The Carnivores

The Carnivores

R. F. Ewer

Comstock Publishing Associates a division of
Cornell University Press | Ithaca and London

First published by Cornell University Press in 1973.
Reissued in cloth and paperback editions, 1985.

Library of Congress Cataloging in Publication Data

Ewer, R. F.
The carnivores.

Reprint. Originally published: Itahca, N.Y. : Cornell University Press, c1973.
Bibliography: p.
Includes index.
1. Carnivora. I. Title.
[QL737.C2E93 1985] 599.74'4 85-12741
ISBN 0-8014-9351-X (pbk.)
ISBN 0-8014-0745-1 (cloth)

Printed in the United States of America

The paper in this book is acid-free and meets the guidelines for permanence and durability of the Committee on Production Guidelines for Book Longevity of the Council on Library Resources.

For

Caroline and Timothy

Contents

Plates

Acknowledgements

The author and the publishers wish to thank the following for providing photographs for this book: C.K.Brain: 4, 10b, 11; Chicago Zoological Society: 7b (photo Leland la France); Bruce Coleman Ltd.: 5a (photo George Schaller), 6a (photo C.A.W. Guggisberg), 8b (photo Leonard Lee Rue), 12a (photo Donald Paterson); Nicole Duplaix-Hall: 9a; H. van Lawick: 6b; H.Patel: 13b; Paul Parey: 15b (photos Lilli Koenig), 1b (photos P. Leyhausen); Fritz Siedel: 15a; C.Wemmer: 9b, 10a; Zoological Society of London: jacket, 1a, 2a and b, 5b, 7a, 8a, c and d, 12b, 13a, 16 (photos Michael Lyster), 14c (photo K.East).

We would also like to thank the following publishers, journals and authors for permission to use drawings.
Professor R.Bourlière for the drawings used in Figure 6.4.
American Zoologist: Schenkel, R. (1967) 7, 319–29, figs. 5a, 6a.
Behaviour: Schenkel, R. (1947) **1**, 81–129, figs. 14, 18.
Biological Bulletin: Scholander et al. (1950) **99**, figs. 2, 5.
Bulletin of the American Museum of Natural History: Van Gelder, R.G. (1959) **117**, 229-392, fig. 7; Allen, J.A. (1924) **47**, 73–281, figs. 8a, 8b, 9b, 16b, 22a, 22b, 23c, 26b, 43; MacIntyre, G.T. (1966) **131**, 119–209, modified versions of figs. 1a, 1b and 2d and fig. 19.
Chicago Natural History Museum: Davis, D.D. (1964) *Fieldiana Zool. Mem.* **3**, figs. 8, 53, 68, 69; Davis, D.D. (1949) *Fieldiana Zool.* **31**, fig. 70, slightly modified version.
Harper, New York: Walls, G.L. (1942) *The vertebrate eye and its adaptive radiation*, three drawings from figs. 71, 85a and b.
Journal of Anatomy: Barnett, C.H. & Napier, J.R. (1953) **87**, 11–21, figs. 6, 10; Welker, W.I. & Johnson, J.I. (1965) **99**, 761–90, one drawing from fig. 14.
Journal of Experimental Biology: Manter, J.T. (1938) **15**, 522–40, fig. 8.
Journal of Mammalogy: Hildebrand, M. (1961) **42**, 84–91, fig. 5, slightly altered version; Matthew, W.D. (1930) **11**, 117–38, two drawings from fig. 2.

Journal of Morphology: Scapino, R.P. (1965) **116**, 23–50, figs. 1, 2, 16a.
McGraw Hill: Hamilton, W.J. (1939) *American Mammals*, fig. 32.
MacMillan, New York: Lull, R.S. (1929) *Organic Evolution*, figs. 44a, b; Gregory, W.K. (1951) *Evolution Emerging*: figs. 20.18d, e, f; 20.19d, e, f; 20.20a, b; 20.21a, b; 20.23a, c, e; 20.27d, e, g; 20.28d, e, g.
Mammalia: Ginsburg, L. (1961) **25**, 1–21, fig. 4c; Didier, R., sixteen drawings from his series of papers 1946–50, vols. 10–14.
Masson et Cie., Paris: Grassé, P.P., *Traité de Zoologie* **16** (1), figs. 270, 506c, 572b, c, 583, 729a, b, 775; Piveteau, *Traité de Paléontologie* **6** (1), figs. 90, 99, 112.
Philosophical Transactions of the Royal Society of London: Matthews, L.H. (1939) **230** (B), 1–78, fig. 3c, d, and two drawings from fig. 2.
Proceedings of the Zoological Society of London: Wood-Jones, F. (1939) **109** (B), 113–29, fig. 10; Anthony, R.L. & Iliesco, G.M. **1926** (2), 989–1015, fig. 9; Pocock, R.I. **1926** (2), 1085–94, fig. 54.
Säugetierkunde Mitteilungen: Weigel, I. (1961) **9**, 1–120, fig. 4 (slightly modified version).
Science: Eisner, T. & Davis, J.A. (1967) **155**, 577–9, figure of mongoose throwing millipede.
Zeitschrift für Tierpsychologie: Leyhausen, P. (1956) *Beiheft* 2, fig. 7; Tschanz, B., Meyer-Holzapfel, M. & Bachman, S. (1970) **27**, 47–72, four drawings from fig. 3.

Preface

APPRECIATION of animals has much in common with appreciation of works of art. In both, what matters is the way the parts are combined to make a meaningful whole: assemble words or notes in a different order and a poem or a symphony has turned to gibberish; match a badger's feet with a leopard's teeth and you also have nonsense. In the one case human skill and judgement have created harmony, in the other it is the power of selection that has produced the harmony of adaptation. The animal, however, is more complex than any work of art and its harmonies more multifarious: structure, physiology and behaviour in all their aspects; relations with the environment, with their fellows and with other species – all fit together and the more we know and understand the more satisfying, both intellectually and aesthetically, is the synthesis we can make.

The carnivores are animals many of whose adaptations are so obvious that, even to a superficial view, there is something admirable about them: it is no mere chance that the two animals man has taken for choice as his house guests are both carnivores, despite the fact that they are more expensive to feed than vegetable eaters.

When I started to write this book, it was my aim to gather together as much factual material as possible and attempt to see what meaning could be extracted from it in terms of adaptive significance and evolutionary history. I hoped, as far as possible, to unite anatomy and physiology with ecology and behaviour; for no one of these can be fully comprehended without all the others. The first part of this undertaking was more difficult than I had anticipated; the information more scattered and difficult to locate. What I have gleaned is certainly incomplete but I have tabulated data wherever possible, for convenient reference. Certain aspects have been neglected. For instance, I have said very little about intraspecific fighting or toilet behaviour and I have not discussed the population dynamics of predator–prey relationships: the former because the information available requires very detailed analysis before it becomes meaningful: the latter largely because time and space are both limited and, in any case, it would tend to take us too far away from the animals themselves and their adaptations, which constitute the main theme of the book. If there are no errors, then this differs from any book I have ever read, except those on the most narrowly specialised topics. For this I can only apologise and offer the same explanation as Dr Johnson did when asked by a lady why he had defined the pastern of a horse as the knee: 'Ignorance, madam, pure ignorance.'

Despite all these things, it is my hope that the book may contain some things to interest those who share with me the belief that the study of carnivores is a rewarding one. Because I believe that such people cover a wide range of zoological expertise, I have tried to keep the use of technical terms to a minimum and, although I believe that structure and behaviour illuminate each other, it is perfectly possible to skip the anatomical section, move from the Introduction to Chapter 5 and refer back to earlier chapters to answer questions as they arise – or to read them at the end instead of at the beginning.

The classification adopted is very largely based on the same authorities as those accepted by Morris (1965). Many groups undoubtedly require revision but that is a highly specialist task for which I am quite unfitted and, in any case, such matters are of little interest to the general reader.

The problem of names is a difficult one. For most of the large species, common names are truly familiar and present no problems – we all know the lion, tiger, wolf or polar bear and the zoologist too knows exactly what species is meant – but what of the cacomistle, the fanaloka or the toddy cat? Such common names are probably familiar only to those who have lived in the animals' natural habitats and convey nothing to the average professional zoologist, still less to the layman. There is the further complexity that the same common name may be used for different species in different areas. The dwarf mongoose in east Africa means *Helogale parvula* but in west Africa, where *Helogale* does not occur, it is sometimes used for *Herpestes sanguineus*. When an animal is mentioned for the first time, I have therefore given both common and scientific names and if a genus contains only a single species I have often, for brevity's sake, used the generic name alone. If, thereafter, I have seemed to use scientific names too frequently I can only apologise to the reader who finds them difficult but I have tried to simplify his problem by making it possible to equate common and scientific names by referring to the index. If you look up, say, toddy cat, you will find that it says 'see *Paradoxurus hermaphroditus*: V, 21' and if you start from *Paradoxurus hermaphroditus* you will find 'common palm civet or toddy cat: V, 21'. The figures mean that by referring to Chapter 10 you can find out the animal's taxonomic position: it is species number 21 in family V (= Viverridae) and falls in the subfamily Paradoxurinae.

The people who have helped me in one way or another are innumerable but there are a few to whom I am particularly indebted: first and foremost to Dr Paul Leyhausen who, with unfailing generosity, has shared ideas, let me watch his animals and see his films, not to mention providing some of the photographs; to the many people who have given me access to their data before publication – Dr C.K.Brain, Mrs Nicole Duplaix-Hall, Dr J.F.Eisenberg, Margaret and Niels Jacobsen, Dr D.Kleiman-Eisenberg, Mr R.L. McLaughlin, Dr A.Rasa, Mr F.E.Sandegren, Mr M.Taylor, Professor Niel Todd, Mr C.Wemmer; last but not least, to Dr Konrad Lorenz, who first made me realise how an animal's behaviour provides the key to understanding everything else about it. I also owe two other debts: one to all those carnivores who have been my friends, from Old Pussy – on whom I was probably imprinted in my cradle – and Jock, the orphaned mongrel with whom I shared puppyhood, to my present kusimanses and civet cats. The other is to my husband and only those who have also shared a life of scientific endeavour can ever know how much that is.

LEGON

R.F.E.

Chapter 1

Introduction

FOR mammals, the digestion of cellulose is a formidable physiological problem but the digestion of flesh is easy. One cannot, however, have it both ways: flesh may be easily digested but it is hard to come by. Moreover, prey species cannot support an equal number of predators, so the business of the killer is a highly competitive as well as a highly skilled occupation. It is therefore no accident that in the history of the placental mammals, although a number of orders have established themselves successfully as herbivores, the predators all belong to a single order, the Carnivora. Much of their fascination lies in the fact that they are so clearly adapted to their mode of life: the otter in pursuit of a fish, the cheetah of a Thomson's gazelle; the lightning-quick pounce of the fox on a vole; even the domestic cat stalking a mouse - each, in its own way, seems to display a grace and perfection of movement which we both envy and admire. This perfection, however, is the product of a long evolutionary history and we cannot understand the carnivores of today without some appreciation of the process which brought them into being.

The evolutionary history of the vertebrates is a story of successive adaptations, offering fresh possibilities whose exploitation led to the development of a host of new forms, each taking advantage of the new advance in its own particular manner. Legs and lungs may first have evolved as adaptations to poorly oxygenated waters, subject to periodic flooding and drying up - but they opened the gateway to the land habitat and their acquisition led to the development of a bewildering variety of amphibians. The reptiles' dry skin and shelled egg gave them a further independence of the water and again a vast variety of new types evolved, exploiting to the full the potentialities of these new evolutionary 'inventions'. Still later the mammals, warm-blooded and hair-covered, enacted once more the drama of expansion and diversification.

The same principle can be seen in operation at the lower taxonomic levels of order and family. Each represents an original adaptation to some particular environment and mode of life, which is manifested in the common characters of the group as a whole. The potentialities of the basic adaptations, however, are manifold and leave room for a large number of variations on the basic theme. Moreover, evolution is an opportunistic process: there is no rule that says carnivores must eat nothing but flesh while rodents must restrict themselves to vegetables. In any large group, there will almost certainly be some species that have become adapted to a rather atypical mode of life. Their basic characters, however, still betray their relationships and their individual peculiarities can be understood only if the original theme has first been grasped.

The fundamental adaptations of the Carnivora relate to their role as predators and the order includes a range of species of very different shapes and sizes, some adapted to dealing with prey as small as insects and some capable of killing a buffalo. It does not, however, follow that because their basic adaptations relate to predation, all Carnivora live on flesh alone. Although many species live mainly on flesh, I do not believe that we know enough to state that there is any member of the order that never partakes of vegetable food: for many species plant foods make an important contribution to the diet and there are even a few that have become secondarily almost pure vegetarians.

The order Carnivora, then, comprises those mammals that originally evolved as predators but whose adaptation within the order has resulted in diversification into a number of distinct lineages - the families - each with its own characteristics related to its particular habitat and general mode of life and its particular food and method of obtaining it. Here again, within each family, evolutionary radiation has occurred and produced a variety of forms, each adapted to a particular mode of life. Sometimes a very similar role has been taken over in one geographical area by a member of one family, in another by one belonging to a different family; or a special food source has been exploited by members of more than one family. We must therefore expect to find that the basic ecological or geographical separation that first brought about the divergence of the families has been somewhat obscured by the adaptations subsequently evolved within the families: we must expect to find parallelism and convergence. Despite such complexities, it is possible to discern in broad outline the typical features distinguishing the families and to see how these reflect the parcelling-out of a series of different ways of life between them. Once the basic adaptational syndrome of each group is clear, we are in a position to appreciate how some genera have departed from the norm for their group and to understand the adaptations of each to its own way of making a living. The preliminary outline which follows is intended to indicate the basic adaptations of the families and to provide a sort of sketch-map which can act as a guide to the sections on structure and behaviour which follow. Subfamily names are given for the more diverse families, since it will be necessary to use these later. A more detailed taxonomic review is given in Chapter 10.

The Carnivora are commonly divided into the following seven families:

Family 1, Canidae – wild dogs, wolves, jackals, foxes, maned wolf, raccoon dog, bat-eared fox

The Canidae are medium-sized carnivores, adapted to swift running on relatively open terrain. They live mainly, but by no means exclusively, on flesh and possess crushing molar teeth suitable for dealing with vegetable food. The canine teeth are fairly large but not highly specialised: they are not particularly sharp and not much flattened and are a good all-purposes weapon, not specifically adapted to delivering a highly oriented death bite. The Canidae are cosmopolitan in distribution, although the Australian dingo was almost certainly introduced by man.

Family 2, Ursidae – bears

Bears are large and heavily built. They walk on the flat of the foot and are adapted to moving about in rough mountainous or hilly country or in forest. They are extremely omnivorous and, despite their large size, they feed largely on fruit, nuts, tubers and insects and do not very often kill prey of any size. The bears are essentially a northern-hemisphere group with a few genera extending southward into India and southeast Asia and into South America. They are unknown in Africa.

Family 3, Procyonidae

Subfamily Procyoninae: raccoons, coatis, potto, cacomistle, kinkajou
Subfamily Ailurinae: pandas

Procyonids are typically forest-dwelling and, like the bears, very omnivorous. They are, however, mostly rather small and a number are arboreal.

The purely vegetarian pandas of eastern Asia are usually regarded as belonging to this family but the typical procyonids are restricted to the New World and most belong to Central and South America.

Family 4, Mustelidae

Subfamily Mustelinae: weasels, martens, polecats, sable, fisher, wolverine, South American tayra and grison, African muishonds
Subfamily Mellivorinae: honey badger or ratel
Subfamily Melinae: badgers
Subfamily Mephitinae: skunks
Subfamily Lutrinae: otters

The Mustelidae are small to medium-sized animals. Typically they are forest dwellers, with long bodies and relatively short legs and are more strictly carnivorous than dogs. The jaws are short and powerful and adapted to delivering an accurately placed death bite. Typically a holarctic group, the mustelids have spread into all the southern continents except Australia. They are extremely successful and have diversified to occupy a variety of habitats. The aquatic otters and highly omnivorous badgers are mustelids that have diverged considerably from the basic pattern of the family.

Family 5, Viverridae

Subfamily Viverrinae: civets, genets, linsangs
Subfamily Paradoxurinae: palm civets
Subfamily Hemigalinae: banded civets and their allies
Subfamily Galidiinae: Madagascar 'mongooses'
Subfamily Herpestinae: mongooses
Subfamily Cryptoproctinae: the Madagascar fossa

The viverrids are mostly rather small animals: they are a relatively ancient group, retaining a number of primitive features and are similar to the mustelids in general adaptation. They are a somewhat heterogeneous assemblage, comprising all the less specialised residual members of the basic stock from which the hyaenas and cats have originated. The two main centres of evolution are the oriental region of tropical Asia and Africa and there are two subfamilies restricted to Madagascar. Viverrids are unknown in the New World.

Family 6, Hyaenidae

Subfamily Hyaeninae: striped, spotted and brown hyaenas
Subfamily Protelinae: the aardwolf

The hyaenas are large animals, essentially flesh-eating and taking carrion readily. The premolar teeth are heavy, the jaws and jaw muscles are extremely strong and the whole head is very massive. Correlated with this, the shoulders are heavy, so that the hindquarters appear disproportionately small. Hyaenas are capable of crushing larger bones than any other carnivore and, although they are competent killers in their own right, they frequently scavenge on the kills of others. The aardwolf, *Proteles*, in contrast, has a reduced dentition and feeds mainly on termites.

Family 7, Felidae – wild cats, ocelots, serval, caracal, lynxes, puma, leopard, jaguar, lion, tiger and cheetah

The cats are strictly carnivorous, vegetable foods playing a very minor role in the diet. The jaws are short and the canine teeth highly specialised for delivering an aimed lethal bite. The usual hunting method involves preliminary stalking and a quick final rush. Although they occur in every type of terrain from desert to dense forest, the typical habitat is woodland or rough country rather than open plains. They are cosmopolitan in distribution, except for Australia.

This division into seven families represents the most generally adopted scheme of classification. Some authorities, however, separate off the mongooses (Herpestinae) as a family on their own, distinct from the Viverridae. They certainly do form a compact and closely linked group, easily distinguished from other Viverridae but most taxonomists have felt that this is sufficiently indicated by according them the rank of a subfamily. Exactly the same is true of *Proteles*: while it may be regarded as the sole representative of a second subfamily of the Hyaenidae, some authorities prefer to place it in a family of its own.

The seven families listed fall naturally into two groups. The Viverridae, Hyaenidae and Felidae have affinities with each other, while the Canidae, Ursidae, Procyonidae and Mustelidae form another group of families, more closely related to each other than to the first trio. The two groups of families were named Herpestoidea and Arctoidea by Winge in 1895. Most later authors have accepted his groups but a variety of other names have been applied to them at one time or another. The historically minded will find a summary of these nomenclatural vicissitudes in Simpson (1945). Simpson argues that the designations Feloidea (for the Felidae, Hyaenidae and Viverridae) and Canoidea (for the Canidae, Ursidae, Procyonidae and Mustelidae) are those most in accordance with the rules of taxonomic procedure. This opinion is, to some extent, a matter of judgement and there are taxonomists who differ. Quite apart from procedural niceties, the designation of the two groups as cat-like (Feloidea) and dog-like (Canoidea) has much to recommend it; since the names refer to familiar species, their meaning is readily apparent and these names are now very generally adopted.

A final point to be considered is the taxonomic status of the seals, sea-lions and walruses. They are certainly related to land carnivores and it used to be customary to classify them as a suborder within the Carnivora. Present usage, however, accords them the rank of a separate order, the Pinnipedia. In the older arrangement, the typically terrestrial families were placed in the suborder Fissipedia and the categories Canoidea and Feloidea had the rank of super-families. With the removal of the Pinnipedia to a separate order the name Fissipedia becomes unnecessary, if we consider only living species but when fossils are taken into account, it retains its usefulness as a term designating all the extant families but excluding their more primitive extinct ancestors not included in these families.

Chapter 2

The skeleton

IT might seem most logical to start a consideration of structure by dealing with external features. It is, however, more convenient to begin with the skeleton and muscles, since outward form cannot be understood without reference to the supporting framework within and to the muscles which move it. It is not, however, proposed to attempt a detailed account of the skeletal and muscular systems of the Carnivora, but rather to concentrate on the adaptive differences that characterise the various groups. Muscles will be mentioned only where some special feature aids in understanding skeletal structure. For more detailed information, the following sources may be consulted. The classical monographs of Mivart (1881) and Ellenberger and Baum (1891) deal with the anatomy of the cat and the dog; shorter and more modern accounts may be found in Taylor and Weber (1951), Sisson and Grossman (1953) and Miller, Christensen and Evans (1964). Windle and Parsons (1897, 1898) give a detailed but purely descriptive treatment of carnivore myology; Mivart (1882b) deals with the musculature of the genet, *Genetta tigrina*, Fisher (1942) with that of the sea otter, *Enhydra lutris* and Langguth (1969) with the maned wolf, *Chrysocyon brachyurus*.

A. Post-cranial skeleton

Efficiency as a predator necessitates the ability to make a wide variety of movements. Prey must not only be sought and captured, but killed and eaten. Carnivores therefore have not specialised in the performance of one particular type of movement to the detriment of others: some are swift runners but most can also climb or dig to some extent and many are good swimmers. As a consequence, the skeleton is very generalised and, indeed, the dog or cat skeleton is frequently chosen to illustrate the typical mammalian structure.

Figures 2.1 and 2.2 show the skeletons of a representative selection of carnivores. Apart from the skull, which will be considered later, the most obvious differences are in general body proportions,

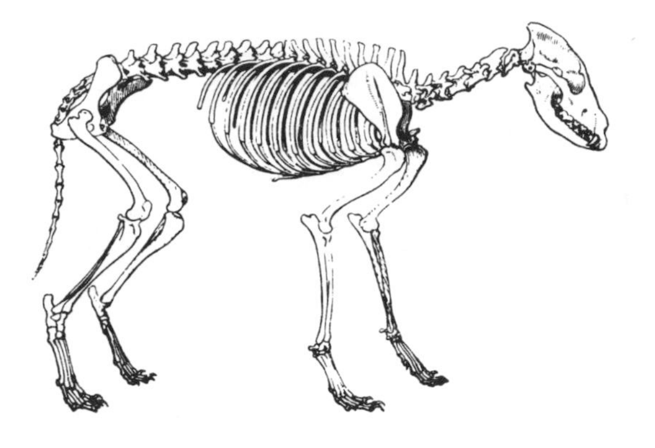

Wolf

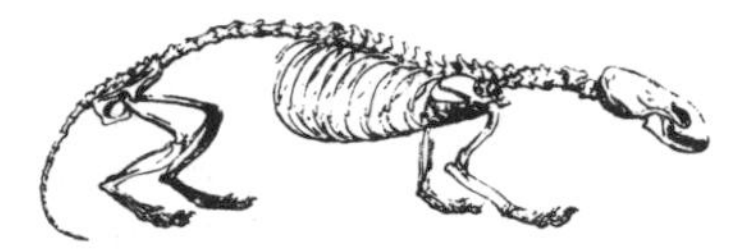

Weasel

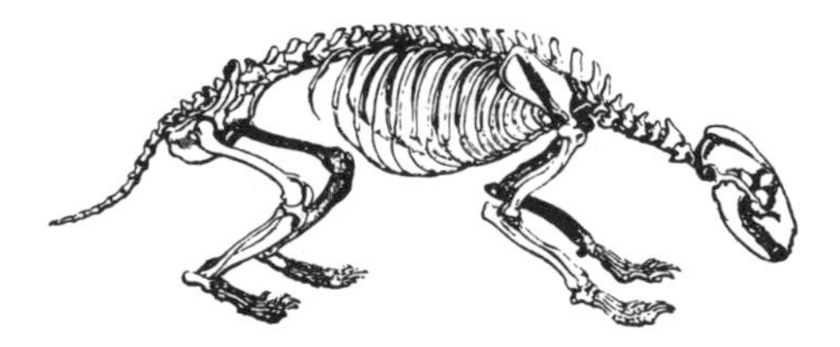

Badger

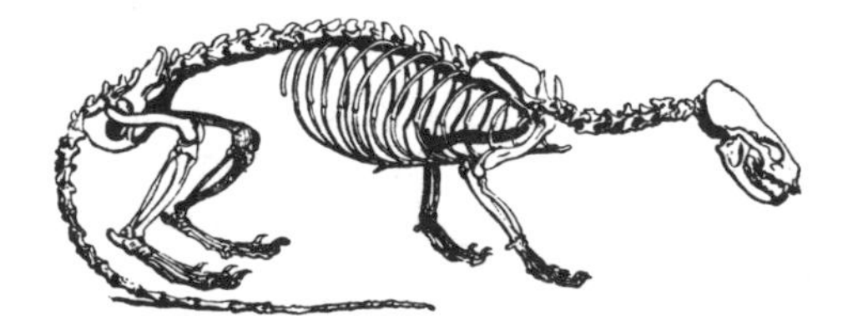

Otter

Figure 2.1. Skeletons of Canidae and Mustelidae

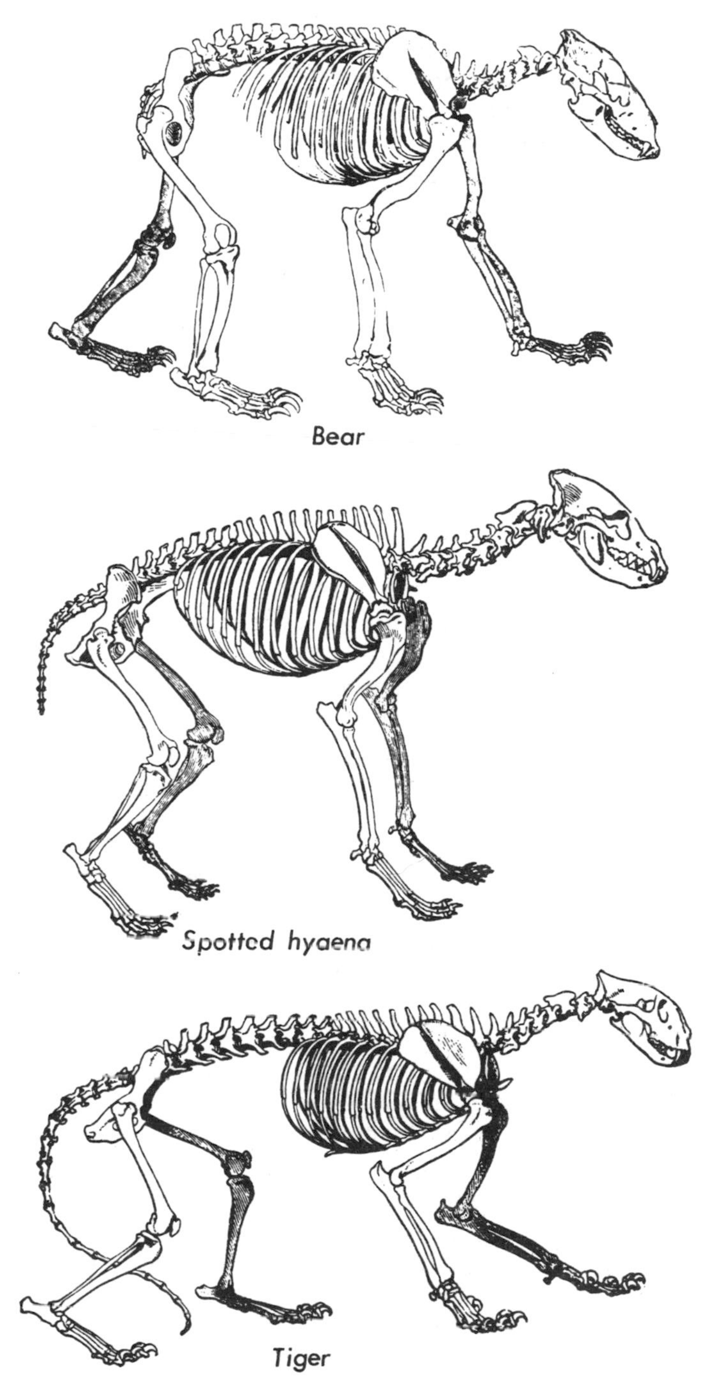

Figure 2.2 Skeletons of Ursidae and Felidae

from the low-slung weasel to the erect wolf and dog, or the lithe cat to the ponderous bear. This is mainly a matter of limb length: the number of body vertebrae does not vary very strikingly from one family to another, although the number in the tail is more variable (table 2.1). There is, however, much overlapping, and vertebral formulae are of little taxonomic value.

Table 2.1
Numbers of vertebrae in the various body regions of carnivores

Family	Cervical	Thoracic	Lumbar	Sacral	Caudal
Canidae	7	13–14	6–8	3–4	14–23
Ursidae	7	14–15	5–6	4–6	9–11
Procyonidae	7	13–14	4–7	3	18–29
Mustelidae	7	13–16	4–6	2–4	15–25
Viverridae	7	13–15	6–7	2–3	15–34
Hyaenidae	7	15–16	4–5	2–4	19–26
Felidae	7	13	7	3	14–28[1]

(Mainly after Grassé, 1967.)

[1] Excluding the Manx cat which may have as few as three caudal vertebrae

In the Indian mongoose, *Herpestes edwardsi*, the lumbar vertebrae, although normal in number are peculiar in structure. The transverse processes are unusually broad and the sacrum too is very wide. Rensch and Dücker (1959) regard this as a protective adaptation, related to the fact that in this species a rather hedgehog-like rolled up defensive posture is often adopted. The related Egyptian mongoose, *H. ichneumon*, does not defend itself in this way and the vertebrae and sacrum are not specially modified.

The length of the tail is, of course, of some adaptive significance. It is long in those species where it acts as an organ of balance, whether in making quick turns when running at high speed, as in the cheetah, or in running along branches or leaping from branch to branch as is done by many of the small arboreal species like the ringtail or cacomistle, *Bassariscus*, the coati, *Nasua*, and the red panda, *Ailurus*. The tail is also long in the kinkajou, *Potos*, and the binturong, *Arctictis*, the only two carnivores in which it is used as a prehensile organ. In a number of cases, such as the bobcat, *Lynx rufus*, the short tail may have little mechanical function but its use as a signal organ has limited its reduction and, apart from the Manx cat, only the non-social bears and the giant panda, *Ailuropoda*, are virtually tailless.

The most striking differences in limb structure from one species to another are correlated with how the animals walk – plantigrade, digitigrade or intermediate (figure 2.3). Digitigrady is, of course, characteristic of the more cursorial species and the limbs show a number of modifications in relation to this habit. Before considering

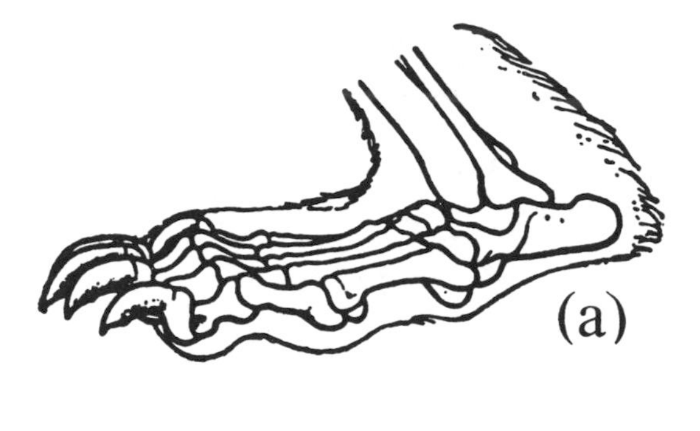

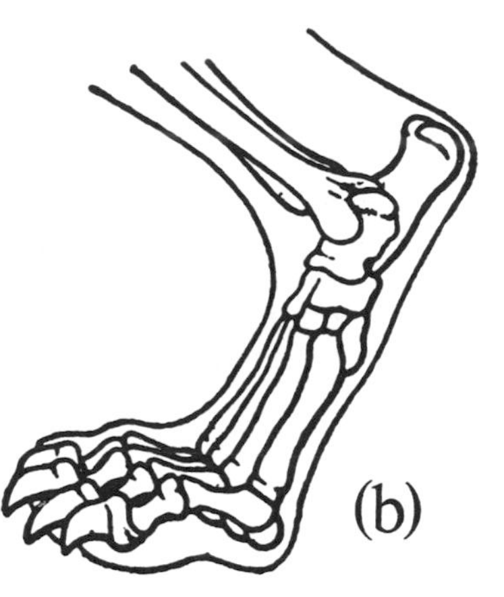

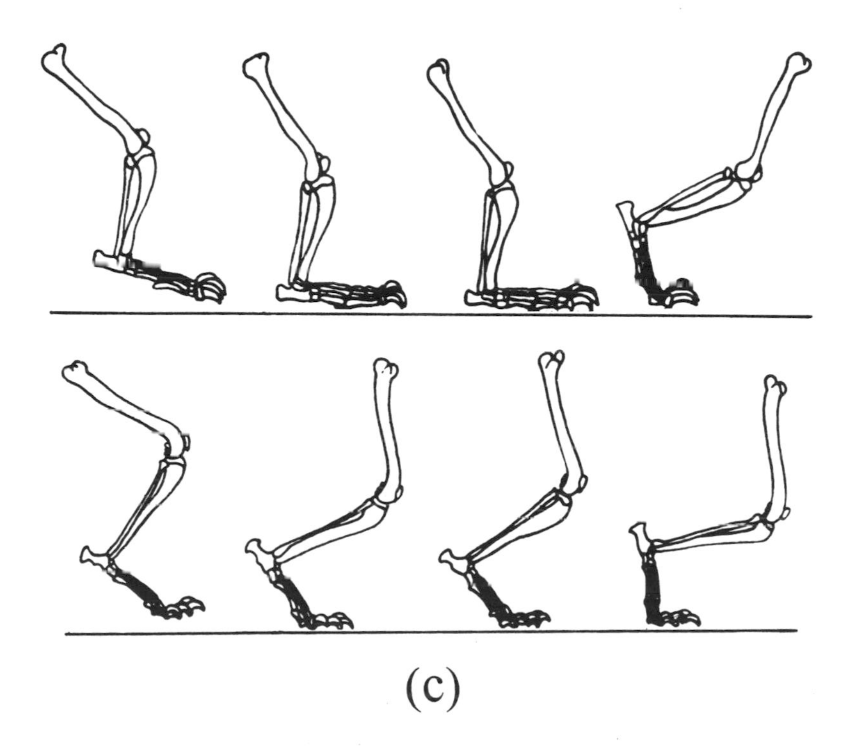

Figure 2.3 Foot of (a) bear and (b) hyaena, showing plantigrade and digitigrade stances [from Lull, 1929]. (c) positions taken up by plantigrade (above) and digitigrade (below) feet as they contact and leave the ground in the walking gait. [After Grassé, 1967]

these modifications in detail, however, it is desirable to understand how a cursorial carnivore uses its limbs in locomotion. Hildebrand (1959, 1961) has studied the cheetah's gallop and he also gives some data on its walking gait. In walking, where speed is not important, economy of effort is the main consideration. In the gallop the reverse is the case; speed is the main desideratum and economy must be sacrificed in its favour. As Hildebrand puts it (referring to the gallop), 'the cheetah does not need to be efficient; it needs to be fast'. Manter (1938) has made a detailed study of walking in the domestic cat; the differences from the cheetah's walking gait are minor and it is convenient to consider Manter's findings first.

The order in which the limbs are moved can most conveniently be represented by a foot-fall diagram (figure 2.4). Here time forms

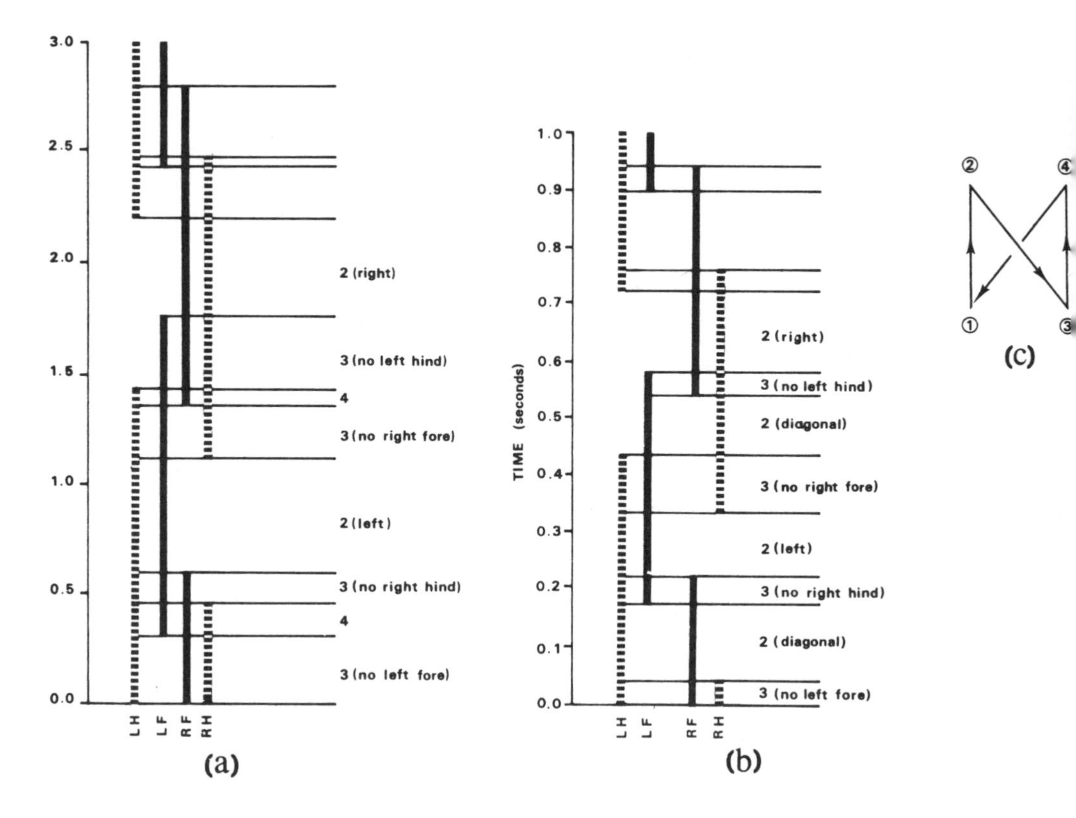

Figure 2.4. Foot-fall diagrams of (a) cheetah and (b) cat, both walking. LH, RH, LF, RF denote left and right fore and hind feet. On the right is shown the number of feet supporting the body during each of the eight phases of the cycle. (c) The 'diagonal' sequence in which the feet contact the ground. [After Manter, 1938 and Hildebrand, 1961]

the vertical axis and the animal should be thought of as walking up the page. The periods during which each foot is in contact with the ground are shown. Of course the foot does not move when it is in contact with the ground and the lines represent time, not distance, the beginning of each stroke giving the moment the foot touches the ground and the end the point at which it is raised again. This walking pattern is often described as 'diagonal', since the step of a fore foot is followed by that of the diagonally opposite hind foot (figure 2.4). There are, however, better ways of describing this gait. From the foot-fall diagram it can be seen that left and right fore feet alternate, overlapping slightly, so that the left is not raised until after the right has been set down and vice versa. The same is true of left and right hind limbs. The fore and hind limbs do not move simultaneously but slightly out of phase, the hind one of each side moving slightly in advance of the fore. A complete cycle comprises eight phases. In Manter's cat, the body is alternately supported by two or by three limbs. Hildebrand's cheetah is walking (relatively) more slowly and the feet remain on the ground for a greater proportion of the total time. There is thus more overlap and in each cycle there are two brief periods when all four paws are on the ground. In neither case, however, is there an unsupported 'floating phase' and there is no point when two fore or two hind limbs alone support the body.

Since elbow and knee are oppositely oriented, it is clear that the movements made by the fore and hind limbs cannot be identical. Figure 2.5 gives a diagrammatic representation of the two phases of motion in each limb: the power stroke, in which the foot is in contact with the ground and retraction of the limb propels the body forward, and the recovery stroke, in which the foot is lifted and the limb protracted in readiness for the next step. In the hind limb, when the paw contacts the ground the femur is protracted and the knee and ankle are slightly flexed. The paw is therefore set down well in front of the acetabulum and a trifle in advance of the knee. The femur is then retracted and at the same time the ankle is slightly extended but there is little movement at the knee joint. The acetabulum (and with it the body) is thus moved forwards. The paw is raised again by flexion at knee and ankle just after the femur reaches the vertical position; the femur is then protracted and the knee and ankle extended for the next step. When the fore foot contacts the ground the humerus is vertical and the elbow and wrist slightly flexed. The paw thus comes down well in advance both of the elbow and of the glenoid. Retraction of the humerus, with extension of elbow and wrist, then follows, driving the glenoid forward. In the recovery stroke, the elbow is first flexed and the humerus protracted; as the latter nears the

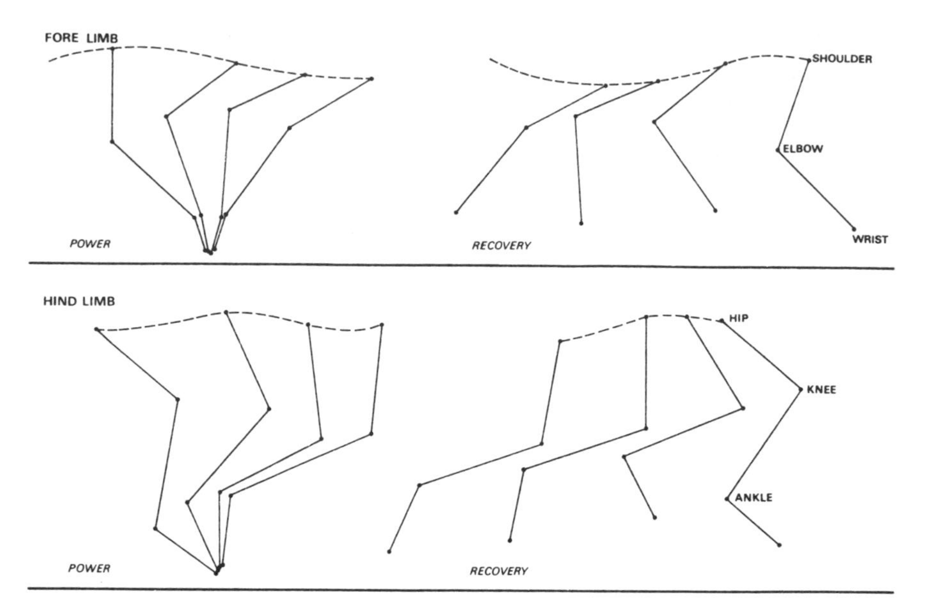

Figure 2.5. Positions taken up by the fore and hind limbs of a walking cat during successive stages of the power stroke and the recovery stroke. [After Manter, 1938]

vertical position, elbow and wrist extend again to set the foot down for the next step.

The major difference between the fore and hind limb actions is thus that in the power stroke the humerus swings through an arc from the vertical position backwards, while the femur works through an arc from the protracted position back to the vertical orientation. In both fore and hind limbs the recovery stroke is made with the distal segments well raised, so that the work done in moving the paw forward is reduced and in both limbs the energy unproductively expended in raising and lowering the body is kept to a minimum by the way in which flexure at the distal joints compensates for the change in the proximal segment from an oblique to a vertical position. In the fore limb, up and down movements of the scapula relative to the thorax also assist in this: the scapula rises when the limb is providing maximal support and falls during the recovery stroke. There is, at the same time, a back and forth pivoting which carries the glenoid forward as the limb is protracted and backward as it is retracted. Hildebrand estimates that in the walking cheetah this action adds about 4½ inches (12 cm) to the length of the stride.

By getting his cat to walk across a sort of three-dimensional weighing machine, Manter recorded the forces exerted by the limbs against the ground (figure 2.6). The limbs, of course, must support the body as well as propel it forward. Since the animal's centre of gravity is nearer the fore than the hind limbs, it is not surprising to find that the former are responsible for more than half of the total vertical (= supporting) forces. Since the limbs slope forwards as

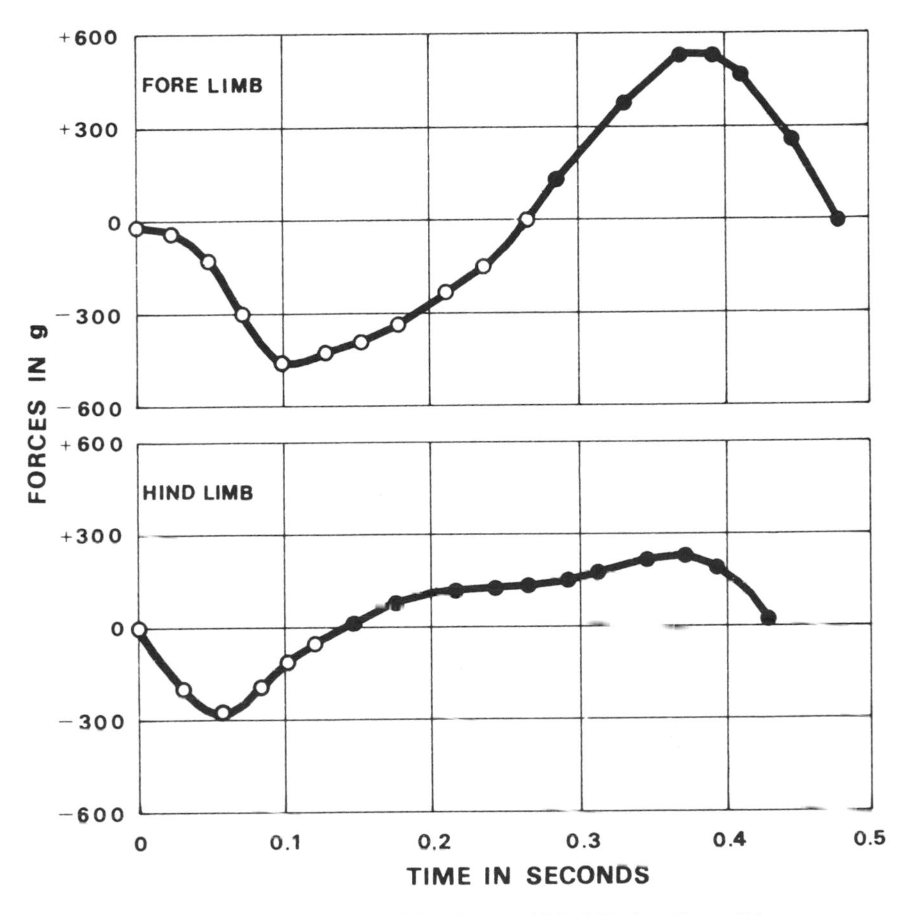

Figure 2.6. Horizontal forces exerted by fore and hind limbs of a walking cat. Positive forces (solid circles) are propulsive, negative forces (open circles) represent retardation. [From Manter, 1938]

they are set down it is also not surprising that they exert a retarding force during the first moments of contact with the ground. What is, perhaps, a little unexpected is to find that it is only in the hind limb that the propulsive force exceeds this retarding action; in the fore limb the retardation is actually greater then the propulsion. This means that the functions of supporting the body and of moving it

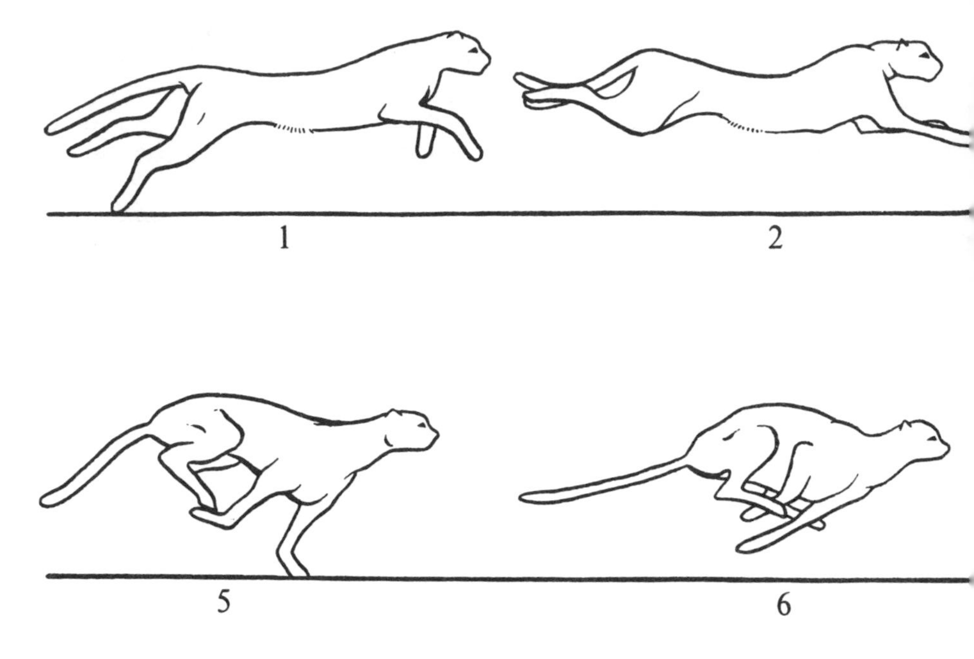

Figure 2.7. Successive positions of galloping cheetah; 2 and 6, the floating phases following the thrusts of hind and fore limbs respectively. In position 8 the animal is about to retract the hind legs and extend the body again, thus returning to position 1 and completing the movement cycle. [After Hildebrand, 1961]

forwards are unequally shared: the hind limb does more than its share of propulsion, while the fore limb takes most of the weight. Barclay (1953) has shown that the same is true of the walking dog.

Another point emerges from the foot-fall diagram. The phasing of fore and hind limbs is such that at the moment when the fore limb strikes the ground and retardation is maximal, the ipsilateral hind limb is further on in its cycle and is exerting a propulsive force. The jerkiness one might expect to be produced by periodic retardation is thus largely ironed out and the animal moves forward relatively smoothly.

Hildebrand (1965), in considering different gaits, distinguishes the symmetrical, where left and right members of a pair (fore or hind) strike the ground at equal time intervals, from the asymmetrical, where they do not. From the foot-fall diagrams it can be seen that the walk is a symmetrical gait, whereas the gallop is asymmetrical. For the sake of clarity, single foot-fall diagrams, showing only one cycle, have been given here: there is, of course, more variation than this

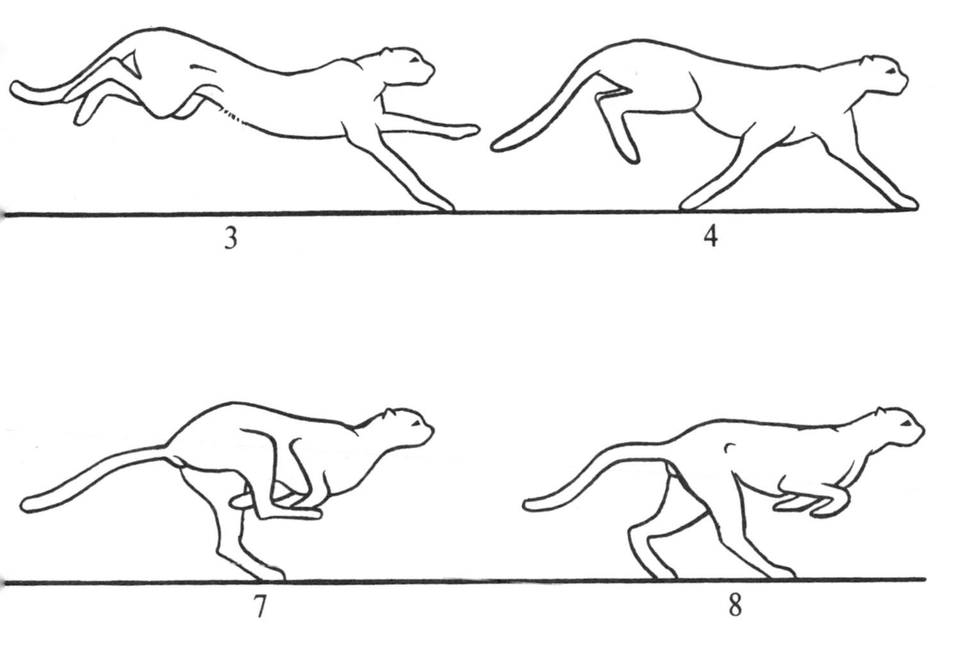

suggests, since no real animal in action is ever maintaining a precisely constant movement pattern over exactly level ground. For a more refined and detailed analysis, which takes this variation into account, Hildebrand (1966, 1968) may be consulted.

At the gallop the cheetah uses the same basic pattern of limb movements as it does at the walk: alternating movements of right and left, with the fore and hind out of phase. The difference lies in the speed with which the limbs are moved and in the increased angle through which femur and humerus are rotated. The faster action provides enough energy to raise the body into the air and propel it forward out of contact with the ground. The fore and hind limbs each create a floating phase of this type, so that there are two such phases per cycle, the one following the thrust of the hind limbs being the longer (see figure 2.7). In addition, the stride is lengthened by extending the body fully as the fore limbs are protracted and flexing it as the hind limbs are brought forward. Hildebrand estimated that this 'caterpillar' action adds about thirty inches (76 cm) to the stride at a relatively slow gallop.

The retarding force exerted by the limbs of the walking cat has already been mentioned. At the gallop, this retardation is greatly reduced, if not eliminated, by two factors. Firstly, the limbs are not merely extended and then set down. They are first protracted very fully without contacting the ground and then brought downward and backward, so that they are already moving fast when they first strike the ground. Secondly, as the fore limbs touch the ground, the flexure of the spine which follows permits the hindquarters to continue their forward travel without any abrupt check.

At a slow gallop the overlap between corresponding right and left limbs, when both are on the ground, is considerable. As speed is increased the limbs move faster, the floating phases increase in length and the overlap of left and right limbs is reduced. This change is clearly seen if we compare the foot-fall diagrams for a carnivore lacking cursorial specialisations (the white-tailed mongoose, *Ichneumia albicauda*), with those for the cheetah at a slow and at a fast gallop (see figure 2.8). The three form a series, with increasing duration of the floating phases and decreasing support by the two limbs of a pair simultaneously. In the mongoose the floating phase before the hind limbs strike the ground is very short and the hind limbs move almost synchronously; in the cheetah travelling at 56 mph (90 km/hr) both floating phases are long and the hind limb overlap is reduced. In the fore limbs overlap may actually be eliminated and Hildebrand was of the opinion that a very brief third floating phase intervened between the strides of right and left fore limbs.

As the cheetah speeds up, the distance covered per cycle increases but the time taken for a complete cycle remains virtually constant at a little under a third of a second. At 32 mph the cheetah covered 13 feet 4 inches per cycle and at 56 mph the distance was 22 feet 10 inches. If the same rhythm of a third of a second per cycle is maintained at higher speeds, it can be calculated that to reach its reputed 70 mph the cheetah would have to cover a little over thirty feet per cycle, roughly four times its own body length.

The cheetah is renowned not only for its fleetness but also for its ability to make rapid turns at high speed. The longitudinal ridges on the pads of the feet described on page 111 are a special adaptation to this habit and are exactly comparable with the treads on a motor-car tyre.

We must now return to our consideration of the limb skeleton. From what we have seen of the gallop it is clear that cursorial adaptation involves more than mere elongation of the bones. When travelling at speed the paw must strike the ground with very considerable force. The digitigrade foot of cursorial species must therefore embody a number of modifications adapting it to withstand the landing shock. The metapodials are elongated and closely bound together; digits 3 and 4 are emphasised so that the structure is paraxonic, a parallel with the cursorial artiodactyls. In the plantigrade foot, the metapodials are shorter and more spreading and the digits

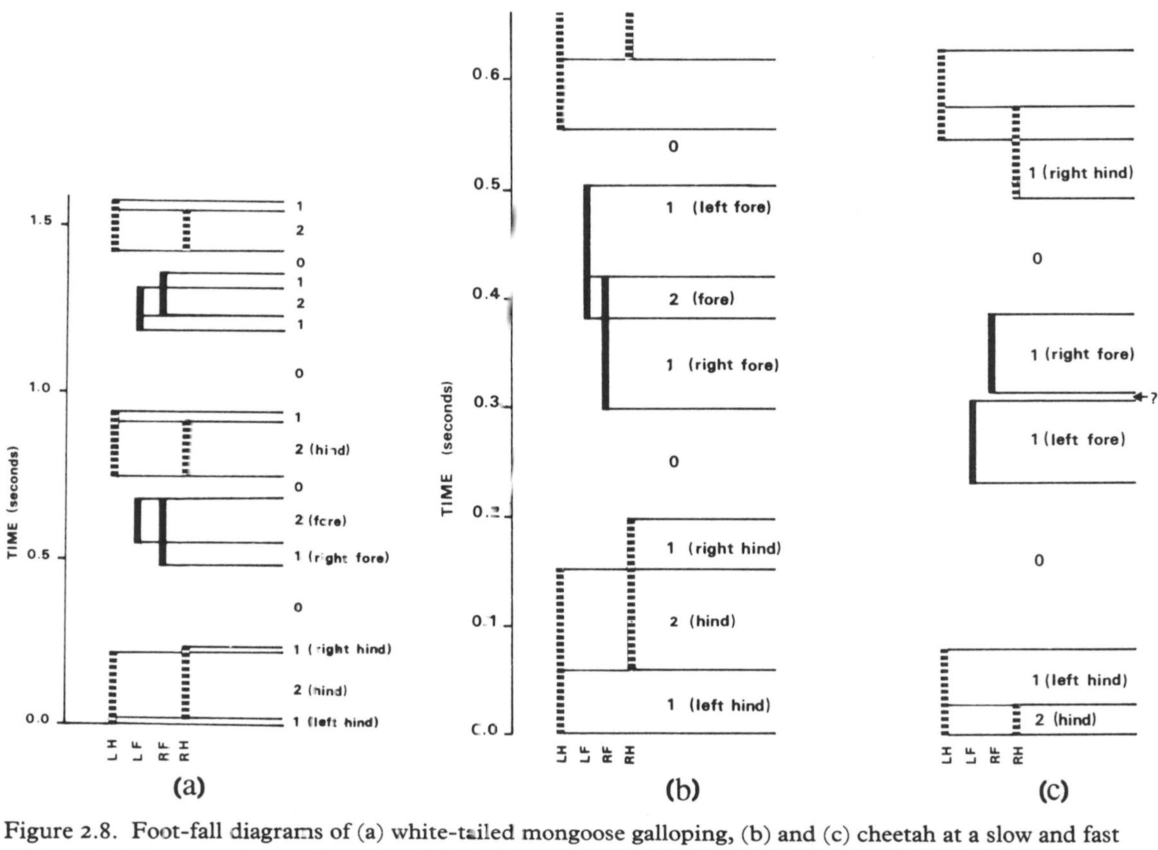

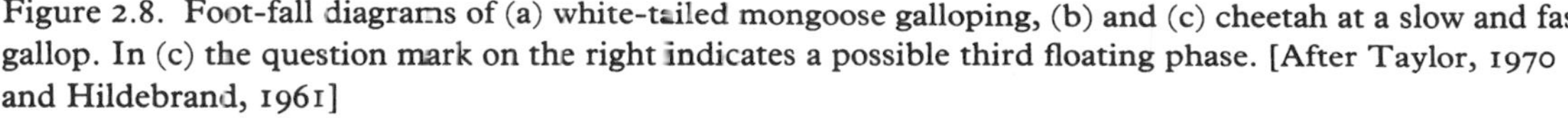
Figure 2.8. Foot-fall diagrams of (a) white-tailed mongoose galloping, (b) and (c) cheetah at a slow and fast gallop. In (c) the question mark on the right indicates a possible third floating phase. [After Taylor, 1970 and Hildebrand, 1961]

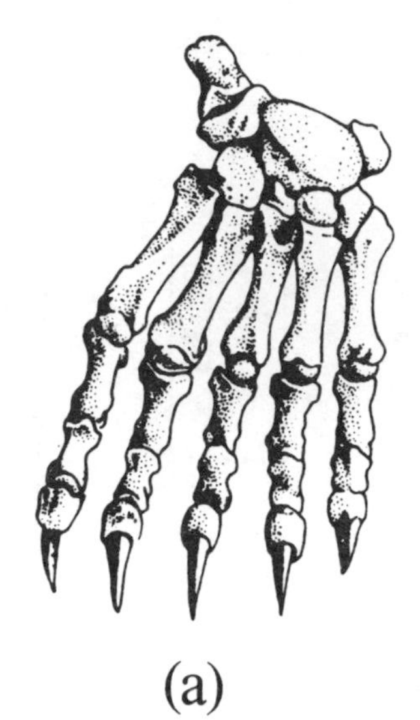

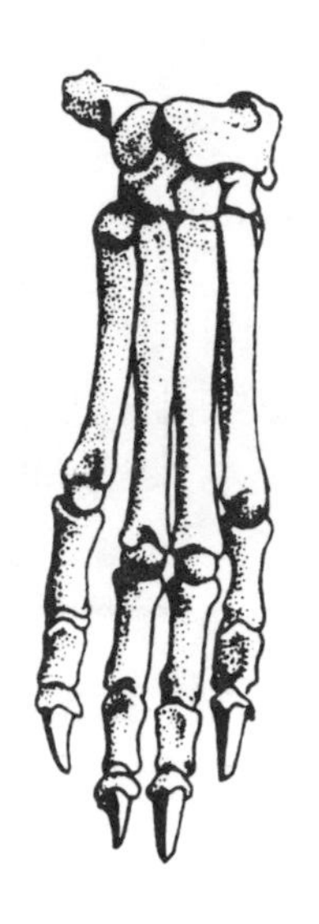

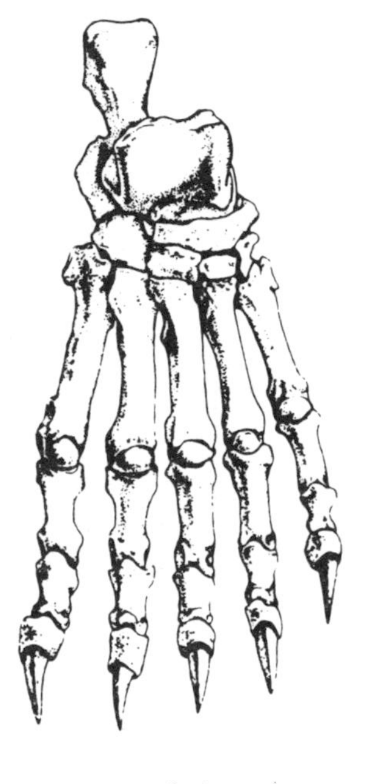

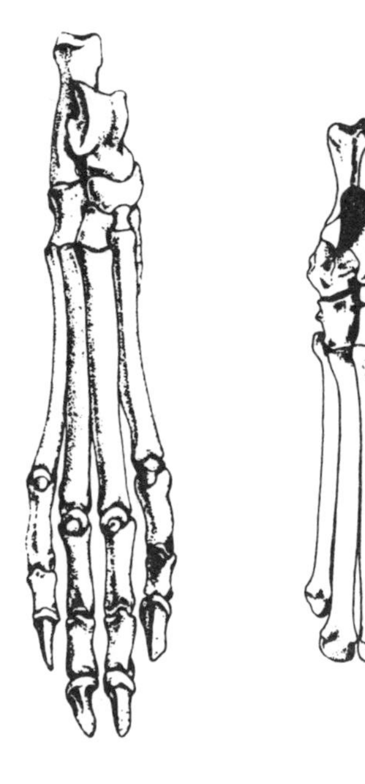

Figure 2.9. Bones of hand and foot in plantigrade and digitigrade carnivores. (a) and (b) right fore paws of bear and hyaena; (c) and (d) right hind paws of the same two animals [from Grassé, 1967]; (e) tarsal and metatarsal bones of right foot of caracal. [From Ginsburg, 1961]

are arranged fanwise (figure 2.9). The other bones of the limbs show correlated differences: they are in general shorter and heavier in plantigrade species and there is greater lateral mobility at wrist and ankle.

Before considering these latter characteristics in more detail, one peculiarity of the carnivore carpus must be mentioned. In all modern carnivores two of the proximal carpals, the scaphoid and the lunar, are fused with the centrale to form a single bone. This gives a firmer arrangement of the wrist and, particularly in the paraxonic cursorial types, where the weight is taken mainly on digits 3 and 4, a firmer bonding between the proximal and distal carpals (see figure 2.9). This fusion, however, is not necessarily associated with cursorial adaptation; it is found in all modern carnivores and has also been independently evolved in species belonging to a number of other orders. Although its function always appears to be to increase the stability of the wrist, the factors making this desirable need not have been the same in all cases. In non-cursorial carnivores, for instance, it is more likely to have been a matter of withstanding the shock when landing from a leap than when running. Yalden (1970) points out that the fused scapholunar also provides a firm basis for flexion at the mid-carpal joint. This can have little importance in normal locomotion, since wrist flexion occurs only during the recovery stroke, when the limb carries no weight. It might, however, be relevant in species that climb or that use their paws in prey capture, both of which actions may involve exerting force on the paw with the wrist in a partially flexed position.

A sesamoid bone, the pisiform, on the ulnar side, is a normal part of the mammalian carpus. Another sesamoid on the opposite, radial side, although not regularly present, is by no means uncommon: a small radial sesamoid is present in all the Canoidea and also in the Felidae. In the red panda the radial sesamoid is somewhat enlarged, while in the giant panda it is larger still and forms what is virtually a sixth digit. The giant panda's radial sesamoid articulates with the scapholunar by a synovial joint but there is some variation in its degree of contact with the first metacarpal. The sesamoid is capable of independent movement and can be brought into partial opposition with the first digit, so that it functions almost after the manner of a thumb (figure 2.10). Four muscles are involved in the movement: the opponens pollicis and abductor pollicis brevis, running between the sesamoid and the metacarpal and first phalanx of digit 1, serve to adduct the former; the abductor longus pollicis, originating high up on the radius and ulna, abducts the sesamoid, while a strong tendon from the palmaris longus muscle, inserting on the lower surface of the sesamoid, pulls it ventrally into a position more effectively opposed to the first metacarpal (Wood-Jones, 1939). Wood-Jones draws

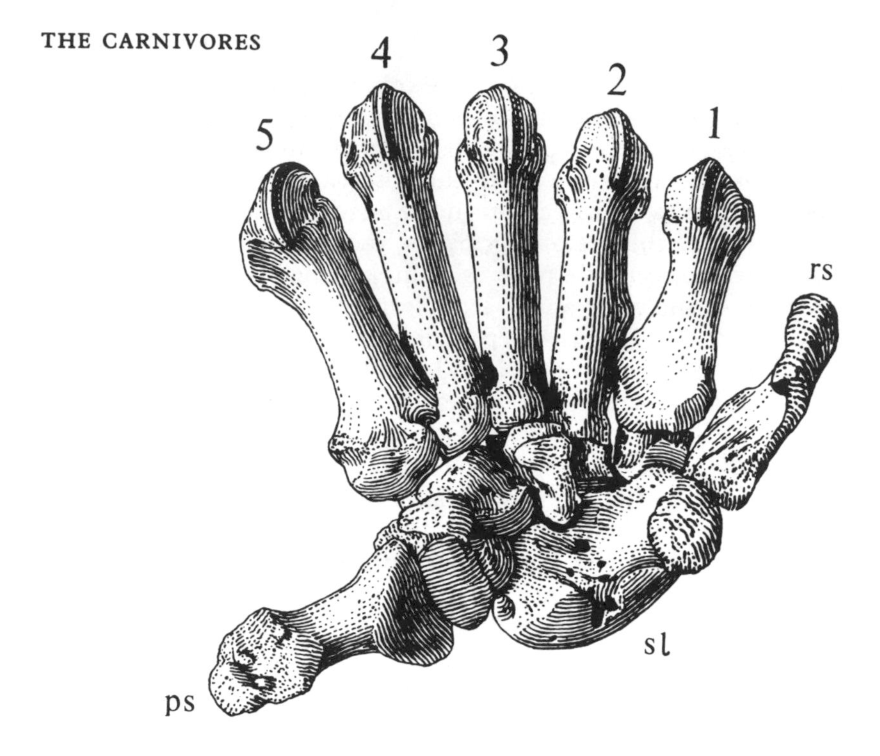

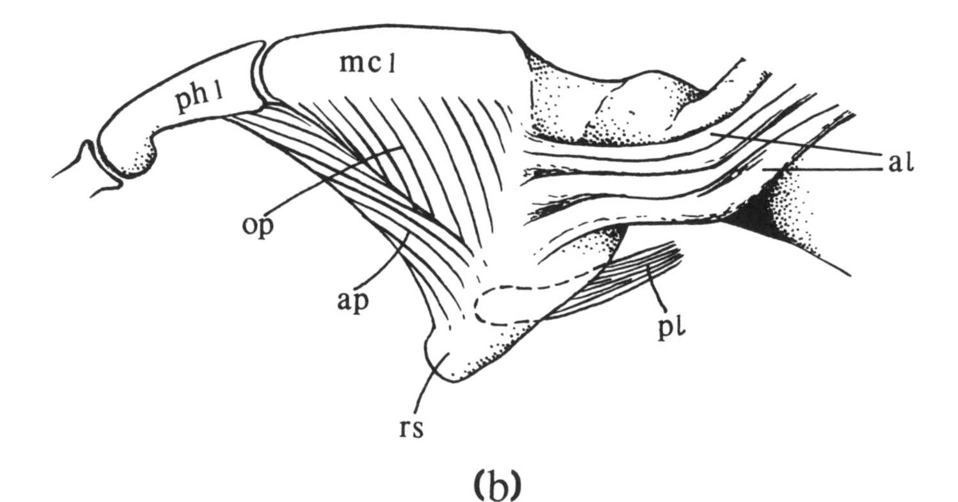

Figure 2.10. (a) Skeleton of right paw of giant panda, ventral view; 1–5, metacarpals of first to fifth digits; rs, radial sesamoid; ps, pisiform; sl, fused scapholunar bones. (b) Side view of radial sesamoid and associated muscles in right fore paw of giant panda. al, abductor longus pollicis muscle; ap, abductor pollicis brevis muscle; mc 1, first metacarpal; op, opponens pollicis muscle; pl, tendon of palmaris longus muscle; ph 1, proximal phalanx of digit 1. [After Wood-Jones, 1939]

attention to the fact that in the panda the pisiform also is extremely large and elongated and has muscle attachments, which could imply that it too is mobile. He suggests that this bone also may have some grasping function but no later workers appear to have investigated this point. In the kinkajou the radial sesamoid of the male is enlarged and pointed. It serves to increase the stimulation of the female during mating, when the male clasps her as a prelude to intromission (Poglayen-Neuwall, 1962).

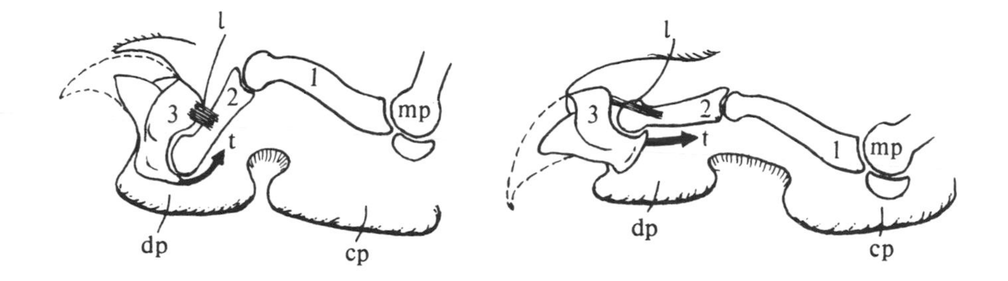

(a) (b)

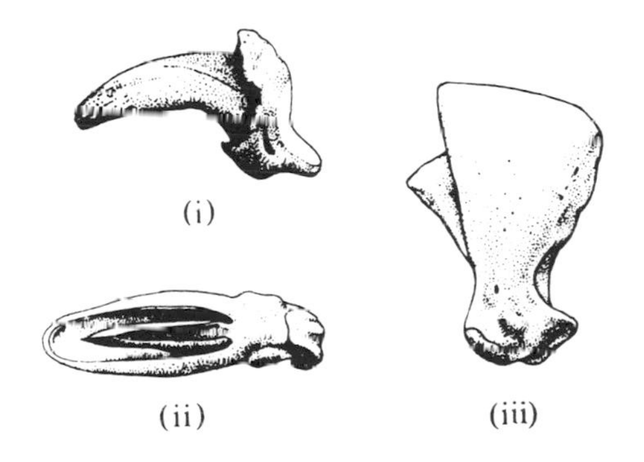

(c)

Figure 2.11. Claw protrusion mechanism in Felidae: 1,2,3, phalanges; cp, carpal pad; dp, digital pad; l, retractor ligament; mp, metapodial; t, tendon of protractor muscle. (a) At rest: the terminal phalanx bearing the claw is held in the retracted position by the ligament and the whole digit is slightly flexed. (b) Claw in use: contraction of the muscle pulling on the tendon t, protracts the claw. At the same time dorsal muscles (not shown) also contract and the whole digit is held firmly in the extended position. (c) Terminal (ungual) phalanx of (i) canid, (ii) and (iii) felid. (i) and (iii), side views; (ii) seen from distal end. [From Grassé, 1967]

The digits of carnivores are typically clawed and even in the exceptional clawless otter, *Aonyx*, there are vestigial claws on some of the toes. Claws may be relatively short and thick, as in the Canidae, or long, compressed and very sharp, as in Felidae and the shape of the terminal phalanx is modified accordingly. In the latter type, the flange at the proximal end of the bone, which protects the base of the claw, is elaborated to form a large hood (figure 2.11). It may be noted in passing that in pandas this bony hood is present but it is not developed in bears (Pocock, 1921a). In the Felidae and in a number of the Viverridae[1] the claws are said to be retractile. It would be more correct to describe them as protractile, since at rest the claw is retracted and is actively protracted by muscular action into the striking position (figure 2.11). In the rest position the digits are slightly flexed and the terminal phalanx is pulled back by a pair of retractor ligaments, originating on the sides of the second phalanx. The tendon of the flexor longus digitorum muscle is attached to the base of the ungual phalanx, so that when the muscle contracts, the phalanx is rotated forwards. Full protraction of the claw, however, requires simultaneous action of the dorsal extensor muscles, which straighten the digit and thus ensure that the action of the flexor is concentrated at the terminal articulation. In striking with the claws, the abductor muscles are also normally brought into play and the digits are spread sideways as well as extended.

In cursorial species like the Canidae and Hyaenidae, the rotatory movement of the terminal phalanx is negligible and the claws are not significantly retracted in the rest position: the tips are therefore normally worn down and the claws are consequently blunt. In the cheetah, too, the claws are only very slightly retracted and on digits 2 to 5 they are worn down like those of a dog. The dew-claw, on the pollex, however, does not come into contact with the ground and is not worn down in the same way: it therefore remains long and sharp and constitutes a formidable weapon which is of considerable importance in prey capture. In captive animals there may be very little wear on any of the claws and in the specimen figured by Pocock (1916e) the other claws were almost as sharp as the dew-claw (see figure 3.17, p. 109).

Many of the mongooses use the non-retractile claws of the fore paws for digging. The claws are narrow and although they become worn down they remain reasonably sharp. In adaptation to this differential use the claws on the fore feet grow much faster than those

[1] Retractile claws have been described in members of three viverrid subfamilies: Viverrinae (*Genetta*, *Poiana*, *Viverra*, *Prionodon*); Paradoxurinae (*Nandinia*, *Paradoxurus*); Cryptoproctinae (*Cryptoprocta*). In the otter civet, *Cynogale bennetti* (subfamily Hemigalinae), the claws are partially retractile.

on the hind feet: a captive animal with restricted opportunities for digging may therefore periodically require a manicure but not a pedicure.

The fore limb is used for a variety of purposes besides locomotion but the hind limb is almost exclusively a locomotor organ and its adaptations are therefore predominantly related to this single function. In this connection the relations of the tibia and fibula are of some interest. Primitively the two bones are separate and the fibula articulates with the tibia at its proximal and distal ends by synovial joints, so that both extremities are movable. The lower articulation is such as to permit rotation of the fibula about its long axis and some degree of supination of the foot is possible. Owing to the different arrangements of elbow and knee in the fore and hind limbs, the tibia and fibula are crossed in supination and lie parallel in pronation, whereas the reverse is true of the radius and ulna in the fore limb. The mobility of the lower end of the fibula permits the foot to adjust itself to uneven ground and may also be of importance

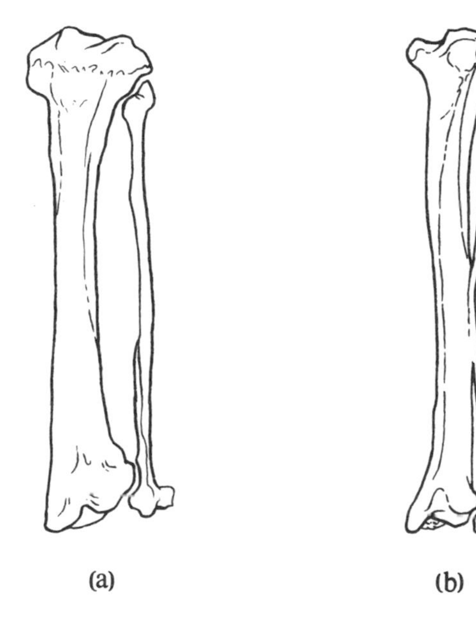

Figure 2.12. Tibia and fibula of (a) Himalayan bear, (b) cheetah.

in climbing. A mobile fibula may be regarded as the basic carnivore arrangement and is typical of the families other than the Canidae and Hyaenidae (figure 2.12). Amongst the Felidae the mobility of the ankle joint is so great in the tree ocelot, *Leopardus wiedi*, that supination through 180° is possible (Leyhausen, 1963) and the animal can thus grasp a branch equally well with fore and hind paws. This facility makes the tree ocelot capable of arboreal acrobatic feats unparalleled in other Felidae (plate 1).

In the more cursorial Canidae and Hyaenidae, movement at the ankle is mainly a hinge action: lateral mobility is restricted and the lower end of the fibula is firmly tied to the tibia by a syndesmosis, giving increased stability at the expense of flexibility. The synovial joint at the proximal end, however, is retained. The same condition is found in the badger but here the stabilising of the ankle joint is presumably related to digging rather than to running. The coati, *Nasua nasua*, is unusual amongst carnivores in that the slender fibula is fused with the tibia proximally but articulates distally by a synovial joint. While the relation of the mobile ankle to the animal's climbing ability is clear, the significance of the fused proximal end is not apparent. In the cursorial cheetah (figure 2.12) the arrangement is unique (Barnett and Napier, 1953). The fibula is long and slender and, although the proximal and distal synovial joints are present as in other felids, the middle of the shaft is closely bound to the tibia by fibrous tissue. The bone itself is sufficiently flexible to permit some independent movement at the two ends. As far as the ankle joint goes, this arrangement may represent a compromise, permitting rather more lateral movement than is present in the Canidae but giving a more stable arrangement than in other Felidae. The upper end of the fibula parallels the condition found in many leaping or hopping mammals, which Barnett and Napier regard as in some way related to the muscle arrangements involved in producing very fast extension of the ankle. If this is correct, then its significance in the cheetah is presumably similar, for a forward leap from the hind feet forms an important part of the cheetah's gallop.

The more proximal articulations of the limbs reflect the same principles. In cursorial species there is a tendency for the action to be limited to movement in the sagittal plane: the head of the humerus is more cylindrical and the articulating grooves between humerus and ulna, femur and tibia, are deeper; whereas in climbers there is more lateral and rotational mobility. One does not usually think of foxes as climbing animals but the grey fox, *Urocyon cinereoargenteus*, is an excellent tree climber (Yeager, 1938; Terres, 1939) and its forearm has greater rotational mobility than that of any other member of the Canidae (Hildebrand, 1954b). In truly arboreal species the action of the limbs involves abduction and adduction even when the animal is merely walking along a horizontal branch. The limb is not simply flexed and brought directly forwards on the recovery stroke, as in the normal walking of terrestrial species, but is swung out laterally in an arc and there is very little flexion. This allows the animal to make a quicker correcting movement if it should happen to slip (Taylor,

1970). Taylor points out that the rolling gait of the experienced sailor exemplifies the use of exactly the same principle to safeguard against possible sudden loss of balance.

The pectoral girdle, as well as the bones of the limb itself, shows modifications related to the way the arm is used. The clavicle is never large in carnivores and in the Canidae it is absent, or reduced to a mere sliver embedded in the cephalo-humeralis muscle. In the Felidae it is a trifle larger but still does not make direct contact with either the sternum or the scapula (figure 2.13). The main function of

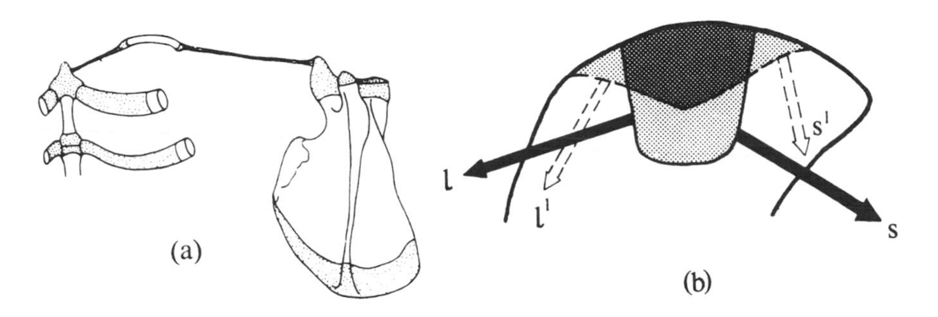

Figure 2.13. (a) Vestigial clavicle of cat, joined only by ligaments to sternum and scapula. [From Grassé, 1967]
(b) Inner surface of scapula, showing area of origin and approximate angles of action of muscles serratus magnus (s, s^1) and levator scapulae (l, l^1) in cheetah (full line and black arrows) and in leopard (broken line and arrows). [After Hopwood, 1947]

the clavicle is to stabilise the lower end of the scapula during abduction of the arm and it also provides attachment for some of the adductor fibres of the pectoralis major and for a part of the deltoideus muscle used in raising the arm. It is therefore not required in cursorial species where abduction and adduction are unimportant and, indeed, would tend to limit movement of the scapula which, as we have seen in the case of the cheetah, contributes to the length of the stride.

Since movements of the scapula are involved in locomotion, it is only to be expected that its form too should be altered in relation to cursorial adaptation. In non-cursorial species the scapula is distinctly fan-shaped, wider above than below but in the cursorial Canidae it is long and narrow, almost rectangular in shape (figure 2.14). The upper border of the scapula can be pulled forward or backward relative to the thorax by the levator scapulae and serratus anterior muscles, which are also important in suspending the thorax from the scapula. The scapula pivots about its mid-point and the lower end, carrying the head of the humerus, is therefore moved backward and forward at each stride by the alternate contractions of these muscles. The elongated scapula of cursorial animals thus increases the effective length of the stride; moreover, forward rotation of the glenoid permits the humerus to be more fully protracted than would otherwise be possible. Amongst the Felidae, the scapula of the cheetah is more elongated than that of any other species and on its inner surface the attachment area for the levator scapulae and serratus muscles is unusually deep and narrow. In the leopard, on the other hand, the muscle attachment is wide and shallow. This difference is related to the different angles at which the muscles work in the two animals (figure 2.13). In the cursorial cheetah their fore and aft component is the major one, whereas in the climbing leopard the vertical component, responsible for raising the thorax, is the more important (Hopwood, 1947).

In the Ursidae (figure 2.14), the scapula differs from that of most other Carnivora in that there is a wide flange, the post-scapular fossa, on the upper part of the posterior margin. The subscapularis minor muscle arises from both lateral and mesial surfaces of this extension and runs down to insert on the head of the humerus (Davis, 1949b). The unusual development of this muscle in bears is correlated with their method of climbing, which involves pulling up the heavy body by the fore limbs. This action tends to pull the humerus away from its articulation with the glenoid and exerts a force which is the direct opposite of the thrust associated with normal quadrupedal locomotion. The function of the enlarged subscapularis minor is to resist this pull. Although many of the Procyonidae are arboreal, their progression is by jumping and running along the branches and not like the climbing of bears. It is, therefore, not surprising that in most of them the subscapular fossa is small. The giant panda, however, climbs very much as a bear does and the scapula has a similar subscapular fossa. This, however, does not necessarily imply any close relationship between the two: it is an adaptation which one would expect to find in any bear-sized carnivore that climbs as a bear does.

Oxnard (1968) used a multivariate analysis method to study the architecture of the mammalian shoulder girdle. His main interest was in primates and their adaptations to climbing but a number of carnivores were included in the study. Of the two canonical variates on which he lays most emphasis, one gives a measure of the degree to which the shoulder is adapted to withstand tension, rather than compression: it might therefore be expected to confirm Davis's findings with respect to the arboreal adaptations of bears and the giant panda. This is indeed the case. The value for *Ailuropoda* falls together with those for the bears, the closest neighbour being the

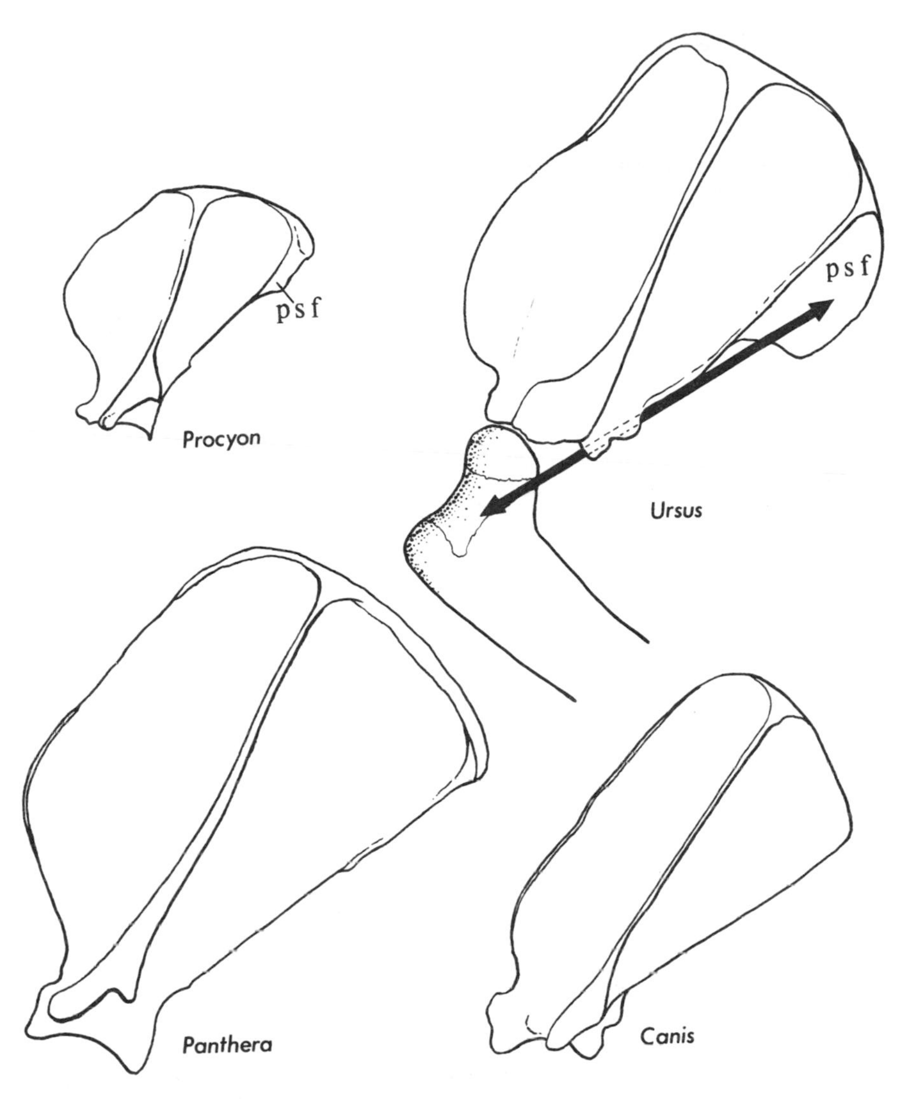

Figure 2.14. Scapulae of various carnivores seen from the side. psf, postscapular fossa. The arrow shows the line of action of the subscapularis minor muscle. [After Davis, 1949]

most arboreal of the Ursidae, *Helarctos*, the Malayan sun bear. One point of some interest is that the value for the polar bear, although the most extreme for the group, lies with the other Ursidae and not with the terrestrial carnivores. This, however, is not really surprising. True, a polar bear does not climb trees, but it swims with a stroke rather like the crawl, pulling itself through the water with its fore

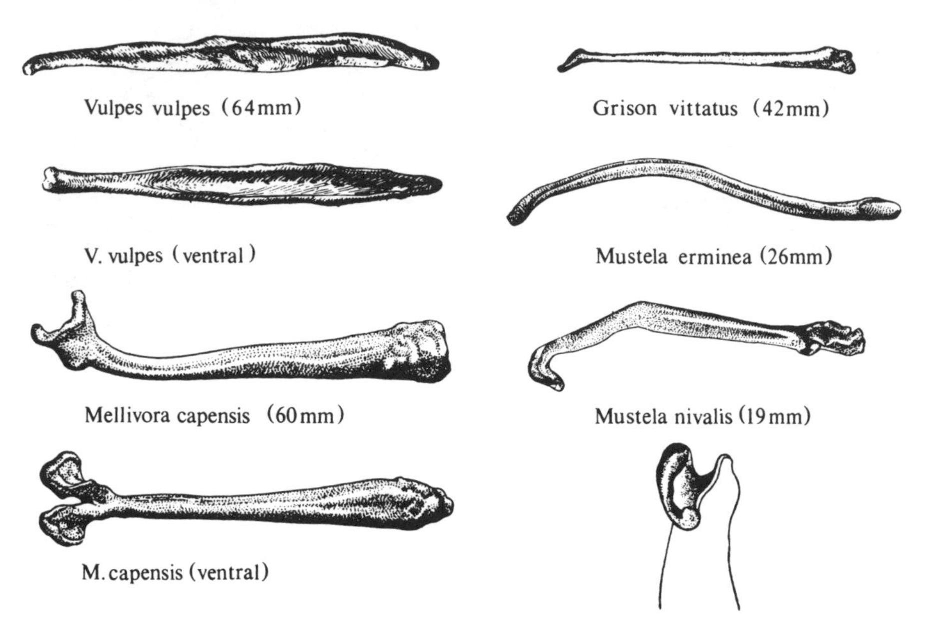

Figure 2.15. Os penis of Canidae and Mustelidae. Lateral views except where otherwise indicated; distal ends to the left. Figures in brackets are the actual lengths of the bones. [From Didier]

limbs while its hind legs trail behind (Flyger and Townsend, 1968). Any swimmer knows that this subjects the arms to a pull, not a thrust and it is therefore natural that the girdle should show adaptations to withstand traction, even if these are less extreme than in the case of the tree climbers. Oxnard's other variates, being chosen for their relevance to other orders, are not very revealing in relation to carnivores.

The proportions of the limb bones remain to be considered. In generalised mammals the segments of the limb decrease in length proximo-distally: the femur is longer than the tibia and the foot shorter; similarly the humerus is longer than the radius and the manus very much shorter. In cursorial adaptation there is a general tendency not only for pes and manus to lengthen but for the proximal segments to become relatively shorter. In both artiodactyls and perissodactyls this adaptation is very pronounced: humerus and femur are short and the main bulk of the musculature is located near the proximal end of the limb. The long distal segments do not carry a heavy load of muscles, for the movements they perform are very restricted. Such an arrangement makes for economy, since the energy

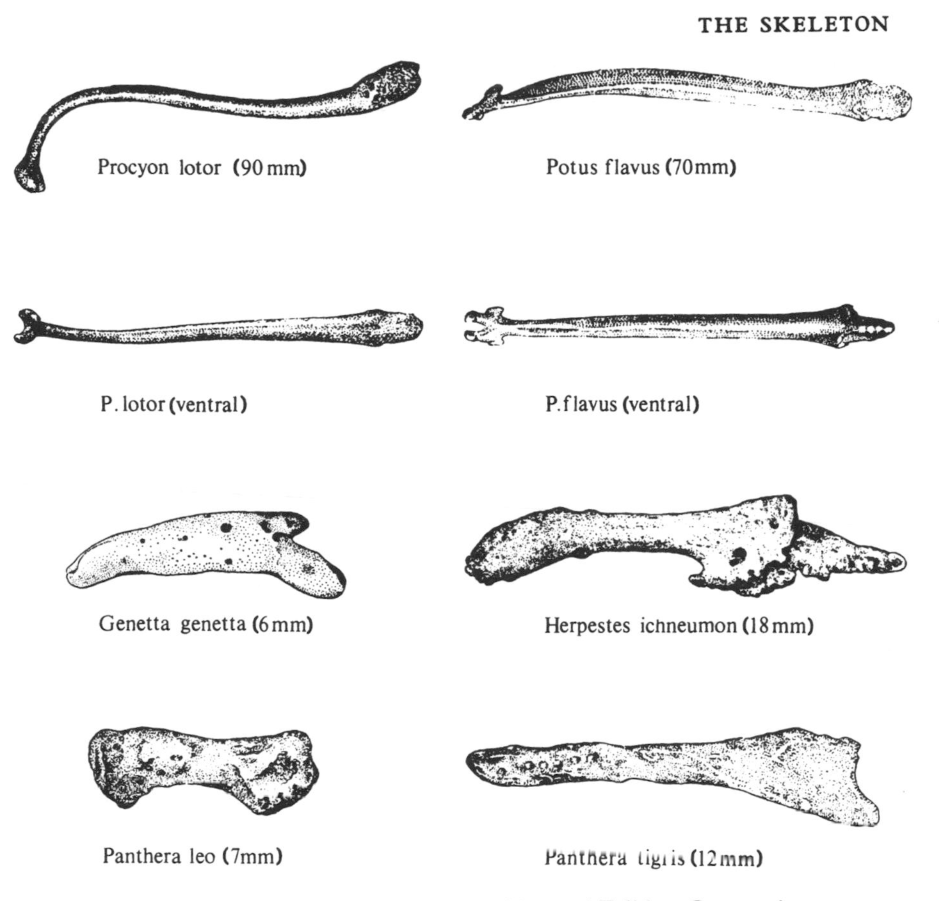

Figure 2.16. Os penis of Procyonidae, Viverridae and Felidae. Conventions as in figure 2.15. [From Didier]

required to swing the limb itself to and fro is minimal. In cursorial Carnivora, however, the trend is not nearly so striking: even in the cheetah and the African hunting dog, femur and tibia, humerus and radius differ very little in length and the same is true of the greyhound. This may be because the limbs of carnivores are never locomotor organs pure and simple. The number of digits is not reduced beyond four, the musculature of the forearm and shank remains relatively complex and the limb proportions may have to be such as to permit the animal to dig a burrow and crawl into it, to swim or to climb a tree, as well as to run.

Before leaving the post-cranial skeleton it is necessary to mention the baculum or os penis (figures 2.15 and 2.16). A penis bone is typically present in all families of the Carnivora except the Hyaenidae. Didier (1946, 1947–8, 1949, 1950) has given descriptions of the bacula of representatives of all the families and Burt (1960) has made a study of those of North American species. In its simplest form the baculum is a rod-like structure, grooved below for the passage of the urethra and the corpus spongiosum. In the Canoidea as a whole, the baculum is of this general type: in the Feloidea, where the penis is much shorter, the baculum is correspondingly truncated and in the Felidae it is very reduced. In the Mustelidae, the baculum is of the elongated canoid type but the distal end is often expanded and assumes a variety of peculiar shapes in different species. In most viverrids a baculum is present but in the banded civet, *Eupleres*, it is very reduced and it is said to be absent in the palm civets *Paguma* and *Paradoxurus*. The fossa, *Cryptoprocta*, is exceptional amongst feloids in having a long baculum and the pandas amongst canoids in having very short ones (Pocock, 1921a).

Earlier studies were mainly concerned with the possible taxonomic value of the os penis. While it is true that the bacula of different species have distinctive characters, the differences do not always reflect taxonomic relationships in any very simple way. Pocock (1918a), for instance, found that the baculum of the weasel, *Mustela nivalis*, is very different from that of the closely related stoat, *M. erminea* and is more like that of the polecat, *M. putorius*. Didier (1947) too remarked on the difference between stoat and weasel. He also found considerable differences between the bacula of two species of raccoon, *Procyon lotor* and *P. cancrivorus*. Amongst the Viverridae, Didier (1948) noted that the marsh mongoose, *Atilax paludinosus*, differs from other mongooses in having an unusually long thin baculum and he also found that the bacula of specimens of the small-spotted genet, *Genetta genetta*, from Europe were quite distinct from those of specimens from Africa identified as belonging to the same species. In this connection it may be noted that, according to Long and Frank (1968), two types of bacula with slightly differently shaped tips occur in the American badger, *Taxidea taxus*. It is not clear from their account whether both types occur within a single population or whether they are characteristic of different areas. Amongst the Felidae, Didier (1949) found that the baculum of the domestic cat resembles that of the European wildcat, *Felis silvestris*, but differs quite considerably from that of *F. libyca*, the African wildcat. He also notes that the penis bones of lion and tiger are distinguishable. Of the Felidae examined to date, the most complex baculum is found in the ocelot, *Leopardus pardalis*.

Deansley (1935) discovered that in the stoat the size of the baculum could be used to distinguish males in their first year of life from older ones. Since then a number of authors have investigated the possibility of using the baculum as a criterion of age in a number of mustelids. Most of them have found, as Deansley did, that after sexual maturity has been reached growth of the baculum is so slow that in practice it can be used only to distinguish immatures from full adults. This has been found to hold for the long-tailed weasel, *Mustela frenata* (Wright, 1947), the mink, *M. vison* (Elder, 1951) and the polecat (Walton, 1968). According to Friley (1949), however, it is possible to divide river otters, *Lutra canadensis*, into four age groups, according to their bacular development.

Relatively little attention seems to have been devoted to the difficult problem of the functional significance of the details of bacular structure; indeed Burt (1960) is inclined to doubt if all the differences found could have adaptive meaning. Long and Frank (1968) have attempted an analysis in terms of the two most likely functions - effecting penetration of the vaginal orifice and providing vaginal stimulation. The relative importance of these two functions may vary from case to case. Long and Frank point out that in many mustelids the male is considerably larger than the female, which may make penetration more difficult. Many Carnivora are known to be induced ovulators and the degree of stimulation necessary to ensure adequate liberation of ova may not be the same in all species.

In the mouse, pseudopregnancy can be induced by mechanical stimulation of the cervix. Diamond (1970) has recently shown that this is most successful if the temporal sequence of stimulation closely mimics the normal sequence of intromissions by the male. He is of the opinion that there is a specific coding of stimulus patterning to which the vagina is most responsive, which may serve as a reproductive isolating mechanism. It is thus quite possible that in the Carnivora, in addition to the timing of the sequence of intromissions, the characters of the baculum, by affecting the quality of the stimulation, may act in the same way. In these terms the bacular differences between closely related species are easily explicable as mechanisms contributing to reproductive isolation.

A further point not considered by Long and Frank is that if the penis is large relative to the vagina, a grooved baculum may serve to prevent occlusion of the urethral canal during copulation. This may also be important in the Canidae, where twisting of the penis occurs during the copulatory tie. In view of all these variables it does not seem unreasonable to suppose that the details of bacular form have significance but, in order to elucidate the problem, each case would have to be considered individually in relation to details of female anatomy and physiology and of copulatory procedure.

In a number of carnivores a small bone, the os clitoridis, homologous with the baculum, occurs in the female. This has been described in representatives of several families: the Canidae (*Urocyon cinereo-*

argenteus, Hildebrand, 1954b); the Procyonidae (*Procyon lotor*, Rinker, 1944 and *Bassariscus* sp., Arata, 1965); the Mustelidae (*Mustela vison*, Long and Shirek, 1970, *Taxidea taxus*, Hoffmeister and Winklemann, 1958 and *Lutra canadensis*, Scheffer, 1939) and the Viverridae (*Cryptoprocta ferox*, Lönnberg, 1902). It may well be present in a great many other species but its function remains obscure.

B. Skulls and teeth

The skull includes not only the housing of the brain and organs of special sense but also the masticatory apparatus. One of the most fundamental features of skull architecture is therefore the type of dentition, with which the form of the jaws, the glenoid articulation and the jaw musculature are adaptively correlated. The size and position of the orbits and the size of the braincase are also factors of importance. Since the temporalis and anterior neck muscles are attached to the outer surface of the braincase, the relationship between brain size and musculature affects the form of the skull. If the muscles require a larger area for attachment than the braincase provides, then this is made good by the addition of bony crests. A median sagittal crest gives more space for the temporalis attachment, while larger neck muscles can be accommodated by the addition of an occipital crest, dividing off the posteriorly facing occiput from the sides and roof of the braincase. Absolute size also affects the development of crests. Nerve cells do not have to be enlarged in proportion to body size and the relationship between brain volume and body volume is not linear, the brain being relatively smaller the larger the animal. Closely related species of widely differing sizes will therefore have skulls that are superficially very different (figure 2.17). In the smaller skull the braincase will dominate the entire structure and its surface, ample for all the muscle attachments required, will be relatively smooth: in the larger animal, the relative importance of the jaws will be greater, the braincase will be relatively reduced and, if the jaws are powerful, its surface will be complicated by the development of bony crests.[1]

Since some of the temporalis muscle fibres originate from the orbital ligament and the connective tissue closing off the back of the orbit, there is a relationship between the size of the eyes, the importance of the temporalis and the size of the postorbital processes on the frontal and on the zygoma which support the orbital

[1] Skulls from captive animals fed on soft foods are often highly abnormal. With little work to do, the jaw muscles do not develop their usual size and strength and the skull may therefore lack crests which would be present in any normal wild individual.

ligament. If the eyes are large and the temporalis muscle is strong, then the postorbital processes are generally large and may even meet to form a postorbital bar. If the eyes are relatively small, then even with a strong temporalis the processes may not be particularly large: similarly, even if the eyes are large, the processes may be poorly developed if the temporalis is weak.

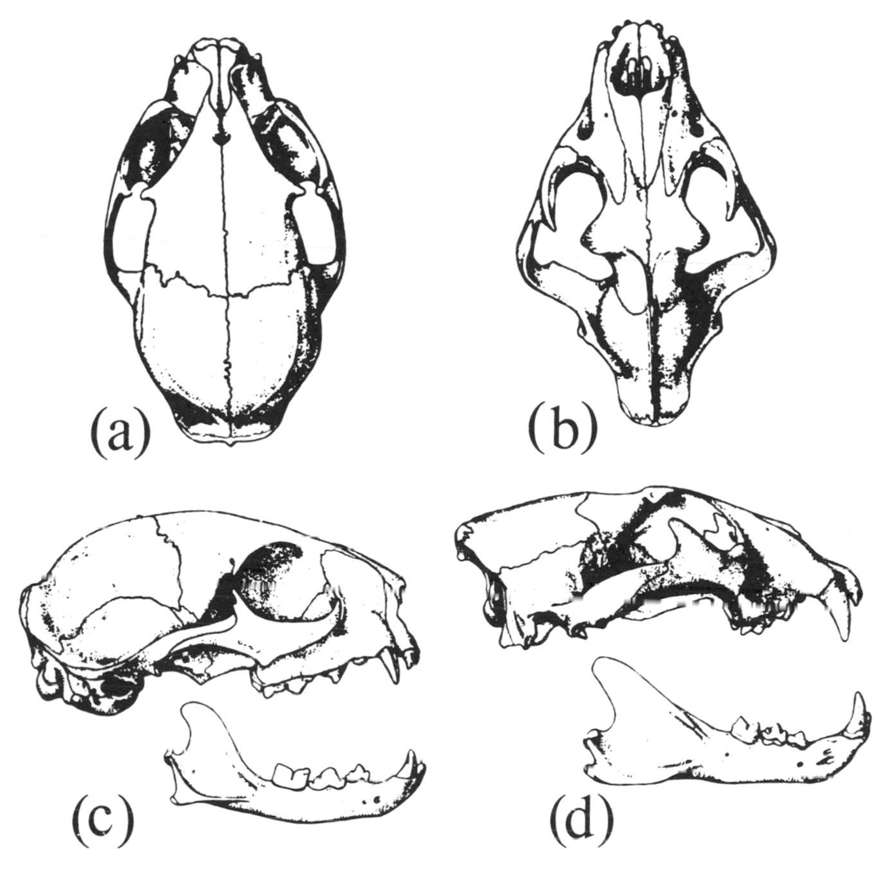

Figure 2.17. Relation of absolute size to skull proportions in Felidae. The relatively large braincase of the small species [jaguarondi, (a) and (c)] provides ample space for jaw and neck muscles without the development of bony crests: in the large species [tiger, (b) and (d)], with a relatively smaller braincase, this is not so and large crests are required. [From Grassé, 1967]

Since the masticatory apparatus is so important in determining skull structure, it is necessary to deal with dentition and jaw musculature in some detail. For convenient reference, dental formulae are summarised in table 2.2, at the end of this chapter.

The feeding habits of the extant Carnivora range from almost exclusive flesh eating to secondary vegetarianism and the teeth and jaws show corresponding modifications. It is therefore desirable to start by considering a type of dentition which is not highly specialised and which is capable of dealing with a variety of foods. For this the Canidae are the most suitable and the dog will serve as an example (figures 2.18 and 2.19). The dental formula, as in the majority of Canidae, is $I\frac{3}{3}$, $C\frac{1}{1}$, $P\frac{4}{4}$, $M\frac{2}{3}$, making a total of forty-two teeth. The canines are large but neither very sharp nor very flattened. The lower canines are set a little in advance of the uppers and small diastemata in front of the upper and behind the lower allow the teeth to interlock when the jaw is closed. The incisors are moderately large and set on a slight curve, so that they and the canines form a mechanism capable of gripping and tearing.

It is characteristic of Carnivora that the carnassial teeth, the last premolar in the upper jaw (P^4) and the first molar in the lower (M_1), are specially adapted to cutting through flesh with a scissor-like action and the dog, chewing sideways at a bone, is using the carnassial shear to slice away the meat. The posterior two cusps of P^4 and the anterior two of M_1 are laterally flattened and, as the jaws close, the blades shear past each other. The two constituent cusps do not form straight lines but are arranged so that each blade has the shape of a wide open V. This increases efficiency by preventing the meat from slipping out forwards and makes the action really more comparable with that of pruning shears than of ordinary scissors.

The cheek teeth in front of the carnassials are simpler. Each has a single, somewhat flattened, main cusp and smaller accessory ones set in line fore and aft: the tooth is thus sectorial, although less so than the highly modified carnassials. Behind the carnassial blades the teeth are modified in the opposite direction, for crushing not for slicing. In the upper jaw the molars constitute the crushing apparatus. However, since the upper and lower teeth are not directly one above the other but interdigitate, with the lower slightly in advance of the corresponding upper, the lower carnassial is necessarily a dual-purpose tooth. The anterior part forms the lower blade of the carnassial scissors, while the posterior region, occluding with M^1, belongs to the crushing apparatus. This part of the tooth forms a basined heel, with cusps on the outer and inner margins. The two cusps fit into the double-basined inner lobe of the triangular M^1 and the last upper and lower molars interlock in a similar way. The dentition as a whole is thus adapted not only for killing prey and slicing off the flesh but also for crushing up vegetable foods.

The jaw action involved in using this dentition is quite complex. Use of the anterior teeth requires only a simple hinge-like jaw closure, with sufficient force to drive in the canines. The carnassial shearing, however, demands lateral movement of the jaw as it closes, so that the upper and lower blades are kept in proper adjustment all the time. Lateral movement could also increase the efficiency of the molars by adding a shearing component to their direct crushing or pulping action. The transversely elongated glenoid articulation acts as a strong hinge joint and at the same time permits some lateral movement. In addition, the mandibular symphysis is not a firm bony fusion but a mobile union involving a complex arrangement of ligaments and fibro-cartilage (Scapino, 1965). This allows some degree of independent movement of the two mandibular rami but the significance of this extra mobility is by no means clear. How much movement actually does occur during chewing could be determined only

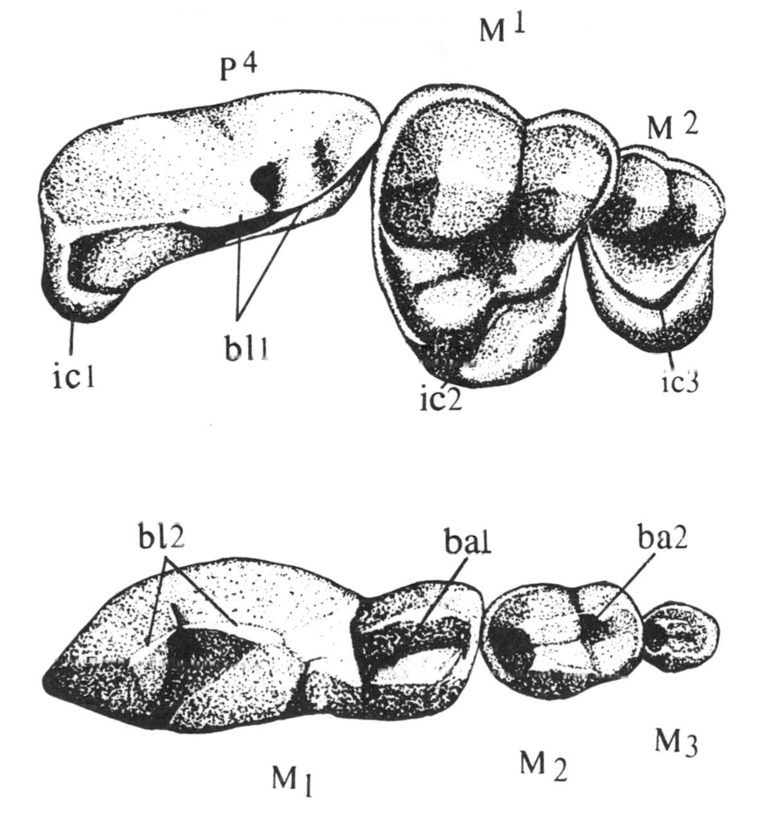

Figure 2.18. Upper and lower carnassial and post-carnassial teeth of dog. The upper and lower carnassial blades (bl1, bl2) shear against each other; the inner cusps of the upper molars (ic2, ic3) fit into the basins (ba1, ba2) on the lower carnassial (M_1) and on M_2 and form a crushing mechanism. ic1, inner cusp (protocone) of upper carnassial. [After Scapino, 1965]

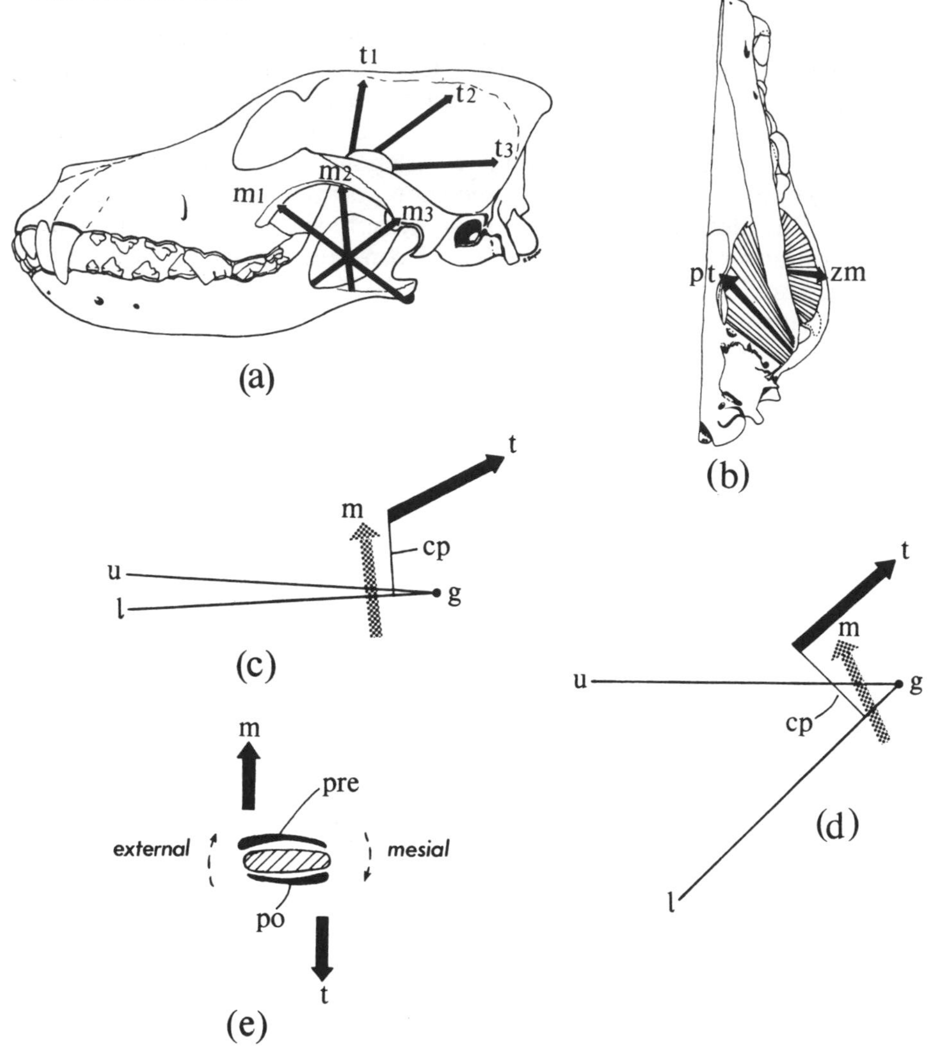

Figure 2.19. (Above) Skull of dog. The arrows show the lines of action of the jaw closing muscles. (a) Side view: t1, t2, t3, anterior, middle and posterior fibres of temporalis muscle; m1, m2, m3, anterior middle and posterior fibres of masseter muscle. (b) Ventral view: pt, pterygoideus muscle; zm, zygomatico-mandibularis muscle. [After Scapino, 1965] (Below) Lines of action of m, masseter and t, temporalis muscles; (c) with the jaws almost closed and (d) wide open. cp, coronoid process; g, glenoid articulation; l, lower and u, upper jaws. The posterior fibres of the temporalis are most effective when the jaws are widely open, the masseter when they are almost closed. (e) Diagrammatic horizontal section through the condyle of the mandible (hatched), to show the forces exerted on it by the masseter (m) and temporalis (t) muscles. The two form a couple tending to twist the condyle in the direction shown by the broken arrows. To resist these forces, the preglenoid process (pre) is enlarged externally, the postglenoid process (po) mesially.

by X-ray filming and this has not yet been attempted. Moreover, neither the most specialised carnassial slicers (the Felidae) nor the most specialised molar grinders (the Ursidae) have a mobile symphysis.

Closure of the jaws is brought about by the action of three sets of muscles, two inserting principally on the outer and one on the inner surface of the mandible (figure 2.19). The temporalis arises from the lateral surface of the braincase and from the ligament behind the eye. Its fibres run down to insert on the anterior border and upper part of the coronoid process. Its anterior fibres thus pull directly upward, while the more posterior ones pull upward and backward. Since the jaw lies lateral to the braincase, the temporalis fibres must also have some tendency to pull the jaw mesiad.

The masseter muscle fibres originate from the lower border of the zygoma and run down to insert on the angle of the lower jaw and in the masseteric fossa. The most superficial fibres, arising from the anterior part of the zygoma, wrap round the angular process and are attached to its inner surface: their action therefore includes a considerable forward pulling component. The fibres of the next layer, arising from the central region of the zygoma, run down approximately vertically to insert on the outer surface of the angular process: they therefore exert a simple closing action. The deepest fibres arise from the posterior part of the zygoma immediately in front of the glenoid and run forward to insert in the masseteric fossa, thus exerting a strong backward pull on the mandible. Since the zygoma lies lateral to the mandible, the masseter fibres must also exert some lateral pull. In addition to these, there are fibres which originate on the mesial surface of the zygoma and insert on the outer surface of the coronoid process: their pull is therefore mainly lateral. This group of fibres is not easily separable from the deep layers of the masseter below nor from the temporalis above and many authors do not recognise it as a separate muscle. Toldt (1905), however, regarded it as functionally, if not necessarily morphologically, distinct and referred to it as the zygomatico-mandibularis muscle, a usage which, although not favoured by classical anatomists, has been adopted by a number of modern authors concerned with functional analysis of jaw mechanics.

The pterygoideus muscles originate on the side of the skull immediately below the orbit and run out and back. The more superficial fibres insert on the inner surface of the angle of the mandible and the deeper ones just anterior to the condyle. The pterygoideus therefore exerts a forward and mesiad pull.

The possession of large canine teeth poses a mechanical problem. To use them effectively a powerful bite must be delivered with the jaws widely open, whereas the carnassials are in operation with the jaws almost closed. The masseter muscle is so arranged that it acts with maximal efficiency in the latter situation but its line of action is brought progressively nearer the condyle the wider the jaws are opened and its mechanical advantage consequently decreases. The same, however, is not true of the temporalis. Downward rotation of the coronoid as the jaw opens actually brings the line of action of the main fibres into a position where their action is more effective: it also moves the insertion of the zygomatico-mandibularis down, so that this muscle acts more effectively as a jaw closer. The anterior portion of the temporalis and the zygomatico-mandibularis are therefore the most important muscles as the jaws start to close from the wide-open position: the masseter and the posterior part of the temporalis become progressively more effective as the bite is carried home and can exert their full force when the carnassial shear is in operation. The jaw muscles are thus arranged in such a way that a strong bite can be delivered with the jaws in any position (figure 2.19).

In addition to the strong closing forces exerted by the temporalis and masseter, the jaw muscles also produce components which could be used to pull the mandible forwards (superficial masseter and pterygoideus), backwards (deep masseter and posterior temporalis), laterad (zygomatico-mandibularis) and mesiad (pterygoideus). As the animal chews, there should thus be no difficulty in making whatever adjustments are necessary in relation to the size and toughness of the food that is being sliced by the carnassial blades. One would expect rather small side-to-side movements of the mandible to be all that is required and, indeed, the shape of the glenoid is such as to prevent any significant fore and aft movement. Although it does have this effect, the form taken by the glenoid – a deeply concave transverse groove – is probably not an adaptation specifically to prevent fore and aft movement so much as to resist the torque which is produced by the combined action of the posterior temporalis and superficial masseter. The former pulling backwards on the inner surface of the jaw and the latter forwards on its outer surface will produce a couple, tending to twist the condyle out of joint. The pre- and post-glenoid processes, showing their maximal development respectively on the outer and the mesial regions of the articulation, are so situated as to resist this torque most effectively (figure 2.19e).

In species where there is a specialised carnassial shear, the side-to-side movements of the mandible are not extensive but in vegetarian species this movement may be increased to produce a grinding action. The Carnivora are the only mammals in which grinding is accomplished in this way: in both the rodents and the ungulates the mandible executes a sort of rocking movement in which one condyle moves forward while the other is held back.

Opening of the jaw is brought about by a single muscle, the digastric, which originates on the paroccipital process just behind the auditory bulla and runs down to insert on the posterior part of the lower border of the mandible.

It is now necessary to deal with the variations in dentition and associated features of skull architecture which characterise the different carnivore families. Dental formulae are summarised in table 2.2 (p. 69).

(i) Canidae

Within the Canidae, departures from the typical form of dentition which has been described above are neither very numerous nor very radical. The three genera, *Cuon*, *Lycaon* and *Speothos* (the Indian dhole, African hunting dog and South American bush dog) share a peculiarity of the lower carnassial. The inner cusp of the talonid is missing, so that this part of the tooth does not form a basin but a subsidiary blade. The first upper molar is correspondingly modified and the projecting inner lobe forms a single basin not divided into two in the usual way by the protocone-metacone ridge; into this basin fits the single hypoconid cusp of the lower carnassial. In all three genera, the importance of the crushing post-carnassial molars is reduced. *Lycaon* retains the normal number but the teeth are not very large, while in the other two there is a reduction in number as well as in size. These features of the dentition suggest a highly predacious habit with diminished importance of vegetable foods in the diet. They could very easily have been independently evolved in the three genera and do not therefore constitute any very cogent grounds for uniting them in a single subfamily, as is currently done by some taxonomists.

A few species show the opposite adaptation, with the carnassial shear reduced and the molar crushing battery increased in importance. In the raccoon dog, *Nyctereutes*, vegetable foods and insects are an important part of the diet and only small vertebrate prey is taken. The carnassial blades are short, the molars large and an extra upper molar is not uncommon, making a total of forty-four teeth. In the bat-eared fox, *Otocyon*, which is largely insectivorous, this emphasis on the molars is much more pronounced. The carnassials are no longer blade-like and the molars are increased to three or four uppers and four or five lowers (plate 2 and figure 2.21).

The lower border of the mandible of *Otocyon* is peculiar in that a large bony flange is present below the angular process. *Nyctereutes* shows the same characteristic but the flange is smaller. This flange was first described by Huxley (1880), who named it the sub-angular lobe but did not discuss its function. Since insects form the main food of *Otocyon* and bulk large in the diet of *Nyctereutes,* it seems likely that the sub-angular lobe is associated with this diet. The teeth of *Otocyon* are high cusped and there is considerable interlocking. One would therefore suspect that the jaw action is a quick chopping ('intercuspidation' of Gaspard, 1964) rather than grinding. This supposition is supported by the details of the insertions of masseter and pterygoideus muscles, which do not show the arrangement characteristic of species in which there is emphasis on grinding (Gaspard, 1964). The digastric muscle inserts on the sub-angular lobe and its line of action is therefore altered in those species possessing the lobe. The insertion is shifted back to a point vertically below the glenoid, which greatly increases the muscle's efficiency as a jaw opener (figure 2.20). In relatively slow chewing through tough meat,

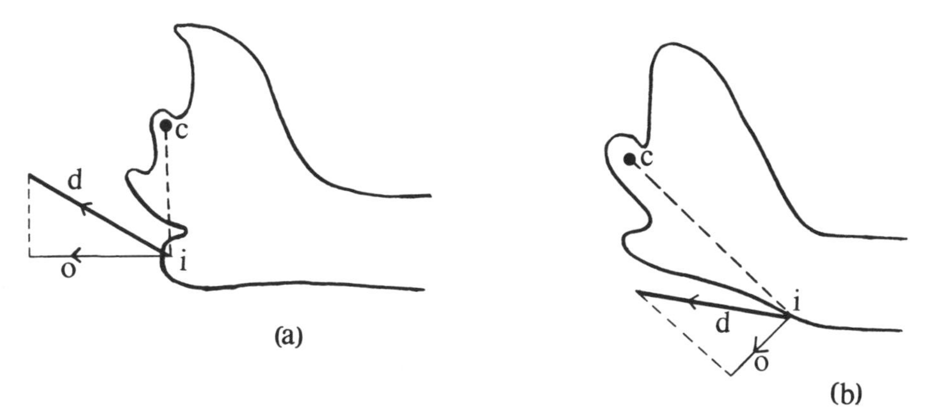

Figure 2.20. Action of depressor mandibulae muscles in (a) *Otocyon* and (b) *Vulpes*. c, condyle; i, insertion and d, line of action of depressor mandibulae muscle. o, the component of d at right angles to ci, responsible for opening the jaw, is larger in (a) than in (b). Moreover, i will move through a greater distance per unit shortening of d, so that the jaw action is faster as well as more powerful in (a).

fast opening of the jaw can be of no importance but in a quick repetitive chopping it clearly could be advantageous. *Otocyon* in fact does chew with remarkable rapidity. I have watched a tame *Otocyon* eating both insects and chicken meat and found that the chewing rate was not less than three bites per second, while some zoo animals which I watched eating a rather wettish mixture containing finely

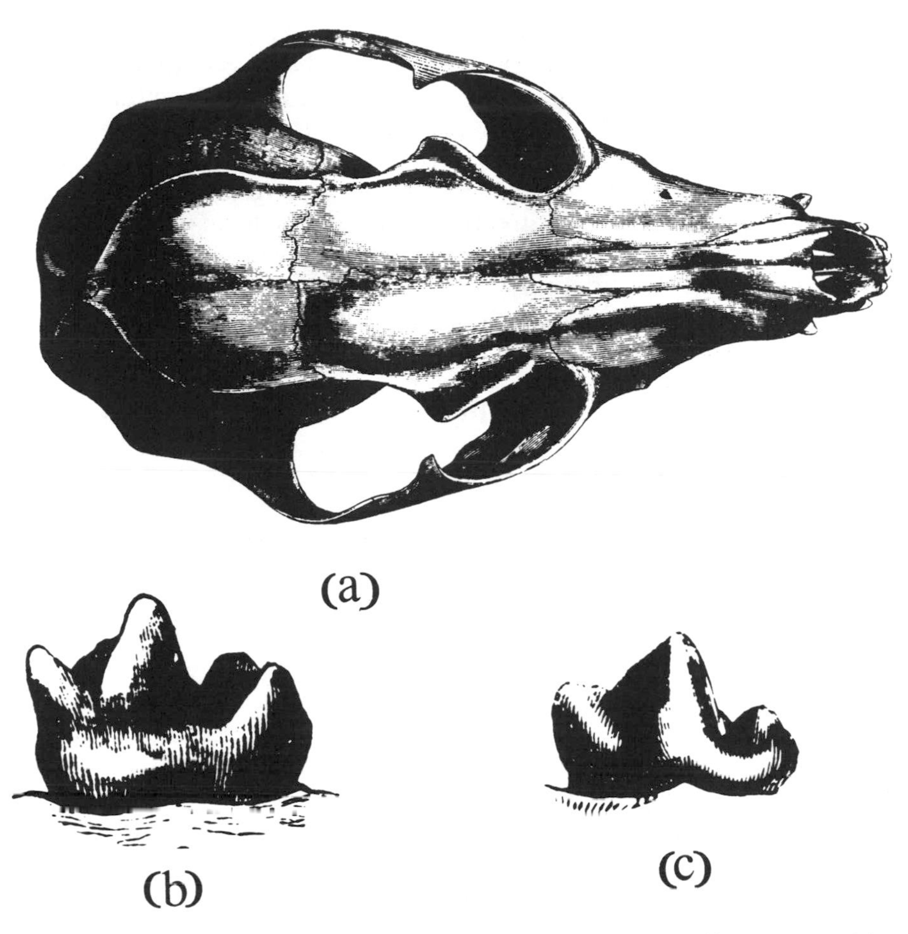

Figure 2.21. *Otocyon*, (a) dorsal view of skull, (b) mesial view of lower carnassial. (c) *Speothos*, mesial view of lower carnassial. [From Huxley, 1880]

chopped meat and milk were like ducks straining their food from the water rather than normal canids chewing and the movement was too fast for me to count.

As is natural in an animal that does not kill large prey, the jaw muscles of *Otocyon* are not very powerful and the skull shows corresponding characteristics. The zygoma is relatively slender and the small temporalis muscles do not meet above the braincase, so that instead of there being a sagittal crest the top of the braincase is formed by a broad smooth area devoid of muscle attachments (figure 2.21).

There has been some discussion as to whether in the dentition of *Otocyon* extra molars have been developed at the back of the jaw, or whether the normal molars have been reduplicated (Wood and

Wood, 1933; Guilday, 1962). This is not a very profitable way to approach the question. As long ago as 1939, Butler pointed out that in the ontogeny of a mammalian dentition the individual teeth are not fully independent and sharply discontinuous entities. Rather it appears that during development there are morphogenetic processes at work which affect the tooth germs over a certain length of the jaw. Genes probably do not directly determine the form of individual cusps or even individual teeth but exert their effects by altering the characteristics of the differentiation fields. In the case of *Otocyon* we can best describe the situation as one in which the molarisation field has been expanded (van Valen, 1964). This results not only in the development of extra molar teeth but also produces a distinct molarisation of the carnassials. In *Speothos* the reverse process has occurred: the molar field is reduced, molars disappear at the back of the jaw and M^1 and the talonid of M_1 are less molarised than usual (plate 2).

As might be expected, there is considerable individual variation in tooth number in these anomalous genera. In *Otocyon* the commonest molar formula is $M\frac{3}{4}$ but four uppers and five lowers are also found, while in *Speothos* the norm is $M\frac{1}{2}$, but individuals with $M\frac{2}{2}$ or $\frac{1}{1}$ are not uncommon (van Valen, 1964). Even amongst the normal Canidae, variations in both directions from the usual $M\frac{2}{3}$ are not infrequently recorded (Wood and Wood, 1933; Hall, 1940; Paradiso, 1966; Sealy, 1968). In view of all this variation, selection for either a reduction or an exaggeration of the molarisation field might be expected very easily to shift the norm in either a positive or a negative direction.

(*ii*) *Procyonidae and Ursidae*

The typical procyonids are rather omnivorous, with the emphasis on the crushing rather than the slicing functions of the dentition. The molars are broad and the carnassials are not specialised as cutting teeth but are also somewhat molarised. The temporalis muscle is not particularly powerful, sagittal crests are not developed and the postorbital processes are small. The jaw action is mainly crushing and chopping without a great deal of sideways movement and there is nothing very noteworthy about skull architecture.

The pandas show elaboration of the crushing features of the dentition. The molars are broad, flat, multicusped teeth; P^3, P^4 and P_4 are well developed and also somewhat molarised (figure 2.22). The jaw action is not a simple crushing one but a definite sideways grinding. The zygoma is very deep, reflecting an enlargement of the zygomatico-mandibularis muscle, which is mainly responsible for the sideways movement and the glenoid is very deep, which prevents

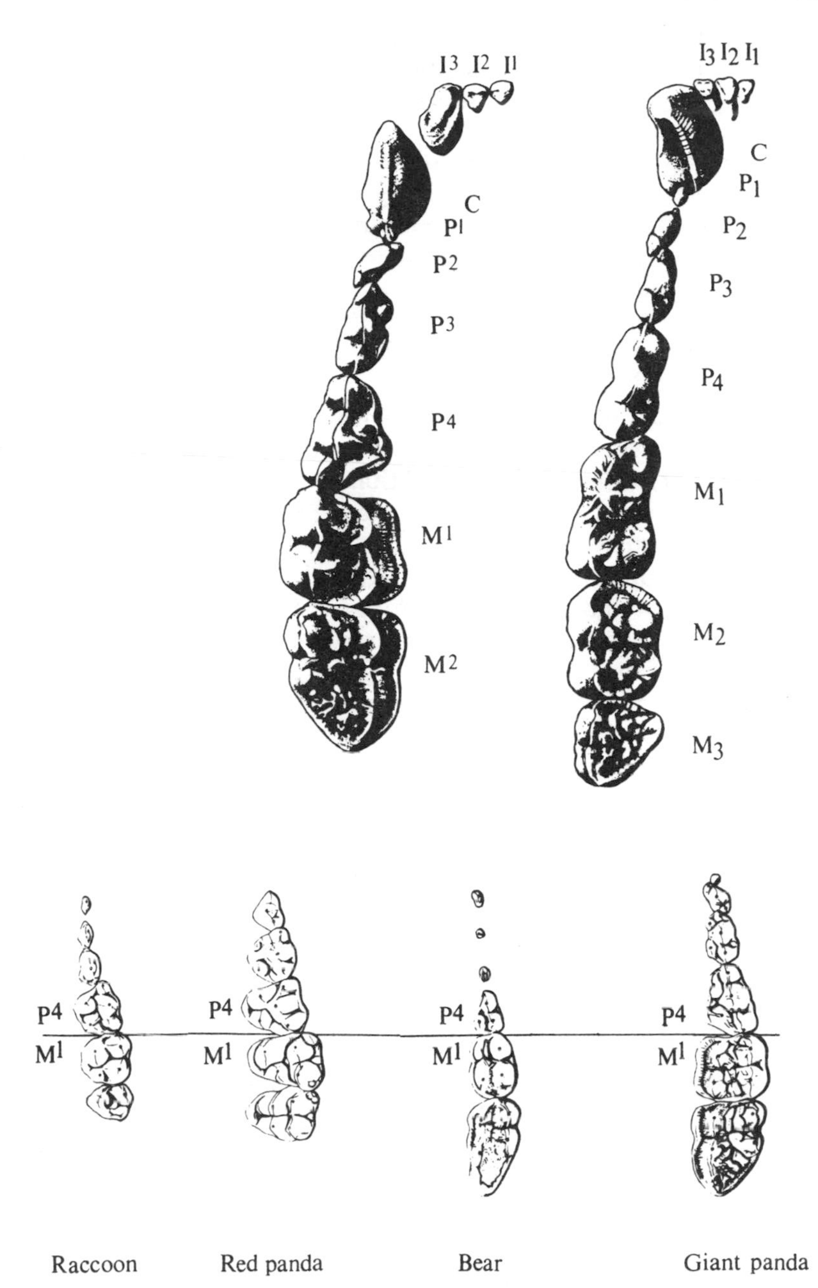

Figure 2.22. (Above) Occlusal view of teeth of giant panda. (Below) Upper cheek teeth of red and giant pandas compared with those of a procyonid and a bear. [After Davis, 1964]

back-and-forth movement of the jaw (Sicher, 1944). The temporalis muscle is relatively small and there is little in the way of a sagittal crest. The bears show comparable adaptations, most marked in purely vegetarian species and least in the secondarily more carnivorous polar bear. The skull of the South American spectacled bear, *Tremarctos ornatus*, is highly aberrant but in many of its peculiarities there is a strong resemblance to the giant panda and the characteristics of the masseteric fossa are remarkably similar. Little is known of the habits of this species but according to Cabrera and Yepes (1960) it is highly vegetarian and feeds extensively on palm nuts, a type of food which presumably requires as much crushing as the giant panda's bamboo stems.

The ursid dentition is broadly comparable with the pandas', in that the emphasis is on crushing and grinding and not on cutting. The molars are so highly modified that attempts to trace homologies between the cusp patterns are not very rewarding but there is one obvious difference between the two groups. In both, the molars provide the main grinding surface but in the pandas, as noted above, the posterior premolars are also important. In bears, the emphasis on the molars is greater and the premolars are much more reduced (figure 2.22). This would seem to suggest that the similarities in the dentitions are convergent, rather than that the pandas are derived from the bears.

Before leaving the bears, one peculiarity in the dentition of the sloth bear, *Melursus ursinus*, must be mentioned. Termites form an important component of the diet of this species and the long tongue and very mobile lips are adapted for licking up small insects. In correlation with this habit, there are only two pairs of upper incisors separated by a median diastema, through which the tongue can be protruded.

(*iii*) *Mustelidae*

The mustelids are a group showing a wide variety of dental adaptations. The subfamily Mustelinae are typically specialised predators with elongated blade-like carnassials, well developed canines and short but powerful jaws. We are accustomed to regard the large felids as the killers *par excellence* but, on a lesser scale of magnitude, the stoat killing a rabbit is no less worthy of respect than a lion bringing down a wildebeest. Although both depend for their efficiency on sharp canine teeth and cutting carnassials, the shape of the skull in the two forms is very different. In the felid, the braincase is high and wide but relatively short, the eyes are large and the zygoma strong. In the mustelid, the braincase is long and low and extends far back behind the glenoid, the eyes are relatively small and the zygoma rather weak. In the felid, both masseter and temporalis muscles are powerful but in the mustelid the weak zygoma indicates that the major share of the work must be done by the temporalis. This is borne out by Schumacher's (1961) finding that in the otter, *Lutra lutra*, the temporalis constitutes almost 80% of the total mass of jaw musculature and the masseter less than 20%. Judging from the skulls, the stoat or weasel would show an even greater temporalis domination.

It has already been explained (page 40) how in carnivores with a more orthodox skull shape the anterior fibres of the temporalis and the zygomatico-mandibularis muscles are responsible for exerting the main force when the jaws are widely open and the canines are being driven home, whereas the masseter and the posterior part of the temporalis are most effective when the carnassial shear is being

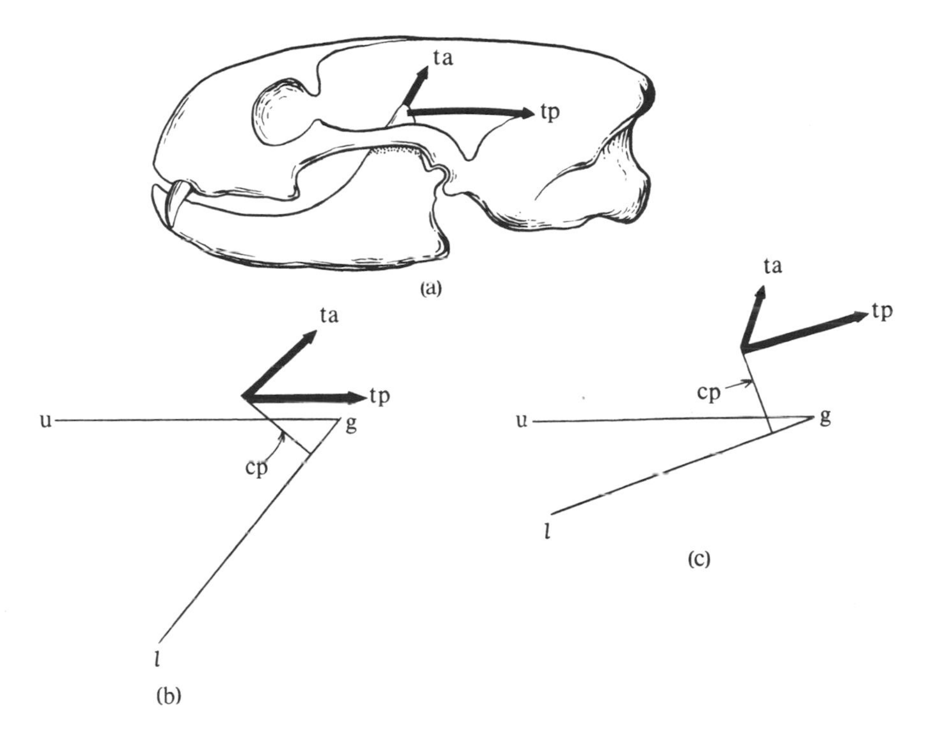

Figure 2.23. (a) Side view of weasel skull to show direction of main fibres of anterior, ta, and posterior, tp, fibres of the temporalis muscle. (b) and (c) diagrams show, respectively, lines of action of the same muscles with the jaws widely open and nearly closed. cp, coronoid process; g, glenoid articulation; u, upper and l, lower jaws. The anterior fibres are most effective when the jaw is widely open, the posterior fibres when it is more nearly closed.

used and the jaws are more nearly closed. The peculiar shape of the weasel's skull reflects a modification of this system in which the posterior part of the temporalis has taken over most of the work normally performed by the masseter and the latter is correspondingly reduced in size. This can best be understood by reference to figure 2.23. As can be seen, there is in the mustelid a great increase in the size of the posterior part of the temporalis, whose horizontally running fibres work to best advantage when the jaws are nearly closed. There must thus be an exaggeration of the normal progressive transition from anterior to posterior temporalis fibres as the jaws close, while the part played by the masseter must be less important. Such an arrangement of the temporalis muscle increases the torque exerted at the articulation when the carnassials are in use. The pre- and post-glenoid processes are therefore enlarged and, indeed, in some mustelids are so developed as to make it impossible to remove the mandible from a skull otherwise than by slipping it out sideways. This condition of the glenoid is sometimes taken to indicate that the bite of the animal is particularly powerful but in fact it is a reflection of an unusual arrangement rather than of any remarkable strength in the jaw muscles. This can easily be appreciated by comparing the skull of a weasel with that of one of the extinct sabre-tooth cats (see figure 9.5, page 373). In the latter the skull is short and high and the attachment area for the posterior temporalis fibres is small while that for the masseter is large. The contrast with the long, low weasel skull, with enlarged posterior temporalis and reduced masseter is extremely striking. Since in the sabre-tooth the temporalis pulls principally upwards on the lower jaw, rather than backwards, the torque produced at the articulation is small: the glenoid therefore need not be so deep nor the pre- and post-glenoid processes so large as in the mustelid. The jaw can therefore be opened much more widely in the sabre-tooth – as it has to be to permit the extremely long upper canine teeth to be used effectively. Thus it is clear that the arrangement of the sabre-tooth jaw muscles is related to the long upper canines, but it is less easy to understand the significance of the mustelid skull shape. Possibly the long, low and narrow skull is related to the habit of pursuing prey down burrows, which is characteristic of the smaller mustelids. Certainly it is the smaller species which show this arrangement in its most extreme form whereas the skulls of the larger species are more 'normal' in shape.

The subfamily Mephitinae (skunks) are omnivorous, making considerable use of seasonally available fruit and insect food but they are also efficient killers of small prey. The carnassial shear is not highly developed; P^4 has a large inwardly projecting protocone and a short blade (figure 2.24). Although the molars are few in number

($M\frac{1}{2}$), M_1 has a large crushing talonid and M^1 is also large, with a wide inner lobe, giving it the rectangular shape characteristic of mustelids, in contrast to the more triangular molars of the Canidae. The three living genera form a series, *Spilogale* being the most predacious and possessing the longest carnassials, *Mephitis* inter-

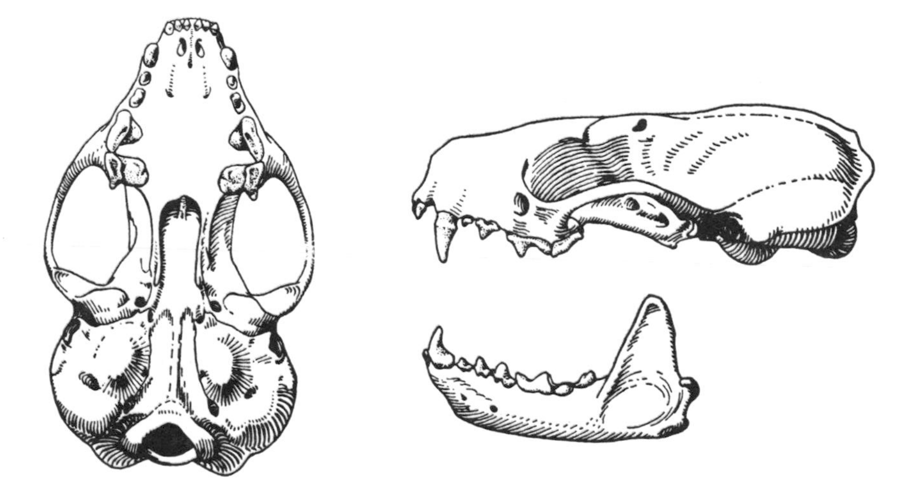

Figure 2.24. Skull of spotted skunk, *Spilogale putorius*. [From van Gelder, 1959]

mediate and *Conepatus* the least predacious form, with poorly developed carnassials and an elongated snout, adapted to rooting in the ground for invertebrate prey.

The Mellivorinae (honey-badgers), Melinae (badgers) and Lutrinae (otters) are subfamilies in which the crushing function of the dentition is emphasised in relation to particular forms of diet. In the Mellivorinae this has not proceeded very far: the molars are rather broad and the protocone of P^4 is large and heavy. In the other two subfamilies, the adaptations to crushing are much more marked. The molars are broad, flat, multicusped teeth and the carnassials are also modified for crushing, with an enlarged protocone and reduced blade. Both temporalis and masseter muscles are well developed and the power of the bite is considerable. Maxwell (1960) gives eloquent testimony to the power of an otter's bite and even a baby badger has a surprisingly strong grip. Although the three subfamilies show dental adaptations which are of the same general type, the detailed differences in cusp patterns make it clear that this has happened independently. Indeed, Pocock (1920b) regarded the teeth of the American badger, *Taxidea*, as so different from those of other badgers as to justify putting it in a separate subfamily. Later workers,

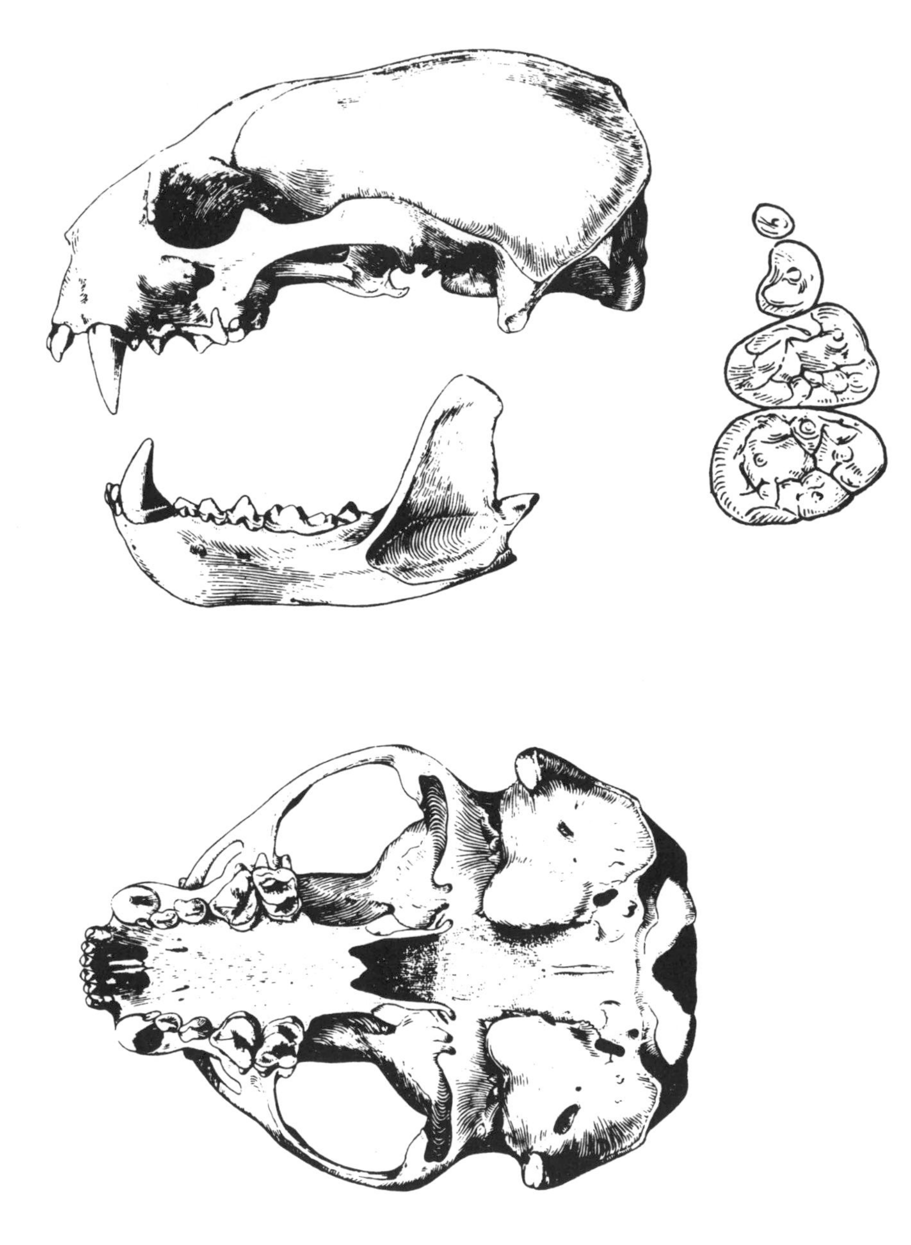

Figure 2.25. Skull of Cape clawless otter, *Aonyx capensis*, and on right, upper cheek teeth of sea otter, *Enhydra lutris*. [From Allen, 1924 and Gregory, 1951]

while not adopting such an extreme view, are agreed that despite superficial similarities the American and European badgers are not very closely related.

The most extreme development of crushing molars is found amongst the otters, in those genera that have specialised as eaters not of fish but of shelled invertebrates. *Aonyx*, the African clawless otter, lives mainly on Crustacea and *Enhydra*, the sea otter, feeds extensively on molluscs, crabs and sea urchins. In both genera the teeth are extremely broad and the carnassials highly molarised, the sea otter showing the more extreme specialisation (figure 2.25). Another peculiarity of the sea otter is that there are only two pairs of lower incisors instead of the normal three. According to Fisher (1941), it is the third incisor that is missing. This I find difficult to credit, in view of the condition in *Aonyx*, where the symphysis is narrow, I_3 is the largest incisor, I_1 is extremely small and the three teeth are staggered, I_2 being set back a trifle behind I_1 and I_3. Loss of the vestigial I_1 would produce exactly the condition found in *Enhydra*. Reduction of the incisors is paralleled but goes much further in the seals, where the Phocidae have $I\frac{3}{2}$, Monachinae $I\frac{2}{2}$ and the Cystophorinae only $I\frac{2}{1}$ (King, 1964). In both groups the reduction probably reflects the diminished importance of the teeth in aquatic carnivores, in tearing the flesh of their prey.

Hildebrand (1954a) points out that in *Enhydra* the lower incisors become more heavily worn than the uppers and interprets this as resulting from use of the lower teeth to prize molluscs loose from the rocks. This is very improbable: sea otters have been seen to remove molluscs by pounding them with a stone but never by using their teeth which, in any case, are far too weak and not conveniently situated for such use. The wear is probably produced by scraping out the contents of sea urchin tests, as described by Kenyon (1959).

(*iv*) *Viverridae*

Like the Mustelidae, the Viverridae cover a considerable range of dental adaptation and are classified into a number of subfamilies. Of these, the most familiar are the Viverrinae (civets and genets) and the Herpestinae (mongooses). The genets are very cat-like in general appearance but the skull and dentition are much less specialised. The carnassials have good cutting blades but the protocone on P^4 and the talonid on M_1 are large and there are two upper molars. The jaws are therefore considerably longer than those of the Felidae and the whole skull is long and low. The civet skull (figure 2.26) is similar but more heavily built and the carnassials are less blade-like.

Amongst mongooses, the least specialised forms, like *Herpestes*

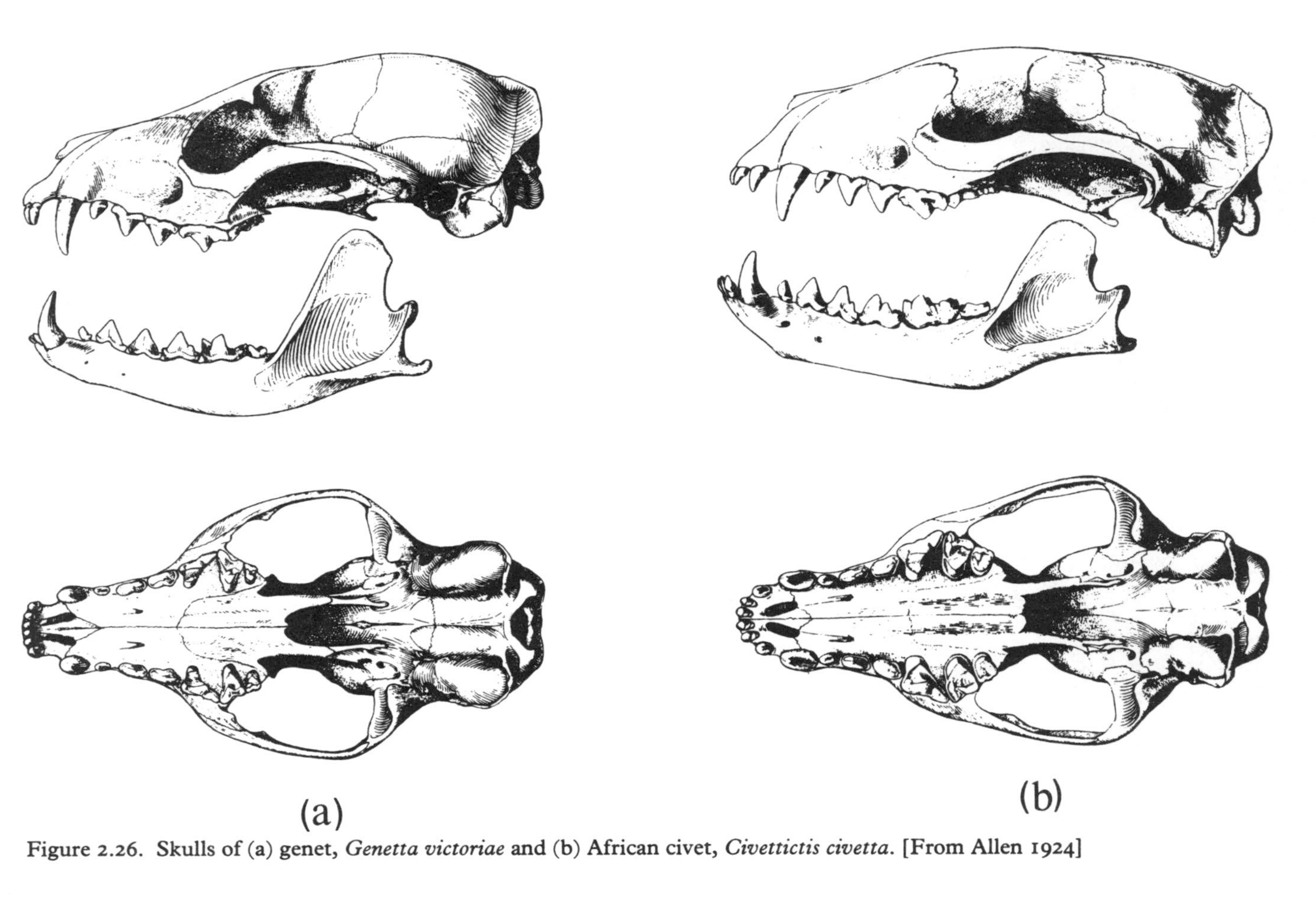

Figure 2.26. Skulls of (a) genet, *Genetta victoriae* and (b) African civet, *Civettictis civetta*. [From Allen 1924]

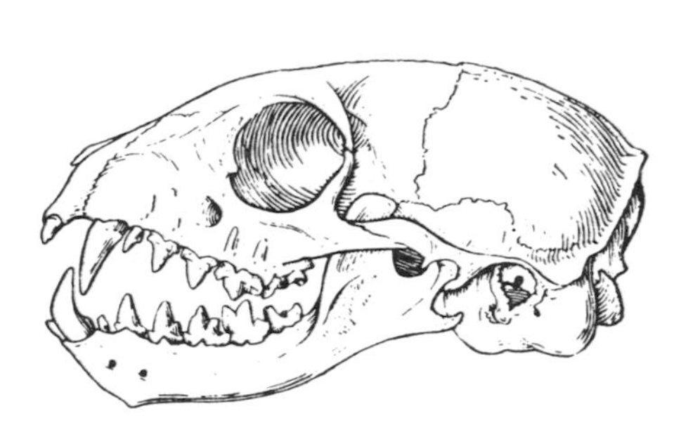

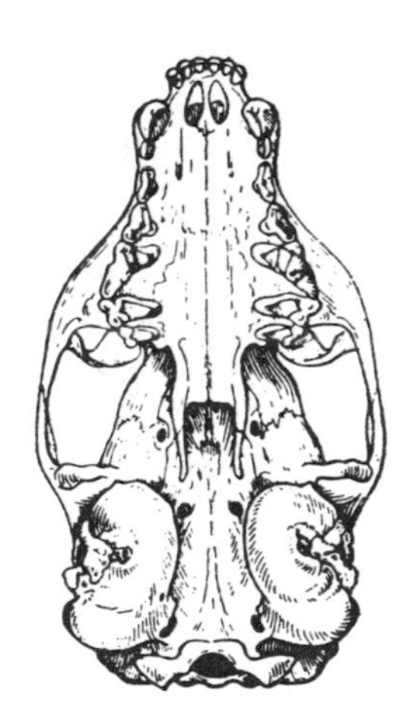

Cynictis

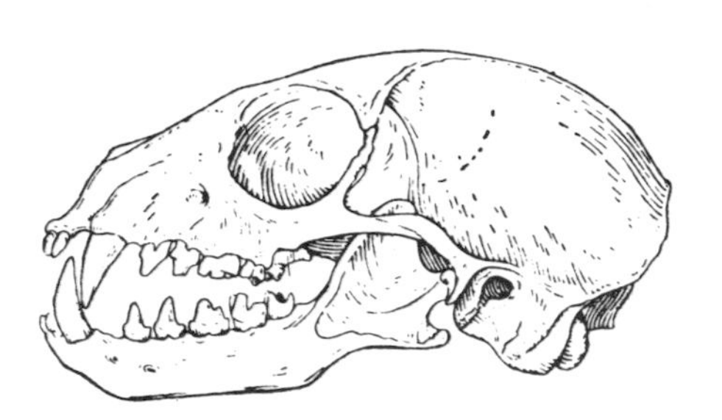

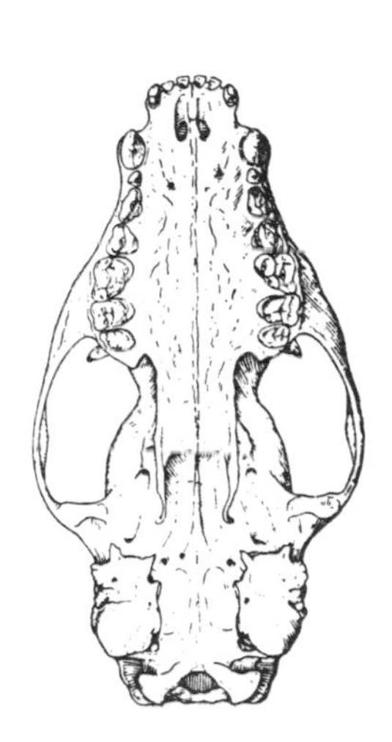

Suricata

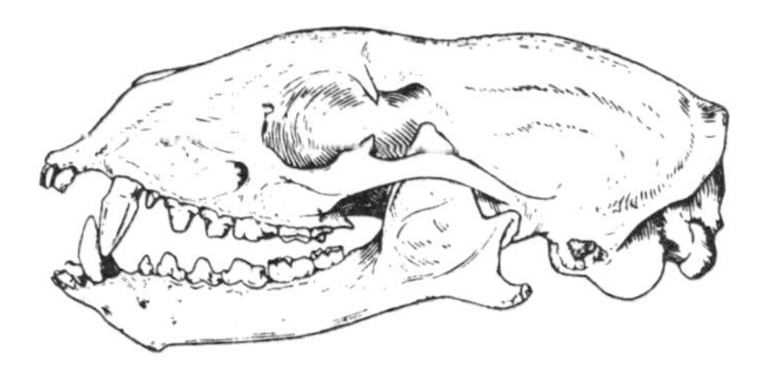

Bdeogale

Figure 2.27. Skulls of mongooses showing dentition of basic carnivorous type (*Cynictis*), insectivorous type (*Suricata*) and crushing omnivorous type (*Bdeogale*). [From Gregory, 1951]

have an all-purposes dentition. The carnassial blade is moderately well developed and two upper molars of the typical viverrid triangular shape are present. The skull is less elongated and the jaws rather more robust than in the civets and genets. The postorbital processes are large and in some genera they meet to form a complete postorbital bar. Although mongooses all have a somewhat varied diet, the relative importance of different types of food differs from one species to another and the teeth are modified accordingly (figure 2.27). *Cynictis*, like *Herpestes*, is an efficient killer of small vertebrate prey and both have long canines and a relatively well developed carnassial shear; *Suricata*, depending largely on insects, has sharp cusped interlocking teeth and the carnassial shear is very poorly developed; in *Atilax* the teeth, particularly the posterior premolars, are very robust and capable of breaking through the shells of the crabs that are included in the diet; while in *Rhyncogale* and *Bdeogale* the wide, blunt-cusped molars and molarised carnassials suggest a predominance of vegetable food in the diet. Information on the feeding habits of the latter two genera is very scanty but Thomas (1894) quotes the collector of his *Rhyncogale* specimens as stating that 'wild fruits are always found inside the stomach of this mongoose'. No

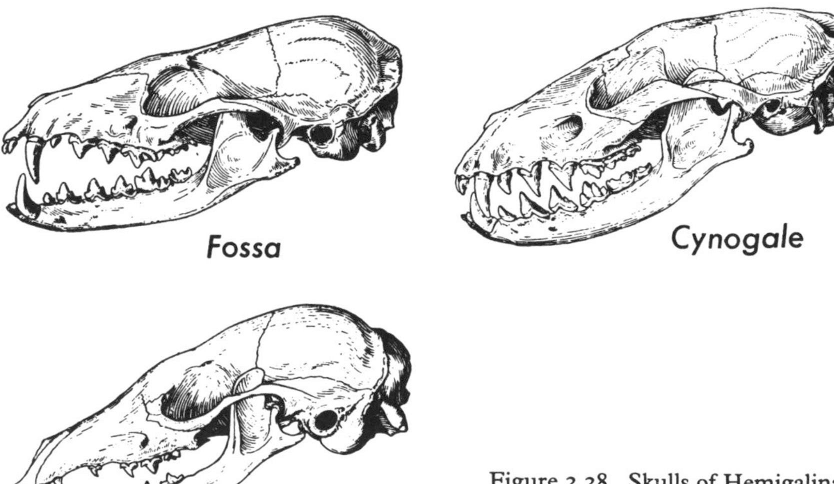

Figure 2.28. Skulls of Hemigalinae. [From Gregory, 1951]

mongoose, however, is a food specialist. All will kill vertebrate prey if the opportunity offers and their jaws are strong and the jaw muscles powerful. *Suricata* may have teeth predominantly adapted to insect prey but a single bite suffices to crush a mouse's skull and it requires quite surprising force to prize open the jaws.

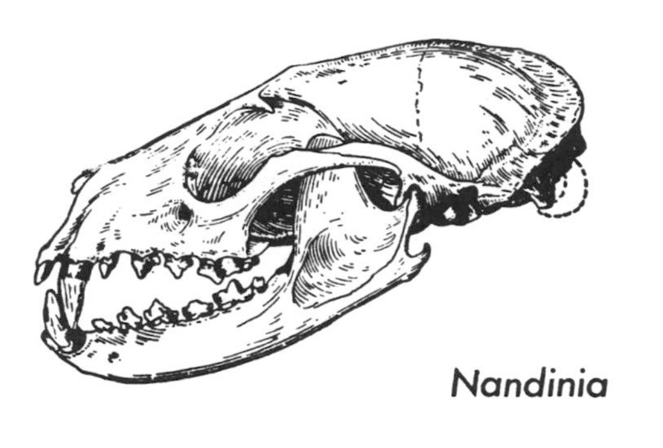

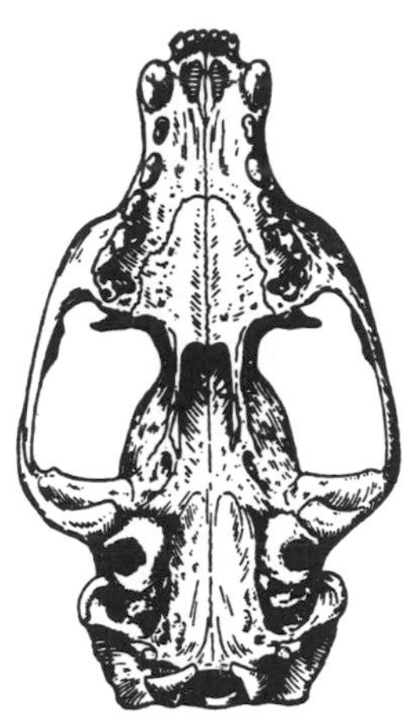

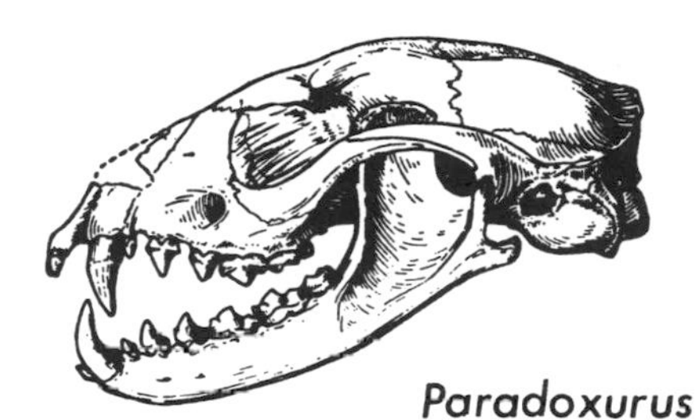

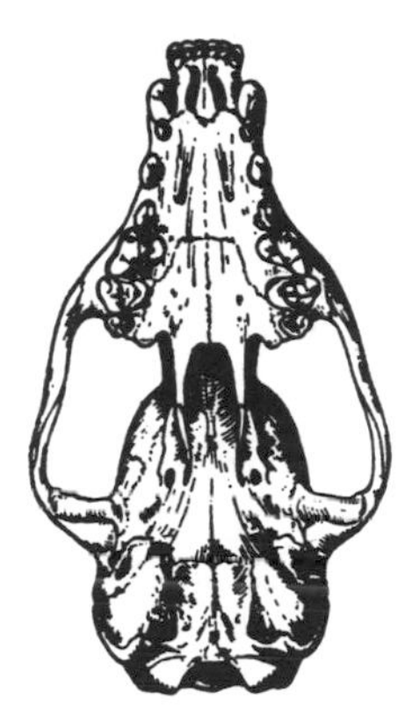

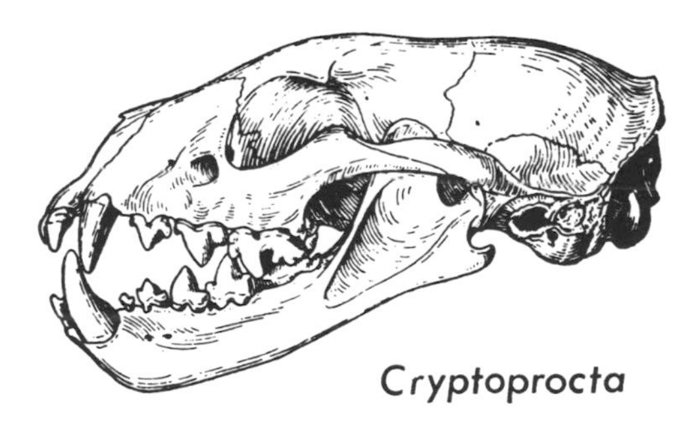

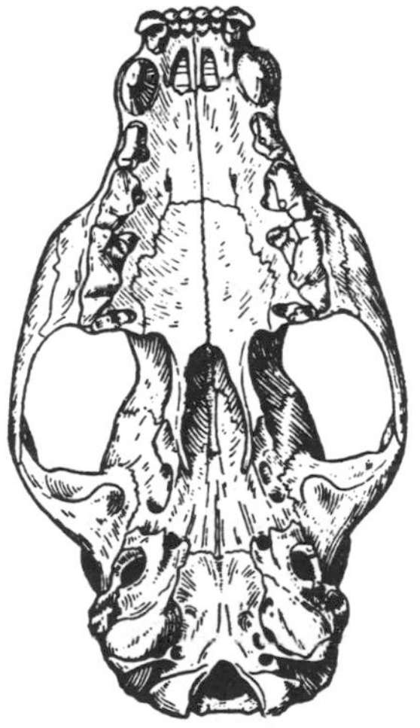

Figure 2.29. Skulls of Paradoxurinae (*Nandinia* and *Paradoxurus*) and Cryptoproctinae (*Cryptoprocta*). [From Gregory, 1951]

The Malagasy mongooses, although classified in a separate subfamily, Galidiinae, resemble the true mongooses in the general characteristics of their teeth and skulls. In the Hemigalinae (figure 2.28) the skull, longer and lower and with less powerful jaws and smaller postorbital processes, is more genet-like but the modifications of the teeth in adaptation to different diets parallel those of the Herpestinae. The fanaloka, *Fossa fossa*, has an all-purposes dentition with moderately well developed carnassial blades and triangular upper molars. The insectivorous *Eupleres* forms an interesting contrast with *Suricata*. The teeth are very reduced, the lower jaw flimsy and the jaw muscles small: clearly *Eupleres* has become a true insect-eating specialist and has lost the ability to deal with other types of prey which *Suricata* has retained. The dentition of *Cynogale*, the otter civet, is quite remarkable. The premolars are sharp and blade-like and the molars broad and crushing, so that the teeth are suited to cope both with small fish and with Crustacea. The contrast between anterior and posterior cheek teeth is so striking and the transition between them so abrupt that if presented with them separately one would hardly credit that they could belong to the same animal.

In the Paradoxurinae (palm civets) the skull and dentition show the same type of robustness as in the mongooses (figure 2.29). *Nandinia*, the two-spotted palm civet, possesses carnassials with reasonably well developed blades but the other genera all show adaptation to a largely vegetarian diet in the blunt cusped molars and molarised carnassials.

The last subfamily of the Viverridae, the Cryptoproctinae, containing only the single genus *Cryptoprocta* (rather confusingly known as the fossa), has paralleled the Felidae to a remarkable degree. The dentition is of an advanced carnivorous type, with enlarged cutting carnassials and only a single very reduced upper molar. The skull, with its large orbits and somewhat shortened braincase, is also suggestive of a primitive felid (figure 2.29).

(*v*) *Felidae*

Of all the Carnivora the cats are the most specialised predators. In the dentition the emphasis is on the large, somewhat flattened canines and the long blade-like carnassials, responsible respectively for killing the prey and slicing through its flesh. The protocone of P^4 is small and situated far forward, so that almost the whole length of the tooth forms the cutting blade. In the lower carnassial, loss of the metaconid and talonid achieves the same thing; the elongated protoconid-paraconid blade runs the whole length of the tooth (plate 3). The jaws are short and powerful and the strong, bowed-out zygoma provides a firm attachment for the masseter muscle as well as ample space within for the passage of the temporalis fibres. As already noted, the attachment area for the latter is enlarged by the development of a strong sagittal crest in the larger species and the wide, high braincase also provides a generous attachment area for neck muscles. The latter are extremely important, since the shock produced when the canines encounter sudden resistance must be transmitted back through the skull to the body. Skull architecture is in fact so arranged that stresses applied to canines or carnassials are transmitted back along smoothly curving strengthened bony arches. These strengthened arches are also so placed as to resist the forces applied to the skull and mandible when tension is developed in the jaw muscles.

According to Leyhausen (1965b), the typical felid kill results from one of the canine teeth entering between two neck vertebrae, forcing them apart and breaking the spinal cord. He points out that the laterally compressed canines are admirably adapted to act in this way and much less so to bite directly on bone; indeed, he is of the opinion that the canines in different species are specifically adapted in size and shape to deal in this way with the neck vertebrae of their principal prey species. Such a technique at first sight seems incredible, since the chances against the canines striking the right place would appear to be very high. It seems too much to ask that, as it bites, the animal should search for the right spot and adjust its grip before biting home. Nevertheless, this may well be exactly what does happen. In the domestic cat, at least, mechano-receptors related to the teeth are abundant, especially around the canines (Jerge, 1963) and the jaw muscles have an exceedingly short contraction time (Taylor and Davey, 1968). Eccles and her co-workers (1968) also mention that the cat's muscle afferent nerves conduct extremely fast. The animal therefore possesses the physiological basis both for 'feeling' with its canines and for biting home with great speed once the teeth have found the correct position.

Although there are minor differences in skull proportions and in the size and degree of flattening of the canine teeth, the skulls of the Felidae are remarkably uniform. The most aberrant is that of the cheetah. The cheek teeth are extremely narrow and blade-like, adapted almost exclusively for meat slicing, which suggests that the cheetah must be incapable of dealing with bones of any size. Brain (1970) in a recent study of the remains left after cheetahs have finished a meal has shown that this is in fact the case (see plate 4). The canines are small by felid standards and not very much flattened. This is probably correlated with a killing technique of biting at the throat (Kruuk and Turner, 1967; Eaton, 1970b, 1970d), which does not demand quite such specialised weapons as the method of breaking the spinal cord. Leyhausen has drawn my attention to the way the nasal aperture is bounded on either side by the roots of the upper canines and points out that a reduction in the size of the teeth permits the aperture to be enlarged. For an animal capable of such a burst of speed as the cheetah, increased air intake may be more valuable than enlarged canines. Strange though it may sound, it may therefore be true to say that the cheetah has small canines because it runs so fast.

The contour of the cheetah's skull is unusual, with the highest point above the eyes and the braincase sloping down from there to the occiput (plate 3). This suggests that the temporalis muscle fibres must act at a slightly different angle from those of other large felids, with a higher proportion of the fibres pulling more nearly horizontally. The fibres must also be shorter, which would imply that the cheetah cannot open its jaws as widely as a lion or leopard. This seems reasonable, in view of the shorter canine teeth. As we have already seen when considering the Mustelinae, it is the anterior, most nearly vertically-directed fibres of the temporalis which are most effective with the jaws widely open. The emphasis in the cheetah on the more horizontally directed fibres is thus correlated with the relatively short canines and consequent smaller gape of the jaws as the bite is delivered. Another characteristic of the cheetah skull is that the infraorbital canal is extremely small, a point noted by Pocock (1916e) and also more recently by Mazak (1968). Neither author makes any suggestion as to the significance of this feature but since the nerves from the tactile receptors at the bases of the vibrissae run through the canal, its narrowness may be associated with the reduced vibrissae (Pocock, 1916e), which can hardly be of the same value to the cheetah as to the more nocturnal species.

(*vi*) *Hyaenidae*

In the hyaenas the carnassials are slightly less specialised as cutting blades than those of the Felidae. The protocone of P^4 is larger and there is a small talonid on M_1. In *Hyaena* a small metaconid is also present but in *Crocuta* this is lost. Like the Felidae, the hyaenas are true predators but their major dental specialisation concerns their ability to break up larger bones than any other carnivore can cope with. This has permitted them to supplement their rations by scavenging on the kills of other species. The bone-crushing adaptations relate mainly to the premolars. The anterior and posterior cusps are reduced and the central cusp enlarged and widened, so that the tooth is converted from a blade-like structure to a heavy conical hammer. In the extant hyaenids, P^3 and P_3 form the principal hammers. In *Hyaena*, with its less specialised carnassial shear, P_4 and the anterior part of P^4 are also modified for crushing and the dominance of the third premolars is less striking than in *Crocuta* (Ewer, 1954).

Bone crushing obviously requires not only hammer-like teeth but also strong muscles and the temporalis attachment on the skull is

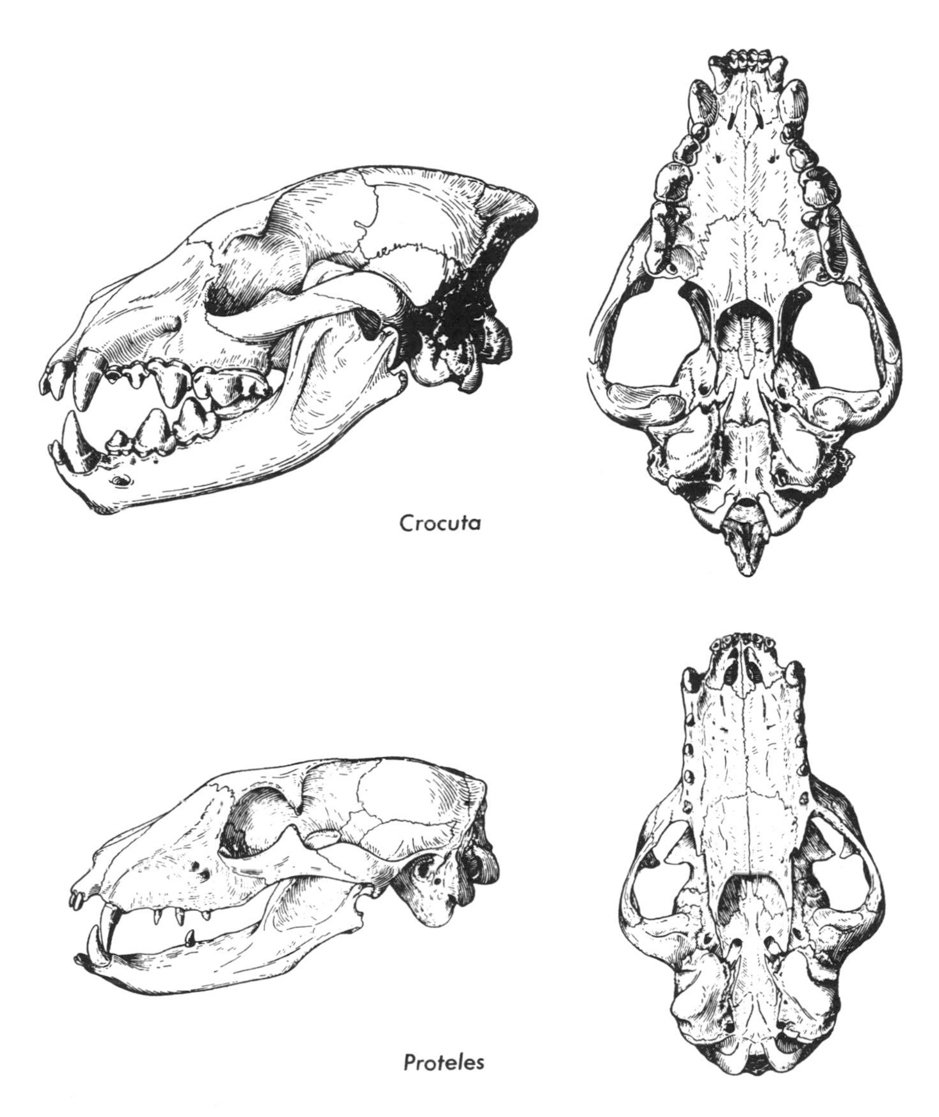

Figure 2.30. Side and palatal views of bone-crushing (*Crocuta*) and termite-eating (*Proteles*) hyaenids. [From Gregory, 1951]

enlarged by a strong sagittal crest (figure 2.30). Although the neck muscles must be strong to carry the weight of the skull, there is no need for them to be increased to the same extent as the jaw muscles and the occiput is therefore relatively small. It is very narrow dorsally and more or less triangular in shape, with the sides not curved outwards as they are in the Felidae. Amongst the Mustelidae, the wolverine, *Gulo gulo*, also an effective scavenger, capable of cracking heavy bones, shows the same thing in more exaggerated form. The sagittal crest projects well above the area of attachment of the neck muscles and in a large animal it extends back far behind the level of the condyles to provide attachment for the relatively enormous temporalis muscles (Shufeldt, 1924).

The extremely reduced dentition of *Proteles* has already been mentioned. The skull and jaws appear disproportionately strong in comparison with the almost vestigial cheek teeth and the postorbital processes are surprisingly large (figure 2.30). The canines, however, although not very large, are by no means negligible and the relatively strong jaw and temporalis attachment may reflect an ability to use these weapons - possibly only defensively.

The milk dentition and its replacement remain to be considered. I do not propose to discuss the theoretical morphological question of whether the molar teeth should be regarded as late erupting members of the same tooth family as the milk teeth, or whether all the permanent teeth belong to one family and the milk teeth to another but rather to deal with functional aspects.

The milk dentition provides the weanling with the necessary equipment for dealing with solid food but, since the jaws are still far from adult size, it is clear that further teeth must develop later. Since elongation of the jaw takes place at the back, it is also clear that the addition of new teeth must take place at the posterior end of the tooth row. The Carnivora possess a highly complex dentition, in which different teeth perform different functions. The milk dentition and the permanent set must therefore each be a functional unit and, what is more, the transition from one to the other must take place without disrupting the functioning of the dentition as a whole. The milk teeth are therefore not merely smaller versions of the permanent premolars that will replace them later: instead the milk cheek tooth most closely resembles the tooth which in the permanent set will erupt one place behind it. Thus Dm^3 and Dm_4 differentiate as carnassials, resembling P^4 and M_1, while Dm^4 is a molariform tooth resembling M^1.

The milk teeth, however, do not exactly resemble their permanent functional counterparts but may show more primitive characteristics. Thus in Felidae, the milk lower carnassial has a distinct talonid and in the upper carnassial the protocone is not so near the front of the tooth (plate 3). Milk canines too often show a tendency for the development of subsidiary cusps (Lydekker, 1912; Broom, 1949).

In the replacement of the milk series by the permanent teeth, functional considerations demand that the animal should not be left at any stage without effective canines. Similarly, in those species where the carnassial shear is important, there must be no period when functional carnassials are not present. The pattern of replacement has been studied in only a few species but these two problems are solved in the same way in each case. Permanent canines do not erupt directly beneath their milk precursors but a little in front or alongside them. There is thus a period when the animal has a double set of canines and the milk tooth is not shed until after the permanent one is sufficiently erupted to be functional. The same is true of the carnassial shear. The permanent carnassials erupt behind the milk representatives before the latter are shed and there is a period when two sets of carnassials are in operation (plate 3). In the lower jaw, where the milk carnassial is the last tooth, this presents no great difficulty. M_1 must simply erupt before Dm_4 is replaced, just as our own first molars do before our milk cheek teeth are shed. In the upper jaw, however, a dislocation of the normal antero-posterior eruption sequence is required: P^4 must erupt to replace Dm^4 before P^3 does the same for the milk carnassial Dm^3. This is exactly what does happen in all the cases that have been investigated. The process is most easily understood if the eruption sequences for the permanent teeth in the upper and the lower jaws are considered separately. These sequences are given in table 2.3 (p. 71) for the few species for which data are available.

The formula for the milk dentition can usually be deduced by deleting the molars from the permanent formula but this procedure is not invariably reliable. Teeth suppressed in the adult may have milk representatives, as in the sea otter, where there are three lower milk incisors. Conversely, there are cases where the milk precursors of permanent teeth have been suppressed. This is true, for instance, of the first premolars of the dog and Linhart (1968) found the same in the fox. The mink has been said to possess only a single pair of deciduous incisors (Kainer, 1954) but Auerlich and Swindler (1968) found that the other incisors are present but vestigial. They are small, erupt very late and are very soon replaced by the permanent teeth. In the ferret too the milk incisors are peculiar. In the laboratory colonies studied by Berkovitz (1968) the incisors were small and non-functional and the lower ones never actually broke through the gum; an extra fourth upper tooth was also very commonly present. In the striped skunk, *Mephitis mephitis*, both upper and lower milk incisors are in this condition: the vestigial germs never produce functional teeth and the first incisors to erupt fully are those of the permanent dentition (Verts, 1967).

It would be interesting to know whether the first milk molar is missing in all the Canidae and it would also be of interest to find out whether the inverted eruption sequence of P^4 and P^3 occurs in Carnivora lacking specialised carnassials as well as in those possessing them.

The skull characteristics which have just been discussed are all directly related to the dentition and jaw mechanics: their adaptive significance in relation to feeding habits is therefore obvious. There are, however, other details of skull architecture whose significance is less apparent but which are nevertheless of importance taxonomically. Amongst these are the detailed architecture of the skull in the auditory region and of the auditory bulla itself. The former has proved to be of considerable value in taxonomic and palaeontological studies but space does not permit more than a consideration of the characteristics of the bulla itself. Further details may be found in papers by Segall (1943) and Hough (1948).

The bulla is well developed in the majority of carnivores but in primitive mammals it is not present and the tympanic bone (derived from the reptilian angular) merely forms a ring which supports the tympanic membrane. In many mammals, however, an inflated bulla, continuous with the auditory meatus, roofs over the membrane and the area of the petrosal housing the inner ear. The bulla may be formed entirely from the tympanic, or a second bone – the entotympanic – may contribute significantly to its formation. The origin of the entotympanic is obscure: its cartilage precursor may be derived from the hyoid arch or it may be a mammalian neomorph, possibly arising from the cartilage of the meatus (Starck, 1964, 1967). However that may be, it seems clear that the evolution of the bulla occurred independently in the Canoidea and the Feloidea, for the contribution of the entotympanic is different in the two groups.

In the Feloidea the tympanic and entotympanic develop separately, almost as though each were about to form an independent bulla. Where the two meet, a bony septum is formed between them, partially separating an antero-external chamber formed by the tympanic, from a postero-internal one formed by the entotympanic. In the Felidae the anterior chamber is small and the septum is situated far forward. In the Viverridae there is considerable variation and the two chambers may be subequal in size or the anterior may even be the larger. *Nandinia* is exceptional in that the entotympanic does not ossify but remains cartilagenous and is therefore usually missing from dried skull preparations.

The Hyaenidae were originally considered to be exceptional amongst the Feloidea in having a bulla which is not divided by a septum. Pocock (1916f, 1928a), however, showed that this is not the case. In felids and viverrids the septum runs more or less at right angles to the lower wall of the bulla. One has therefore merely to cut through the lower wall in order to see the septum - or, indeed, it can usually be seen through the opening of the meatus or felt by inserting a probe. In the Hyaenidae, however, the partition runs more or less parallel to the lower wall of the bulla and is therefore easily demonstrated only if the bulla is sectioned vertically.

In the Canoidea the cavity of the bulla is typically undivided and appears to be formed from the tympanic alone. This, however, may be deceptive. In certain cases at least, an entotympanic element is present in the early stages of ontogeny, although its separate identity is later lost and it fuses completely with the tympanic. According to Weber (1927), this is the case in the badger, *Meles meles*, and Starck (1964) found the same to be true of the dog, the wolf and the brown bear, *Ursus arctos*. In a number of Canidae a small horizontal septum is present in the mature bulla; it is particularly obvious in the maned wolf, *Chrysocyon*. Flower (1869a) regarded this as homologous with the septum of the Feloidea but later workers have disagreed with this and interpret the septum as a secondary outgrowth of the tympanic alone, the entotympanic not being involved. In the Mustelidae, the inner wall of the tympanic is drawn out into a series of crests, which in some of the Mustelinae unite to produce a cellular structure (Weber, 1927).

The function of the bulla has not been investigated in Carnivora. Webster (1962), however, has studied the function of the extremely large bulla of the kangaroo rat, *Dipodomys*. The bulla increases the air space behind the tympanic membrane and so reduces the damping of the chain of auditory ossicles. This permits resonance effects to occur and the sensitivity of the ear to the resonant frequencies is increased. It is not yet clear what structures determine the resonant frequencies but in the kangaroo rat two peaks are involved. Their frequencies correspond to the sounds made by a rattlesnake as it strikes or an owl as it swoops on its prey. The kangaroo rat is often able to evade these predators by a quick vertical leap at the last moment. If the bullae are blocked with wax, which reduces the sensitivity to the resonant peak frequencies, no locomotor abnormalities are detectable but the animals do not show this type of escape behaviour and are therefore easily captured by the predators.

In view of these findings it does not seem unreasonable to suggest that in Carnivora too the enlarged bulla produces a high sensitivity to particular frequencies, resulting from resonance phenomena. In their case, of course, the adaptation would be offensive rather than defensive. Leyhausen (1956) has shown that high pitched rustling sounds are very effective in alerting the cat to the presence of prey and the bat-eared fox can locate insects underground, apparently by hearing the sounds made by their movements (Smithers, 1966a). Presumably the enlarged bulla of the fennec reflects the same ability. Carnivores in which the bulla is particularly inflated are all species in which one might expect to find acute hearing, especially for high pitched sounds. In the vegetarian pandas, where there is no apparent need for such sensitivity, the bulla is extremely reduced and in the omnivorous bears it is also small.

While it thus appears reasonable, in general terms, to suggest that the carnivore bulla, like that of *Dipodomys*, has the function of enhancing sensitivity to certain frequencies, no analysis has yet been made of the role played by the septum and it is not clear what significance should be attached to the variations in its position that occur in different species.

Other features of some taxonomic value which affect the posterior region of the skull are related to the details of the blood supply to the brain. The brain receives its blood from vessels arising from the circle of Willis (cerebral arterial circle) and from the median basilar artery.

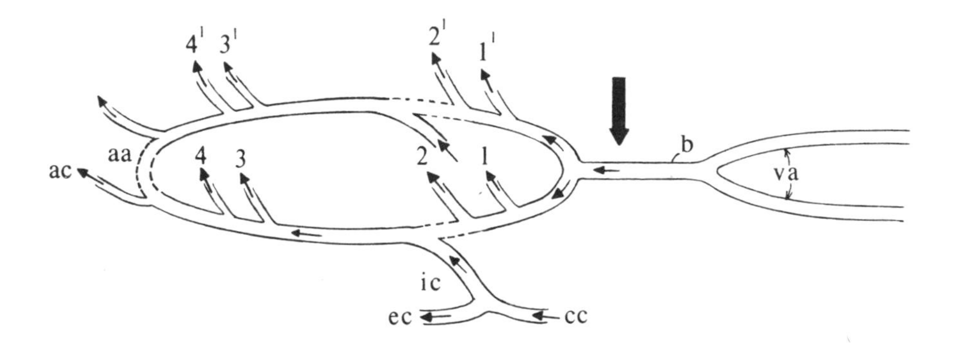

Figure 2.31. Formation of the circle of Willis: diagrammatic view from side and slightly from above. aa, anterior anastomosis; ac, anterior cerebral artery; b, basilar artery; cc, common carotid; ec, external carotid; ic, internal carotid; va, vertebral arteries; 1–4, arteries running up from the circle to supply the brain. The large arrow is at the level of the back of the skull. In some species (e.g. domestic cat) the anterior anastomosis is absent and the circle is incomplete anteriorly: in many mustelids the right and left anterior cerebral arteries unite to form a single median vessel: in many viverrids the basilar artery is very small and the circle is supplied mainly by the internal carotids.

The latter is formed by the union of right and left vertebral arteries which enter the skull through the foramen magnum. The posterior part of the circle of Willis is formed by bifurcation of the basilar artery to form right and left limbs, from which the arteries supplying the hind brain and cerebellum originate. The anterior portion of the circle on either side is formed principally from the internal carotid, running forward to give off vessels supplying the cerebral hemispheres and linked to the posterior part of the circle by a backwardly directed anastomosis. In some species, left and right internal carotids do not unite but in most carnivores an anterior anastomosis, joining left and right sides, completes the circle (see figure 2.31). Although the arrangement is basically very simple, blood from the carotid artery actually reaches the circle by rather complex and devious pathways. The common carotid runs along the outside of the base of the skull and becomes overlaid by the auditory bulla. In some viverrids its course within the bulla is marked by a groove in the tympanic bone but in the Canoidea and Herpestinae this becomes roofed over to form the carotid canal. In the Felidae the canal lies between the entotympanic and the basioccipital. The carotid canal is longer in the Canoidea than in the Feloidea, a characteristic regarded as of some systematic importance.

The internal carotid branches off the main vessel close to the anterior end of the carotid canal and bends back to enter the skull through the foramen lacerum medius, where it participates in the formation of the circle of Willis. The main trunk continuing its way forwards is usually referred to as the external carotid, although it is not exactly homologous with the external carotid of lower vertebrates; it is also sometimes referred to as the internal maxillary artery. This vessel runs forward to the orbit, where various branches enter the skull to contribute to the circle of Willis. Jewell (1952) gives details of these vessels in the dog and Chapuis (1966) provides a comparative account with particular emphasis on viverrids.

The point of importance in the present context is that on its pathway forward from the auditory region to the orbit the external carotid lies close to the skull wall. Its course across the alisphenoid may be marked by a slight groove, or this may be roofed over to form the alisphenoid canal. This opens anteriorly close to the foramen rotundum, through which the maxillary branch of the trigeminal nerve leaves the skull. The presence or absence of the canal is characteristic of certain groups and it therefore has some taxonomic value. An alisphenoid canal is present in the Canidae, Ursidae and some Viverridae but is absent in Felidae and Procyonidae, with the exception of *Ailurus*. In Hyaenidae the situation is a trifle peculiar. The blood vessel remains outside the bone and there is no true alisphenoid canal but the foramen rotundum opens at the base of a blind-ending tube in the alisphenoid. Pocock (1916h) suggests that this is a secondary condition, derived from an arrangement of the type found in *Genetta* (figure 2.32). Here there is a normal alisphenoid canal but the foramen rotundum opens deep within it instead of in the more usual position, near its anterior end. If some change were now to occur in the position of the external carotid, so that it no longer passed close against the skull, then the canal might not vanish, since it would still act as the passageway for the nerve fibres; its posterior end, however, might well become closed, resulting in the formation of an alisphenoid tube of the hyaenid type.

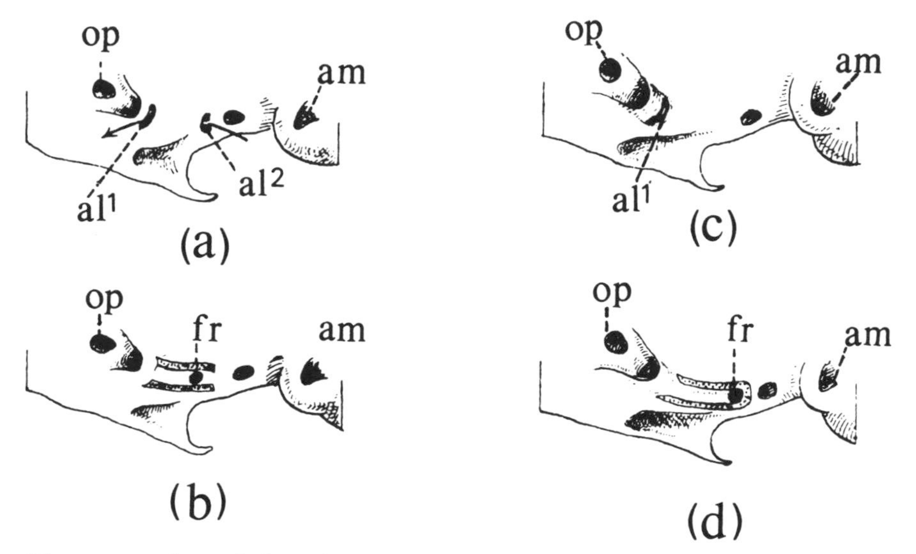

Figure 2.32. Lateral view of skull in region of alisphenoid canal. (a) *Genetta*, (b) same with roof of canal cut away (cut edges of bone shown stippled). Canal (shown by arrow) with normal two openings, transmitting external carotid artery; foramen rotundum opens deep within the canal instead of in the usual position anterior to it. (c) *Viverricula*, (d) same with roof of canal cut away. The external carotid passes on the outer surface of the skull; the posterior opening of the canal is closed to form a tube, at the base of which the foramen rotundum opens al^1, al^2, anterior and posterior openings of alisphenoid canal. am, auditory meatus; fr, foramen rotundum; op, foramen for emergence of optic nerve. [From Pocock, 1916h]

The small Indian civet, *Viverricula*, shows exactly the same arrangement as the hyaenids but in other viverrids the situation is variable: in many species the normal alisphenoid canal is present; in some genera (*Cynogale*, *Eupleres*, *Crossarchus*[1]) a short canal may or may not be present as an individual variation, while in others (*Galidia*, *Galidictis*, *Salanoia*) it is absent.

In the Carnivora the region of the skull between eye and ear is involved in the shortening which accompanies reduction in the post-carnassial dentition. It is therefore hardly surprising that there should be considerable variation in the exact relations between the blood vessels and the underlying bone in this area and consequently in the details of the alisphenoid canal. Although the function of the canal is not discussed in standard anatomical works, the most obvious suggestion is that it protects the carotid from occlusion when the pterygoideus muscles contract. This would accord with the fact that it is in those groups with a reduced post-carnassial dentition that the canal is absent. In them the orbit lies so close to the auditory bulla that the external carotid reaches the orbit almost immediately after emerging from the carotid canal. This takes it clear of the pterygoideus muscle and there is therefore no need for the protective canal. However, as is so often the case with taxonomic characters, attention has been directed almost exclusively to anatomical description and no functional investigations have yet been made.

The turbinal bones are rarely well preserved in fossil material and even in museum skulls of extant species they are often incomplete. They do, however, provide some taxonomic characters of value, including one which distinguishes Canoidea from Feloidea. The turbinals comprise a single maxillary, a nasal and a primary series of ethmoturbinals, the latter differentiated originally from the median nasal septum. These are sometimes collectively referred to as endoturbinals. They are supplemented by later developing outgrowths from the side walls of the nasal passages, forming an ectoturbinal series. The number of elements in this external series is extremely variable. The Ursidae have about forty, many more than are present in any other family; in Mustelidae there are twenty-six and in Procyonidae only fourteen. The numbers in the primary endoturbinal series are lower: seven to eight in Procyonidae and five to six in the other families (Anthony and Iliesco, 1926). In the Feloidea the maxilloturbinals are relatively small and their branchings are not very complex. In the Canoidea the maxilloturbinals are enlarged and thrown into extremely complex foldings (figure 2.33).

The olfactory epithelium extends only over the more posterior turbinals towards the back of the nasal cavity. The size of the turbinals is therefore no measure of olfactory acuity. The main function of the turbinal bones is believed to be that of warming and moistening

[1] Pocock (1916h) says that a short canal is present in *Crossarchus* but specimens of *C. obscurus* from west Africa which I have examined include a few skulls lacking the canal.

the incoming air. This is obviously desirable in terms of protecting the tissues of the lungs but it could also be valuable in terms of heat and water conservation. One might therefore expect the turbinals

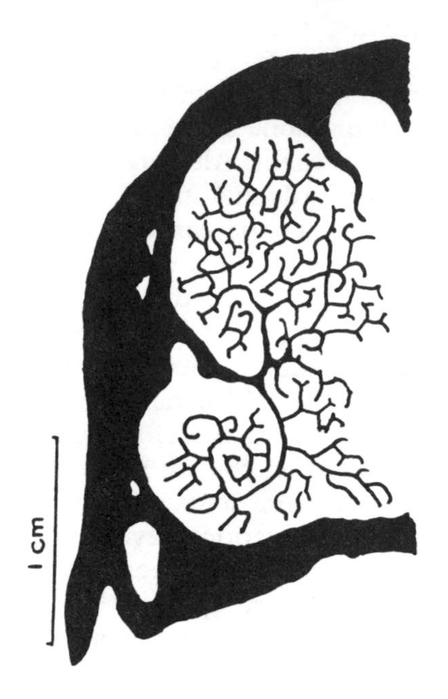

Figure 2.33. Transverse section of nasal region of badger skull to show complex ramifications of the maxillo-turbinal characteristic of the Canoidea. [From Anthony and Iliesco, 1926]

to be maximally developed in very active animals, where the ventilation rate must be high, or in those that live in very dry or very cold environments. However, I am not aware that any correlations of this kind have been established. Unfortunately we are therefore again left with bare anatomical facts, devoid of the meaning with which only a functional and physiological analysis can endow them.

Table 2.2
Dental formulae in the families of Carnivora

(Common dental anomalies are summarised by Hall (1940))

1. CANIDAE

    Typically: $I\frac{3}{3}$, $C\frac{1}{1}$, $P\frac{4}{4}$, $M\frac{2}{3}$ = 42
    Cuon: $I\frac{3}{3}$, $C\frac{1}{1}$, $P\frac{4}{4}$, $M\frac{2}{2}$ = 40
    Speothos: $I\frac{3}{3}$, $C\frac{1}{1}$, $P\frac{4}{4}$, $M\frac{1}{2}$ = 38
    Nyctereutes: a third upper molar is sometimes present, making a total of 44 teeth.
    Otocyon: $I\frac{3}{3}$, $C\frac{1}{1}$, $P\frac{4}{4}$, $M\frac{3-4}{4-5}$, with a maximum number of 50.

2. URSIDAE

    Usual formula: $I\frac{3}{3}$, $C\frac{1}{1}$, $P\frac{4}{4}$, $M\frac{2}{3}$ = 42 but the anterior premolars are very reduced and are often shed early, so that in the adult dentition only $P\frac{2}{2}$ remain, giving a total of 34 teeth.

    Melursus is unusual in having $I\frac{2}{3}$.

3. PROCYONIDAE

    Usual formula: $I\frac{3}{3}$, $C\frac{1}{1}$, $P\frac{4}{4}$, $M\frac{2}{2}$ = 40
    Potos: $I\frac{3}{3}$, $C\frac{1}{1}$, $P\frac{3}{3}$, $M\frac{2}{2}$ = 36
    Ailurus: $I\frac{3}{3}$, $C\frac{1}{1}$, $P\frac{3}{4}$, $M\frac{2}{2}$ = 38
    Ailuropoda: $I\frac{3}{3}$, $C\frac{1}{1}$, $P\frac{3-4}{3}$, $M\frac{2}{3}$ = 38–40
    (The small P^1 is variable in occurrence)

4. MUSTELIDAE

    Typically: $I\frac{3}{3}$, $C\frac{1}{1}$, $P\frac{2-4}{2-4}$, $M\frac{1}{1-2}$ = 28–38
    The variation in dental formulae within this family makes it desirable to treat the subfamilies separately.

    (i) Mustelinae
    The commonest formula, found in *Mustela*, *Poecilictis*, *Ictonyx*, is $I\frac{3}{3}$, $C\frac{1}{1}$, $P\frac{3}{3}$, $M\frac{1}{2}$ = 34
    Gulo, *Martes*: $I\frac{3}{3}$, $C\frac{1}{1}$, $P\frac{4}{4}$, $M\frac{1}{2}$ = 38
    Grammogale: $I\frac{3}{3}$, $C\frac{1}{1}$, $P\frac{3}{2}$, $M\frac{1}{2}$ = 32
    Lyncodon, *Poecilogale*: $I\frac{3}{3}$, $C\frac{1}{1}$, $P\frac{2}{2}$, $M\frac{1}{1}$ = 28

    (ii) Mellivorinae $I\frac{3}{3}$, $C\frac{1}{1}$, $P\frac{3}{3}$, $M\frac{1}{1}$ = 32

    (iii) Melinae
    Usual formula: $I\frac{3}{3}$, $C\frac{1}{1}$, $P\frac{4}{4}$, $M\frac{1}{2}$ = 38 but in some species P^1 is very small and may be shed early, so that there appear to be only 36 teeth.
    Taxidea: $I\frac{3}{3}$, $C\frac{1}{1}$, $P\frac{3}{3}$, $M\frac{1}{2}$ = 34

    (iv) Mephitinae
    Usual formula: $I\frac{3}{3}$, $C\frac{1}{1}$, $P\frac{3}{3}$, $M\frac{1}{2}$ = 34
    Conepatus: $I\frac{3}{3}$, $C\frac{1}{1}$, $P\frac{2}{3}$, $M\frac{1}{2}$ = 32

    (v) Lutrinae
    Usual formula: $I\frac{3}{3}$, $C\frac{1}{1}$, $P\frac{3-4}{3}$, $M\frac{1}{2}$ = 34–36
    (The small P^1 is variable in occurrence)
    Enhydra: $I\frac{3}{2}$, $C\frac{1}{1}$, $P\frac{3}{3}$, $M\frac{1}{2}$ = 32

5. VIVERRIDAE

    Most commonly $I\frac{3}{3}$, $C\frac{1}{1}$, $P\frac{4}{4}$, $M\frac{2}{2}$ = 40
    Since there is considerable reduction in some groups, the subfamilies are listed separately.

    (i) Viverrinae
    Usually the full complement.

Poiana, Prionodon: $I\frac{3}{3}$, $C\frac{1}{1}$, $P\frac{4}{4}$, $M\frac{1}{2}$ = 38 A vestigial M^2 is sometimes present in *Poiana*.

(ii) Paradoxurinae

Paradoxurus, Paguma, Macrogalidia, Arctogalidia: full complement of 40 teeth but in *Macrogalidia* the first premolars (upper and lower) and in *Paguma* both these and the last molars are very small and may be shed early.

Nandinia: $I\frac{3}{3}$, $C\frac{1}{1}$, $P\frac{4}{4}$, $M\frac{1}{2}$ = 38

Arctictis: $I\frac{3}{3}$, $C\frac{1}{1}$, $P\frac{4}{3}$, $M\frac{2}{2}$ = 38

(iii) Hemigalinae

$I\frac{3}{3}$, $C\frac{1}{1}$, $P\frac{4}{4}$, $M\frac{2}{2}$ = 40

(iv) Galidiinae

Usually $I\frac{3}{3}$, $C\frac{1}{1}$, $P\frac{3}{3}$, $M\frac{2}{2}$ = 36 but a small P^1 may be present in some species.

Salanoia: $I\frac{3}{3}$, $C\frac{1}{1}$, $P\frac{4}{3}$, $M\frac{1}{2}$ = 36

(v) Herpestinae

Ichneumia, Rhyncogale, Xenogale, Bdeogale, Cynictis, Paracynictis, Liberiictis; typically $I\frac{3}{3}$, $C\frac{1}{1}$, $P\frac{4}{4}$, $M\frac{2}{2}$ = 40 but the first premolars are very small in a number of species and may be absent or shed early.

Herpestes: $I\frac{3}{3}$, $C\frac{1}{1}$, $P\frac{3-4}{3-4}$, $M\frac{2}{2}$ = 36 to 40 in different species of this genus.

Mungos, Crossarchus, Atilax, Helogale, Suricata:

$I\frac{3}{3}$, $C\frac{1}{1}$, $P\frac{3}{3}$, $M\frac{2}{2}$ = 36

(vi) Cryptoproctinae

$I\frac{3}{3}$, $C\frac{1}{1}$, $P\frac{3}{3}$, $M\frac{1}{1}$ = 32 but a vestigial P^1 may be present but early shed.

6. HYAENIDAE

Hyaeninae: $I\frac{3}{3}$, $C\frac{1}{1}$, $P\frac{4}{3}$, $M\frac{1}{1}$ = 34

In *Proteles* the cheek teeth are extremely reduced and irregular and there may be as few as 24 teeth in all.

7. FELIDAE

Typically $I\frac{3}{3}$, $C\frac{1}{1}$, $P\frac{3}{2}$, $M\frac{1}{1}$ = 30. Anterior premolars may be missing as an individual anomaly, giving a minimum of 26 teeth.

P^2 is regularly absent in a number of the short faced species, including *Profelis, Caracal, Lynx rufus, Prionailurus rubiginosa, Mayailurus iriomotensis* and *Otocolobus manul* and also *Acinonyx.*

Table 2.3
Eruption sequences of teeth of permanent dentition

Where two teeth erupt more or less simultaneously, they are linked by a (+) sign. The first tooth to erupt is shown on the left and arrows indicate time intervals.

Panthera uncia (snow leopard) (data from Pocock, 1916d)

Dental formula: $I\frac{3}{3}$, $C\frac{1}{1}$, $P\frac{3}{2}$, $M\frac{1}{1}$

Upper jaw sequence: $I^1 + I^2 \rightarrow P^2 \rightarrow I^3 \rightarrow C^1 + P^4 \rightarrow P^3$
$\qquad\qquad\qquad\qquad\qquad +$
$\qquad\qquad\qquad\qquad\quad M^1$

Lower jaw sequence: $I_1 + I_2 \rightarrow I_3 \dashrightarrow C_1 \rightarrow P_3 \rightarrow P_4$
$\qquad\qquad\qquad\qquad\qquad \searrow M_1 \nearrow$

Mustela vison (mink) (data from Auerlich and Swindler, 1968)

Dental formula: $I\frac{3}{3}$, $C\frac{1}{1}$, $P\frac{3}{3}$, $M\frac{1}{2}$

Upper jaw sequence: $I^1 + I^2 \rightarrow I^3 \rightarrow C^1 \searrow \quad \nearrow P^2 \rightarrow P^4 \rightarrow P^3$
$\qquad\qquad\qquad\qquad\qquad\qquad M^1$

Lower jaw sequence: $I_1 \rightarrow I_2 \rightarrow I_3 \dashrightarrow C_1 \rightarrow P_2 \dashrightarrow P_3 \rightarrow P_4$
$\qquad\qquad\qquad\qquad\qquad \searrow M_1 \nearrow \dashrightarrow \searrow M_2 \nearrow$

Vulpes vulpes (red fox) (data from Linhart, 1968)
Linhart's observations are rather scanty and the incisor eruption times were not determined with certainty. As far as could be judged from weekly observations, the sequences for the other teeth were as follows:

Dental formula: $I\frac{3}{3}$, $C\frac{1}{1}$, $P\frac{4}{4}$, $M\frac{2}{3}$

Upper jaw sequence: $P^1 \dashrightarrow P^4 + C^1 \qquad \nearrow P^2 + P^3$
$\qquad\qquad\qquad \searrow M^1 \nearrow \dashrightarrow \searrow M^2 \nearrow$

Lower jaw sequence: $P_1 \dashrightarrow C_1 \dashrightarrow P_2 + P_3 + P_4 \searrow$
$\qquad\qquad\qquad \searrow M_1 \nearrow \dashrightarrow \searrow M_2 \nearrow \dashrightarrow M_3$

Paradoxurus hermaphroditus (Indian palm civet or toddy cat)
Pocock (1934) gives data for premolars and molars only:

Dental formula: $I\frac{3}{3}$, $C\frac{1}{1}$, $P\frac{4}{4}$, $M\frac{2}{2}$

Upper jaw sequence: $P^1 \rightarrow P^2 \dashrightarrow P^4 \rightarrow P^3 \searrow$
$\qquad\qquad\qquad\qquad \searrow M^1 \nearrow \dashrightarrow M^2$

Lower jaw sequence: $P_1 \rightarrow P_2 \dashrightarrow P_3 \rightarrow P_4$
$\qquad\qquad\qquad\qquad \searrow \quad \nearrow \qquad +$
$\qquad\qquad\qquad\qquad\quad M_1 \dashrightarrow M_2$

Mellivora capensis (honey badger). Lydekker (1912) does not give an eruption sequence but he figures a skull in which dm^3 is still in place and P^4 is fully erupted behind it.

Chapter 3

Anatomy of the soft parts

A. The pelage

CARNIVORE furs have always been highly prized. In medieval England the right to wear the furs of ermine (the winter coat of the stoat), sable, marten and genet was restricted to royalty and the high nobility - commoners had to make do with rabbit or lambskin. The same predilection for carnivore furs is still with us: in Canada the nine furs bringing in the greatest total revenue in 1960 included six carnivores - mink, lynx, otter, fox, ermine and marten. The majority of the most esteemed furs come from cold-climate species, whose winter coats must needs be thick and warm - lynx, ermine, arctic fox, sable, marten, mink - but some, like leopard, jaguar and ocelot, owe their popularity mainly to their beautiful markings. Since furs have become less and less necessary with the development of other means of keeping warm in winter, the latter, selected for appearance rather than insulating qualities, have become increasingly sought after. It is horrifying to learn that the United States fur trade in 1969 is estimated to have dealt with some half a million skins of the more lovely of the Felidae, none of which today exists in sufficient numbers to support wholesale slaughter on this scale.

Zoological textbooks, however, usually pay scant attention to coat colour or markings; they do not even point out that while the Canoidea are typically plain coloured, the Feloidea are typically marked with blotches, spots or stripes, although within each family there may be some species that are not thus patterned. This difference suggests a basic divergence in early phylogeny into forest or woodland ancestral feloids and open country ancestral canoids.

Where it has been adequately investigated, coat colour in mammals has been found to be under complex genetical control (Searle, 1968) and mutations affecting coat colour turn up quite frequently in experimental populations. In domestic species of carnivores, as of other mammals, it has always been possible to produce variations in colour by appropriate selection and crossing. This is true not only for the dog and cat which have a long history of domestication behind them but also for those that have been bred for their furs for only a relatively short time, such as fox and mink. Every now and then melanistic, albino or erythristic individuals are reported in the wild for various species of carnivore (see, for instance, Cross, 1941; Ulmer, 1941; Green, 1947; Neill, 1953; Allen and Neill, 1956; Funderberg, 1961; Roest, 1961; Matheson, 1963; Johnson, 1968), the black panther, which is simply a melanistic leopard, being the best known. The famous Indian 'white tigers' are actually a semi-albino strain in which pigmentation is reduced but not absent and are more correctly designated as chinchilla mutants (Robinson, 1969a). They were originally bred from the progeny of a single male showing the unusual colouring. True albino tigers have been reported only once (Thornton *et al.*, 1967). The most spectacular case of aberrant colouring in the wild, however, is the curious cheetah (plate 13) with ocelot-like streaks and blotches instead of the normal spots, which Pocock (1927b) originally described as a new species.

Despite the fact that aberrant colour mutants appear regularly, there are only a few cases in which there seems to be a relatively stable polymorphism, with more than one colour variety maintained in the population. This occurs in the African golden cat, *Profelis aurata*, with dark, light and a variety of intermediate forms (van Mensch and van Bree, 1969), in the jaguarondi, *Herpailurus yagouaroundi* (with a dark reddish form and the normal greyish buff), and also in the red fox in North America and in the arctic fox (the former with a range of shades from normal red to melanic, for which the details are given by Cross (1941); the latter with normal and 'blue' varieties). In the leopard, too, the black and the normal spotted forms exist together in densely forested regions of southeast Asia in what appears to be a stable polymorphism (Robinson, 1969b). As a rule, however, colour aberrations are rare in natural conditions, a testimony to the fact that they are disadvantageous compared with the normal pelage and the mutant genes are therefore eliminated fairly rapidly. It is quite clear, however, that, should conditions alter, changes in coloration could be evolved quite rapidly. It is therefore not surprising to find that although a certain colouring may be typical of a family or subfamily, there are always some unorthodox, nonconformist species. Nevertheless, if we are studying a relatively small taxonomic unit we may expect to find that details of colouring have some classificatory value. Table 3.1 (page 87) gives a summary of the patternings typical of the various carnivore families or subfamilies.

The primary function of coloration is concealment and predators are no exception. Concealment may be either defensive, offensive or both at once, for a small carnivore may need to avoid a larger predator and at the same time remain undetected by its own potential prey. Most carnivores are just as cryptic as their prey: desert species are light, sandy coloured; forest or woodland species darker, often with mottling or other markings which are difficult to see in an uneven pattern of light and shade; the polar bear matches the arctic landscape and several northern carnivores have a seasonal colour change to a white winter coat. In a zoo a tiger may look very conspicuous but according to Schaller (1967) tigers are surprisingly cryptic in their natural habitat and I am continually amazed by the way my civet cats can blend into the most seemingly inadequate cover and become virtually invisible.

Other factors besides concealment, however, are involved in carnivore colour patterns. The simplest and most obvious of these is the principle of warning coloration. It may be advantageous for the possessor of some special defence or weapon to advertise his unpleasant or dangerous qualities, rather than to attempt to avoid detection. A small carnivore with a bold pattern of black and white is best left unmolested: he almost certainly possesses stink glands (see table 3.1, p. 87, for details).

Warning colours constitute an inter-specific signalling device, carrying the simple message 'danger - don't touch', but special colour marks may also have the function of assisting in much more complex intra-specific communications. As Schenkel (1947) showed in his classical study of the wolf, the face, including the ears, and the tail

Figure 3.1. Contrasting facial expressions in the wolf: (a) aggressive threat, (b) anxiety. [From Schenkel, 1947]

are the main organs of expression (figure 3.1) and it is therefore only to be expected that they should be the main areas distinguished by special colour coding.

We are still, however, very far from understanding exactly what part is played by the markings which characterise so many species. The dark colour and long tufts of hair on the caracal's ears undoubtedly make ear movements more obvious and I have seen a pair of these animals exchanging ear twitches - a movement by one evoking an answering twitch by the other - but the social role played by the movements remains to be elucidated. Possibly they are simply a means of maintaining social contact; a visual equivalent of the contact calls which have this function in a number of species. Similarly, the bobcat's tail is very distinctively marked; moreover it is held curved up so that the marking is visible from in front and it is frequently moved from side to side. Here, too, a detailed study would be required to work out the significance of the semaphore system.

In a few cases it is possible to put forward a definite suggestion about the function of special marks: the light chest marks in certain bears, for instance, probably serve to accentuate a bipedal threat posture. The hair at the end of the cheetah's tail is longer than elsewhere and the black and white bands are very distinct. I have seen the tail tip apparently functioning as a 'follow me' signal from mother to young. A female and her cubs were resting in long grass: when the mother finally decided to move off she raised her tail as she did so, holding it almost clear of the grass. It was exceedingly obvious, much more so than the rest of her body, and the young fell in behind and followed her. Had I been in their place, I would have found the tail tip markings extremely helpful in keeping track of mamma. Schaller (1967) quotes Leyhausen as believing that the white patch on the under-side of the tail of the Asian golden cat, *Profelis temmincki*, is used in this way and he himself suggests that the brilliant white ear spots of the tiger may play the same role as 'follow me' signals. The tail tip and the back of the ears are, of course, the ideal places for a signal which is to be received from the rear, so this suggestion is extremely plausible and the occurrence of ear spots in a number of other felids is not surprising. It would be interesting to know whether the young of species possessing ear spots can be induced to follow a model with ear spots more readily than one not so marked and whether this is not the case in species that do not have ear spots.

Facial markings may merely serve to emphasise expressive movements of mouth, eyes or ears. Similarly, open-mouthed threat is emphasised by the black gum line in many felids, contrasting with light surrounding fur and, in the marsh mongoose, by the pink nose tip and lips which stand out against the dark face. The markings, however, may sometimes have a more specific role. Kleiman (1967), for instance, believes that the facial markings of the raccoon dog and bat-eared fox are related to social grooming, which is particularly common in these two canids. The face is the area most frequently groomed and she suggests that the markings serve to direct the groomer's responses to this particular region. The same may be true of the dark 'tear marks' on the face of the cheetah, for in this species too social grooming is common and is concentrated on the face (Eaton, 1969b).

The idea that markings serve to orient responses - not necessarily only friendly ones - between individuals may be of more general applicability than has hitherto been recognised. The play fighting of young African civets, *Civettictis civetta*, is very prone to turn somewhat vicious. I have repeatedly been struck by the fact that as soon as this happens the actions become more stereotyped: the animals fence for position with their noses and bite predominantly at the side of the neck, just where there is a very distinctive black and white marking. The same sort of 'neck fencing' is also characteristic of the preliminaries to mating and when adult, the female would prevent the male from stealing her food by snapping at his neck stripe (see also plate 14). The orientation of the bite is so definite that it is impossible to

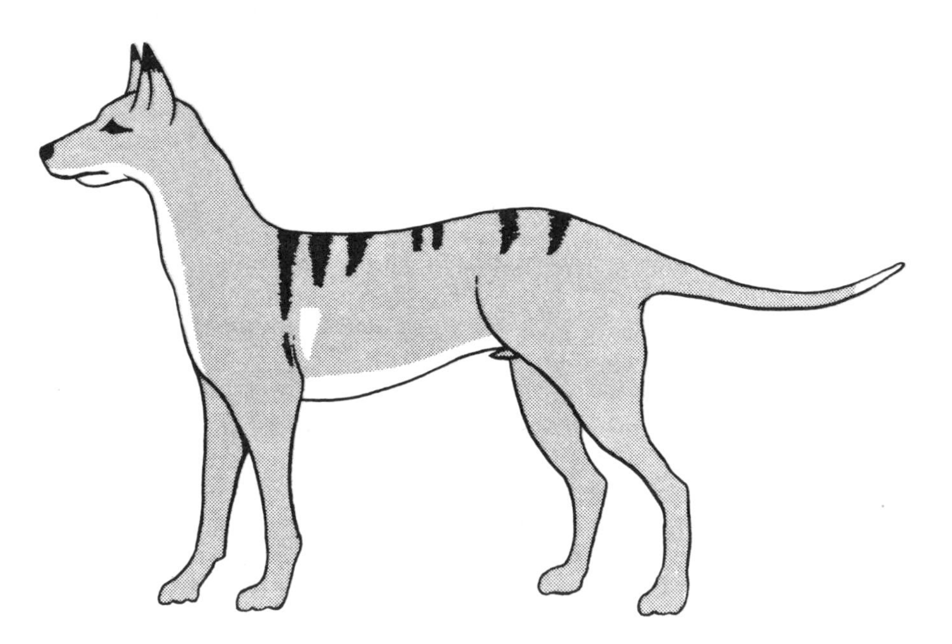

Figure 3.2. Body markings related to social behaviour in a hypothetical prototype canid. Expressive movements of ears, eyes, mouth and tail are emphasised by dark or light colouring; the white belly accentuates the submissive posture while dark stripes on shoulder and rump make piloerection more striking; the white stripe on the shoulder and the white cheek may serve to direct bites to relatively non-vulnerable areas. [After Fox, 1969]

resist the assumption that the function of the mark must be to direct the bites to a specific region where there are no vital structures near the surface. Fox (1969a) has independently suggested that in the Canidae, markings on the shoulder or cheek may have the function of orienting the bite in ritualised intra-specific fighting (figure 3.2).

It may not be without significance that nose-fencing, combined with neck biting, of the sort described in the civets, is characteristic of the fighting of a number of species of mongoose and that distinctive markings on the side of the neck are present in three species, the crab-eating mongoose, *Herpestes urva*, the striped-necked mongoose, *H. vitticollis* and the Gambian mongoose, *Mungos gambianus*; I am indebted to Mr L. W. Robbins for the information that *Liberiictis* also has a neck stripe, much like that of *M. gambianus*. Pocock (1911) suggested that in *H. urva* the neck stripe should be regarded as warning coloration, since the secretion of the anal glands is said to be 'aqueous, horribly foetid and projectile to a great distance'. The same argument applies to *H. vitticollis*, which can also expel a malodorous anal gland secretion. However, a warning function of the kind Pocock suggests, directed to other species, is not incompatible with a second, bite-directing role in intra-specific encounters and I have seen a pair of *H. urva* indulging in a squabble in the same manner as has been described for the civets. In the kusimanse, *Crossarchus obscurus*, an extremely stylised form of fighting with bites directed almost exclusively at the neck often accompanies courtship. In this species there is no definite neck stripe but the hair in this region is somewhat lighter than its surroundings, forming what could be either a rudimentary or a vestigial neck stripe.

The region of the side of the neck sometimes shows another peculiarity. In a few viverrids the hair in this area does not slope backwards in the normal manner but instead is directed forwards. Pocock (1933, 1934) records this in *Macrogalidia musschenbroeki* and in *Paradoxurus zeylonensis* and *P. jerdoni* but offers no suggestion as to its significance. It seems possible that it may be related to a social grooming pattern in which the side of the partner's neck is licked in a forward direction. It is not at all difficult to imagine that an original tendency to nose-fencing and neck biting, evolved with the function of minimising hurt in quarrels, might be taken one step further and the very area which was once the target of bites become the special grooming place. Reversed neck hair also occurs in another viverrid, the banded palm civet, *Hemigalus derbyanus*, and also in one genus of the Felidae, *Leopardus*, where it must obviously have been evolved quite independently.

In a few species markings are so variable that a human observer can use them to distinguish individuals. This is true, for instance, of the ventral white patterning of the weasel (Linn and Day, 1966) and individual tigers or cheetahs can be recognised by differences in their facial markings (Schaller, 1967; Eaton, 1970c). Whether the animals themselves use such differences for individual recognition is doubtful in the case of the weasel and tiger but seems highly probable in the social, highly visual African hunting dog, *Lycaon pictus*. In *Lycaon* (plate 6) not only are the body markings variable but so are the marks on the face and the distribution of black and white on the tail (Stevenson-Hamilton, 1914). In the only pack I have observed, every one of the twenty-one individuals could be distinguished both by face and by tail. In view of their group hunting activities, it is easy to appreciate that individual recognition could be as important from the rear as in a face-to-face encounter. Wolves, although not distinctively patterned like *Lycaon*, show a considerable range of individual variation in coat colour and even within a litter there may be darker and lighter, greyer and more rufous animals (Mech, 1970). These differences may well have some value in facilitating individual recognition; otherwise it is difficult to see why such variability should persist.

The mane of the lion is in a category of its own - the only really striking example of sexual dimorphism in the Carnivora. This is presumably correlated with an unusual social structure, for in lions the normal unit is a group of females and juveniles, together with a smaller number of adult males. Although the mane may have some protective function in fighting and may play a role in threat, its primary value is probably as a more general indicator of social status. Since its size and colouring are highly variable, even within a single habitat, it may also assist in individual recognition.

All these aspects of coloration involve secondary functions of the pelage: the primary one, of course, is to provide the insulation without which the metabolic cost of homoiothermy would be prohibitive. Carnivores are found from the equator to the Arctic regions and the majority of them remain active the year round. Arctic species are able to withstand extremely low environmental temperatures almost entirely because of the magnificent insulating qualities of their fur. Their basal metabolic rates and body temperatures do not differ from those of temperate and tropical species of comparable size (Scholander *et al.*, 1950c) and yet they are able to maintain normal body temperature in extremely cold environments without having to increase metabolic heat production (figure 3.3). Scholander and his co-workers (1950b) attempted to determine the critical temperature at which such an increase does occur for a number of Arctic species. Unfortunately they underestimated the capacities of their subjects and designed their apparatus to function only down to −30°C. At

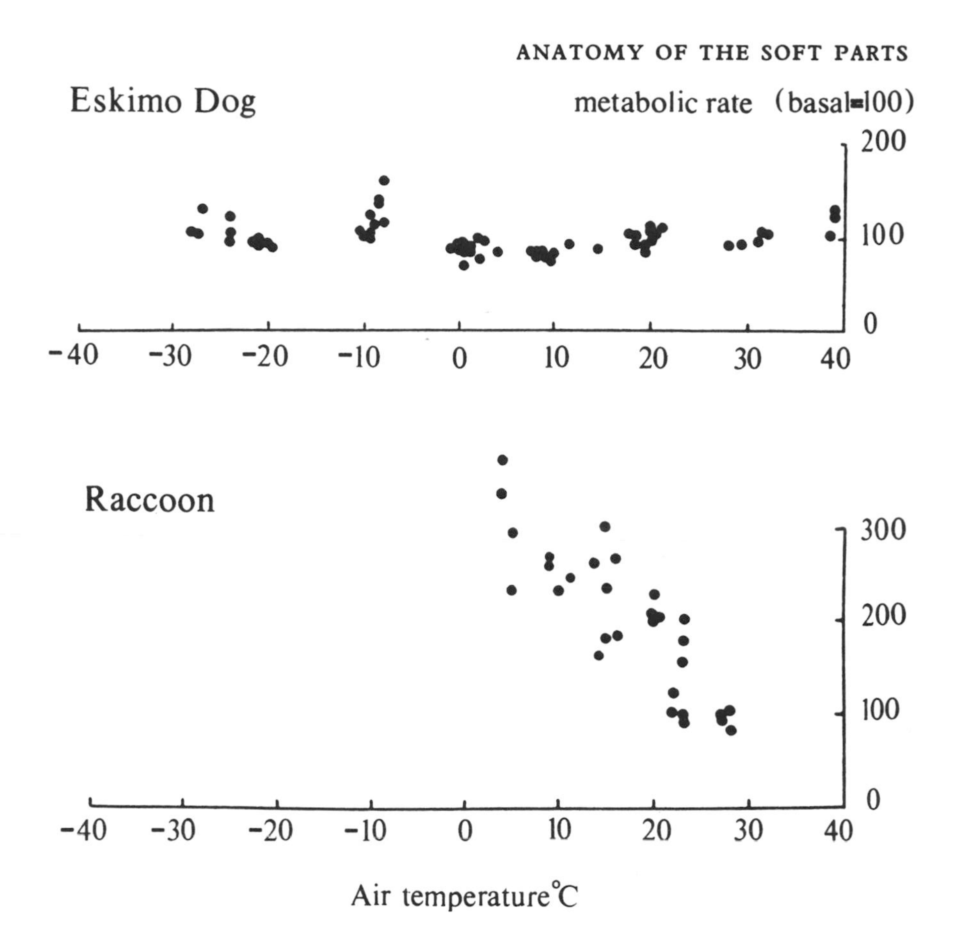

Figure 3.3. Metabolic rate in relation to external temperature in an arctic carnivore (Eskimo dog) and a tropical one (raccoon). [From Scholander *et al.*, 1950]

this temperature the husky dog and arctic fox showed no metabolic increase and it is clear that they were nowhere near their limits, for it was observed that the dog could sleep at −45°C and one arctic fox slept at −60°C. Husky dog puppies, however, were just reaching their critical limit at −30°C and polar bear cubs (age unstated) reached theirs at −10°C. In contrast to these figures, the values for the tropical crab-eating raccoon, *Procyon cancrivorus*, and coati, *Nasua narica*, were found to be in the region of +20°C (figure 3.3).

The performance of the arctic species implies an ability to adjust effective insulation in relation to external temperature. If adjustment were not possible and metabolic rate remained unchanged, then the arctic fox that slept peacefully at −60°C would boil if the environmental temperature rose to freezing point. Insulation can be altered in three ways. The limbs and nose are relatively thinly haired:

heat loss is therefore decreased if the animal curls up so that these areas are not exposed. By fluffing up or sleeking down the hair, its effective thickness and hence its insulating power can be altered. Finally, a physiological mechanism may come into play if these adjustments alone do not suffice to cut down heat loss sufficiently to maintain body temperature: the peripheral circulation may be reduced and the skin temperature allowed to fall. This is equivalent to increasing the thickness of the layer that intervenes between the cold outside world and the warm inside of the animal. The superficial tissues of arctic species are adapted to withstand this chilling and no damage results.

Scholander and his co-workers (1950a) devised a simple method of measuring the insulating capacities of the furs of the different species they studied. Two pieces of skin, with the fur side outwards, were clamped one on either side of an electric heating plate. The external

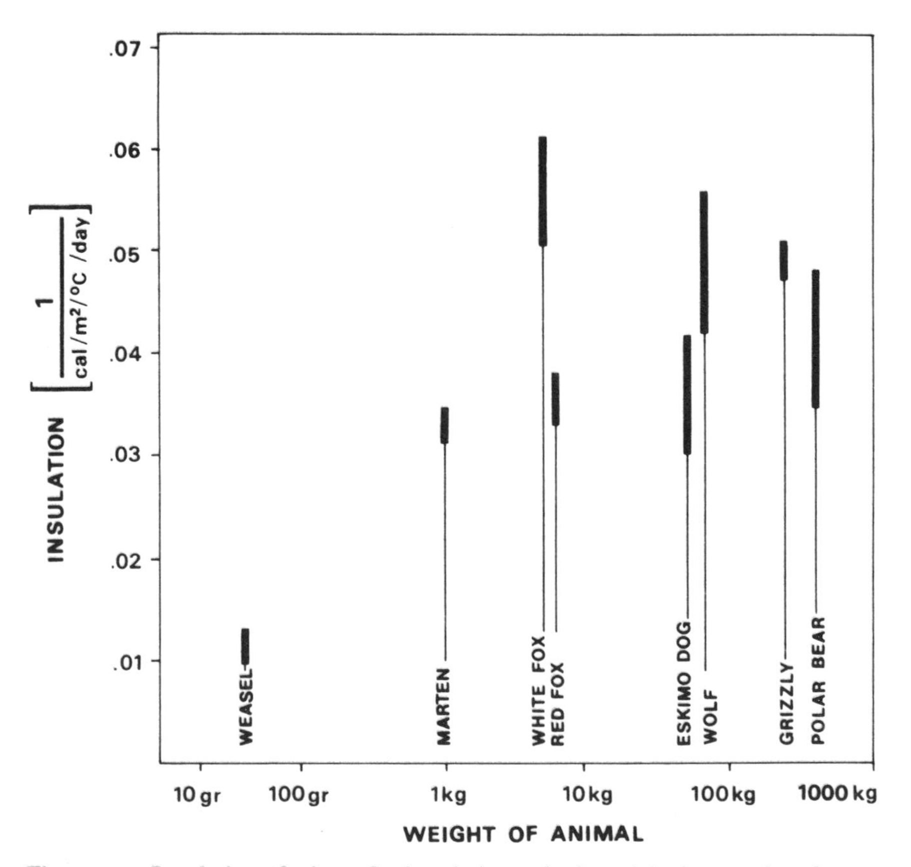

Figure 3.4. Insulation of winter fur in relation to body weight for a series of arctic carnivores. [From Scholander *et al.*, 1950a]

temperature was maintained at 0°C and the power required to keep the hotplate at 37°C (approximately normal body temperature) was measured. Knowing the area of the heating plate, this could be converted to watts per m^2 per 1°C temperature difference, or expressed as calories per hour per m^2 per 1°C. The reciprocal of this value then gives a measure of the insulation. The results they obtained (figure 3.4) were in satisfactory agreement with the critical temperature measurements: the arctic fox had the best insulation, followed closely by wolf, grizzly bear and polar bear, with husky dog and red fox a little lower, while the tropical raccoon and kinkajou gave very much lower values.

While on the subject of the furs of northern species, an interesting fact about the fur of the wolverine may be mentioned. This fur is the Eskimos' first choice for lining the hoods of their parkas. The reason for this is that the frost formed by condensed moisture from the breath can very easily be brushed off the wolverine's rather coarse and very smooth hairs; a flick with the mittened hand suffices. With other furs the frost cannot be thus dislodged but must be melted off, with consequent wetting of the fur. If there is a slight rise in external temperature, the warmth of the breath may start to melt the frost *in situ*, which is decidedly uncomfortable for the wearer (Quick, 1952).

The arctic summer may not be long but it can be quite warm. Irving (1964) gives a seasonal temperature range of from −60°C to +20°C as characteristic of arctic climates. It is therefore not surprising that species living in such conditions should have different summer and winter pelages. In most of the species that have been investigated, both from temperate and from Arctic regions, this is achieved by two fairly rapid cycles of moulting and regrowth; one in spring, when the heavy winter coat is shed and replaced by the lighter summer one and the other in autumn, to restore the *status quo*. No investigations have been made on tropical carnivores but by combing my tame female kusimanse every evening and weighing the accumulated combings each week, I have found that she too has a spring and an autumn moult, despite the fact that she lives at a latitude of only 5° north.

Some of the Canidae, however, manage with a single rather prolonged cycle. Shedding of the winter coat occurs in spring, and the summer coat which replaces it consists mainly of short guard hairs, with very little under-fur. It is not until the end of summer that the under-fur comes in thickly and the guard hairs at the same time grow longer, so that the warm winter coat is re-established. A cycle of this type is found in the coyote, *Canis latrans* (Whiteman, 1940), and the red fox (Bassett and Llewellyn, 1948) but it is not universal in the Canidae, for the arctic fox has the usual two moults (Johansson, 1960). What happens in the wolf is not clear. According to Mech (1970), Goldman believes that there is a single cycle, as in the coyote, but Novikov says that there are two moults. Since they worked in different areas, it is of course possible that both are right and that the moulting cycle is not uniform over the whole range of the species. Unfortunately, so little is known about other families that one cannot say whether the single moult cycle is or is not restricted to the Canidae, or even how common it is in the latter family. Until a great deal more information has been acquired, it is not

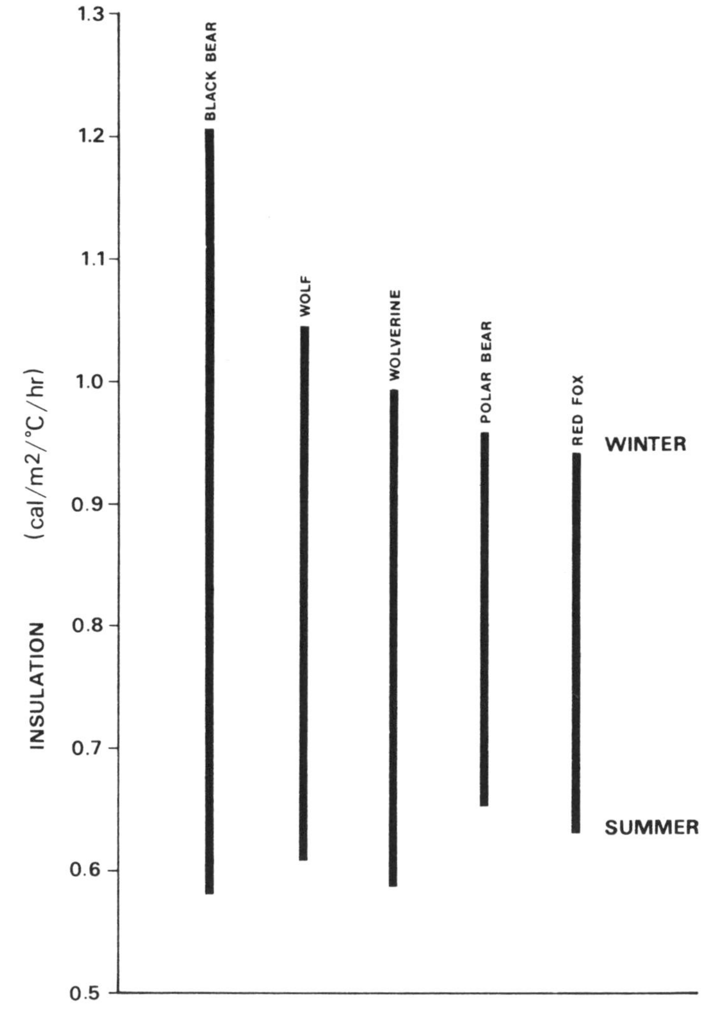

Figure 3.5. Insulation of winter and summer coats of a series of arctic carnivores. [Data from Hart, 1956]

possible to come to any definite conclusion as to why there should be two types of moulting cycle but the single cycle, involving as it does a delay between shedding and full regrowth of the replacing hairs, would appear to be a secondary condition. Whatever its significance may ultimately prove to be, it has the same result as the two-moult cycle: the animal can have a lighter coat in summer and a warmer one in winter.

Hart (1956) has measured the changes in insulation which result from the seasonal pelage change for five species of carnivore and his data are shown in figure 3.5. The black bear, *Ursus americanus*, with the heaviest winter coat, shows the biggest seasonal change - an approximate halving of insulation in summer. The insulating values of the summer coats of the other species do not differ very much from that of the black bear but, with less effective winter insulation, the seasonal change is less.

In a number of northern species the winter moult involves not only a thickening of the pelage but also a change in colour (figure 3.6). The winter coat is white, making the animal inconspicuous against the background of winter snow. This happens in a number of mustelids and also in the normal brownish grey colour variety of the arctic fox. The 'blue' colour phase of this species does not turn white in winter but moults to a slightly paler grey. Johansson's (1960) breeding studies have shown that the difference between the two colour varieties is due to an incompletely recessive autosomal gene for 'white', which affects the pigmentation of the summer coat as well as the colour change that occurs at moulting.

In some species the change to white occurs only in the northern part of the range or at high altitude, while in the warmer, more southerly or more low-lying areas, the winter coat is dark. This is true of the stoat in Europe and the long-tailed weasel in America (Cott, 1940; Rothschild, 1944, 1957). The factors that control the colour change are still not fully understood. It is obvious that although an animal may moult without changing colour, it does not change colour without moulting. It is therefore necessary to consider first the factors that control moulting. In the ferret (Bissonnette, 1935; Harvey and Macfarlane, 1958) and the stoat (Rust, 1965) moulting is controlled by day length, acting through the pituitary. In the fox, *Vulpes vulpes* (Bassett *et al.*, 1944), the long-tailed weasel and the stoat (Bissonnette, 1943), moulting can be induced by manipulation of the daily light-dark cycle; presumably here too the pituitary is involved. The responses to light relate to the duration of the photoperiod, not to the fact that day length is changing, since moulting can be initiated by a sudden shift in the light/dark cycle to a new but constant regime.

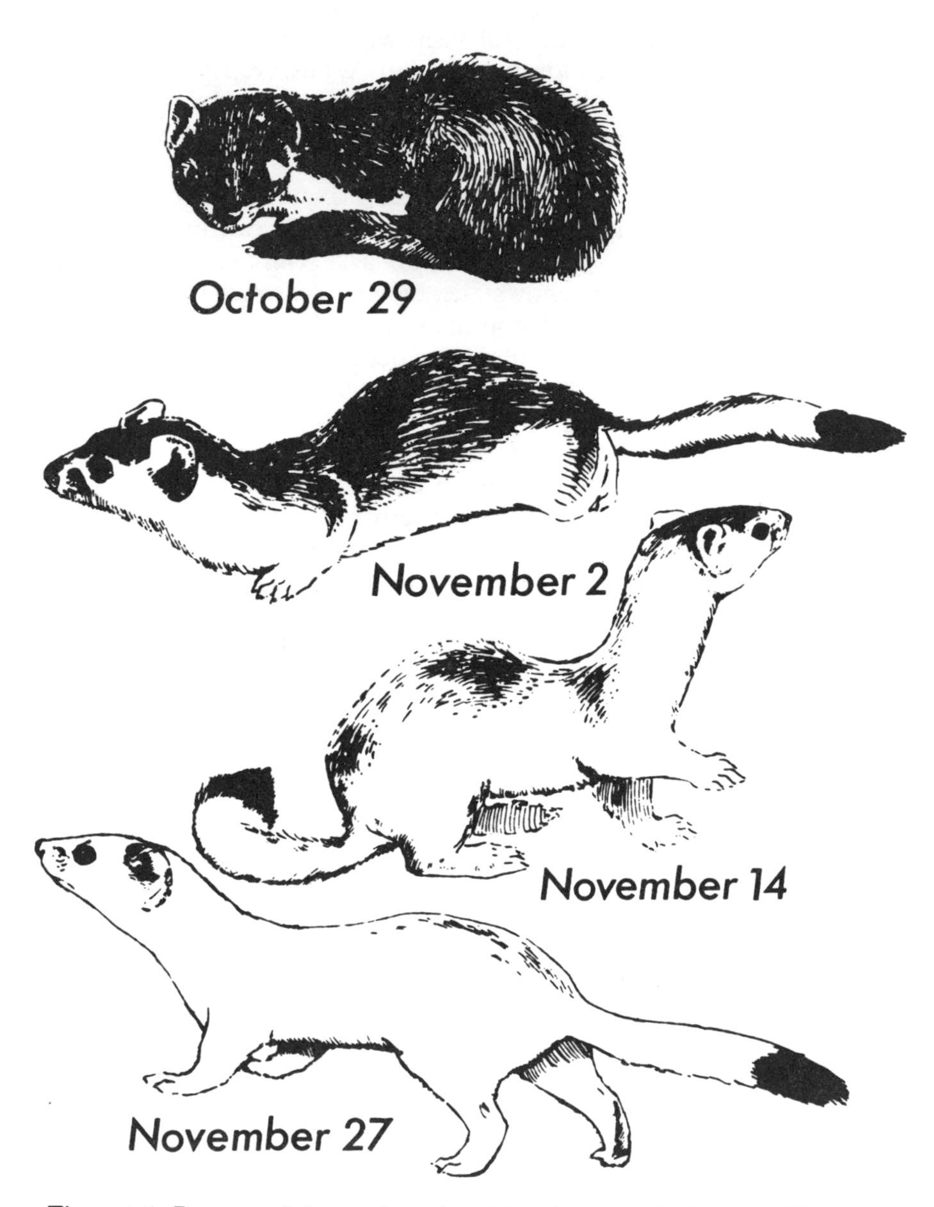

Figure 3.6. Progress of the moult to the white winter coat in the stoat. [From Hamilton, 1939]

In the stoat, Rust (1965) has shown that the normal cycle of pigmentation as well as that of moulting is under the control of the pituitary. Intact animals, whether in brown or in white pelage, will moult to brown if held under a twelve hour dark/twelve hour light regime but hypophysectomised animals will grow only white hair. The involvement of the pituitary suggests that the pigmentation cycle too is determined by the day length conditions. However, since a moult can be in either direction - white to brown or brown to white - it is clear that the light requirements for the two cycles are not identical. In the normal winter moult the conditions required for inhibition of pigmentation are fulfilled before moulting occurs; in the spring moult they are not.

Most workers have found that temperature changes have little effect on the pigment cycle. The most striking case of temperature independence is the regular moulting to white of the arctic foxes in the London Zoo, even in a winter climate very much milder than that of their normal habitat. Rothschild's (1942, 1944) experiments on stoats, however, gave anomalous results. Her animals were caught at Peterborough, in the south of England, where a change to white at the winter moult is exceptional. She found that if kept at low temperature the autumn moult was to white. In spring, on the contrary, even if kept cold, the animals moulted to brown.

Rothschild's results may be brought into line with Rust's findings if two factors are taken into account. Firstly, there is genetic diversity: there is no need to assume that the light requirements are identical within a species over the whole of its range: indeed, it would be very surprising if this were the case in a species existing over a wide range of latitude. The second factor is the effect of low temperature on the moulting, as distinct from the pigmentation, cycle. Rust (1962) kept winter-trapped Wisconsin stoats under artificially increased day length, some at 70°F and some at 20°F. All moulted to brown much earlier than they would have done in natural lighting but the animals kept at the lower temperature required a longer exposure to the increased lighting before moulting was initiated. Since the Peterborough stoats do not usually moult to white in winter but the Wisconsin ones do, we must assume that in the English animals the lighting requirements for pigment inhibition have not normally been fulfilled by the time the moult occurs, whereas in the American ones they have. If in Rothschild's animals cold slowed down the initiation of moulting, as it did in Rust's experiments, then this could have permitted the processes leading to inhibition of pigment formation to 'catch up' and so have caused a white moult instead of a brown one. This hypothesis would also account for the fact that Rothschild found a remarkable variability in the response to cold, even among animals that were littermates. If two processes are involved, running, as it were, neck and neck, then it is not very difficult to believe that very minor variables might suffice to decide which is leading - moulting or pigment inhibition.

In a climate where winters are usually mild and prolonged snowy conditions are rare there is obvious adaptive value in having the responses to light of the moulting and the pigment cycles so arranged that temperature may swing the balance either way. On the other hand, if hard winters are the rule, one might expect the responses to be so arranged as to ensure that by the time the winter moult occurs pigment formation will already have been inhibited. It is therefore not surprising that the arctic foxes go white in the English winter: they have no reason to be adapted to cope with such mild conditions. The same is true of the stoat which Rothschild (1957) caught in Switzerland at an altitude of 2320 m and which at Peterborough moulted to white in a winter too mild to cause colour change in the local animals.

Clearly it is desirable that the lighting requirements for moulting and for pigment inhibition should be precisely determined but this has not yet been attempted. A little, however, is known about the pituitary-mediated endocrine changes involved. Rust (1965) has found that in hypophysectomised stoats kept on a twelve hour light/twelve hour dark schedule, the new hair which grows to reclothe a plucked area is white. If, however, the animals are injected with corticotrophin or with melanocyte-stimulating hormone, then the new hair is pigmented. Changes in the levels of these hormones may therefore be responsible for controlling pigmentation.

The vibrissae remain to be considered. These are enlarged hairs whose bases are richly supplied with sensory nerve endings, so that they function as tactile organs. They are most numerous on the face but many carnivores also have a bunch of vibrissae on the postero-external surface of the wrist. The facial vibrissae show a characteristic arrangement into a number of tufts. Pocock (1914b) describes the typical carnivore arrangement and in a series of papers[1] he gives details for the various families of carnivores. Figure 3.14 (page 104) shows his nomenclature, which is generally adopted.

In bears the vibrissae are vestigial and they are somewhat reduced in a number of fossorial species. In aquatic species they are generally rather few in number but are very stiff. Pocock (1915a) points out that this stiffness is essential if the vibrissae are to remain standing out from the body while the animal is in water. The most striking departure from the basic arrangement, however, is met with in the

[1] Pocock 1914c; 1915a–f; 1916a, b, e, g; 1920b; 1921a, b, c; 1926; 1927a; 1928b, c.

Felidae, where the inter-ramal tuft is absent (Pocock, 1914b). One may argue that while vibrissae in this position are of use to a short-legged animal, or to one that finds its food by nosing about in soil or litter, they would be useless in the felids, whose chins are not often close to the ground. The same argument would apply in the Canidae but to a lesser extent. Although long-legged, the canids do use the nose for ground snuffling and in food burying and in them the inter-ramal tuft, although generally rather small, has not been lost. Moreover, there is some correlation with body build and, presumably, with the use of the nose, for the tuft is vestigial in the long-legged maned wolf, *Chrysocyon*, and exceptionally large in the equally exceptionally short-legged bush dog, *Speothos* (Pocock, 1927c) (see plate 7). It is difficult to judge whether the loss of these vibrassae in the Felidae bears any relation to the development of hypertrophied skin glands on the chin and to the affectionate chin rubbing that betokens their use. The cat's chin glands, however, are only a minor example of a phenomenon of considerable importance in the Carnivora - the development of special scent glands - to which we must now turn.

Table 3.1
Carnivore colour patterns

1. CANIDAE

Typically uniform except for markings on face and tail tip, which are not uncommon.

Lycaon: anomalous in having the whole body marked with irregular blotches and patches of black.

2. URSIDAE

Generally uniform: *Tremarctos* has white markings on face and chest and a light chest mark is present in *Selenarctos*, *Helarctos* and *Melursus*.

3. PROCYONIDAE

Body typically uniform but tail marked with dark rings: facial markings also common. The giant panda is anomalous, being white with black limbs and shoulders, ears and eye patches.

4. MUSTELIDAE

Typically uniform but there are white head stripes in the badgers and black and white stripes or blotches in the following species possessing stink glands:

Mephitinae: all three genera (the American skunks).

Mustelinae: *Vormela*, *Ictonyx*, *Poecilictis*, *Poecilogale*[1].

Lyncodon, the Patagonian weasel, also has stink glands and is conspicuously coloured: grey above and black below with a broad white band on the head.

Melinae: *Mydaus javanensis*, the stink badger, has the usual badger head striping extended as a white line all along the body.

[1] Pocock (1908) doubted whether *Poecilogale* possesses effective stink glands and suggested that its colouring might be an example of Batesian mimicry of *Ictonyx*. He was wrong: *Poecilogale* may not stink as badly as *Ictonyx* and does not discharge its glands unless severely upset; but when it does so the effect is both nauseous and persistent (Alexander and Ewer, 1959).

5. VIVERRIDAE

(i) Viverrinae: typically with a pattern of dark spots tending to be arranged in lines or to join up to form longitudinal stripes dorsally; tail with dark rings.

Prionodon: spots larger with no tendency to form lines; if anything, they tend to form cross banding dorsally.

Osbornictis: (plate 1) anomalous in having no spots: white facial markings, tail dark, without rings; body a uniform chestnut-brown; juvenile pelage also without spots, somewhat darker than adult (Verheyen, 1962).

(ii) Paradoxurinae: very similar to Viverrinae but markings tend to be less distinct.

Paguma and *Arctictis*: facial markings but no body patterning or tail rings.

(iii) Hemigalinae: typically with transverse bandings dorsally.

Fossa: spots arranged in longitudinal rows; markings more distinct in juvenile (Mivart, 1882a) (see plate 10).

Hemigalus and *Cynogale*: unmarked.

Eupleres: young have dark shoulder stripes but these are lost in the adult.

(iv) Galidiinae: typically with dorsal longitudinal stripes.

Galidia: body striping absent, tail ringed.

Salanoia: body obscurely spotted.

(v) Herpestinae: typically brindled or grizzled; tip of tail often distinctively coloured.

Suricata and *Mungos mungo*: dorsal transverse banding, particularly marked in the lumbar region.

Herpestes urva, *H. vitticollis*, *Mungos gambianus* and *Liberiictis*: unusual in showing stripes of contrasting colour running from the angle of the mouth back to the shoulder.

(vi) Cryptoproctinae: the one species, *Cryptoprocta ferox*, is uniformly coloured.

6. HYAENIDAE

Typically with somewhat irregular transverse striping on body and cross banding on legs.

Hyaena brunnea: stripes only on legs.

Crocuta crocuta: irregular blotches, not arranged in any particular pattern.

7. FELIDAE

The felids have the most complex and varied coloration of all the carnivores. They are typically blotched, spotted or striped. Weigel (1961) described the basic pattern as composed of dark markings with a tendency to fuse together to form bands or stripes. These are predominantly longitudinal on the dorsal surface, more nearly transverse below the shoulders and on the flanks; the upper parts of the limbs too are usually cross banded. Breaking up of the markings to form smaller spots, or emphasis on fusion to give more distinct striping, result in modifications of the basic pattern. Paling of the centre of the marks produces rosettes while further paling may reduce the patterning virtually to vanishing point (figure 3.7).

Only a few species, however, are more or less uniformly coloured and in some of these, like the lion and the puma (Young and Goldman, 1946), there are markings in the young. Blonk (1965) has described an unusual puma from South America which at three years old still showed distinct spotting on the body. The Asian golden cat, *Profelis temmincki*, has a relatively uniform body colour although facial markings are present but the African *P. auratus* is extremely variable. The ground colour ranges from golden brown to dark grey and the spotting varies from pale rosettes to very distinct black spots which tend to fuse together in the mid-dorsal region to form longitudinal lines. The caracal, possibly closely related, also shows considerable variation. Although uniformly coloured dorsally there is often some patterning ventrally, which may range from faint irregular blotches to perfectly distinct spotting. The jaguarondi is the least marked of all the small cats; Weigel (1961) describes the patterning as reduced to a few markings on the face and some faint splotches on the belly but he does not mention the characteristics of the juvenile pelage.

The genus *Felis*, as defined by Pocock (1951) is characterised by a pattern of solid spots, never rosettes, which show a marked tendency to form transverse stripes on the shoulders but never horizontal lines across the flanks. White ear spots are never present.

Panthera: typical patterning of rosettes; white spots on the back of the ears present. The tiger is anomalous in having a very clear pattern of transverse stripes but the typical ear spots are present. Loss of patterning in the adult lion has already been mentioned.

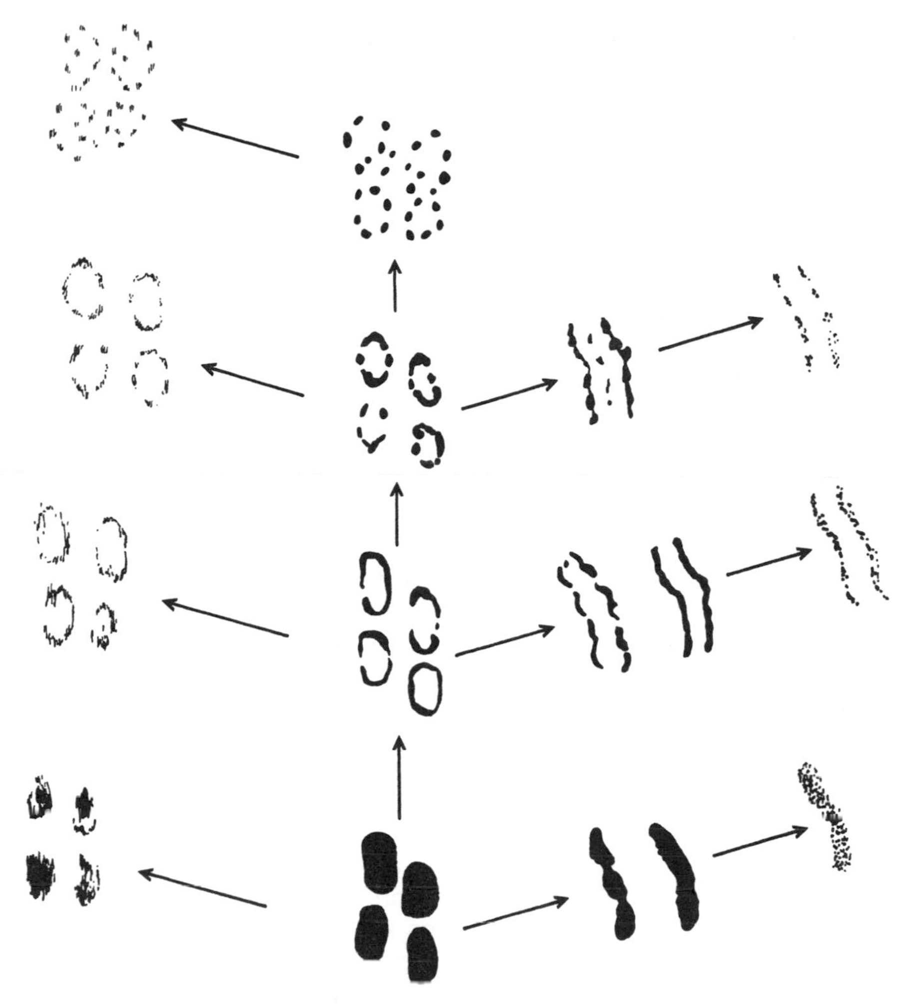

Figure 3.7. Diagram to show possible derivation of the various coat patterns of the Felidae from a basic pattern of dark spots by various combinations of three processes: 1, break-up of individual spots (central vertical column); 2, fusion to form stripes (to right of mid line); 3, paling of spots or stripes (to left of mid line and far right). [After Weigel, 1961]

The ocelots, *Leopardus*, are characterised by spots with a lighter centre and darker outline, tending to fuse to form streaks and by the presence of white ear spots.

Ear spots have probably originated independently a number of times. They are also present in a number of other species – the serval, *Leptailurus serval*, the fishing cat, *Prionailurus viverrinus*, the Bengal and rusty spotted cats, *P. bengalensis* and *P. rubiginosa* and in *Mayailurus iriomotensis*.

The cheetah is as distinctive in colouring as it is in other characters. No other felid has so conspicuous a pattern of clear distinct spots,[1] normally without any tendency to form streaks or bars (but see plate 13); moreover, the ontogenetic change is anomalous, being from a less distinctly marked juvenile pelage to a very clearly spotted adult coat.

B. The skin glands

The skin glands, apart from the mammary glands, are primarily concerned with thermoregulation and with providing a suitable hair dressing to keep the pelage in good condition. In the majority of mammals, however, they have taken on a secondary function in relation to olfactory communication. In certain areas the skin glands have been elaborated and secrete substances whose function is to act as a means of exchanging information. The behaviour associated with the use of such glands and the social role they play will be discussed later. Here it is necessary only to deal with their anatomical location and their distribution within the families of carnivores, so far as these are known. In a number of cases the presence of specialised glands has not been demonstrated histologically but is assumed because the animals have the habit of rubbing the area concerned on each other or on inanimate objects. It is very likely that in fact areas specialised for smell production are far commoner than is at present known.

In the Herpestinae enlarged sebaceous glands producing a substance with a honey-like odour have been found at the bases of the genal vibrissae in *Herpestes edwardsi* (Rensch and Dücker, 1959). Zannier (1965) assumes on behavioural grounds that similar glands occur in the dwarf mongoose, *Helogale parvula* and I have found that both *Atilax paludinosus* and *Crossarchus obscurus* rub their cheeks on various objects in contexts which suggest that smell marking is involved. Schaffer (1940) reports glands round the lips of the fox, while in the cat, as already mentioned, enlarged sebaceous glands are present on the chin. Prescott (in press) has recently shown that in the cat specialised skin glands are also present in a number of other regions of the body.

Glandular patches are found on the ventral surface in a few mustelids. In *Taxidea* Pocock (1925) described an elongated depression in the skin of the abdomen just in front of the penis, into which a series of little glandular pockets opens on either side: the gland is said to be present in the male only. Hall (1926) found a glandular area, marked externally by shorter hair, in much the same position in the American marten, *Martes americana* and the wolverine but could find none in dried skins of fisher, mink, striped and spotted skunks and the river otter. Amongst the procyonids, the kinkajou has

[1] Varaday (1964) once counted the spots on his tame female cheetah; there were 1,967.

paired manidular glands and in addition two patches, one sternal and one abdominal, where the hair is scanty and the skin glands enlarged (Pocock, 1921a; Poglayen-Neuwall, 1962). The only similar structures known in the Feloidea are the glandular areas on the throat found in *Galidia elegans* and *Fossa fossa* (Albignac, 1969b, 1970) and the chest gland of *Cryptoprocta ferox*. Vosseler (1929b) found that in the latter species the gland becomes enlarged only when sexual maturity is reached, is larger in the male than in the female and secretes most actively during the mating season.

Enlarged skin glands may also be associated with the root of the tail. Schaffer (1940) reports them in this position in the badger, the otter and also the cat. In the latter he found that, as in *Cryptoprocta*, the glands do not become enlarged until sexual maturity is reached, are most active during the mating season and are larger in the male than in the female. Most of the Canidae possess a rather more elaborate glandular area on the dorsal surface of the tail – the 'violet gland', so named because in the fox, where it was first described, the secretion is said to have an odour resembling violets. According to Fox (1971a) the smell of both arctic and red fox secretions is 'ambrosia-like', presumably meaning sweet and pleasant. Wolf and coyote glands have a similar but weaker odour while that of the grey fox he describes as smelling strong and musk-like.

Schaffer (1940) describes the histology of the gland in some detail. He found that in a pair of foxes killed during the breeding season the tail gland of the female was larger than that of the male. In a pair killed before the onset of breeding, the gland was distinctly less active in both sexes. Grassé (1955) incorrectly quotes Schaffer as saying that the gland is larger in the male. Pocock (1914c) noted the presence of the tail gland in the arctic fox and Thompson (1923) described the glands of the coyote and wolf and of the red, arctic, kit and grey foxes. The gland is largest in the latter and is marked externally by a patch where the under-fur is absent and the guard hairs are tipped with black: a similar dark patch is also very obvious in the fennec, *Fennecus zerda*. Hildebrand (1952) too noted that the gland is particularly large in the grey fox. Its presence has also been recorded in the raccoon dog (Seitz, 1955) and the bush dog (Langguth, 1969). In domestic dogs there may be some enlargement of the sebaceous glands on the dorsum of the tail but they do not form a typical violet gland. According to Schaffer, *Lycaon* is the only wild canid in which it has been established that the violet gland is absent. Langguth (1969) could find no external trace of the gland in the maned wolf but he did not make any histological examination.

Sweat glands are present on the interdigital membrane of the paws in a number of species and these may also have some function in setting scent. In the Canidae, little glandular pockets are also present on the dorsal surface of the membrane, near the bases of the digits. The arctic fox (figure 3.16, page 108) is unusual in possessing a pair of deep glandular pockets on the lower surface between the foot pads which, according to Pocock (1914a), give off a strong 'foxy' smell.

It is in the ano-genital region, however, that the most complex smell producing glands occur. The skin round the anus and just internal to it contains numerous small skin glands and in addition to these, complex aggregated glandular structures are present. Schaffer (1940) has attempted to classify the different types of gland that may occur and has devised a terminology of which the part relevant to the Carnivora involves five different categories. In my opinion his classification is unnecessarily complicated and serves to confuse the picture rather than to clarify it. His terms, however, may be retained to describe two quite distinct types of glandular structure that occur in Carnivora: the anal sacs and the perineal or perfume glands.

Although anal sacs are more widespread, it is convenient to deal first with the perineal glands. These are composed of a compact mass of glandular tissue, typically lying between the anus and the vulva or penis and opening on to a naked or sparsely haired area, which may be infolded to form a storage pouch. Perineal glands occur only in the Viverridae. They are absent in Herpestinae and Cryptoproctinae but present in the other three subfamilies.

In the Paradoxurinae (Pocock, 1915f) the perineal glands are simple, consisting of a patch of glandular tissue, usually divided into right and left halves, and emptying by numerous small openings on to an overlying area of naked skin. This area may be almost flat (*Paradoxurus hermaphroditus*) or folded to form lappets on either side, thus making a sort of storage pocket (*Paguma larvata*, *Arctictis*, *Nandinia*). The position of the naked area varies from species to species. In *Arctictis* (figure 3.8) the gland of the male lies entirely posterior to the penis but in the female it surrounds the vulva and extends well forward in front of it on either side (Vosseler, 1929a; Story, 1945). Although Pocock's original description (1915f) differs from this, his later amended account (1939) is in agreement. In *Paradoxurus* the gland of the female extends forward slightly on either side of the vulva while in *Arctogalidia* the posterior end of the rather small glandular area of the female surrounds the vulva but the main part extends forward in front of it and the gland is absent in the male (Pocock, 1939). In *Nandinia* the gland in the female lies entirely anterior to the genital opening and the same was true of the only adult male that I have examined. Pocock (1915f), however, says that in the male the gland surrounds the penis and extends beyond it both posteriorly and anteriorly. *Nandinia* is unusual in that there is no clear division of the glandular mass into right and left halves. From the above descriptions, it is clear that Schaffer's attempt to erect a special category of pregenital glands, distinct from perineal glands lying between anus and genital opening, is not only meaningless but in practice impossible.

In the Hemigalinae the condition is not very different from the Paradoxurinae (Pocock, 1915a, b). There is a glandular area, lying between anus and genital opening, delivering its secretion into a median longitudinal depression in *Hemigalus* (figure 3.8) and into a

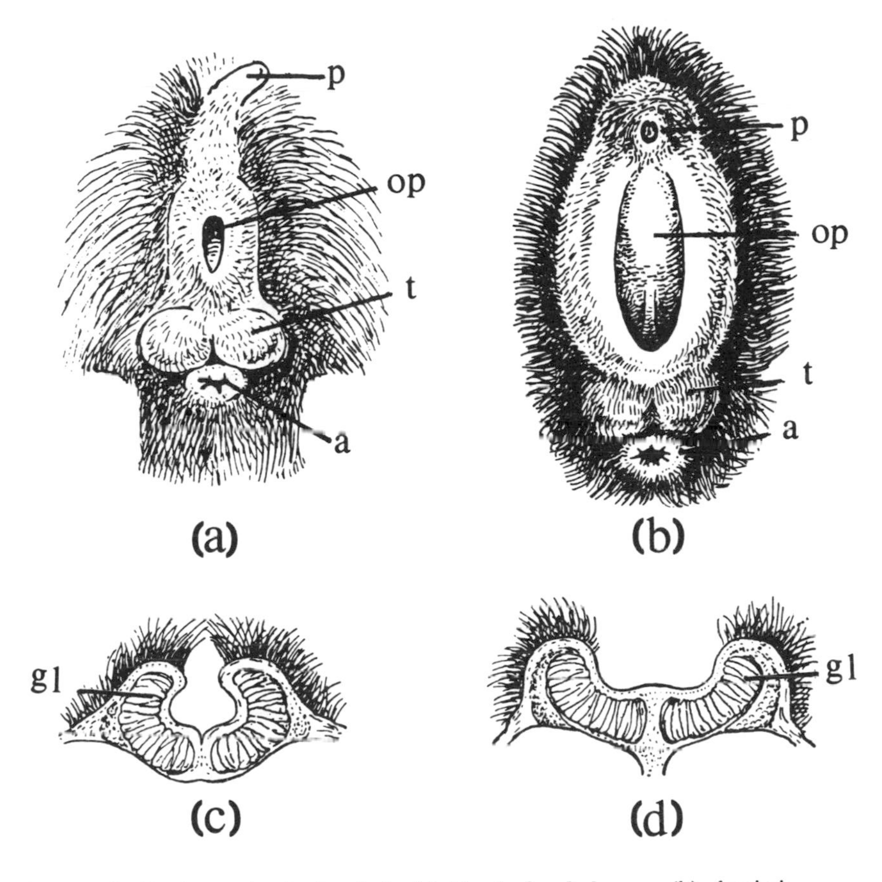

Figure 3.8. Perfume gland of male in (a) *Hemigalus derbyanus*, (b) *Arctictis binturong*, both with the lips of the gland slightly opened to show the pocket beneath. (c) Transverse section of *Arctictis* gland with lips closed in rest position. (d) Same with lips pulled back, ready to deposit scent. a, anus; gl, mass of glandular cells; op, opening of pocket; p, penis; t, testis. [From Pocock, 1915b and f]

number of shallow pits which are not surrounded by labia in *Cynogale*. According to Pocock, the gland is absent in *Fossa* and *Eupleres*. His description for the Galidiinae (1915d) is not very detailed but the structure does not appear to be very different and in the female of *Galidia* the gland apparently extends forwards a little beyond the vulva.

In the Viverrinae there is a sharp contrast between the Prionodontini (linsangs), in which there is no perfume gland (Pocock, 1933), and the Viverrini (civets and genets), in which it reaches its highest development (Pocock, 1915e). In the latter the secretory tissue forms the usual glandular mass, lying between the anus and the genital opening. The overlying skin forms sparsely haired folds or lappets on either side which are complicated and invaginated to varying degrees

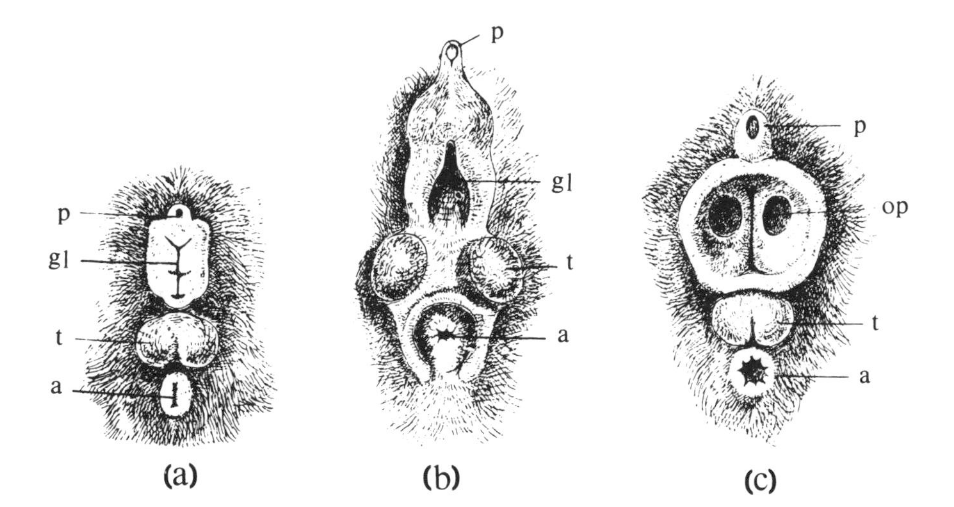

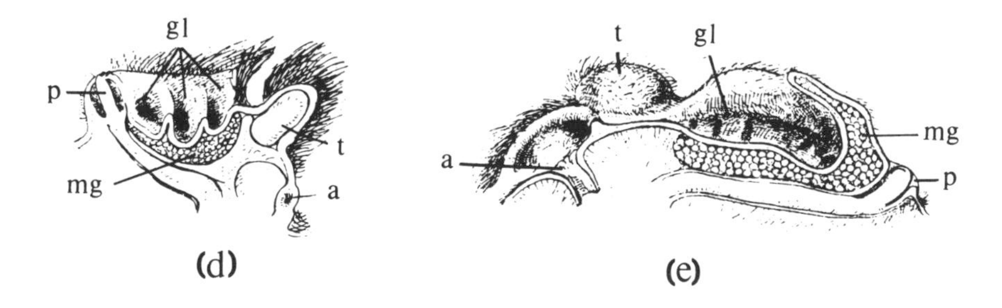

Figure 3.9. Perfume gland of male in (a) *Genetta*, (b) *Viverra*, (c) *Civettictis*. (d) and (e) glands of *Genetta* and *Viverra* in longitudinal section. a, anus; gl, opening of gland; mg, mass of glandular cells; p, penis; t, testis. [From Pocock, 1915e]

to form a storage organ for the secretion. In *Genetta* (figure 3.9) this consists of little more than a number of subsidiary foldings of the lappets, the details differing from species to species. No data are available for *Poiana* and *Osbornictis*.

In the large Indian civet, *Viverra zibetha* (Pocock, 1915e), the labia are separated posteriorly but fuse anteriorly some distance behind the penis (figure 3.9). They enclose an invagination which extends forwards under the fused portion to form a deep pocket which is very slightly subdivided into right and left halves by a low median longitudinal ridge. The glands open on the internal faces of the labia by a series of small openings. *Viverra* is unusual amongst viverrines in that the anal area is surrounded by a depressed area of naked skin, suggestive of a rudimentary anal pouch (*vide infra*).

In *Civettictis civetta* (figure 3.9) the perineal glands are somewhat more complicated and Pocock's account does not agree fully with the photograph given by Allen (1924). The latter shows an arrangement very much as in *Viverra*, whereas Pocock shows the labia separate right forward to the base of the penis. The difference may possibly be related to age: Allen's animal was very old, with heavily worn teeth; Pocock's specimen is described as adult, but may well have been only just mature and the gland may not yet have reached its maximal development. According to Pocock the glandular mass within the labia is invaginated on either side to form a large pocket, communicating by a wide opening with the space between the labia. Allen gives a diagram which suggests the presence of this inner pocket but the details are obscure and he gives no accompanying description. It seems clear, however, that the storage pouch is not constituted merely by a prolongation of the space between the labia but includes a pair of inner pockets, one on either side, and both authors are agreed that the gland empties its secretion into each of these latter by one major aperture.

The anal sacs are vesicular cutaneous invaginations opening by a short canal or duct, one on either side of the anus or just internal to it. Aggregated glands, often of more than one histological type, open either into the vesicle or into the canal, or the wall of the vesicle may be composed of secretory epithelium. The whole is invested with a distinct muscular coat, particularly well developed in those species capable of expelling the secretion to some distance (figure 3.10). Confusion arises because these structures are often referred to simply as 'anal glands', a term which Schaffer (1940) restricts to the diffuse skin glands opening directly on to the surface of the skin in the region of the anus or just internal to it.

Anal sacs are almost universally present in Carnivora but in the Ursidae and the aquatic Lutrinae they are very reduced and are said to be completely absent in some bears (Pocock, 1921a), in the raccoon dog (Seitz, 1955) and in the sea otter (Jacobi, 1938). In the Canidae and Felidae there is nothing of particular note about their structure. They are relatively small, those of the lion being about an inch long,

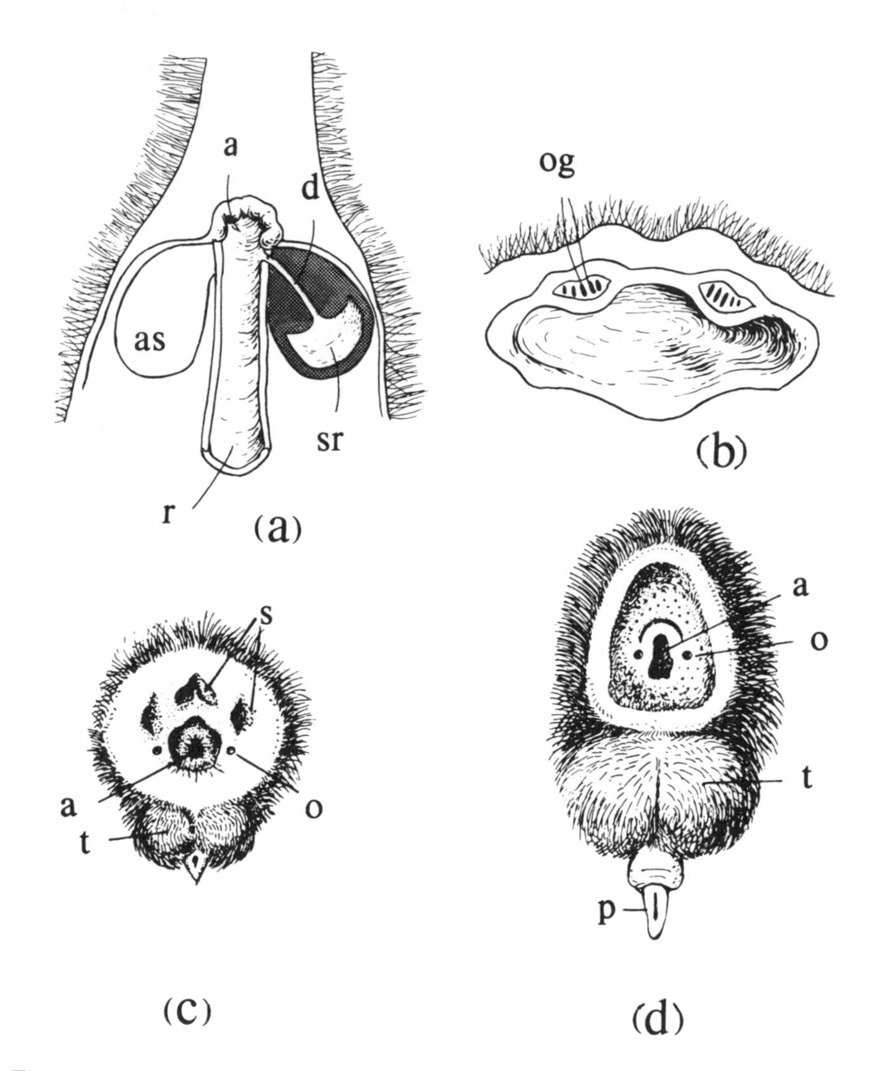

Figure 3.10. (a) *Lyncodon* anal region dissected to show the anal sacs, as. The rectum, r, and the anal sac on the right have been cut through horizontally. a, anus; d, duct; sr, storage reservoir; stippled area, secretory part of anal sac. (b) *Nasua* anal region pulled open to display the openings of the dorsally situated glands, og. Anal pouches of male in (c) *Helogale* and (d) *Atilax*, pulled open to display the openings of the anal sacs. a, anus; o, opening of anal sac; p, penis; s, subsidiary pockets; t, testis. [From Pocock, 1916g]

and the ducts open on either side close to the anus. Most of the Procyonidae have normal anal sacs, much like those of the Canidae, but the kinkajou and coatis are exceptional in lacking them. In the kinkajou their function has presumably been taken over by the sternal and abdominal glands. In the coati (figure 3.10) there is a glandular area situated along the dorsal margin of the anus, consisting of a series of pouches opening by four or five slits on either side. No other carnivore has exactly comparable glands; the nearest approach is the unspecialised pouch glands found in some herpestines. In the red panda the sacs are of normal type and open on an area of naked, slightly invaginated skin around the anus, which resembles a rudimentary version of the anal pouch characteristic of mongooses (Pocock, 1921a). The condition in the giant panda does not appear to have been described but presumably the sacs are present, since the panda marks by rubbing the anal region against objects (Morris and Morris, 1966).

In the Herpestinae (figure 3.10) the skin around the anus is invaginated to form an anal pouch, usually closing to form a transverse line. Into this pouch the ducts of the anal sacs open and when the secretion is to be applied the pouch is everted to expose the openings, which may lie on either side of the anus or distinctly above it. In some species the lining of the pouch itself also contains enlarged cutaneous glands. These are uniformly distributed in *Atilax* but aggregated into patches in *Mungos*. In the latter genus, in *Helogale* and in *Suricata* the lining of the pouch is folded to form a series of subsidiary pockets, which presumably store the secretion of these accessory glands.

In *Cryptoprocta* also an anal pouch is present, lying mainly dorsal to the anus. A pair of diverticula open into the pouch just below the anus: these presumably are normal anal sacs, although Pocock (1916b) appeared to be in some doubt as to their nature.

In the Mustelidae the anal sacs open just within the anus and the honey badger, *Mellivora*, is unusual in possessing an eversible anal pouch very much like that of the mongooses (Pocock, 1920a). In a number of species the anal sacs are very large – those of the striped skunk are said to be the size of a pigeon's egg – and are specially modified in relation to the storage and ejection of the secretion as a means of defence. The distal part of the sac is dilated to form a storage reservoir and the muscle coat is unusually well developed. Blackman (1911), in a paper dealing mainly with histology, gives a vivid description of the glands of the striped skunk:

Each gland is a secretory sac in a muscular tunic and is furnished with a duct to convey the secretion to a little teat-like pipe near the verge of the anus. Contraction of the muscular investment compresses the sacs and spurts the fluid in two jets. The action is really the same as that of a syringe with a compressible bulb.

In the Hyaenidae the anal sacs also open into an anal pouch, somewhat similar to that of the Herpestinae. It differs, however, in that the anus opens into the bottom of the pouch and the large anal sacs more dorsally. Harrison Matthews (1939) has described the condition in the spotted hyaena, *Crocuta crocuta*. In both sexes the anal sacs consist of lobular glands surrounding a central cavity in which the

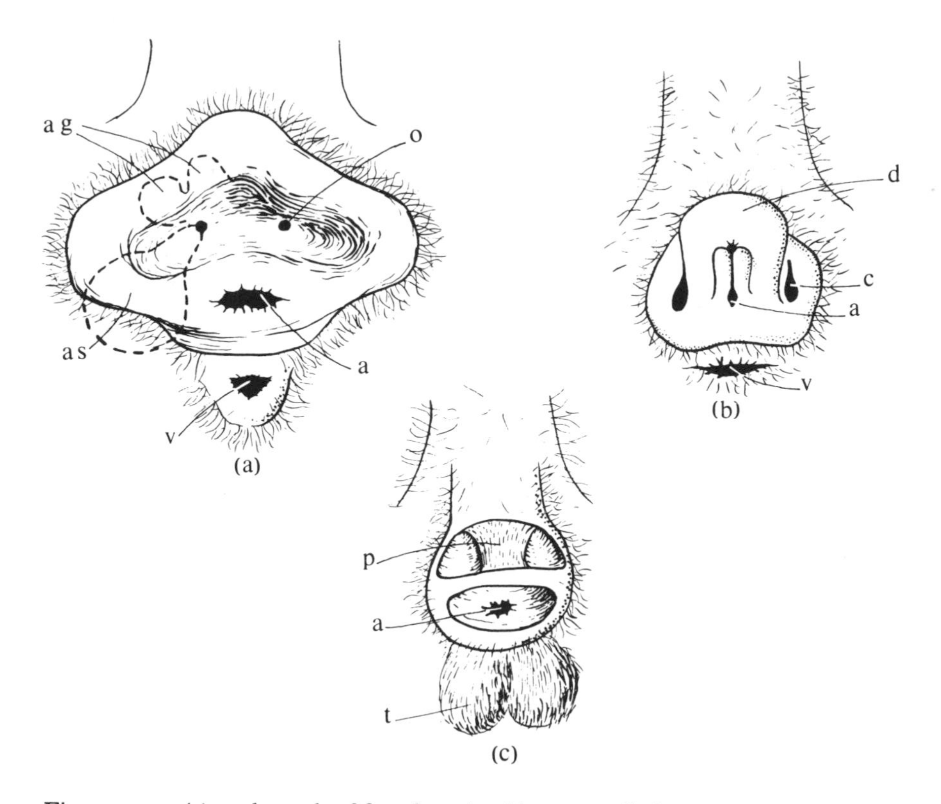

Figure 3.11. (a) anal pouch of female striped hyaena pulled open to show the openings of the anal sacs, o. The positions of the sacs, as, and of the auxiliary glands, ag, are shown on the left in broken lines. a, anus; v, vulva. [After Pocock, 1916a] (b) anal region of young female striped hyaena, actively protruded by the animal itself. a, anus; c, crypt containing opening of anal sac; d, dorsal protrusion; v, vulva. [Drawn from photograph in Fox, 1971] (c) Anal region of male badger (*Meles meles*) with anal and subcaudal pouches pulled open. a, anus at bottom of anal pouch; p, partition separating the two pockets of the subcaudal pouch; t, testis. [After Pocock, 1920b]

secretion is stored and they open by short ducts into the anal pouch. Between the anal sacs lie a number of smaller glands which open directly into the upper part of the pouch, very much like the pouch glands found in *Atilax* and *Mungos* amongst the Herpestinae. The arrangement in the aardwolf, described by Flower (1869b) is essentially the same but in the striped hyaena, *Hyaena hyaena* (Pocock, 1916a) the pouch glands are aggregated to form on either side a pair of auxiliary glandular masses opening independently into the pouch by a number of scattered apertures (figure 3.11). In the brown hyaena, *Hyaena brunnea*, the auxiliary glands appear to be still further differentiated and, according to Murie (1871), each opens into the pouch by a single duct. Unfortunately de Boom (1957), in his description of a female brown hyaena, does not give any description of the supplementary glands.

In *Crocuta* the picture is complicated by the fact that the region between the anus and the tail is greatly swollen, due to the presence of the underlying glands and the skin is thrown into a pocket-like fold immediately beneath the tail. According to Harrison Matthews (1939), this pocket is not glandular and should be regarded as a mere incidental, not as a structure of any importance in its own right. Schaffer (1940), however, compares it with the supra-anal pocket of badgers (*vide infra*); a procedure which, to judge from Matthews' account, does not seem justifiable.

The condition in *Meles* (figure 3.11) is described by Pocock (1920b). The ducts of the normal anal sacs open in the usual mustelid position, just internal to the anus, which is sunk in a slight depression. Above this and distinct from it the skin is invaginated to form a pocket, partly divided at its base into right and left halves. The skin lining the pouch is hairy and highly glandular and secretes copiously. A sub-caudal pocket of this type is also present in *Arctonyx* but is absent in *Taxidea* where, as already mentioned, an abdominal gland is present (Pocock, 1920b).

Schaffer, with his passion for complexity, regards this sub-caudal pocket as something distinct and separate from any of the structures so far described. It seems to me, however, that the circum-anal pouches of Herpestinae and *Cryptoprocta*, the largely supra-anal but entirely comparable pouch of hyaenids, the badger supra-anal or sub-caudal pocket and the anal pouches of *Mellivora* and *Ailurus*, although they clearly have more than a single evolutionary origin, are all basically similar. They are invaginations of the skin in the anal region, in the walls of which the normal circum-anal glands may be hypertrophied to varying extents. There is little justification for distinguishing further between them. It is thus necessary to draw a distinction only between the almost universal anal sacs proper, the viverrid perineal or perfume glands and these peri- or supra-anal pouches, present in varying forms and elaborated to varying degrees in Herpestinae, Cryptoproctinae and Hyaenidae and occurring sporadically in a few members of some of the other families.

In addition to the characteristics of the fur coat and its associated glands, other external features are of some importance, both in relation to their adaptive significance and their taxonomic value. These include the structure of the area of naked skin around the nostrils forming the rhinarium, the ears and the paws. These will be dealt with briefly. Detailed descriptions for a great many species may be found in Allen (1924) and in the series of papers by Pocock dealing with external features of carnivores which are cited in the references.

C. The rhinarium

The rhinarium (figure 3.12) is modified in various ways in relation to the use of the snout and the degree of mobility of the lips. In general, where there is much mobility the rhinarium tends to be prolonged downwards as a grooved extension, the philtrum, which contacts the margin of the mouth (plate 11). The sloth bear, *Melursus*, with its exceedingly mobile lips, has a highly peculiar rhinarium which makes contact with the mouth over a wide area and Pocock (1918b) was of the opinion that the nostrils could probably be closed as a protection against bees, as they can in aquatic species to prevent entry of water. In most aquatic species the rhinarium is rather reduced and in *Cynogale*, the otter civet, it is specially modified so as to bring the nostrils to the dorsal surface.

In species where the snout is elongated and used in digging, the rhinarium is shallow and lacks a philtrum. Further details may be found in the series of papers by Pocock mentioned at the end of the last section.

Hill (1948) points out that in some mammals the detailed pattern of sculpturing on the naked skin of the rhinarium closely resembles the patterning of ridges on the human finger tips and, like the latter, may show individual characteristics. He believes that ridging of this type may be related to highly developed tactile sensitivity of the snout. Carnivores, however, do not show such complex patterning and the surface sculpturing is a simple 'pebbling', giving a granular appearance (see plate 10). Although there are specific differences in the size and degree of separation of the granules, these do not provide the basis for individual identification by means of 'nose prints' in the same way as people can be identified by their finger prints.

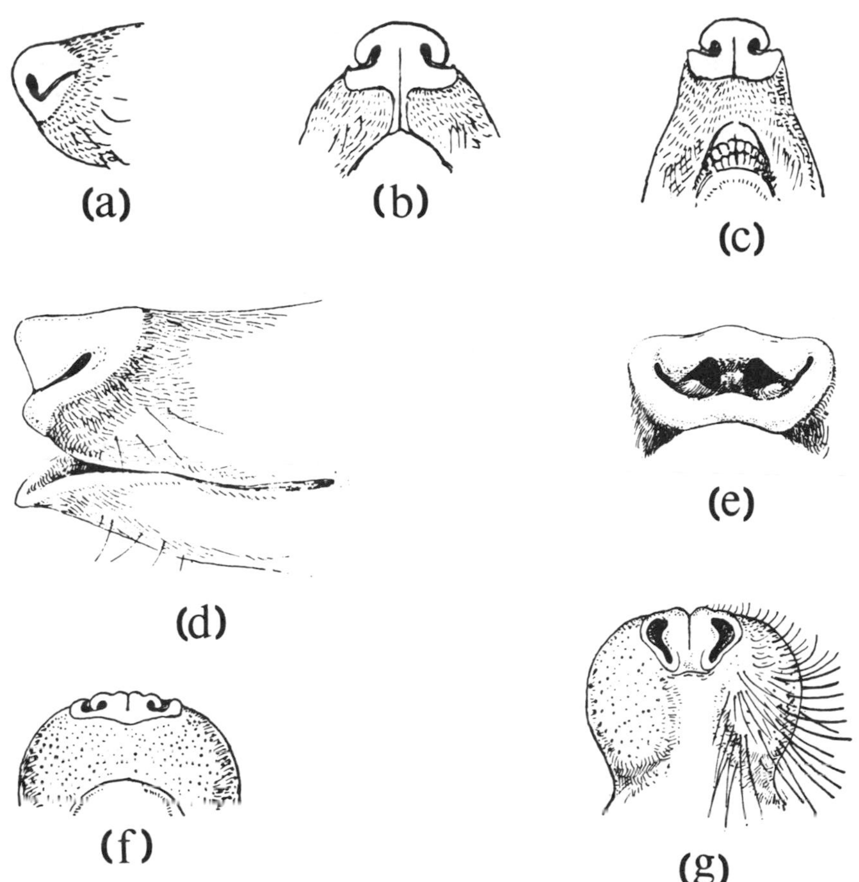

Figure 3.12. Rhinaria of (a) and (b) *Mungos*, side and ventral views; (c) *Crossarchus*, ventral; (d) and (e) *Melursus*, side and front views; (f) and (g), *Cynogale*, ventral and dorsal views. [From Pocock, 1914c, 1915a, 1916g]

D. The ears

The importance of ear markings or decorations has already been mentioned. The use of the ears as organs of expression has also had its effects on the size and shape of the pinna and may be one of the factors deciding whether the ear should be pointed or rounded. Apart from considerations related to the primary function of the pinna as a sound wave collector, a number of other factors are involved in determining the size of the ear. One of these is temperature: an animal living in a very cold climate cannot afford the increased heat loss which large ears would involve and no arctic species has large ears – witness the small ears of the arctic fox as compared with those of the common red fox (plate 8). On the other hand, in hot climates the increased heat-radiating surface provided by large ears may be an added advantage and all the really large-eared species do live in warm climates – for instance the serval (plate 10) amongst the Felidae and *Lycaon* (plate 6) amongst the Canidae. The bat-eared fox and fennec have already been mentioned. Here the extremely large pinnae seem to be related to the special method of prey catching, which involves acute hearing and highly directed listening, as well as to temperature regulation. It is worth noting that *Proteles* has larger ears than the other hyaenids. I know of no description of prey capture in this species but it would clearly be advantageous not to trouble to open a termite mound unless the inhabitants are near the surface. The enlarged pinnae suggest that *Proteles* may listen before it digs.

The pinna which collects sound waves so effectively could equally well act as a funnel to introduce undesirable substances, such as earth or water, in species which burrow or swim. It is therefore not surprising that in burrowing species, like the mongooses, or aquatic ones such as otters, the pinna is reduced. Another factor which may tend to limit ear size is the frequency of intra-specific fighting and the tactics involved. There would be little to gain from enlarged ears if they were bound to end up in tatters before very long. The large ears of *Lycaon* may be partly a reflection of the relatively peaceful social relations that characterise this species.

When we come to look at the details of ear structure, however, it becomes much more difficult to understand the selective forces involved. Details of structure do, however, provide characters of taxonomic importance and it is therefore necessary to deal briefly with them. Figure 3.13 illustrates both the classical terminology of Mivart and Pocock's (1915f) simpler version. The bursa is one of the most enigmatic structures. It is a typical feature of the carnivore ear and is absent only where there has been some reduction in the size of the pinna. All the Felidae and Canidae possess a bursa. Amongst the Viverridae it is absent only in the Herpestinae, all small-eared diggers or burrowers: in *Cynictis* (figure 3.14), which has the largest ears of any mongoose, a small depression at the posterior margin may represent a vestigial bursa (Pocock, 1916g). A similar depression is the nearest approach to a bursa in the Hyaenidae, a fact which suggests that the enlargement of the ears of *Proteles* is a secondary adaptation. Amongst the Mustelidae the bursa is absent in otters, most badgers and also in a number of other species that inhabit burrows (Pocock, 1927a): it is also absent in the rather small-eared bears. In default of any better explanation it has been suggested that the bursa is related in some way to the mobility of the ear, permitting the pinna to be folded down or rotated without at the same

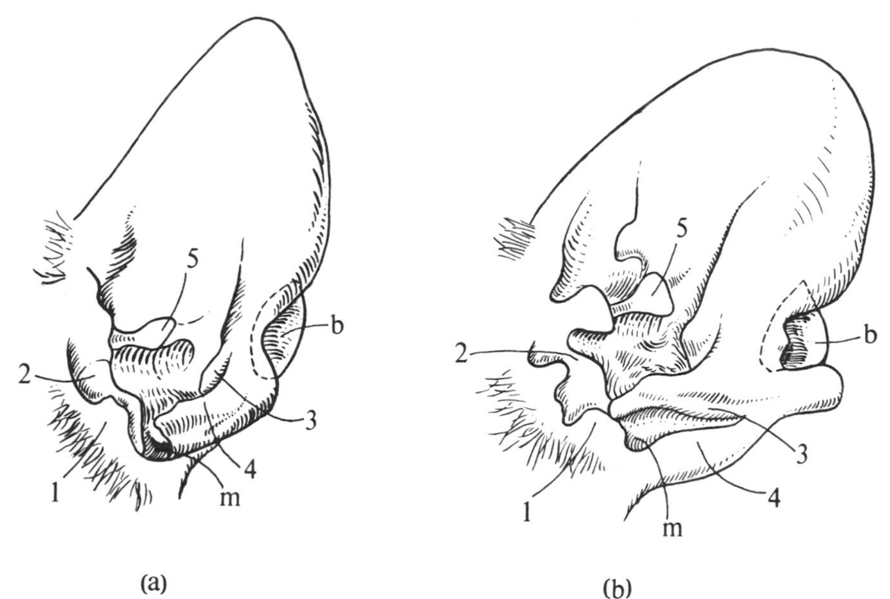

Figure 3.13. (a) Ear of *Paradoxurus*. m, opening of auditory meatus; b, bursa. The terms used by Pocock and by Mivart to designate the various ridges are as follows, Pocock's name being given first in each case:

1 anterior external ridge: tragus
2 anterior internal ridge: post-tragus
3 posterior internal ridge: } anti tragus
4 posterior external ridge: }
5 superior ridge: supra-tragus

(b) Ear of Linsang, showing unusual complexity of ridges. Symbols as in (a).

time occluding the meatus. No attempt, however, has yet been made to establish any detailed correlation between the structure of the bursa and the movements of which the ear is capable.

When it comes to the significance of the ridges at the base of the ear we are in no better case. The primary function of the superior ridge would appear to be simply to act as a buttress, ensuring that the meatus remains open, despite movements of the pinna as a whole. The anterior and posterior ridges may be primarily protective but may also be involved in the localisation of sound (see Chapter 4). However, until experiments have been carried out to study the effects on sound reception of removing or altering various parts, little can be said about the significance of variations in their shape. What, for instance, apart from the fact that it underlines their distinctness from the genets, can be said of the peculiar structure of the linsang ear (figure 3.13), with its wealth of extra flaps and ridges?

In aquatic and burrowing species the problem is obviously less one of keeping the meatus open than of being able to close it when necessary and the ridges and flaps are so constructed as to permit this. In most mongooses closure is effected by moving the posterior ridges forward and the superior ridge downward so that both contact the anterior ridge (figure 3.14). In *Crossarchus*, however, the superior

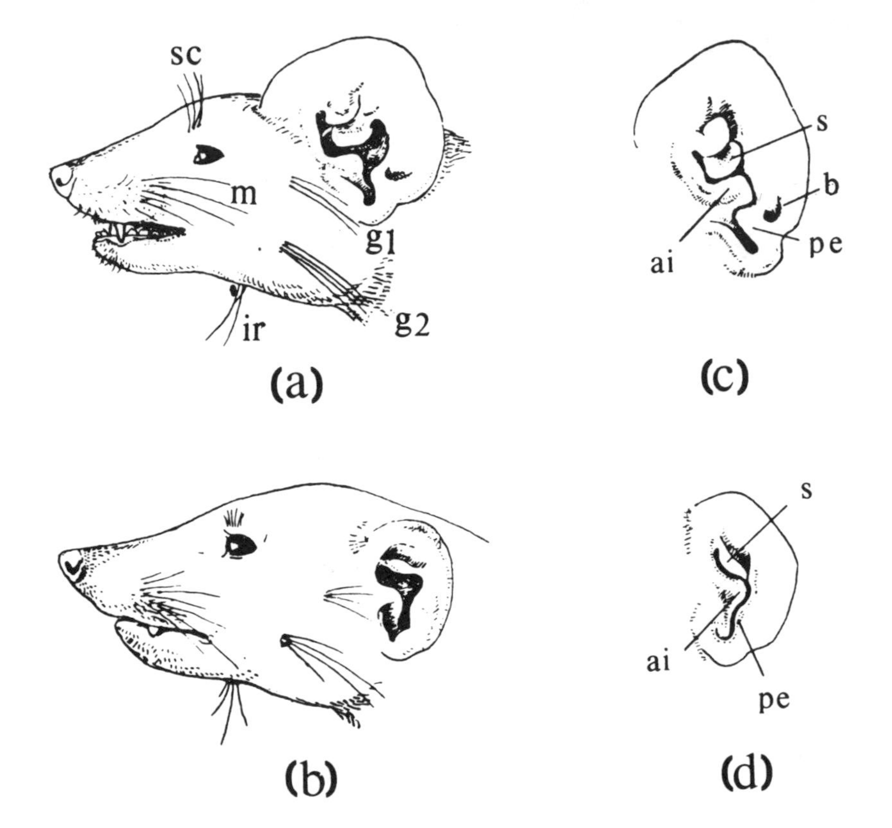

Figure 3.14. Heads of (a) *Cynictis*, (b) *Suricata* with ears open. (c) and (d), with ears closed for burrowing by apposition of antero-internal (ai), postero-external (pe) and superior (s) ridges. b, vestigial bursa. Both show the normal carnivore arrangement of vibrissae, somewhat reduced in the more fossorial *Suricata*. g1, g2, genal tufts; ir, inter-ramal tuft; m, mystacials; sc, superciliary tuft. [From Pocock, 1916g]

ridge is placed almost vertically and although in my tame female I can see the posterior ridges move forwards as her nose sinks deeper into the hole she is excavating and back again as she raises her head, I have never been able to detect any movement of the superior ridge. In burrowing mustelids, too, closure is the result of forward movement of the posterior ridges (Pocock, 1926, 1927a) and in the otter civet as well as the otters the ear can be closed to keep out water by apposition of the posterior and anterior ridges (Pocock, 1915a, 1928c).

E. The paws

The skeletal structure of the paws has already been dealt with but the outer coverings, particularly the foot pads, are also of some interest. A pad, with its thick cushion of fibrous connective tissue, tends to hold together the overlying bones and limits their independent mobility. The pads must therefore be subdivided in such a way as to permit movement where necessary. In species lacking cursorial specialisation the digits are somewhat spreading and capable of considerable movement relative to each other; moreover, during locomotion there is movement not only at wrist or ankle but also at the more distal articulations. Even in the most plantigrade species the foot does not leave the ground all at once but progressively, so that the final thrust is given by the toes. The arrangement of the pads generally regarded as primitive (figure 3.15), allows for both these sorts of mobility. There is a pad at the end of each digit, forming a cushion under the main point of pressure, and a series of four more beneath the phalangeo-metapodial joints, often more or less united to form a plantar pad. The arrangement of the plantar pads reflects a peculiarity of the mammalian hand and foot; namely that the first digit has only two phalanges whereas all the others have three. This reflects a basic functional distinctness of the first digit which is much older than the mammals themselves, for the two-jointed first digit can be traced back in their reptilian ancestry as far as the early therapsids. Its original function would appear to be that of providing a buttress to give lateral stability to the limb, while the main propulsion was provided by the other, longer digits. This basic skeletal difference has been retained in the mammalian lineage; there is a tendency for the first digit to be slightly offset from the others and it often has some degree of independent mobility. Correlated with this, the corresponding foot pad is often separate from those at the bases of the other digits and if there is reduction of the first digit the pollical or hallucal pad may be lost. Behind the plantar pads, towards the proximal end of the metapodials, lies a pair of metacarpal or metatarsal pads. In the fore paws the shortness of the carpus means that these pads can act effectively as a 'heel pad', cushioning the carpus as a whole. In the more elongated foot, however, they are not beneath the heel; their position seems slightly peculiar and possibly reflects a greater mobility in primitive forms at the tarso-metatarsal joint. In extant carnivores there is a tendency for the metatarsal pads to be

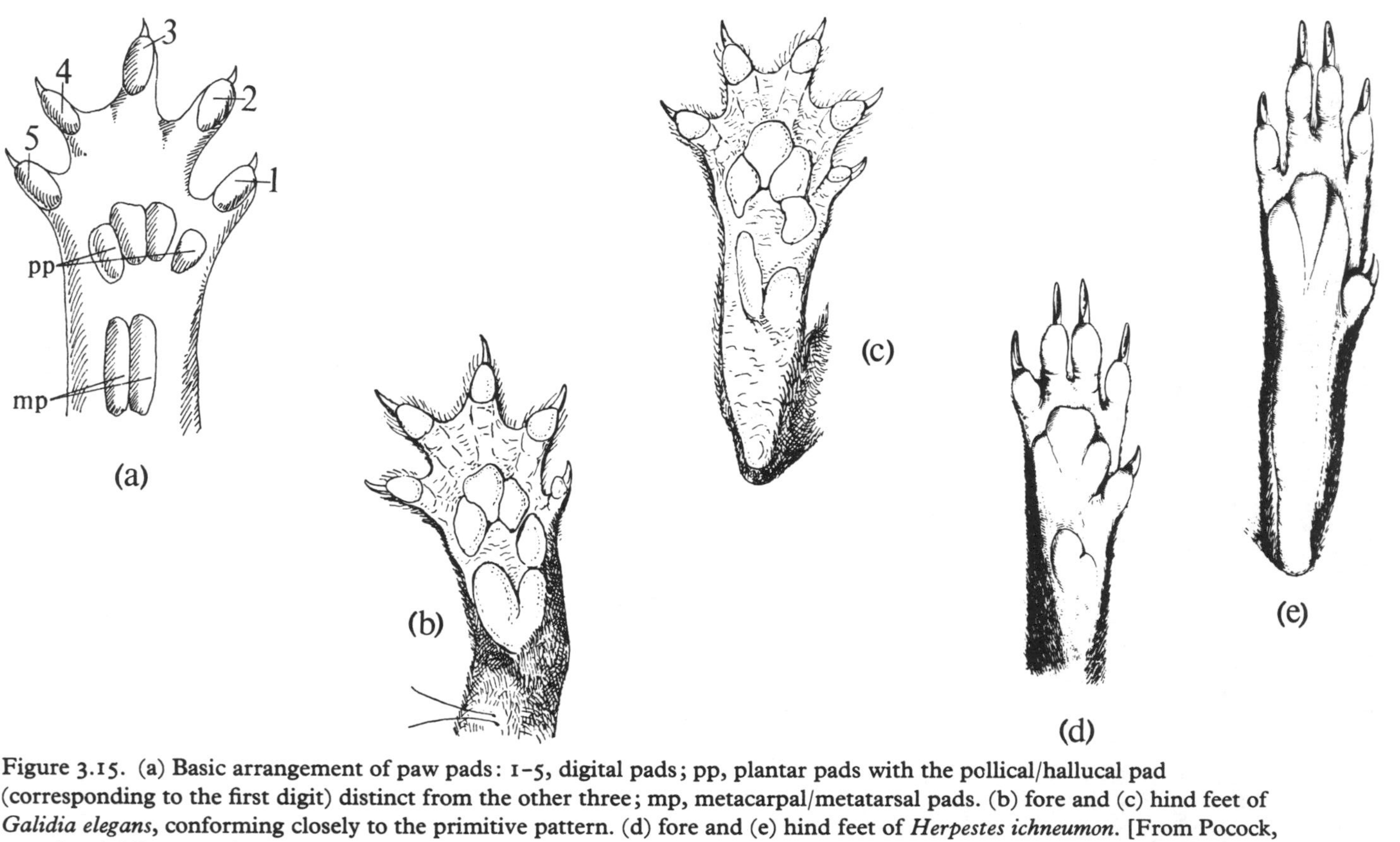

Figure 3.15. (a) Basic arrangement of paw pads: 1–5, digital pads; pp, plantar pads with the pollical/hallucal pad (corresponding to the first digit) distinct from the other three; mp, metacarpal/metatarsal pads. (b) fore and (c) hind feet of *Galidia elegans*, conforming closely to the primitive pattern. (d) fore and (e) hind feet of *Herpestes ichneumon*. [From Pocock, 1915d and Allen, 1924]

elongated backwards towards the heel in the plantigrade forms or to be lost entirely with the development of even partial digitigrady.

Pocock's papers give details of paw structure for representatives of all the families of carnivores. He regarded the presence of hair on the surfaces between the pads as advanced and naked soles and palms as primitive. This may be true in general but must be interpreted with some caution. Certainly the plantigrade foot is commonly naked, whereas with the development of digitigrady, where the thick, closely apposed pads are the only part that contacts the ground, all the other surfaces are hairy. Hair, however, may be specially developed to protect the feet against extremes of temperature. The polar bear and the Tibetan sand fox, *Vulpes ferrilata*, both have extremely hairy feet as an adaptation to walking in snow, although in other respects the bear's foot is primitive and the fox's advanced. Both the sand cat, *Felis margarita*, and the fennec have similarly well clad feet in adaptation to walking on hot desert sand. A naked sole, on the other hand, may be an adaptation to digging or to swimming and is not necessarily primitive. Readers of the Jungle Books will, no doubt, be saddened to know that 'red dog' (*Cuon alpinus*) does not differ from the wolf in having hair between the toes. This piece of misinformation Kipling gathered from Blandford's *Natural History of India* but in fact the feet of the two species are equally hairy.

The primitive condition – a broad foot with the palmar pads not fused, the lobes of the pollex and hallux large, the metapodial lobes well developed and with little hair – is met with in several families. Amongst the Canoidea it is typical of more generalised members of both Procyonidae and Mustelidae. In the Feloidea the same basic structure characterises a number of the viverrid subfamilies – the Paradoxurinae, Galidiinae (figure 3.15) and Hemigalinae – and it is only slightly departed from in *Cryptoprocta*. The Viverrinae and Herpestinae show some slightly more advanced features. In many of the former the foot is longer and narrower, the pollical and hallucal lobes are reduced and the surface is extensively haired. In the fully terrestrial Herpestinae (figure 3.15) the foot is very narrow, the digits closely bound together and the first digit reduced or even lost, as in the pes of *Cynictis* and in both manus and pes in *Paracynictis*, *Bdeogale* and *Suricata*. The arrangement of the pads corresponds: the hallucal and pollical lobes are missing, the metapodial lobes small and indistinct and there is little development of hair; at most a covering of the posterior half of the foot in the more digitigrade forms – *Ichneumia*, *Cynictis*, *Helogale*, *Bdeogale* and *Xenogale*.

In bears there is a tendency for the pads to fuse into large cushions extending right across the foot; one formed from the interdigitals and one from the metapodial pads. The degree of hairiness varies

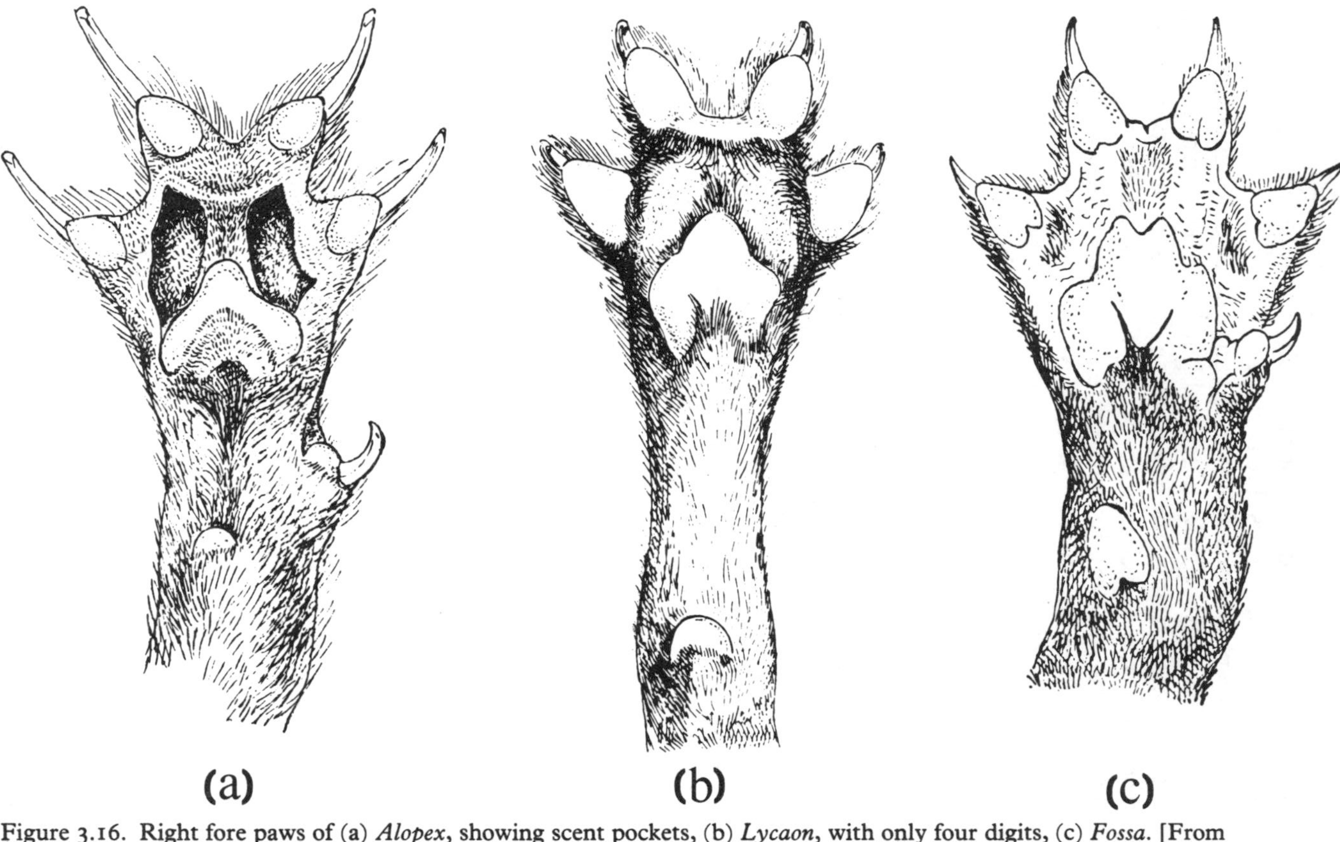

Figure 3.16. Right fore paws of (a) *Alopex*, showing scent pockets, (b) *Lycaon*, with only four digits, (c) *Fossa*. [From Pocock, 1914c, 1915c]

with habitat from the almost complete covering of the polar bear to the entirely naked sole and palm of the tropical *Melursus*.

The Canidae, Felidae and Hyaenidae show the typically digitigrade pad arrangement with which we are familiar in our cats and dogs. The hallucal and pollical pads are absent and the three remaining interdigitals are fused to form a single shamrock-shaped plantar pad. The digits are carried somewhat flexed, so that the digital pads lie close beside the plantar pad and the whole virtually forms a single supporting structure. In the manus, the pollex with a small digital pad lies well back from the other digits and does not contact the ground, while the single carpal pad lies still further proximally. The latter cannot play any part in normal locomotion and its function is by no means apparent: possibly it plays some role in braking when a sudden stop is made or prevents skidding when landing from a jump. In the hind foot the hallux is absent and there is no metatarsal pad. *Lycaon* (figure 3.16) is exceptional amongst the Canidae in having only four digits in the fore foot. In the Hyaenidae *Proteles* is unique in possessing the pollex: both *Crocuta* and *Hyaena* have only four digits. I

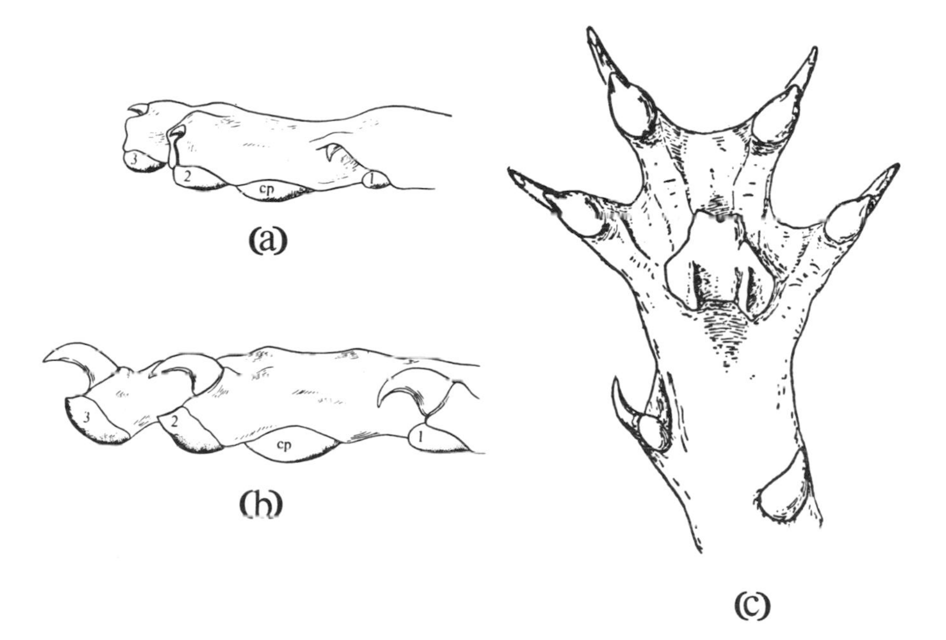

Figure 3.17. Side view of right fore paw of (a) ocelot, (b) cheetah (a captive animal with unnaturally sharp claws). 1, pollical pad of dew claw; 2, 3, digital pads of first and second digits; cp fused carpal pads. (The hair has been clipped short to make the structure clearer.) (c) Plantar view of paw of cheetah, showing ridges on carpal pad. [From Pocock, 1916e]

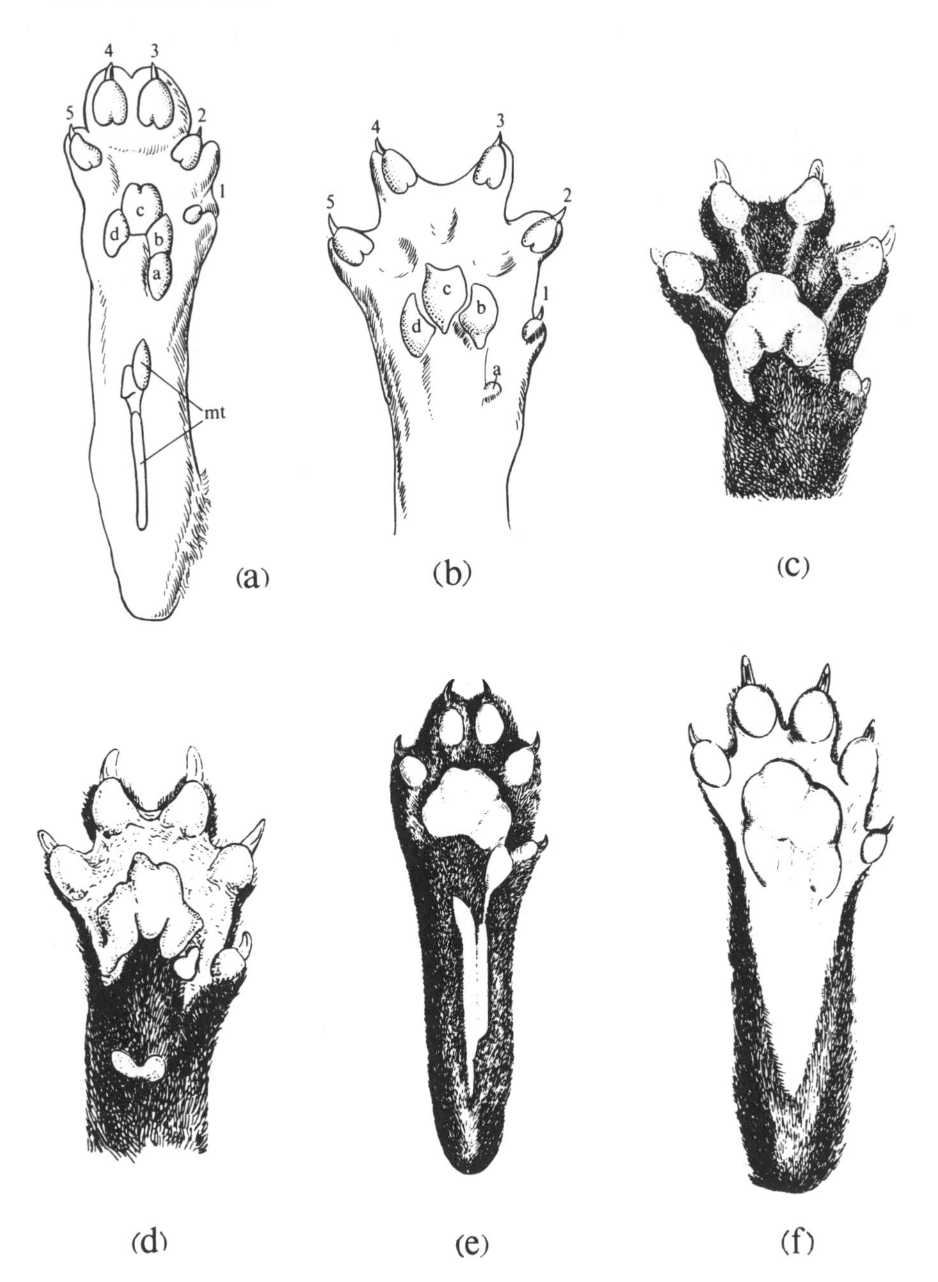

Figure 3.18. Right hind paws of (a) *Poiana richardsoni*, (b) *Linsang linsang*: 1–5, digits; a, hallucal pad; b–d, tarsal pads; mt, metatarsal pad (absent in *Linsang*). Right hind paws of (c) *Viverra*, (d) *Civettictis*, (e) *Genetta*, (f) *Osbornictis*. [From Pocock, 1915e and Allen, 1924]

know of no detailed description of how *Proteles* digs, so it is not possible to judge whether the presence of the pollex is a positive adaptation for digging or simply the retention of a primitive character which has been lost in the more cursorially specialised Hyaeninae.

In the Felidae the retracted claws are protected to a greater or lesser degree in different species by loose skin folds. These folds are absent in the cheetah, where the claws are but little retracted. The cheetah's foot shows a number of other modifications (figure 3.17). The digital pads and also the metacarpal pad are extremely hard and pointed at the front, possibly an adaptation to sudden braking. The palmar pads bear a pair of longitudinal ridges instead of the more usual slight depressions – the functional equivalent of tyre treads, serving in the same way as anti-skid devices.

The typical digitigrade pad pattern has obviously been acquired independently in Felidae, Hyaenidae and Canidae. It is therefore no surprise to find an almost identical arrangement in yet another case, to wit *Fossa fossa* (plate 10 and figure 3.16), the most digitigrade of the viverrids. Here, however, the condition is a little less advanced: the palmar pads are less fully united; small hallucal and pollical lobes are present and the hair on the interdigital membrane is restricted to little patches between the digits (Pocock, 1915c).

From the descriptions given above it is clear that the various families or subfamilies have characteristic types of pad arrangement. This means that the taxonomic value of such characters is not restricted to distinguishing between closely related species but may have slightly wider applications. Thus the foot pad arrangement is one of the features which show that *Poiana* is an aberrant genet and not, as it was originally called, an 'African linsang', and the characters of the footpads were amongst those used by Allen (1919) to justify placing *Osbornictis* in a separate genus from *Genetta* and by Pocock (1915e) for separating *Civettictis* from *Viverra* (figure 3.18). Unfortunately the pads are not very helpful in the problem of the relationships of the giant panda. The pads are somewhat bear-like but no more than one might expect as the result of convergence (figure 3.19). The main difference is that the pes of the panda is not fully plantigrade, the heel does not contact the ground fully and there is no heel pad. This could, however, be interpreted as a secondary adaptation to more arboreal and less terrestrial habits.

Family characteristics may, of course, be modified in relation to special adaptation. The mustelids show the most extreme examples both of fossorial adaptation – in the feet of the badgers with their strong claws and short stout digits – and also of aquatic adaptation, in the otters. Here the pads are nearly obliterated and the interdigital webs enlarged to produce a paddle-like structure. The most extreme

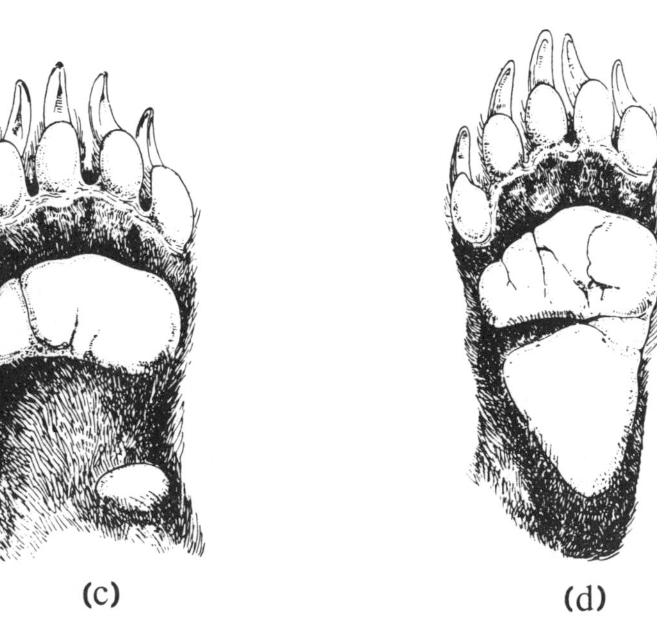

Figure 3.19. Left fore (a) and hind (b) paws of giant panda and of bear, *Ursus americanus*, (c) and (d). [From Davis, 1949]

case is the sea otter, where the hind feet resemble the flippers of a seal. For one type of aquatic predation, however, paddle-like paws are not suited. An animal that does not depend on fish but feeds largely on Crustacea caught in shallow water need not be a highly skilled swimmer, but must be able to feel for its prey in crevices and under stones. For this, thin mobile fingers not united by a web and with a fine sense of touch are required. It is striking that members of three different families adopting this technique of food capture have paws very similarly modified - the raccoon, the clawless otter and the crab-eating and marsh mongooses (figure 3.20).

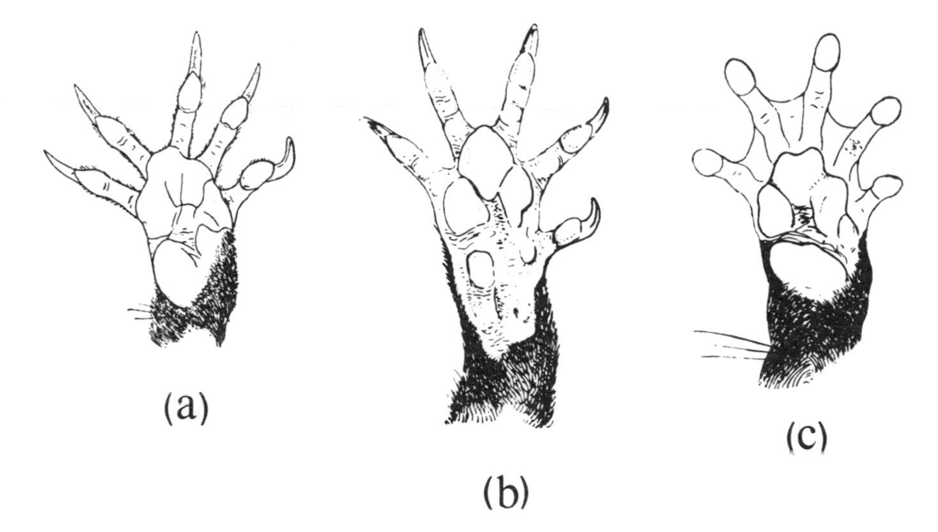

Figure 3.20. Right fore paws of (a) *Procyon lotor*, (b) *Atilax paludinosus*, (c) *Aonyx capensis*. [From Pocock, 1916g, 1921a, b]

Verheyen (1962) notes that in an area in the Congo near Katshungu *Osbornictis* is reputed to live largely on Crustacea. This is very surprising since neither its broad feet nor its rather delicate sectorial teeth seem in any way adapted for securing or devouring such prey but look more suited to a diet of frogs and fish. Mammals, however, are exceedingly adaptable and, if the report is true, this may be a case of a species turning to an unorthodox food because the latter is abundant and its usual prey is scarce. It is this same adaptability in the past that has produced the wealth of diverse mammalian species that we see around us today. If the raccoon's ancestors had never caught crayfish, however inexpertly to start with, there could have been no selective advantage in modifications increasing its efficiency. Its paws would have remained unmodified and the fishing pattern would never have been built into its brain as an essential part of its prey-catching behaviour.

F. Viscera

The visceral anatomy of carnivores does not present any particularly striking features and it is not necessary to do more than mention a few points in relation to the alimentary and reproductive systems.

The tongue in mammals is characterised by the presence on the dorsal surface of somewhat cornified non-gustatory papillae. In the Feloidea these are particularly well developed, with sharp, hard, backwardly directed points, and they are often arranged to form a distinct spinous patch near the front of the tongue: they are clearly visible in plate 11. In the larger predators these papillae spinosae are of use in rasping flesh from bones and must also be of some importance in grooming the fur. The size and shape of the spinous patch and also the shape of the individual spines differ from group to group. In the Canoidea the papillae are generally smaller, a special spinous patch is not developed and there is much less variation than in the Feloidea. Details for a number of species are given by Sonntag (1923, 1925).

The ability of carnivores to regurgitate, either to provide food for the young or simply to get rid of excessive or unsuitable food, is well known. This ability may be related to the fact that the striated muscle coat of the oesophagus is not, as in ourselves, limited to the proximal end but extends all the way to the stomach, as it does in the cud-chewing ruminants (Weber, 1927).

As is to be expected in a primarily flesh-eating group, there is a general tendency for preponderance of the fore gut with reduction in the caecum and hind gut and also a tendency for the distinction between colon and rectum to be lost. The total length of the gut is approximately five times the body length in dogs and shorter – rather less than four times the body length – in the more strictly carnivorous Felidae. The sea otter, although strictly carnivorous, has a much longer gut, about ten times the body length (Jacobi, 1938), but this is hardly surprising in view of the high proportion of indigestible material included in the food. In the dog it has been shown that, as in herbivores, bacterial fermentation in the large intestine produces fatty acids, which are absorbed into the blood stream. Although the concentration of fatty acids per unit dry weight of gut contents is much the same as in herbivores, the large intestine is so short that the total amount of fatty acid absorbed makes only a very small contribution to the animal's energy metabolism (Phillipson, 1947). Since bacterial fermentation thus appears to be of little importance in the digestive physiology even of the somewhat omnivorous dog, it is not surprising that in carnivores as a whole the size of the caecum bears no simple relationship to the relative importance of vegetable food in the diet; a fact which Mitchell (1905) interpreted as reflecting a general tendency for the reduction of a functionally unimportant organ.

The Felidae, the most advanced fully carnivorous family, show a correspondingly advanced type of gut structure: the caecum, although present, is small and the hind gut is short with virtually no distinction between colon and rectum. The fact that the gut of the domestic cat is significantly longer than that of the European wildcat may represent a secondary adaptation of the former to the more varied diet it receives in its association with man. It would, however, be desirable also to have comparative figures for the African wildcat, usually regarded as ancestral to the domestic cat, before accepting this conclusion.

In the Hyaenidae the gut is much more primitive: the caecum is relatively large and the colon distinctly marked off from the rectum. In the Viverridae the caecum is highly variable in size (Mivart, 1882b). It is relatively large in a number of Herpestinae but very small or absent in *Arctictis* (Garrod, 1873, 1878) and absent in *Nandinia* (Flower, 1872). The situation in the Canoidea is less simple. The caecum is present in the Canidae and is larger than in any other carnivore family, which, in view of their general tendency towards a mixed diet, seems quite simply adaptive (Flower, 1879, 1880). In the Mustelidae, Procyonidae, and Ursidae, however, the caecum is absent (Mitchell, 1905). This suggests that the very considerable reliance on vegetable foods shown by at least some members of these three families is a secondary habit, adopted only after the caecum had already been lost in the ancestral forms.

The gut of the giant panda, which is of particular interest in relation to its highly specialised vegetarian diet, has been described by Raven (1936) and by Davis (1964). The stomach has an elongated, extremely muscular pyloric end which is almost gizzard like. The caecum is absent and the gut as a whole is short, only about five times the body length; roughly the same as in dogs. The proportions, however, are very different, the small intestine being relatively short and the large intestine long. Davis interprets this as meaning that the food is still largely in an unassimilable form when it reaches the small intestine. A long passage through this part of the gut would therefore be of no value and it is more advantageous to pass it on as soon as possible to bacterial processing in the large intestine. This, however, does not seem to be particularly efficient. Morris and Morris (1966) say that enormous quantities of droppings are produced, containing much undigested material and Sneldon (1937) estimated that the animal must spend between ten and twelve hours a day in feeding in order to obtain sufficient nourishment. With this one may contrast Schaller's (1969c) finding that in favourable circumstances a lion need be active no more than four hours out of the twenty-four.

As in other orders, the kidney is lobulated in species that are marine, like the sea otter (Barabash-Nikiforov, 1962), or that attain large size, like bears and the giant panda. The polar bear, being both large and marine, has the highest number of lobules recorded for any carnivore, namely sixty-five (Raven, 1936).

The main features of interest in the male reproductive organs are the reduction of the accessory glands and the peculiar structure of the penis in the Canoidea. There is never a vesicula seminalis and in the Canoidea the bulbo-urethral (Cowper's) glands are also absent. Small glandular ampullae are present at the distal ends of the vasa deferentia in the Ursidae and in the giant panda but absent in Procyonidae and Canidae, according to Davis (1964): Grassé (1955), however, states that they are present in the Ursidae, Mustelidae and Procyonidae; Raynaud (1969) lists them as present in Canidae, Mustelidae and Procyonidae, while Sisson and Grossmann (1953) say that in the dog small ampullae are present. The truth seems most likely to be that the glands are present in all families of the Canoidea but may be very small in some species. In the Feloidea these glands are absent but there are bulbo-urethral glands. The prostate is almost universally present but is small in the Ursidae and *Ailurus* and, according to Davis, is absent in *Ailuropoda*.

The penis in the Canoidea is highly specialised. Erectile tissue is reduced and its functional stiffening role has been largely taken over by the baculum. The penis is long and the skin covering it is attached along its length to the abdominal wall so that the tip is forwardly directed. Long (1969) gives some details of the structure for a number of American Mustelidae. In the Canidae the reduction of erectile tissue is less extensive than in other families and proximally it forms a ring, encircling the base of the glans. This swells during copulation and prevents withdrawal of the penis until detumescence has occurred, thus producing the 'locking' characteristic of canid mating.

In the Feloidea the baculum is smaller and erectile tissue plays a correspondingly more important role. The penis is shorter and the attachment of the prepuce to the body wall is less extensive than in the Canoidea but it is not correct to state that at rest the tip of the penis is directed backwards: in fact it points downwards and slightly forwards in the usual manner. The ability of the Felidae to project a stream of urine directly backwards is not the result of an unusual orientation of the penis but is achieved by curving the tip of the glans backwards as the urine is voided.

In the domestic cat the basal part of the glans is covered with backwardly-directed horny spines (figure 3.21). These do not reach full size until the age of six or seven months and their growth is dependent on androgen production by the testes. They regress in castrates and grow again within two months if androgen is administered (Aronson and Cooper, 1967). The various functions that have been suggested for the spines are not mutually exclusive: they may act as a retention device during the brief period of intromission, they may provide added sexual stimulation for the male or they may be of importance in stimulating the female sufficiently to cause ovulation. The latter is probably the most important since, although the available data are scanty, there is a suggestion of a correlation between induced ovulation and the presence of spines. Spines have been reported in a number of species besides the cat, namely *Cryptoprocta* (Lönnberg, 1902), the ferret, mink, marten and raccoon (Zarrow and Clark, 1968), and the spotted and striped hyaenas (Wells, 1968): small spines are also present in *Proteles* (Flower, 1869b; Wells, 1968). With the exception of the hyaenids, for which there is no information, all the others mentioned are known to be induced ovulators. In the dog, on the other hand, the glans is smooth and ovulation is spontaneous.

The clitoris is moderately well developed in the Carnivora and the presence of an os clitoridis in a few species has already been mentioned. In the Hyaenidae the remarkable development of the clitoris of *Crocuta* (Matthews, 1939; Davis, 1949a; Deane, 1962) to form an organ resembling a penis is well known and is responsible for the common beliefs that spotted hyaenas are hermaphrodite or that they change their sex. The male appearance of the female external genitalia is further enhanced by the presence of a pair of swellings composed of fibrous tissue which resemble a scrotum (figure 3.21). The external genitalia of *Hyaena* and *Proteles* are quite normal and these peculiarities are found only in *Crocuta* (Flower, 1869b; Wells, 1968). In a juvenile female the urinogenital canal traverses the clitoris and is no wider than that of the male. As sexual maturity is reached, however, the opening elongates and finally forms a slit lying beneath the clitoris, which is sufficiently large to permit both copulation and parturition to occur (Matthews, 1939). Davis (1949a) states that the urinogenital canal does not traverse the clitoris but merely lies close beneath it: however, since he did not have access to juvenile material he did not realise that this was a secondary condition.

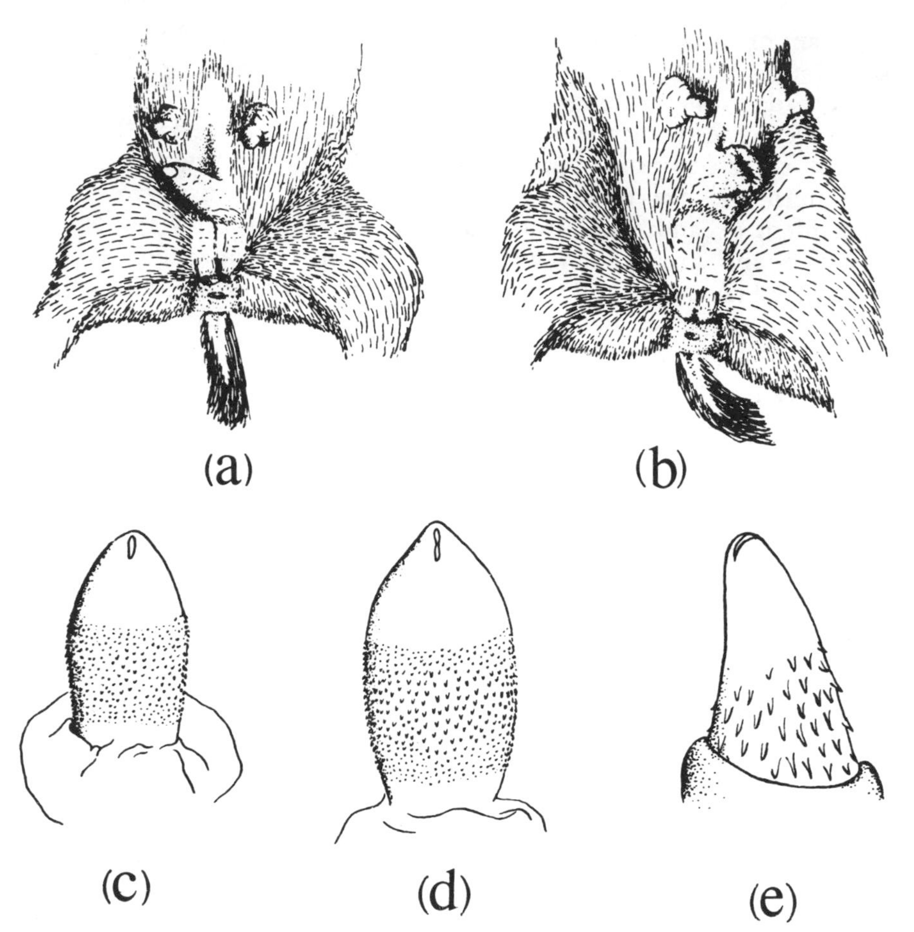

Figure 3.21. (Above) External genitalia of adult female spotted hyaena (a) before coming on heat, (b) during oestrus. (Below) glans penis of (c) adult female, (d) adult male spotted hyaena and (e) of adult male domestic cat (not to same scale). [(a) to (d) from Matthews, 1939]

It is known that in a rat or guinea pig testicular grafts will cause a penis-like growth of the clitoris. Moreover, in *Crocuta* the glans is covered with spines in both sexes. In view of the androgen dependence of the cat's penial spines these facts all suggest that the ovary of the female *Crocuta* may be capable of an unusually high level of androgen production. This in turn may be related to the unusual social relations that characterise this species. The female is not only slightly larger than the male, she is also socially dominant over him. In most other carnivores, although a male may withdraw temporarily if a female repulses him during the early stages of courtship or when defending her young, he is at other times dominant. The significance of the unusual situation in *Crocuta* is not yet clear.

Amongst the viverrids also there are species in which the female genitalia have a somewhat male appearance. In *Arctictis* (Story, 1945) the clitoris is somewhat peniform but in *Cryptoprocta* the male features are more marked. The elongated peniform clitoris is supported by an unusually long os clitoridis and scrotum-like swellings are present (Lönnberg, 1902). Unfortunately nothing is known about the social relations between the sexes in these two species.

Chapter 4

The special senses

HIGHLY efficient sense organs are essential for a predator: prey must be located from a distance and the final attack often demands very accurate orientation. Hearing, sight and smell need not, however, be equally important. Some predators hunt largely by smell; others depend more on sight, while for those whose prey is small or who hunt in dense cover, hearing may play the dominant role. Almost everyone who writes about carnivores kept as pets gives some estimate of the relative importance of these three major senses and often a rough comparison with their own abilities. Analytical experimental studies, however, are rare: apart from the domestic cat and dog, detailed information is rather scanty and no one worker has investigated all the sensory abilities of a single species. It is therefore most convenient to summarise what is known of the various special senses in turn.

A. Vision

The requirements for adequate vision by night and by day are not identical. The sensitivity necessary for vision in dim light requires firstly a large pupillary aperture, so that as much light as possible enters the eye. A large pupil means a large lens and cornea but the rest of the eye must not be increased in the same proportion or the extra light will merely be spread over a larger retina and there will be no gain in sensitivity. If the lens and cornea alone are enlarged, their curvature will have to be increased so as to focus the light on the relatively nearer retina. The eyes of nocturnal species therefore usually have a relatively large anterior chamber with a large and much curved lens and a highly convex cornea (figure 4.1). The second desideratum is a highly sensitive retina, which implies one containing many rods. Sensitivity is further increased in nocturnal mammals, including carnivores, by the presence of a reflecting layer, the tapetum lucidum, outside the receptor layer of the retina. Light which has passed through the receptor layer without being absorbed is

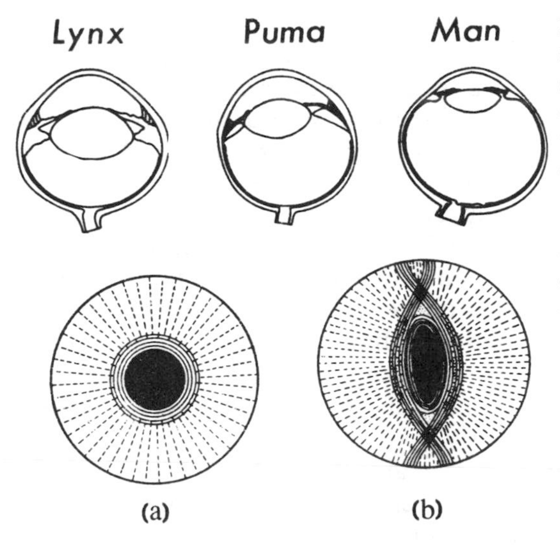

Figure 4.1. (Above) Vertical sections of eyeballs of felids adapted for nocturnal (lynx) and round-the-clock (puma) vision, with human eye for comparison. (Below) (a) circular pupil, closed by circular sphincter: to close the eye completely the muscle fibres would have to be able to shorten to zero length. (b) Slit pupil, closed by interlacing fibres, capable of closing the pupil completely. [From Walls, 1942]

reflected back again and so has a second chance of stimulating a receptor. When a bright light is shone on the eye the resulting 'eyeshine' is produced by reflected light which has missed being absorbed on the return journey and so is transmitted back to stimulate the eye of the observer. The cells forming the tapetum contain very little cytoplasm but are packed full of regularly arranged refractile rodlets. In species with very good night vision, like the cat, there may be as many as fifteen cell layers in the tapetum. A tapetum is almost universally present in carnivores, the only species said to lack it being the mongooses *Cynictis* and *Suricata* (Walls, 1942). Walls does not say how many species of mongoose have actually been examined and it may well be that the same applies to some of the other diurnal species. *Atilax paludinosus* and *Ichneumia albicauda*, both nocturnal, have very obvious eyeshine, whereas in *Crossarchus obscurus*, which is largely diurnal but will also forage after dark, eyeshine is present but is less striking.

Vision in bright light has the opposite requirements. The receptor cells must be cones, adapted to function at high intensities and it must be possible to close the pupil down to a narrow aperture. Many carnivores, probably the majority, have eyes adapted to function both by night and by day. For such round-the-clock vision the retina must contain both rods and cones, in proportions depending on the relative importance of night and day vision. Usually there is a central area of greatest acuity, where the proportion of cones is higher than elsewhere. If the sensitivity is high (i.e. the proportion of rods is high) then the pupil must be capable of closing virtually to zero in bright sunlight. The domestic cat is one of the few species for which measurements of the lower threshold for vision have been made. Gunter (1951) trained six cats to select an illuminated panel, behind which they received their food, in preference to an unlit one. The cats could still make the discrimination at a level of illumination only one sixth of that required by man. No animal can 'see in the dark' but that a cat can see in what is 'the dark' to a human observer was very vividly illustrated in Gunter's experiments when his cats consistently selected the illuminated panel which he himself could not see.

No such tests have been carried out with *Suricata* but my tame ones certainly had much poorer night vision than I have – on the other hand they would stare up at a hawk against a sunlit sky so intensely bright that I could not maintain my gaze. Presumably, if any rods are present in the retina of this species they must be a very small minority in a predominantly cone retina.

High sensitivity depends on the ability of the eye to add the responses from a large number of receptors, no one of which is sufficiently activated to trigger off an impulse in an optic nerve fibre. The eye that gives high sensitivity cannot therefore be expected to have high acuity, since acuity (or resolving power) depends on being able to distinguish the activation of one receptor from that of its neighbour. It is therefore of some interest to know how the acuity of the cat's eye compares with that of the much less sensitive human eye. Acuity can be measured by finding out how narrow a series of black and white bars can be made before they cease to be distinguishable. This would obviously differ with distance from the eye but this can be allowed for by imagining a circle with its centre at the optical centre of the eye and expressing our result as how many cycles of black bar–white bar it takes to fill one degree of arc. Since the length of the arc increases with distance from the eye, the effect at the retina remains constant. For the cat's eye, values of five to six cycles per degree have recently been obtained by Berkley and Watkins (1971), using an electrophysiological method based on recording the potentials evoked in the cerebral cortex. For man, the value is 50 to 55 cycles; so, although the cat's eye is vastly more sensitive than our own, its resolving power is correspondingly less.

Other species that have been investigated, using training methods, are the polecat (Neumann and Schmidt, 1959) and the small-clawed otter, *Amblonyx cinerea* (Balliet and Schusterman, 1971). Both gave values of the same order of magnitude as the cat, approximately four cycles per degree and the otter's acuity was much the same in air and under water. No investigations, however, have been made on species such as the diurnal mongooses, in which one might expect to find a higher acuity and lower sensitivity than in the cat.

Since the cat's retina is so sensitive, it is clear that a highly efficient pupillary closure is required. This, of course, is the function of the slit pupil. A circular sphincter can never give complete closure but the arrangement of the muscle fibres in the slit pupil permits this (figure 4.1). The larger Felidae have broadly oval pupils, only slightly elongated vertically, which appear circular when they are widely open and amongst the Felidae a true vertical slit pupil is restricted to the genus *Felis*, as defined by Pocock. A vertical slit pupil is also found in some viverrids: it occurs, to my knowledge, in *Nandinia*, *Paguma* and *Arctogalidia* and there may well be other species showing this characteristic. According to Walls, the palm civet, *Paradoxurus*, is exceptional amongst carnivores in having a horizontal slit pupil. Complete closure, excluding all light, would, of course, be pointless: one might as well just shut the eyes. The cat and viverrid vertical pupils in fact close completely in the middle but leave a minute pinhole aperture at either end and in *Paradoxurus* the middle of the iris is notched, so that one central pinhole remains when the rest is closed.

A horizontally elongated oval pupil is found in *Cynictis* (Walls, 1942), *Helogale* (Zannier, 1965) and also in *Suricata* and in *Crossarchus*. This has the advantage of extending the visual field in the horizontal plane, which may be of value to short-legged terrestrial creatures like mongooses and the horizontal pupil probably occurs in other species besides the ones mentioned: it may well be characteristic of the Herpestinae as a whole.

In man and the higher primates the ability to change the shape of the lens is highly developed and this makes it possible to alter the focal length to suit the distance of the object which is being watched. In many other orders this power of accommodation is very poorly developed but the Carnivora are unusually well endowed in this respect. In the otter, whose eye must function both in air and in water, the iris sphincter is unusually powerful and acts as an auxiliary device, changing the shape of the lens by constricting the anterior portion (Walls, 1942). Another peculiarity of the eye of *Lutra* is that the retina is folded up into a series of ridges. Pilleri (1967) found a similar arrangement in the retinas of two other aquatic mammals, the white whale and the hippopotamus but not in those of any of the terrestrial species he examined. He therefore regards the folds as an adaptation to underwater vision but how they function remains obscure.

One of the most striking features distinguishing human eyes from those of other mammals, including carnivores, is that only in man is a large area of white sclerotic surrounding the iris visible. This is related to eye movements. If the eye moves extensively, then the lids must be arranged so as not to cover the pupil when the eyeball is moved to left or to right; this can be achieved only if the area on either side of the iris is exposed when the eye is looking directly forwards in the rest position. The extent of eye movement is, in turn, related to the position of the eyes and the size of the visual field. In man, the forwardly directed eyes give a large field of binocular vision and hence accurate distance judgement, but the total field is correspondingly reduced. This is compensated for by the extensive eye movements. In Carnivora, too, accurate distance judgement is a desideratum and binocular overlap, although less than in man, is considerable. Overlap is maximal in the Felidae, where the angle between the optic axis and body axis ranges from 4° to 9°. In the cat the total visual field is 287° with a binocular overlap of 130°. In dogs the divergence of the eyes from the body axis is somewhat greater: 15° to 25° and the maximum for carnivores is approximately 50°. The descussation of the optic nerve fibres reflects the degree of binocular overlap. In man, with frontally directed eyes, the proportion of uncrossed optic nerve fibres is half. For the cat the proportion is a third and for the dog a quarter.

Another feature of the carnivore eye is the well developed nictitating membrane, which in many species can be pulled right across the eye and acts as a device for protecting or cleaning the surface of the conjunctiva. Burrowing or digging species, such as mongooses, have this ability and so have the polar bear and the giant panda. In the latter two species, Walls (1942) suggests that the membrane may be used to protect the eyes against glare, as we use dark glasses to prevent snow blindness.

Colour vision is mediated by cones but from what has already been said it should be apparent that the presence of cones in the retina does not necessarily imply colour vision: it may mean no more than that the eye is adapted to function at high levels of illumination. For many years it was believed that, apart from man and the higher primates, all mammals were colour blind. This, however, is not the case. Dücker (1957) and Gewalt (1959), working with viverrids and mustelids respectively, were able to show that certain species in each family had the ability to distinguish colours from any shade of grey. Dücker's tests were more extensive: she used sixty stages from black through grey to white, while Gewalt used twenty steps. Of the species studied, the Indian mongoose, *Herpestes edwardsi*, had the most highly developed colour vision and was able to distinguish from any shade of grey not only red, yellow, green and blue but also orange and violet. Table 4.1 summarises the findings of the two investigators. Gewalt's results are quoted from his own paper: in Dücker's summary (1965b) they have been slightly simplified.

Dücker's remarks about *Viverricula* are of interest. Although this

Table 4.1
Responses to colours in Viverridae and Mustelidae
Data from Dücker, 1957 and Gewalt, 1959

	Red	Yellow	Green	Blue
Viverridae				
Herpestes edwardsi	+	+	+	+
Viverricula malaccensis	+	–	+	–
Genetta tigrina	–	–	–	–
Mustelidae				
Mustela lutreola	+	(+)	+	(+)
M. putorius	+	(+)	(+)	+
M. erminea	+	(+)	(+)	(+)
M. furo	+	–	–	–
Martes martes	–	–	–	(+)

+ clear and definite discrimination
(+) slight sensitivity, or some results contradictory
– not discriminated from grey

species ultimately proved to be able to distinguish red and green as colours, it normally made little use of this faculty and responded preferentially to brightness cues. Special training was required to induce the animal to make a discrimination based on colour. A cat which Dücker tested for comparison with her viverrids always responded to brightness alone and gave no evidence of colour vision. A number of investigators have in the past come to the conclusion that the cat is totally colour blind and none of the early investigations reaching the opposite conclusion were adequately controlled to eliminate brightness cues. More recently, a number of workers (Mello and Peterson, 1964; Sechzer and Brown, 1964; Meyer and Anderson, 1965), using careful training techniques with satisfactory controls, have found that it is possible, with very prolonged training, to get a cat to make a discrimination truly based on colour. A good review of the relevant literature is given in Meyer and Anderson (1965).

Although the cat's retina possesses a cone-rich central area, the results of most modern investigations have indicated that there is only one type of cone. Granit's earlier work (1947), suggesting three cone processes, is susceptible of another interpretation. On the basis of a single type of cone, with its maximal sensitivity at a different wave length from that of the rods, some degree of colour discrimination should be possible, provided the intensity of illumination chosen is suitable – high enough to stimulate the cones but not so high as to put the rods out of action. Daw and Pearlman (1969), recording from single units in the optic tract and lateral geniculate nucleus, at different wave lengths and different intensities of illumination, have defined the ranges over which the two sorts of receptor are active.

This makes it possible to find out whether the cat's ability to discriminate colours is restricted to the mesopic range and vanishes at higher levels of illumination. Contrary to expectation, experiments designed along these lines showed that a cat trained to discriminate colours at an intensity of illumination allowing both the rods and the cones to function could still do so when the intensity was raised sufficiently to saturate the rods (Daw and Pearlman, 1970). Following this surprising result, Daw and Pearlman resumed the search for evidence of the presence of more than one type of cone by recording from the cells in the cat's lateral geniculate body, the first processing station on the way from the retina to the cerebral cortex. Out of 386 lateral geniculate cells whose responses were recorded, four gave evidence of differential responses according to the wave length of the light stimulus used.[1] This can only mean that they are linked with receptors of two types. One of these corresponds with the previously known cones; the other is sensitive to much shorter wave lengths. The four two-cone cells identified were all located in the B cell layer of the geniculate body. The total number of B layer cells tested was 87, so that even within this single cell layer the proportion of two-cone cells was rather low and the number of the corresponding short wave cones in the retina would therefore appear to be small.

Daw and Pearlman have thus provided clear evidence for the existence of a second, somewhat rare, type of cone in the cat's retina, with maximal sensitivity in the short wave length range of the spectrum but their experiments should not be taken as proving that no other type of cone exists: it remains possible that a third type of cone, with its maximal sensitivity towards the long wave end of the spectrum, may be present in even smaller numbers. Since the characteristics of the normal (green-absorbing) and the new (blue-absorbing) cones have been established, it should now be possible to carry out tests designed to find out whether any behavioural evidence can be produced for the existence of a third (red-absorbing) type of cone and hence whether the cat is a dichromat or a trichromat.

The prolonged training that is required to induce a cat to make a colour discrimination indicates that responses to colours are no part of a cat's normal behaviour. Meyer and Anderson (1965) are of the opinion that the cat simply does not pay attention to colours and, so to speak, cannot believe that anything so negligible as colour could be the real cue to where the food is to be found. I would suggest a slightly different interpretation: possibly the cat does not normally process the information coming from its retina in such a way as to make a hue discrimination. Most biologists are aware that one can 'teach' one's

[1] In the preliminary account of this work (Daw and Pearlman, 1970) the figures given are 3 out of the 118 geniculate cells which had been examined at that time.

brain to process the information coming from the eyes in new ways – the most familiar example being learning to use a monocular microscope with both eyes open and simply to suppress the information coming from the eye that 'sees only the works', as a student once expressed it to me. Maybe the cat also has to train its brain to use the information from its eyes in a new way and this is why such prolonged training is required. Dücker's *Viverricula*, on the other hand, although it took some time to learn to respond to colours, was in a different category from the cat and would seem to be a case where what the animal had to learn was indeed to attend to a cue which it normally regarded as unimportant.

All this, however, tells us nothing about what the cat is normally doing with the information from its cones. Why does it require two sorts, if it does not distinguish between the messages they send to the brain? The obvious suggestion is that the function of the short wave cones is simply to increase the range of wave lengths to which the light-adapted eye is sensitive. The cat may not care whether the grass in which a mouse is hiding is bluish green or yellowish green – but it may care very much whether it can or cannot see clearly towards the bluer, short wave length range.

Whatever may be the true explanation of why it is so difficult to train a cat to discriminate colours, it is clear that colour normally means nothing to a cat and the same is probably true of other species in which training is accomplished only with difficulty. Dogs have not been studied as carefully as cats but a number of early workers found evidence of some slight degree of colour discrimination. Their results, however, all suggest that to dogs, too, colour is of little relevance. Cole and Long's (1909) claim to have trained a raccoon to make a choice based on colour rests on inadequate brightness control. Michels *et al.* (1960) could find no evidence of colour vision in this species, thus confirming Davis's (1907) earlier results. Any species possessing a reasonable proportion of cones in addition to rods might well prove capable of some degree of colour discrimination in the mesopic range, granted sufficient persistence on the part both of experimenter and animal. Biologically significant colour vision, however, is to be expected only in species that search for food by day and amongst these is most likely to occur in those whose diet is varied and includes such things as fruits and insects, where colour assists in recognition. A number of mongooses fall into this category and it is highly probable that other species besides *Herpestes edwardsi* will prove to have good colour vision. Dücker (1959) notes that the retina of *H. ichneumon*, like that of *H. edwardsi*, contains a central area composed entirely or virtually entirely of cones and suggests that it too probably has colour vision.

B. Hearing

Next to man, the domestic cat is the animal whose auditory physiology has been most extensively studied. It has been recognised for many years that the sensitivity of both cats and dogs to sounds extends well above the human upper limit but no detailed investigations were made before 1955, when Neff and Hind found that cats could be trained to respond to tones up to 60 kHz – approximately two octaves higher than the normal human limit of 15 to 16 kHz. The cat's sensitivity was high up to 50 kHz but declined rapidly above this frequency. Wever *et al.* (1958) used the physiological method of recording the cochlear potentials evoked in the ear of an anaesthetised animal. Their results differ from those of Neff and Hind. The range of high sensitivity found was approximately the same but the upper limit was much higher: a response was still given at 100 kHz, the upper limit of their sound producing apparatus. The discrepancy is probably due to technical difficulties. The production of high frequency sounds at high intensity is not a simple matter and Wever *et al.* are of the opinion that the sound source used by Neff and Hind was simply not producing a sufficient intensity to cause stimulation when working at the highest frequencies. Since it is unlikely that in nature the cat ever encounters intense stimulation at these very high frequencies, one can take it that the point at which sensitivity falls off rapidly, namely about 50 kHz, approximates to the normally useful upper limit. One interesting point shown by Wever and his co-workers is that in cats, as in man, there is a decline with age in sensitivity to high notes: their four and a half year old cats were distinctly less sensitive than their yearlings.

Peterson *et al.* (1969) have made experiments similar to those of Wever on a few species from each of the families of carnivores. They give values for the upper frequency limit and also for the upper limit of the high-sensitivity range. The latter is taken as the point where a standard response of 1 microvolt potential is produced by a sound pressure amplitude of 20 db over that required at the frequency to which the ear is most sensitive. One may object to this on the grounds that if two species give identical responses to a certain frequency, this may be regarded as within the sensitive range of one but above it for the other merely because the maximum sensitivity at some lower frequency is greater for the second species than for the first. One could avoid this by regarding the 'useful upper limit' as the point at which a sound pressure of 1 dyne/cm^2 suffices to produce the standard 1 microvolt potential. This point can be read off the sensitivity curves provided and table 4.2 gives both types of useful upper limit, 'relative' being Peterson's value and 'absolute' the value as just defined.

Table 4.2
Upper limits of hearing in various carnivores

Species	Upper limit (kHz)	Upper useful limits Relative (kHz)	Absolute (kHz)
Paradoxurus hermaphroditus (palm civet)	100	70	70
Herpailurus yagouaroundi (jaguarondi)	100	55	70
Bassariscus astutus (ringtail)	100	45	70
Felis catus (domestic cat)	100	35	65
Nasua nasua (coati)	95	45	60
Procyon lotor (raccoon)	85	35	50
Tayra barbara (tayra)	85	40	40
Canis latrans (coyote)	80	30	30
Canis familiaris (dog-greyhound)	60	15	35
Selenarctos thibetanus (Asiatic black bear)	—	30	30
Crocuta crocuta (spotted hyaena)	—	15	30
Vulpes vulpes (red fox)	65	6	20

In all these species not only the upper limit but also the upper useful limit, as defined by the second criterion, falls well above the human upper limit. The civet, jaguarondi, ringtail, coati and cat all take small prey: high-pitched squeaks and rustles are therefore important to them and all are sensitive to very high frequencies. For the bear and hyaena such an ability is much less vital and their upper limits are lower, while raccoon, tayra, coyote and dog are intermediate and may be regarded as giving an idea of the range for terrestrial carnivores without particular auditory specialisation. Evans (1969) quotes preliminary results indicating that hearing in the river otter, *Lutra canadensis*, falls within the range of this latter group.

The results for the fox are puzzling. A much higher upper limit would have been expected in such an expert hunter of small rodents in long grass. Had the results depended on measurements made on a single animal one would have been tempted to suggest that it must have been abnormal but six foxes were studied and gave concordant results. The technique used involved destruction of the auditory bulla and it is therefore possible that the resonance characteristics of the latter may be such as to increase the sensitivity to high frequencies but for the meantime one can only record the results as surprising.

For some obscure reason Peterson and his co-workers appear to have expected the characteristics of the auditory system to be related to taxonomic position rather than to how the animal makes its living.

They therefore did not select the species they studied on any functional criteria. It would be extremely interesting to compare related species with widely differing modes of life – *Proteles*, for instance, might be expected to be very sensitive to high frequencies, in contrast to *Crocuta*, while *Otocyon* would surely give a very different result from the red fox.

A lower threshold is not very meaningful, since vibrations may be felt as a pulse below the level at which a definite tone is audible. There is, however, a point at which the sensitivity becomes distinctly less as the frequency is reduced and which presumably gives some idea of the lowest frequencies which are normally significant in the life of the animal. These are surprisingly high: about 100 Hz in fox, coyote and hyaena and from 200 to 300 Hz in other species. Middle C has a frequency of 264 Hz, so sounds significantly lower than this would seem to be relatively unimportant to most of the species studied. The range over which hearing is acute thus does not exceed our own by as much as the upper limits alone might suggest. Rose (1968) estimates that a cat hears well over a range of approximately ten octaves, as compared with eight and a half in man. With at least forty thousand fibres in its auditory nerve as against our own thirty thousand, this is not surprising.

Since sensitivity varies with frequency one cannot make a simple statement comparing the acuity of hearing in two species. The sensitivity of cats and dogs does not differ very much from that of man at frequencies up to approximately 500 Hz but at higher frequencies both have vastly more acute hearing than man (data from Spector, 1956).

Pitch discrimination has been investigated by training experiments in both cats and dogs. Pavlov's (1928) classical conditioning experiments showed that dogs are capable of distinguishing two notes differing by one-eighth of a tone. Since the tests were done using a frequency of 1 kHz, which is within the sensitive range for a dog, they should give a measure of the animal's maximal discriminatory powers. Neff and Diamond (1958) showed that a cat also can discriminate one-fifth or even one-tenth of a tone within its sensitive range but at lower frequencies this sinks to a semitone.

The ability to locate sounds is also obviously of importance to a carnivore. Neff and Diamond (1958) found that cats could discriminate with 75% accuracy between two sound sources separated by an angle of 5°, a performance of the same order of accuracy as in man. That the cat is not superior to man may at first seem surprising, since the mobile pinna is often popularly believed to aid in localisation. This is not so: the pinna serves to collect sound waves and, no doubt, adds considerably to auditory acuity while its movements allow the animal to scan the environment or to direct its attention to some particular sound source. The movements, however, do not aid in localisation of sound: they may even complicate the problem by adding another variable, since any asymmetry in the positions of the two pinnae must enter into the computation which the central nervous system must perform in order to locate the source of a sound. Nevertheless, the presence of a pinna, as distinct from its movements, is responsible for the ability to locate sounds at some point in space outside the body (Whitfield, 1971). Differences in the time of arrival and intensity of sound received by the two ears can give information about the direction of the source but cannot alone lead to localisation. With complex sounds, however - and most natural sounds are complex - the presence of a pinna introduces variations in the quality of the sound received in relation to the position of the source and it is these which permit of true localisation. The shape of the pinna, so much more elaborate than a simple collecting device like an ear trumpet, is presumably related to the production of these variations. The ability to use these subtle differences in the sound reaching the two ears may depend on learning, just as we learn to use secondary visual cues in our visual judgement of distance but it is the presence of the pinna which provides the basic information which makes this learning possible.

Species in whose lives vocal communication plays an important role must be able to distinguish between different sounds and it is therefore not surprising that many of them are capable of learning to respond to a variety of vocal signals from man. Most cats and dogs, without special training, learn to respond to their own name and to a call signifying that food is ready and most dogs acquire a considerably larger vocabulary. This ability is, of course, an essential basis for the training of a sheep dog. Burns (1969) lists eight vocal commands which a sheep dog must recognise as a basic minimum but notes that the usual vocabulary is considerably larger. Furthermore, many commands can be given either verbally, when the dog is close at hand, or in a code of whistles, when he is at a distance from the shepherd.

C. Smell

Both behaviourally and physiologically, smell is one of the most difficult senses to investigate; partly because of the difficulty of controlling and measuring the stimuli and partly because olfactory communication plays so small a part in our own lives. We can, nevertheless, distinguish a great many smells and a little training can greatly improve our powers, although, in general, smells are rarely of great moment to us.

In the carnivore world, things are very different. Odours are important not only in tracking prey but in a variety of other ways. As long ago as 1897, E.T. Seton had realised that dogs use their urine as a means of information exchange and referred to special urinating spots as 'scent telephones'. Scent plays a role in individual recognition, in assessment of sexual status and in the interactions between mother and young. The multiplicity of odour-producing glands found in carnivores is testimony enough to the fact that even in species that do not hunt by scent, odours are vastly important. Most investigations of olfaction, however, deal with dogs, largely because of the use that is made of their tracking abilities, both in sport and in police work.

Kalmus (1955) was mainly concerned with the ability of dogs to distinguish between individual human odours. He showed that the dog could discriminate between the odours of members of a single family and was even able to distinguish those of identical twins, provided the two odours were presented together. The odour of one twin, however, would be accepted in place of the other provided the latter was not present. The individual odour characteristics are not greatly influenced by the region of the body from which they emanate - palm of the hand, armpit, sole of the foot. The regional odour differences are quite perceptible to man but they do not confuse the dog in his identification of the individual.

King *et al.* (1964) found that dogs could still detect the smell trace left by human fingerprints on a glass slide up to six weeks after it had been handled if the slide was kept indoors, but when the slides were allowed to weather out of doors the smell ceased to be reliably detected after one or two weeks. The experiments of Moulton *et al.* (1960) on the olfactory thresholds of dogs to pure compounds (formic, acetic, propionic and other longer-chain aliphatic acids) are of less biological interest. The dog's nose was not found to be strikingly more sensitive to such substances than the human nose and there is no particular reason why it should be. Where the dog excels is not in the sensitivity of the individual olfactory receptors but in the variety of smells he can detect; in his ability to make discriminations which must be based very largely on quantitative variations in a number of simultaneously present constituents and to remember them as we remember a face or a voice.

Kalmus's investigations included some tracking tests but the most extensive investigations of the tracking abilities of trained dogs remains that of Budgett (1933). He not only showed that a dog can follow the trail of an individual person but that if his quarry dons rubber boots or even mounts a bicycle, the dog, after a check and some casting about, will still follow the trail. This Budgett attributes to the dog's ability to follow the scent of bruised grass: on losing the original scent, he follows the only detectable trail. Budgett also investigated the conditions in which scent does or does not lie well and concluded that conditions for tracking are optimal when the ground temperature is a little higher than the air temperature. This condition is usually fulfilled in the early evening, when air temperature falls faster than ground temperature and Budgett points out that this is the favourite hunting time of many of the more keen-scented carnivores.

One might expect the spectrum of smells to which the nose of any particular species is responsive to bear some relation to its mode of life, in particular to its feeding habits: a vegetarian species, for instance, might be more responsive to plant odours and flower scents than a flesh-eater. The sensitivity of cats to catnip is exceptional and rather a special case. Their responses to this particular plant odour exactly resemble the behaviour shown by the female in oestrus (Palen and Goddard, 1966). It therefore seems almost certain that the scent contains some component identical with, or closely resembling, an odour produced by the animals themselves. The major constituent of the plant's essential oil is cis, trans-nepetalactone (Waller *et al.*, 1969) but I know of no attempt to find out whether a related compound is produced by cats. Unfortunately comparative studies on odour sensitivities of different species have not been made but there is abundant behavioural evidence that some mammals secrete and respond to marking substances which have no detectable odour to the human nose.

The vomero-nasal, or Jacobson's organ, is a structure about whose function in mammals very little is known. It consists of a pouch lined with receptor cells very like those of the olfactory organ and is situated anteriorly in the roof of the mouth. It opens by a duct into the mouth and in many species there is also an opening into the nasal passage. In view of its location it seems very likely that Jacobson's organ may be capable of savouring the food in the mouth, as suggested by Negus (1956). Negus believes that it may play some part in reflex stimulation of gastric secretion, but in view of the speed with which many of them swallow their food this seems rather improbable, at least as far as carnivores are concerned. Unfortunately in many of the higher primates, including man, Jacobson's organ is not functionally developed and introspection can therefore give us no clues as to its action. There are, however, indications that it may be of importance in relation to activities other than those connected with feeding. Many mammals, including a number of carnivores, make the curious grimace known as flehmen in which the lips are pulled up, the nose is wrinkled and drawn back, the head raised and in some species breathing is stopped for a moment. Flehmen may be evoked by a number of strongly smelling substances but is normally seen in response to the smell of conspecific urine. Knappe (1964) has suggested that flehmen serves to bring odours into contact with Jacobson's organ. This view is supported by Verbene (1970), who has made an extensive study of flehmen in the Felidae and gives descriptions of the details of the process in a variety of species. Flehmen is widespread, possibly universal in the felids, which all possess a functional Jacobson's organ: it is, however, manifested only rather weakly in the domestic cat, where the organ is rather small. Amongst the mustelids, flehmen occurs in the tayra (Brosset, 1968) and I have seen it in the viverrids *Suricata* and *Civettictis* but have no information about the condition of Jacobson's organ in any of these species. It is generally believed that the Canidae do not flehm but Kleiman (personal communication) has seen the coyote, the side-striped jackal and the bush dog do so in response to the smell of urine. As far as is known, canids have a reduced Jacobson's organ and, indeed, that of the dog is said to contain no olfactory receptors (Barone and Lombard, 1966). It would therefore clearly be of interest to investigate the organ in those canids that do flehm.

Since flehmen normally occurs in connection with mating activities and is most commonly performed by the male, it may be that Jacobson's organ has some special sensitivity to the odours emitted by the oestrous female. It would be interesting to find out whether a cat would still respond to catnip if the openings leading to Jacobson's organ were blocked. Adrian's (1955) finding that in the rabbit, smells which activate the nasal olfactory receptors are without effect on Jacobson's organ, certainly suggests some highly specific sensitivity but unfortunately he did not test the effect of the smell of an oestrous female rabbit. Winans and Scalia (1970), also working with the rabbit, have recently succeeded in tracing the nerve fibres from Jacobson's organ to their destination in the hypothalamus and their findings support the idea that the organ is concerned both with feeding and with sexual behaviour. The hypothalamus contains what one may describe as major motivational centres, where information received from the external world and from within the animal itself is brought together and correlated. Motivation for feeding, for example, is thus made to depend not only on how tempting the available food is but also on how full the animal's stomach happens to be at the time. The two areas which receive input from Jacobson's organ are the medial hypothalamus (together with the medial preoptic area) and the ventro-medial nucleus. The former areas are concerned with sexual activities and damage to them interferes with copulatory behaviour: the latter is concerned with the regulation of feeding behaviour, in particular with bringing a period of feeding to an end. The main pathway from the nasal olfactory organ relays separately to more lateral regions of the hypothalamus. The connection of Jacobson's organ with sexual activities and with feeding thus seems reasonably certain but further analysis is clearly necessary before any very detailed ideas about its function can be put forward.

D. Taste

In eating, taste always acts in intimate cooperation with smell. This makes it impossible to dissociate the two in behavioural tests, which are therefore not suitable for any detailed study of the sense of taste. It is, however, a common observation that a highly developed 'sweet tooth' is much more characteristic of dogs than of cats. Since 'sweet' is one of the four major human gustatory categories it seems reasonable to assume that dogs, like ourselves, are well provided with receptors responding to sweet substances, whereas cats are not. Biologically this is reasonable: sugars are of significance to any species whose diet includes fruits but can have little relevance to one that is predominantly a flesh-eater. It is therefore not surprising that in captivity it is the more omnivorous carnivores that tend to become very fond of sweets.

The physiological investigations that have been made suggest that the dog's taste receptors are more sensitive to sweet substances than those of the cat. Appelberg (1958) studied the responses of fibres in the glossopharyngeal nerve and summarises the results of earlier workers, who recorded from the chorda tympani. In the dog, fibres responding to salt, bitter, acid and sweet substances are present in both nerves but in the cat no definitive sweet fibres have been found. Cats, however, do have some sensitivity to sugar. Beidler *et al.* (1955), recording from the chorda tympani nerve as a whole, found some response to sucrose although the threshold was high and Pfaffman (1955), recording from single units, found some which, although most sensitive to salt or to acid, also responded to sucrose.

The picture is complicated by the fact that cats are said to be unable to digest sugar and consequently to suffer from diarrhoea if they eat much of it (Bartoshuk *et al.*, 1971). The latter workers found that cats that had suffered in this way learned to avoid drinking a dilute salt solution which also contained sugar. Possibly, therefore, the cat's rather low taste sensitivity to sugar has the function of permitting it to avoid sweet foods, rather than to choose them out. Nevertheless, I have known one exceptional cat that had a passion for chocolate, raisins and even golden syrup and suffered no digestive troubles as a result of eating such delicacies. Possibly individual cats differ genetically both in their ability to metabolise sugars and in their responsiveness to sweet tastes, just as people differ in their sensitivity to the bitter taste of phenyl thiourea. On the other hand it is equally possible that, as in the case of human lactose intolerance, the ability of cats to digest sucrose depends more on whether they have eaten sugar from an early age or not than on any inherited differences. However that may be, under domestication, where many sweet foods not normally encountered by any wild felid are available, there may be some advantage in being able to metabolise and to taste sugars: cats with a sweet tooth may therefore be becoming commoner.

E. Tactile senses

In the majority of carnivores the part of the body most responsive to tactile stimuli is the muzzle and the vibrissae are hairs specially modified to increase this sensitivity. A number of species, however, show considerable manipulative dexterity and one would expect them to have paws with an unusually high degree of tactile sensitivity. Of these, the raccoon, *Procyon lotor*, is the species on which the most extensive experiments have been carried out. The 'intelligence tests' which the animals trained by Cole (1907) and by Davis (1907) were able to master required considerable manual skill but were not designed specifically to test this ability. Rensch and Dücker (1963), however, carried out experiments with the object of testing tactile discrimination. Their raccoon was trained to distinguish between two objects which he could not see, by feeling them with his paws: when he chose correctly, he received a food reward. His powers of tactile discrimination proved to be of the same order as our own. A smooth sphere could be distinguished from a rough one when the latter was covered with tubercles 0·3–0·4 mm in height but at 0·10–0·15 mm this was no longer possible. A similar threshold was found when students were asked to make the same discrimination. The raccoon and the students were also able to tell the difference between spheres of 2·5 and 2·64 cm diameter, a difference of only 1·4 mm or 0·53%. After eleven months the raccoon still remembered which were the correct objects to choose to obtain his reward.

The physiological basis of this discriminative ability has been investigated at various levels. Zollman and Winkleman (1962) have shown that the raccoon's paw has a very high density of touch receptors, while Welker and Seidenstein (1959) have found that in the somatic sensory area of the cerebral cortex the representation of the paw is gigantic in comparison with any other part of the body and much greater than in the cat or the dog. Welker and Johnson (1965) showed that the same is true of the processing station en route to the cortex in the ventrobasal nucleus of the thalamus (figure 4.2). Welker *et al.* (1964) give a general account of the raccoon somatic sensory system and include some comparison with another procyonid, the coati. The coati is less adept with its paws than the raccoon but uses its snout more in locating and extracting insects from the soil. Cortical sensory representation is correspondingly adjusted, the paw occupying a much smaller and the rhinarium a much larger area.

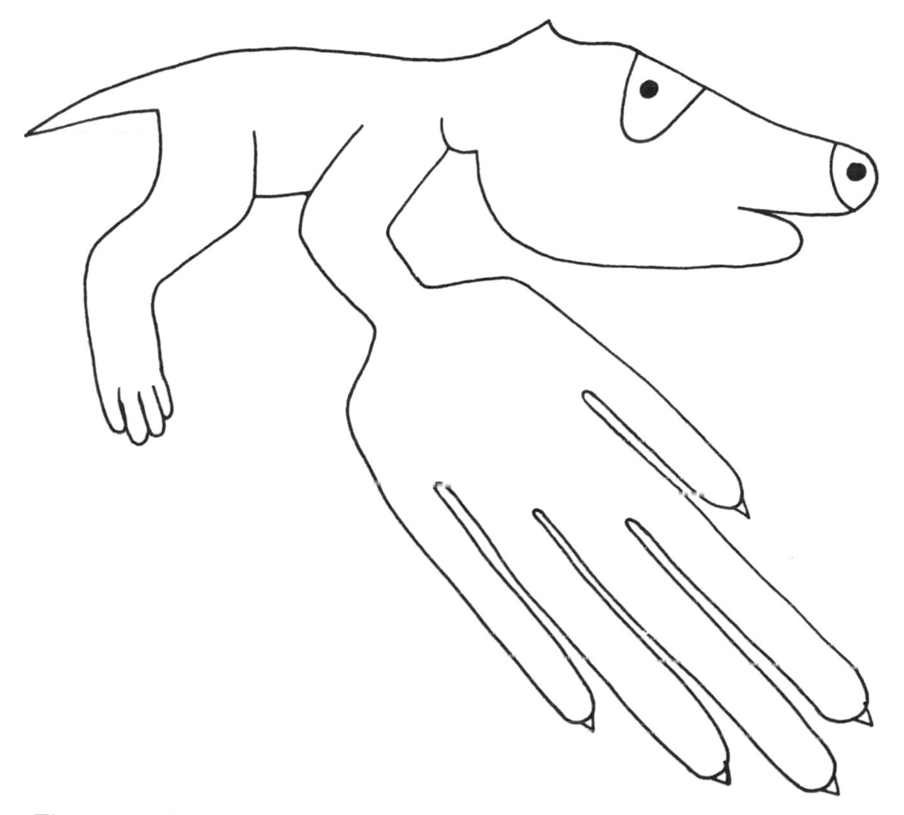

Figure 4.2. Raccoon: outline of body with the proportions distorted so as to correspond with the volume of nervous tissue in the thalamus concerned with relaying sensory information from the skin of the various parts of the body to the cerebral cortex. [After Welker and Johnson, 1968]

Rensch and Dücker (1969) trained a kinkajou to undo a series of different types of fastenings and found that this species too possesses considerable manual dexterity but no investigations on the physiological basis of the skill have been made.

Both the Mustelidae and the Viverridae include species that, like the raccoon, use their paws with skill in capturing aquatic prey. Amongst the former are the otters, *Aonyx*, *Amblonyx* and *Lutrogale* and, despite the shortness of its fingers, the sea otter, *Enhydra*, and amongst the viverrids, the crab-eating and marsh mongooses, *Herpestes urva* and *Atilax paludinosus*. The giant panda's expert handling of bamboo shoots is yet another case of skilled manipulation but the abilities of all these species still await detailed study.

Note added in proof

Dr Bernau (Z. Säugetierk., **34**, 223–6, 1969) has found that the mongoose, *Suricata suricatta*, also possesses colour vision. Her animals could distinguish red, yellow, green and blue as colours and histological examination showed that the retinal receptors were predominantly cones.

Chapter 5

Food and food finding

IF one had to give a general answer to the question, 'What do the Carnivora eat?' it would be a very simple one – 'What they can get'. Carnivores are very adaptable in feeding habits as in other ways and very few are restricted to or even largely dependent upon a single food. The existence of any species that is thus dependent is precarious and the highly successful carnivore is the one that combines sufficient specialisation to make it an effective predator on its chosen type of prey with flexibility enough to permit it to utilise a variety of foods in relation to changing seasons or varied habitats.

'Availability' is, of course, not a simple matter of abundance: a species that is better concealed or better able to defend itself may be less available to the predator than a less numerous but more vulnerable one. Even within a species, all classes of individual are not necessarily killed by a predator in proportion to their numbers in the population. The young, the aged and the sick or injured may be preferentially taken and prey behaviour may sometimes produce results that are at first glance surprising. The lions in Lake Manyara National Park, for instance, preying mainly upon buffalo, kill a disproportionate number of adult males. Although individually bigger and stronger than females and juveniles, the fact that many of the males are solitary lays them more open to attack than the weaker females and young who have the protection of group defence (Makacha and Schaller, 1969).

Two further factors may operate to make the predator's kill not dependent solely on the relative abundance of the various types of food in his habitat. The first of these is palatability. One cannot assume that all types of meat are equally attractive to a particular species of predator and, indeed, we know perfectly well from the behaviour of captive animals that this is not the case. There is little information on the operation of this factor in the wild but although a hungry animal cannot afford to be finicky, it would be surprising if preferences based on palatability played no part in prey selection.

The second factor is experience. Where making a kill requires some skill, experience leads to marked improvement in technique. Different prey species require different techniques and an animal that has become an adept killer of a particular type of prey may therefore tend to concentrate on this species and kill it preferentially. Domestic cats illustrate the operation of this factor. Most cats are rather inefficient predators of birds but a cat who has once learnt how to deal with them tends to become a bird fancier. Amongst the smaller carnivores that do not normally attack domestic stock, the occasional 'rogue' animals that do so may well be individuals that happen to have perfected an unusual technique, very much like the cat that becomes a bird specialist.

Before one can claim to understand the feeding habits of a single species, it is thus necessary to know what it eats in a variety of habitats covering its entire range and also covering the entire annual cycle. Such complete information is available for relatively few species. Most of those on which attention has been concentrated are species that are of economic importance either because, like the fox, they are accused of depredations on domestic poultry or livestock, or because, like the lion, they play a key role in the tourist industry.

The methods adopted in studying the food of carnivores fall into two categories. Firstly, there is the direct observation of kills. This method is applicable only to the larger species and particularly to those living in relatively open terrain, where kills are easily located. Its shortcoming is that small food items are liable to be missed. The second method is the indirect one of identifying the food remains in stomachs and intestines or in droppings. The former, obviously, has the disadvantage that it entails extensive killing. Once the food remains in gut or scat – to use the convenient American term – have been identified, the problem of how to record the results arises. Two methods are in common use, each with its own limitations. Food items may be recorded merely by percentage occurrence; i.e., out of every hundred stomachs or scats, how many contain the particular item? The obvious difficulty here is that the results do not give a direct measure of importance in the diet: 90% of scats each containing a beetle may mean less than 50% of them each with a rabbit. The second method is to attempt to measure the percentage of the total volume constituted by each category of remains. For a freshly filled stomach this may give a clear picture of relative importance in the diet but the longer digestion has gone on the more unreliable it becomes. On *a priori* grounds one would expect that the beetle would leave in the scats a higher proportion of undigested material than the rabbit and the proportions for remains of young and of old birds and mammals might not be identical. Lockie (1959), working with foxes, has greatly refined the technique for dealing with scats. Once the principal foods have been identified, the proportion of digestible matter in each is determined by feeding them to captive animals and finding out what live weight is represented by the dry weight of the remains recovered in the scats. A correction factor for each food item can then be calculated, by which remains in scats can be converted to live weight digested. As one might expect, beetle remains in scats represent less food value per gram than do vertebrate remains and those of small birds and mammals were found to represent less than large ones. This method is obviously applicable only to species where the necessary observations on captive animals can be readily made but once the correction factors have been determined their use is simple. Such refinement, however, is hardly necessary if the types of prey taken do not differ widely in the proportions of indigestible matter they contain: for instance, Erlinge (1968b), working with captive otters, found that the frequency of occurrence in the scats gave a very good estimate of the relative importance in the diet of various species of fishes and frogs. As one might expect, crayfish, with a higher proportion of indigestible material, were slightly over-represented by this method.

Bothma (1966) uses a combination of percentage occurrence and percentage volume. The various food items are arranged in order of importance first by one method and then by the other and numbered according to rank order. The two figures are then added to give an index number for each food item and these indices are then arranged in a new rank order, which Bothma considers to give a better estimate of the relative importance of the various food items than either the percentage frequency or the percentage volume alone.

The fact that different workers have used different recording methods complicates comparison of their results, but for most practical purposes it is sufficient if the food items can be classified into three major categories: (i) major constituents or staples; (ii) minor but significant items and (iii) items figuring only occasionally or too small to be of calorific importance. In the last category, it must be borne in mind that a small proportion of vegetable food may be negligible in terms of energy supplied but may yet be significant as a source of vitamins.

In what follows the families are dealt with in turn but the amount of information available, both on the foods and the methods of obtaining them, is so variable that no very consistent treatment is possible. I have, however, tried as far as possible to start with the most predacious and proceed to the more omnivorous or more vegetarian species.

A. Canidae

(*i*) *Large group-hunters*

(*a*) *Lycaon pictus,* the African hunting dog Data on *Lycaon* kills have been collected in a number of the East African National Parks (Wright, 1960; Lamprey, 1963; Kruuk and Turner, 1967; Estes and Goddard, 1967), in the Kruger National Park in South Africa (Pienaar, 1969) and in the Kafue National Park in Zambia (Mitchell *et al.*, 1965). *Lycaon* preys for choice on small to medium-sized antelope but if these are not available a *Lycaon* pack is capable of coping with larger prey, and wildebeest, topi, hartebeest, waterbuck, kudu and eland all figure in the diet. *Lycaon* is typically an animal of the open grasslands but also ranges into light woodland, and no species illustrates more clearly the way in which the staple prey changes in relation to availability in different habitats. In East Africa *Lycaon* lives on the plains game, with Thomson's gazelle preponderating in the kills. In the Kruger Park impala are the principal food while in Kafue top place is shared by duikers and reedbuck. Table 5.1 lists the percentage of the total kills in all three areas of the

Table 5.1
Principal prey species taken by *Lycaon* in different areas, expressed as percentage of total kills

Species	E. Africa	Kruger National Park	Kafue National Park
Thomson's gazelle	**60**	—	—
Wildebeest	**26**	1	4
Grant's gazelle	**8**	—	—
Impala	2	**81**	2
Waterbuck	—	**7**	2
Kudu	—	**7**	4
Duikers	—	2	**26**
Reedbuck	1	2	**25**
Hartebeest	(3-topi)	—	**16**

three 'top prey' species in each habitat. Not only is there no overlap between the top three prey in the three areas but of the nine species only three (impala, wildebeest and reedbuck) figure at all in the prey lists in all three areas. The data are not sufficient to show any seasonal trends apart from the fact that in Kafue there is probably more predation on reedbuck in the dry than in the rainy season.

Cannibalism has not been reported and while Kruuk and Turner found that the Serengeti animals sometimes took carrion, Pienaar regarded this as very exceptional for the Kruger Park *Lycaon*. In the Kruger Park, census records of the prey species were available.

Pienaar was therefore able to find out whether the frequencies of the various species in the *Lycaon* kills were in proportion to their abundance and on this basis he calculated a 'preference index'. The species taken at frequencies above their relative abundance were, in order of preference, impala, waterbuck, kudu and reedbuck: warthog, zebra, wildebeest and tsessebe had very low ratings. 'Preference' calculated in this way means simply differential killing and from what has been said above it should be clear that one cannot conclude that the flesh of the various species is necessarily relished in the order shown. The preference is a measure not only of which species the *Lycaon* choose to hunt but also of how often they succeed in making a kill – i.e., in addition to palatability, all the factors which affect ease of capture are involved.

(*b*) *Cuon* and *Speothos*, the Indian dhole and the bush dog   The two other canids possessing highly carnassial dentitions are the Indian dhole, *Cuon alpinus*, and the South American bush dog, *Speothos venaticus*. No studies of the feeding habits of either have been made but from the writings of the older hunter naturalists it appears that the main prey of *Cuon*, like that of *Lycaon*, is small to medium-sized ungulates – deer, ibex, wild sheep and pigs. The habit of pack hunting, however, also makes it possible for larger species, such as buffalo, gaur and banteng to be killed and attacks have even been recorded on tiger, leopard and bears.

In captivity *Speothos* has been found to enter water readily and to swim well – 'with otter-like ease', according to Hershkovitz (1969). This lends support to the information given to Tate (1931), that the semi-aquatic pacas are extensively preyed upon and are pursued into the water when they attempt to escape by swimming. It should, however, be noted that this statement, often quoted as fact, is based on hearsay, and Tate makes clear that he never himself saw a bush dog.

(*c*) *Canis lupus*, the wolf   Studies on wolf predation have been made only in the northern part of the range of the species, principally in North America and there is no detailed information about the food habits of wolves in India. Pullianen (1965) has studied the wolf in Finland and Pimlott (1967) gives a summary of earlier work in North America, together with his own investigations in Algonquin Park, Ontario; Mech (1970) also gives a summary of previous work. The picture that emerges is reasonably consistent. In winter, when beavers are safe below the ice and terrestrial rodents in burrows or in runways beneath the snow, the wolves subsist almost entirely on large ungulates – caribou, elk, moose or deer (*Odocoileus* spp.) according to local availability. In summer, snowshoe hare, together with beaver, marmot and smaller rodents, figures significantly in the diet but in only one locality did species other than ungulates make up more than half of the summer food. This was in the Pakesley area in Ontario, to the west of Algonquin Park and here beaver constituted the main summer food. In winter, however, as in other areas, the Pakesley wolves were dependent on ungulates - deer and moose. Wolves have been known to eat almost every available type of small prey - small mammals, birds, snakes and lizards, fish; even insects and earthworms. Grass and berries too are sometimes eaten but none of these items can be regarded as making a significant contribution to the diet.

Pimlott points out that, within the range of the migratory barren ground caribou, although the wolves are not free during their denning period to follow the migrating herds, the caribou may still constitute an important summer food resource. Not every caribou joins the migration - there are always stragglers who remain behind - and the wolves may continue to prey on these after the main herds have gone. Pimlott quotes Kelsall (1957) as finding that forty-two out of sixty-one scats examined contained caribou remains at a time when caribou had supposedly left the area completely.

Rausch (1967), in a study mainly devoted to population ecology, gives some incidental information on the principal food of the wolves in four areas in Alaska. In the southeastern coastal region, with a relatively temperate climate, the principal prey species are sitka deer, beaver and Rocky Mountain goat, with moose of importance in some localities. In the other regions caribou replaces deer as the main prey, with Dall sheep, Rocky Mountain goat, snowshoe hare, beaver and other rodents as subsidiary resources. Further south, in the southern highlands of Wisconsin, Thompson (1952) found that white-tailed deer was the main prey at all seasons of the year and lagomorphs and rodents were of very minor importance. He also noted that grass appeared in the scats in quantities quite sufficient to show that it was deliberately eaten now and then and not merely ingested accidentally because it was adhering to the food.

The wolves of Isle Royale National Park are a rather special case. Isle Royale, in the northern part of Lake Superior, has an area of 210 square miles, which suffices to support a single wolf pack. Studies on these wolves have been carried on over a number of years and are described by Mech (1966) and by Jordan *et al.* (1967). The main prey available is moose, together with beavers as a summer addition to the diet. Shortage of food during the period when young are being reared appears to be the factor responsible for the high pup mortality which keeps the wolves and moose in equilibrium and permits both species to continue to exist in their largely isolated environment.

The largest prey species killed by wolves are the bison and the musk ox. Both wolves and bison have been exterminated from most of their previously common range and virtually the only place where they now occur together is in Wood Buffalo National Park in Alberta. Mech (1970) quotes investigations made in this area by Fuller (1966), who, from an analysis of scats and stomach contents, found that the wolves were feeding mainly on bison. They did not, however, appear to be able to cope with adult animals in good condition, for the eight kills which he was able to examine were calves and old or sick animals. Wolves occur along with musk ox on Ellesmere Island but apparently are not able to kill them at all easily. Mech (1970) quotes Tener (1954a,b) as finding that in this area the wolves' main prey was arctic hare and only 17% of the scats examined contained musk ox remains.

Of the large, group-hunting Canidae, *Lycaon* is the species on which the most detailed studies of hunting technique have been made. The observations of Kruuk and Turner (1967), working in Serengeti and Estes and Goddard (1967) in Ngorongoro are in very general agreement. Hunting is visual and is carried out by daylight, mainly in the early morning and late afternoon, and is preceded by a series of mood-synchronising activities which ensure that all members of the pack are ready to set out together. This begins with a greeting ceremony, based on the juvenile food-begging pattern. The animals lick and nuzzle at each other's mouths, wagging their tails, and arousal gradually spreads throughout the group. The general excitement may work up into bouts of chasing and fighting-play or may die down and all settle to rest again for a little while. Finally, however, all are alerted and the pack moves off. The only pack I have watched set out with the sun behind them, which also happened to be upwind, giving them optimal conditions for sighting prey undetected. This may have been mere coincidence: according to Kühme (1964) no attention is paid to wind direction but he does not mention whether the sun's position is taken into account.

Lycaon usually hunt where there is little cover available and there is no attempt at stalking but Estes and Goddard record one case where the cover of tall grass was used effectively and also describe an incident when the dogs spread out and moved slowly up to the crest of a slight rise, then broke into a run and swept over the crest at full speed on a broad front. Unfortunately there did not happen to be any prey on the far side of the ridge. Normally, however, the approach to the prey is direct, at a slow trot with occasional pauses. As they close in the dogs slow down to a walk with the head held low, in line with the body. This slow walk continues as long as the prey stand their ground but the moment they flee the dogs launch themselves into top speed pursuit. In a small pack studied by Estes and Goddard, probably consisting of a single family group, one particular dog usually took the lead and made most of the kills but in larger packs it is more usual for a number of dogs to start independent chases, although once one clearly stands the best chance of overtaking his victim, there is a tendency for the rest to join in behind him. Since the prey may circle or zig-zag, the dogs in the rear can often take short cuts and may thus succeed in surrounding it. In a long chase the original leading dog may give up and his place be taken by another who, by remaining well behind to start with, has followed a shorter and less exhausting course but it is probably an exaggeration to regard this as an organised relay system. The kill is made by a leap and bite at the flank which fells the victim, followed immediately by tearing open the belly, the whole pack joining in as fast as they catch up. The prey is devoured with no quarrelling whatsoever: there is no growling, only an excited twittering call and the juveniles are all permitted to join in.

In a kill I witnessed, one lame animal was given his full share and when a crocuta attempted to steal the remains, one of the largest dogs snatched the meat back, carried it a short distance and laid it down for the younger members of the pack to share.

The slinking horizontal walk when approaching the prey is also shown by the wolf and Pruitt (1965) believes that caribou can tell whether the wolf means business or not by the posture adopted. A mongrel terrier of mine also sometimes moved towards another dog using this same gait. He did this in circumstances which suggested that he was not contemplating a serious attack but was playing at prey catching, using the other dog as a substitute for genuine prey.

Wolves, like *Lycaon*, normally capture their prey by chasing and fell it by leaping up and biting at rump, flanks and sides, although with certain types of prey the nose and shoulders are also singled out for attack. Really extended pursuits are unusual and the longest ever recorded by Crisler (1956) and Mech (1970) were five and three miles (approximately 8 and 5 km). Few wolf habitats include such flat and open terrain as the Serengeti plains and the fact that long chases are commoner in *Lycaon* probably reflects the heavier going the wolves have to cope with, particularly in winter when snow cover greatly increases the effort required to maintain speed.

For the wolf, smell plays a more important part than sight in the detection of prey. Although prey may be encountered by chance and is occasionally found by tracking, it is much more usual for it to be located by direct scenting: Mech (1970) records that in forty-five out of fifty-one moose hunts he witnessed, the prey was found in this way. Usually detection occurs at distances of less than 300 m but

Mech once saw the wolves pick up the scent of a moose cow and her twin calves approximately $1\frac{1}{2}$ miles ($2\frac{1}{2}$ km) away. When the scent is noticed the wolves come to a sudden halt and stand alert, sniffing and looking towards the source of the odour. Sometimes they then perform a group ceremony, much like the *Lycaon* greeting ceremony, before changing direction and making for the prey. As they draw near to their quarry the approach becomes more cautious and they attempt to steal up unobserved: in this they are favoured by the fact that since they are following the scent of the prey they will normally be coming on it upwind. Even so, they are usually noticed before they have got closer than 10 m and the quarry is alerted. What follows depends on the behaviour of the potential victim, which in turn varies according to species. Relatively small prey such as deer (*Odocoileus* spp.) have little chance of fighting off a wolf pack if they stand their ground. Their best chance of escape lies in speed and they usually flee at once. With a reasonable start a healthy adult deer can normally outdistance the wolves, who appear to be able to appreciate very quickly when a pursuit is unlikely to succeed and soon abandon the chase. However, if they do succeed in coming up with their quarry the result is a foregone conclusion: the deer cannot keep the wolves at bay and it is speedily pulled down and killed.

With prey as large as a moose the situation is different. A healthy adult animal is capable of defending itself and driving off the attacking wolves. Most of our information about wolf predation on moose comes from Isle Royale. Here, with moose the wolves' staple prey and wolves the moose's only predator, each species has considerable experience of the tactics of the other and both are capable of judging the likely outcome of an encounter. Mech (1970) summarises the various sequels that may follow the detection of a moose by the wolf pack. If the moose refuses to run and stands ready to defend itself when the wolves attack it is capable of beating them off and Mech never saw a kill made in such circumstances. The moose will charge the nearest wolves, then wheel rapidly and lash out with its fore feet at any animals attempting to take it in the rear. If the wolves cannot force the moose to abandon its defence and flee, they leave it after harassing it for only a few minutes. Only adult moose will stand their ground and Mech is of the opinion that the necessary confidence to do so is unlikely to be shown by any but young and healthy adult animals. Certainly many of the animals that flee and are ultimately killed prove to be aged or in poor condition.

If the moose does not stand but flees as soon as it detects the wolves then, provided it gets a good start, it will quickly outdistance them and the wolves abandon the chase very quickly if they are not obviously catching up. Even when they do catch up with a fleeing animal the issue is still uncertain. The moose may now stop and stand at bay and once again Mech saw no case where it did not succeed in driving off its assailants. Of those that continue to run, once the wolves have drawn level, the majority outlast and outrun their pursuers and succeed in escaping. However, if the wolves have sufficient reserves of strength to launch an actual attack on the moose while it is on the run, then in the majority of cases a kill follows. Once the moose has been weakened by bites at rump and flank the wolves close in and several will attack simultaneously, biting and gripping the nose and throat as well as the hindquarters. A moose that does escape after being wounded may be found again by the wolves a day or two later and, if it has been significantly weakened or slowed down, may then fall victim to them.

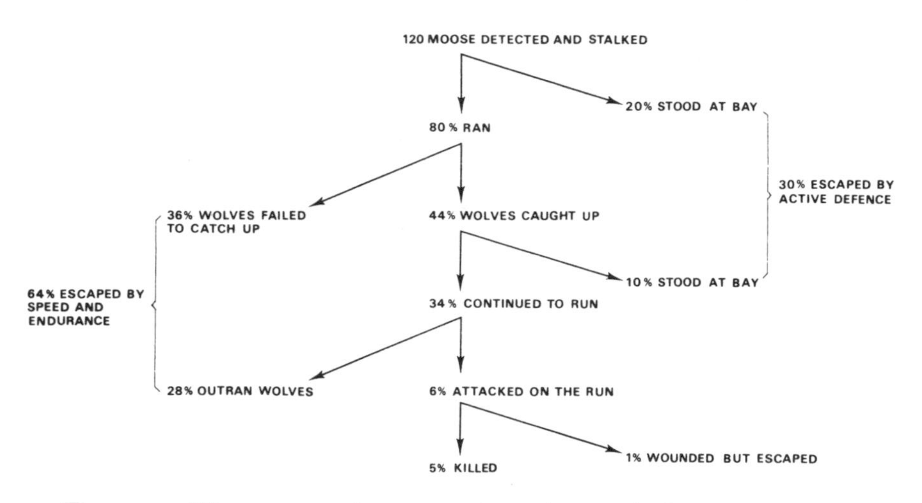

Figure 5.1. The outcome of 120 moose hunts by a pack of 15–16 wolves in Isle Royale National Park. [After Mech, 1970]

As may be appreciated from this description, the percentage of the adult moose located and attacked that are ultimately killed is very low (see figure 5.1) but with calves the success rate is much higher and calves make up a high proportion of the kills. A calf on its own is easily killed but the mother normally defends it pertinaciously and the wolves' problem therefore is to separate the calf from the cow. This is usually achieved by several wolves harassing each animal, so that while the mother is engaged in driving off one group others succeed in reaching the youngster.

Mech (1970) also give details of the hunting of caribou and mountain sheep. In all cases it appears that the success rate of the wolves is low: many attempts fail for every one that succeeds. If this is so, it seems *a priori* likely that the individuals killed are not a random sample of the prey population but are mainly animals that are in some way inferior to or at a disadvantage as compared with their fellows. Studies of the kills made bear this out: they consist mainly of the young and the aged and those killed as young adults are often injured or diseased. Mech is of the opinion that in fact the hunting techniques used by the wolves consist of setting up a sort of test situation which makes it possible for them to pick out the individuals that are slightly inferior in some way. With caribou, a whole herd is set in motion and any animal moving with less than the usual vigour can be detected: in the case of moose and deer the tests are individual; attacks on the animal that runs or defends itself strongly and vigorously are soon abandoned and the wolves seek an easier prey. Van Lawick-Goodall and van Lawick (1970) have expressed exactly the same opinion about *Lycaon.* They believe that once a herd starts to move the dogs single out those individuals that are slightly 'off colour' and just a trifle slower than their fellows. Errington's (1967) extensive studies on muskrat populations in Iowa show that in a territorial prey species the differential effects of predation may be even more striking. He found that the muskrats capable of holding a territory in the most favourable parts of the habitat were virtually immune from predation but that mink and fox levied a heavy toll on the homeless, the transients, the weaker animals forced out to live in sub-optimal habitats and on those suffering from disease or from wounds received in territorial fights with their fellows.

It is not difficult to see why predation should have this differential character. A prey species worth the predator's while to attack would soon be eliminated if every hunt meant a kill and the quarry that always escapes is no longer prey. The fact that in stable predator–prey systems a balance based on differential killing has been achieved is one good reason for not attempting to conserve prey species in isolation but in association with their predators. The natural predator is not a random killer, nor does he choose trophy specimens: his selection, based on what he can most easily get, is comparable with that of the stock breeder who eliminates weaklings from his breeding herd but the predator's success or failure provides a more sensitive test than any game ranger or wildlife management expert could ever hope to emulate.

(*ii*) *Jackals*

The fact that jackals are carrion eaters and scavengers, both on the kills of larger carnivores and on garbage round human habitations, is well known and so is the fact that they will raid cultivated fruit crops. Only a few detailed studies on feeding habits have, however, been made. Observations on the black-backed jackal, *Canis mesomelas*, have been carried out in four different localities: in the Kalahari National Park (Bothma, 1959), in the Transvaal, covering both reserves and farmlands (Grafton, 1965) and in Serengeti and Ngorongoro (Wyman, 1967).

In Serengeti, where Wyman made a comparative study of the black-backed and golden jackals, the latter, *Canis aureus*, is essentially an animal of the open plains, while *C. mesomelas*, although it is seen in the same area, is typically found in bush and in the transition zone bordering the open plains. On the plains Wyman found that both species were living almost exclusively on Thomson's gazelle and on the abundant beetles in the droppings of the rich ungulate fauna. The jackals' scats contained about equal volumes of the two types of food. Most of the gazelles were the jackals' own kills and almost all were very young animals. Scavenging was rare and only about 3% of the scat volume was made up of scavenged larger prey. In the transitional zone insects were much less important in the diet of *C. mesomelas* and their place was taken by small mammals, ground birds and vegetable food. In this area *C. aureus* took fewer gazelle but more insects and about the same proportion of small prey and vegetable food.

In Ngorongoro, where gazelle are scarcer and wildebeest are the dominant ungulate, the jackals are less self supporting, do more scavenging and eat more carrion. Van Lawick-Goodall and van Lawick (1970) found that in this area the placentas of wildebeest were greatly relished by *C. aureus* and during the calving period formed an important food source. They also record that snakes are quite frequently killed and eaten and that fallen fruit and mushrooms are seasonal additions to the diet.

In the Kalahari, where small game and larger predators are both scarce, the jackals can exist neither as self-supporting predators, as they do in Serengeti, nor as scavengers, as in Ngorongoro: here the main bulk of the food in ten *C. mesomelas* stomachs examined by Bothma consisted of insects, mostly locusts. A spring hare (*Pedetes capensis*), a few lizards and scorpions, with some grass and fruits made up the rest.

In the Transvaal, carrion was the largest single food item and made up about a third of the total volume of the contents of 185 stomachs analysed by Grafton (1965). Remains of small mammals and domestic stock, mostly sheep, tied for second place with less than 10% each of insects, birds and vegetable foods. Hirst (1969) records *C. mesomelas* feeding on impala in a Transvaal Nature Reserve but they were not capable of overpowering a healthy adult and killed only lambs or animals in poor condition.

In the Kruger National Park Pienaar (1969) reports that *C. mesomelas* lives on carrion, small mammals (including mongooses), eggs and young of ground-nesting birds, reptiles, insects and fruits. It also takes newborn young of a number of species of antelope and is capable of killing adults of some of the smaller species, such as steenbok and grysbok.

Clearly both the golden and black-backed jackals are capable of existing as independent predators if small game is abundant. Failing this *C. mesomelas* may scavenge, eat insects or become a killer of sheep. For *C. aureus* in areas other than Serengeti there is not a great deal of information. In Kanha Park, in central India, Schaller (1967) found, from an examination of 138 scats, that the two pairs of animals resident in his study area lived mainly on rodents and there was rather little scavenging. He estimated that rodents made up about 80% of the food volume: birds and fish were secondary items and lizard remains were quite frequent (11·6% of scats). Vegetable foods – fruit, grass, seeds and garbage – were present in over 25% of the scats but in this area, in contrast to Serengeti, insects were of negligible importance. In the Lebanon, Lewis *et al.* (1968) found that small vertebrates and garbage were the principal foods and there was no evidence of predation on domestic stock.

For the side-striped jackal, *Canis adustus*, there is virtually no information. It is more timid, solitary and nocturnal than *C. mesomelas*. The molar teeth are relatively larger and the carnassials relatively smaller than those of *C. mesomelas* (Ewer, 1956a), suggesting a greater dependence on insect and vegetable foods and a less predacious habit. The stomach contents of three animals examined by Bothma (1965) were very diverse. One contained nothing but fresh oribi meat, one nothing but fresh green grass while the third, from a juvenile, had grasshoppers, some domestic sheep remains and a little grass.

According to Wyman (1967) both the black-backed and the golden jackals capture Thomson's gazelles mainly by running them down in a direct pursuit, although careful stalking was seen now and then. Once the game has been pulled down, however, the killing techniques differ. *C. mesomelas* kills most commonly by biting at the throat of the prey but *C. aureus* usually tears out the abdomen. *C. mesomelas* pairs commonly hunt together at all seasons but in *C. aureus* co-operative hunting of this sort was common only from January to April. This follows the breeding season and is presumably the period during which the parents have to provide food for the young. Both species carry off and bury any left over food and Wyman is of the opinion that the jackals return within twenty-four hours to finish off the food thus cached.

Eaton (1969c) saw a jackal in Nairobi National Park apparently assist a cheetah to make a kill by running amongst a herd of mixed game and so distracting their attention that the cheetah was able to approach more closely than would otherwise have been possible before being noticed. Both jackal and cheetah had young and the jackal family waited until the cheetahs had eaten and then fed on the remains of the kill. This behaviour was seen several times. Eaton considered that it was a technique learned by the particular individual jackal who some months previously had been seen merely scavenging on the kills made by the cheetah.

(*iii*) *Coyote and foxes*

(*a*) *Canis latrans*, the coyote The coyote, because of its wolf-like appearance, has often been accused of serious predation on livestock and poultry. In fact, it resembles the fox much more closely than the wolf in its feeding habits. The staple prey is lagomorphs and rodents. Carrion is readily taken but killing of large prey, although it does occur, is not common. Birds, although regularly taken, do not constitute an important item of the diet. Insects and plant food are eaten whenever they are seasonally abundant.

Sperry's (1933) pioneer studies, based on stomach contents of coyotes killed in twelve western states of America, covered only the months of September to November and thus missed the season when insects and fruit are most abundant. He found that lagomorphs and carrion were the principal foods and although domestic sheep and goat remains were present in 14% of the stomachs, Sperry was of the opinion that much of this was carrion. His subsequent work, extending the period of observation from two to five years and bringing the number of stomachs examined up to over eight thousand, produced exactly the same conclusions. Sperry (1939) also made a separate analysis of stomachs of coyotes that had lost one paw as a result of trapping injury. There was surprisingly little alteration in the diet of the maimed animals: rodents and lagomorphs decreased but still accounted for one third of the food volume; carrion increased to make good the deficit from 25% to 35% and utilisation of domestic stock also increased.

A. Murie's (1940) extensive study in Yellowstone National Park was based on the analysis of over five thousand scats. He found that rodents were the staple food, with field mice numerically far the most common prey, followed by pocket gophers. Snowshoe hare remains were not very common. O. J. Murie (1945) found that the diet was exactly the same in an area in Wyoming just south of Yellowstone. A little further north, in two areas in Montana, lagomorphs (snowshoe hare, cottontail and jackrabbit) were more important. Bird remains were present in 4 and 12% of the scats from the two areas and plant food and insects in 5% and 18%. Remains of domestic cattle and adult elk (*Cervus canadensis*) were present only as carrion but some elk calf remains represented kills. Further north still, in British Columbia, the snowshoe hare was the staple food. Domestic stock and mule deer were both taken mainly as carrion but there were records of lamb and a few mule deer kills.

Both in British Columbia and in Yellowstone Park, predation on ducks was found to be slight, even in areas rich in waterfowl but Sooter (1946) noted that in a nature reserve in Oregon, the coyotes removed the eggs of various sorts of waterfowl and cached them for future use, very much as foxes do.

The studies of Fichter, Schildman and Sather (1955) covered several areas in Nebraska differing in soil type, vegetation and farming practice and extended over a little more than four years. The results of their analysis of 747 stomach contents and 2,500 scats showed that small mammals constituted the main prey; birds took second place; insects and fruit, while their total contribution to the diet was relatively small, were seasonally important. Insects reached their maximum in summer, fruits in late summer and autumn: in one area the percentage occurrence of fruit in the scats rose from 0% in spring to over 95% in autumn, when scats containing nothing but fruit remains were common. As one might expect, there were quantitative differences in the type of mammal prey taken in the different habitats. Where lagomorphs were abundant these were the main food: elsewhere the coyotes took a higher proportion of small rodents, principally various species of *Microtus* and these formed the staple food. Pocket-gophers appeared to be a second choice food, taken significantly only where the coyotes had difficulty in catching enough lagomorphs or field mice. In one study area muskrats were common and were preyed upon principally in spring, corresponding with their dispersal phase, when they make extensive journeys over land and so are most easily captured. Mule deer figured in the diet in late winter and most of them were believed to have been animals either killed by winter blizzards or so weakened as to render them easy prey. Domestic stock remains figured in 26% of the stomach contents examined and although the authors believed that much of this represented scavenging rather than predation, they did not attempt to assess the relative proportions of carrion and kills.

A. Murie (1951) made an extensive study which was specifically designed to estimate damage to livestock in an area in Arizona (San Carlos Indian Reservation) used as a cattle range. In this area small mammals were not abundant: the coyotes made good this deficiency not by depredations on the cattle but by taking more insect and plant food. No less than 80% of the 3,981 scats analysed contained plant food. Grasshoppers were extensively eaten from May to September and a variety of fruits and berries during summer and autumn. Juniper berries, with their good keeping qualities, were primarily a winter food. Cattle remains figured in only 16% of scats, mostly as carrion but there was some predation on calves. Some of these, however, were animals too weak to follow their mothers and would have died very soon in any case. Murie estimated that the stock losses to coyotes were negligible in comparison with those due to poor nutrition on over-grazed pastures.

Minkley (1966) records a coyote in Mexico catching turtles in water fifteen inches deep. While turtles are obviously not important as a food resource, this does illustrate the coyote's adaptability and readiness to take advantage of whatever food is available.

(*b*) *Vulpes vulpes*, the red fox The food of the red fox has been investigated over a wider geographical extent than that of any other predator, covering its natural range in Great Britain, Sweden and North America as well as in Australia, where the fox was introduced in Victoria in 1868 by sporting interests and soon spread over most of the mainland. All these investigations have reached the same conclusion: the fox is a predator on small mammals, principally lagomorphs and rodents and supplements its diet by insects and fruit in summer and autumn. Carrion is taken if available and ground-nesting birds and their eggs, particularly where they occur at high density, may be seasonally important. Middleton (1954) recorded heavy fox predation on sitting partridges in a preserved area in Hampshire and Cobnut (1955) found the same thing with pheasants in Kent. Tinbergen (1965) found that foxes were regular predators on the eggs and young of the black-headed gull nesting colony at Ravenglass.

Southern and Watson (1941) made a survey of the food of foxes in highland and lowland areas in Wales and the English midlands respectively. In both areas rabbits were the main food and birds were a significant item. In the highlands sheep carrion was important and in the lowlands a higher proportion of insects and rabbits made good the deficiency of carrion. Small rodents, principally voles, were more important in the highlands but in neither area were they a major food constituent.

In America the picture is much the same, with the emphasis on small rodents or on lagomorphs according to local abundance (Errington, 1935; Scott, 1943; Fisher, 1951; Wood, 1954; Dodds, 1955) but in America more wild fruits are available in autumn and these therefore play a more important role in the diet at that time of year than they do in Great Britain. Frijlink (personal communication) finds that the foxes in Algonquin Park scavenge on the remains of wolf kills and is of the opinion that this extra source of food has permitted the foxes to attain an unusually high population density.

Englund's (1965a) extensive study of foxes in Sweden covered a series of different areas from the far north to the more populated southern farming districts and also the island of Götland. Rabbits were common only in Götland and the extreme southern part of the mainland and in these regions they were more important in the fox's diet than were rodents. In the other areas voles, lemmings and mice were the main food, with *Microtus* spp. by far the most numerous.

In Australia, McIntosh (1963a) found that in the Canberra area, where neither small rodents nor small marsupials are common, the staple foods were rabbits and sheep carrion. There was little evidence of true predation on sheep but during lambing time sheep placentas were often eaten and this appears to be a form of food which foxes find extremely attractive. In farming areas, where mice were abundant, these played a much more important role in the diet and rabbit correspondingly less. Insects and plant food showed the normal seasonal fluctuations, with insects at a maximum during the hottest months of December to February and plants a little later. In Australia, however, plant food was never very important and bird predation was also low. In none of the areas studied by McIntosh were there sufficient small terrestrial marsupials for any estimate to be made of the effects of fox predation on their abundance. Marlow (1958), however, expresses the opinion that foxes have in the past been responsible for reducing their numbers and believes that their higher density in Tasmania is partly due to freedom from fox predation.

A number of studies have been devoted to the problem of how the fox manages in an unusual habitat or when his normal food resources are drastically cut down. Heit (1944), investigating a muskrat habitat in a salt marsh in Maryland, found that the foxes not only came into the marsh area to hunt muskrats but some were actually denning in the occasional patches of higher ground in the marshes. He estimated that, although *Microtus* was the most frequent prey in the scats, the larger muskrat was the major food item.

Johnson (1970) has studied the food of foxes on Isle Royale, where the number of prey species available is less than on the neighbouring mainland. Snowshoe hare is the only lagomorph present and is the principal prey species: *Peromyscus*, not usually much relished by foxes, figures rather frequently in the diet, presumably because *Microtus* is absent. In winter the foxes rely mainly on these two species, together with moose meat scavenged from wolf kills. In late summer and early autumn they feed mainly on fruits and berries and in August and September some 80% of scats contain plant food.

Englund (1965a) had data not only for fox foods but also for the abundance of small rodents over a three-year period which included 'bad' and 'good' rodent years. He was therefore able to find out what happened in a bad rodent year in three areas where small rodents were normally the staple food. In two of them the deficit of rodents was made up for by an increased consumption of carrion and garbage. In the third these alternative sources were less readily available and the foxes took more insects and fruit in summer and more birds in winter.

Errington (1937a) in Iowa made a similar comparison of the food of foxes in a normal year and in a drought year when small rodents were scarce. Here the foxes ate many more insects and muskrat appeared in the diet more frequently, presumably because the drying up of streams had made them more accessible. Dexter (1951) records the shooting of a crippled fox with only one fore paw whose stomach proved to contain nothing but plant food, mainly timothy grass (*Phleum pratense*).

The most interesting problem of this type, however, is the question of how the fox was affected by the large-scale destruction of rabbits as a result of control by myxomatosis. The effects of myxomatosis have been studied both in England (Lever, 1959) and in Sweden (Englund, 1965c). Lever divided his post-myxomatosis findings for 1955–7 into lowland and highland areas so that Southern and Watson's (1941) earlier study could serve as the pre-myxomatosis control. In the lowlands decrease in rabbit food was compensated for by increased predation on voles and birds and by a slight increase in insect food. The highland foxes also took more voles and considerably more insects but bird predation showed little change. In Gloucestershire, Burrows (1968) also found that the disappearance of rabbit was followed by increased predation on voles.

Englund's study dealt with the island of Götland and has the advantage that exactly the same area was covered by the same worker before and after myxomatosis and the two studies were separated by only one intervening year, during which myxomatosis had reduced the rabbit population to 5% of its former level (Englund, 1965b). Despite this drastic reduction, the foxes' rabbit consumption declined only to half its previous level, suggesting that the remaining rabbits were being hunted more intensively. The foods that showed compensatory increase were hares, small rodents and game birds, mainly pheasants.

The picture of the fox that emerges from all these studies is of a predator primarily of small rodents and lagomorphs: but he is also an opportunist, taking advantage of whatever is most readily available, whether it be sitting pheasants or garbage and scorning neither carrion, nor insects, nor vegetable foods. Dependent on no single resource, he can compensate for deficiency in one food by taking more of another and can thus occupy a wide range of habitats or survive quite major alterations within a single area.

Such a picture, however, is incomplete. It assumes that the fox selects his food purely on the basis of accessibility and it takes no account of special tastes or preferences. There is, however, considerable evidence that foxes do have special tastes and prefer some foods to others. Their extreme fondness for sheep's placenta has already been mentioned. According to McIntosh (1963a) captive foxes prefer hares to rabbits and Lund (1962) found that, given *Microtus*, *Clethrionomys* and *Apodemus* to choose from, *Microtus* was easily top favourite, *Clethrionomys* was acceptable and *Apodemus* not very attractive. The common preponderance of *Microtus* over other small rodents in the foxes' diet may not therefore be purely a matter of ease of catching: they may be specifically sought after because they taste better. Scott and Klimstra (1955) found that, in an area where *Peromyscus* was five times as numerous as *Microtus*, the latter was still the favourite food. Errington (1935) found only one *Microtus* left uneaten outside a fox den where scats showed that seventy-five had been recently eaten. Only three *Peromyscus* were identified in the scats and two had been left uneaten.

The large consumption of fruit and berries that is recorded wherever these foods are abundant occurs at a time of year when the foxes are not hard put to it to find food, which suggests a genuine liking for such fare. Burrows (1968) remarks on the large consumption of fallen apples and pears by the foxes in a Gloucestershire fruit-growing area and Lewis *et al.* (1968) report that in the Lebanon such cultivated fruits as figs and grapes are raided by foxes. This brings to mind not only the fable of the fox and the grapes but also the biblical 'take us the foxes, the little foxes that spoil our vines', which some commentators have attempted to explain as referring to fruit bats. There is no need for any such sophistry: the fox who was despoiling the fruit crop need be none other than the ordinary red fox, our own familiar Reynard.

(*c*) *Vulpes velox*, the swift fox Cutter's (1958a, b) study of the food of this species in Texas was based on the analysis of scats collected over a period from spring to August, together with a few stomach contents. Lagomorphs were the most important food, with insects, especially grasshoppers, in second place. Domestic poultry and game birds were not taken but remains of small passerines were fairly common and were also seen scattered about near den entrances. Rodent remains were found but were never very common and grass occurred quite frequently.

Kilgore (1969) obtained very similar results, working in Oklahoma, but since his scat collections were made in spring and in August and were analysed separately he was able to provide some information on seasonal changes. The main bulk of the food consisted of small mammals. In spring, lagomorphs accounted for over 70% of the volume of the scats and rodents were a minor item but in August the position was reversed, with lagomorphs scarce and rodents predominating. Cutter mentions that most of the lagomorph remains he identified belonged to young animals, so the seasonal change which Kilgore found probably reflects the changing availability of young lagomorphs. The main rodents which he recorded were *Rheithrodontomys*, *Peromyscus* and *Perognathus*, in proportions matching the frequency with which he was able to catch them in traps. Ground-roosting birds, lizards and plant food were minor items and insect remains, although present in many scats, made up only a small fraction of the total volume, whereas in Cutter's study they were much more abundant. The reason for the difference may well be that Kilgore missed the peak period for insect eating; for Hawbecker (1943) found that in scats gathered in July both insects and plant food were important in the diet. He also found that although antelope ground squirrel (*Citellus nelsoni*) were common in the area studied they did not appear in the scats and kangaroo rats, *Dipodomys* spp., were the main mammalian prey. Burns (1960) records the stomach of a single animal killed on a road in New Mexico as containing 18 *Perognathus*, 3 *Peromyscus*, 1 immature mouse and 2 lizards (*Phrynosoma* sp.).

(*d*) *Vulpes chama*, the silver fox Bothma (1966) investigated the food remains in thirty-seven stomachs of this species, mainly from the Transvaal. The three most important foods taken were rodents, carrion and insects and he found no evidence of predation on poultry or domestic stock. The stomach of a single specimen from the Kalahari proved to contain large numbers of harvester termites and no vertebrate prey (Bothma, 1959).

(*e*) *Alopex lagopus*, the arctic fox The arctic fox makes his living very much as one might expect an ordinary red fox to do, were he capable of living in the far north. Barabash-Nikiforov (1938) has studied the feeding habits of the animals living on the Commander Islands, at the western end of the Aleutian chain. As might be expected, much of the fox's food is gathered on the shores – molluscs, echinoderms and crustaceans, together with general scavenging on whatever is cast up by the tide and the foxes also scavenge on garbage round human habitations. In summer, eggs and young birds are taken from the nesting colonies and fish coming up the rivers to spawn provide another seasonal food source while in autumn, various types of berries are eaten. On Bering Island, *Clethrionomys* figures in the diet but this species is not present on Copper Island. The habit of storing away food when it is abundant, for use at a later date, is very well developed in *Alopex* and must be of considerable importance in enabling it to survive through the long winter.

(*f*) *Urocyon cinereoargenteus*, the grey fox In southern Wisconsin, Errington (1935) found that the foods of grey and red foxes inhabiting the same area were qualitatively very similar, but that there were differences in the proportions of the various items. Cottontail rabbit was the main prey of the grey fox with *Microtus* occupying second place, while in the red fox the relative importance was reversed. This appeared to be because the red fox hunted predominantly in the ploughed or stubble fields where *Microtus* was most abundant but *Urocyon* kept more to the uncultivated hills, where the rabbits were more numerous. Chicken figured in the diets of both species but mainly as carrion or garbage. Gander (1966) also reports that *Urocyon* will take carrion. Wood (1954), working in Texas, also found that cottontails were the main prey of the grey fox with cotton rats (*Sigmodon*) second in importance. Persimmons constituted the main plant food and although acorns were available very few were eaten. For the raccoons in the area, on the other hand, the acorns were the favourite plant food.

(*g*) *Cerdocyon thous*, the crab-eating fox Walker (1964) quotes data from Mondolfi for the stomach contents of nineteen animals killed in Venezuela. In order of importance the foods found were rodents, insects, fruits, lizards, frogs and crabs. *Cerdocyon* is also said to dig up turtles' eggs. In captivity *Cerdocyon* is extremely omnivorous and thrives on a diet including fruits and cereals (Da Silveira, 1968).

(*h*) *Otocyon megalotis*, the bat-eared fox Although this species is well known to be largely insectivorous, detailed records of the food taken are difficult to find. Bothma (1959) recorded the stomach contents of eight specimens from the Kalahari National Park as consisting entirely of insects, apart from one gecko and a small amount of vegetable food. Over 80% by volume of the insects were termites, the rest being beetle larvae together with a few adult beetles and locusts. The termites were workers and both they and the beetle larvae could have been captured only by digging.

The stomach of another animal from the Kruger Park contained mainly carrion, together with one lizard and one small murid (Bothma, 1965) while one from Somalia contained only insects (Azzaroli and Simonetta, 1966).

A fox is too big to pursue mice along their own runways. Their capture where there is any form of cover thus presents a problem, for a direct snap with the jaws would most likely result only in a mouthful of grass. The fox however, has a special technique to cope with this problem. He first locates the prey and then performs the characteristic 'mouse-jump' (figure 5.2). The forequarters rise high in a rearing leap and descend vertically on the victim, the two fore paws and the nose contacting the ground virtually simultaneously. Smith (1944), watching a fox using this technique to capture field mice, saw it catch four mice consecutively without once missing its strike. The mouse-jump has also been seen in a slightly less highly developed form in captive grey and arctic foxes and also in the coyote (Fox,

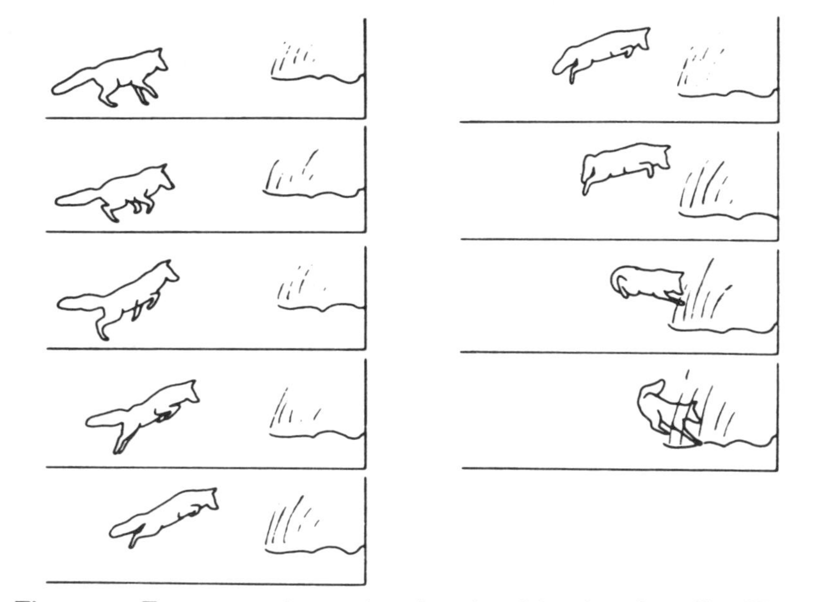

Figure 5.2. Fox executes 'mouse jump'; series of drawings from film. [From Leyhausen, 1956]

1969b). The latter is of some interest, as it underlines the previously mentioned similarity of the coyote's feeding habits to those of foxes, rather than wolves. Seitz's (1959) hand-reared golden jackal also did the mouse-jump; van Lawick-Goodall and van Lawick (1970) have seen jackals in the wild do so; however, as might be expected from its desert habitat, this behaviour is apparently absent from the repertoire of the fennec. Since so many sorts of fox prey upon small rodents, it is likely that the mouse-jump is also used by a number of other species.

Fox did not see his red or arctic foxes shake their prey but his grey foxes did so with great vigour; both red and grey often seized the prey in their jaws and, with a quick flick of the head threw it to one side or vertically up in the air and then attacked it again. According to him, grey, red and arctic foxes seize the prey not by the neck but somewhere near the middle of the body. Tembrock (1957a), on the other hand, says that the red fox usually seizes small mammals by the neck and describes not only throwing but also shaking of the prey. He also quotes accounts of foxes killing larger prey by a bite on the throat and describes the stalking of prey.

As far as is known, all foxes are capable of digging burrows but for those that prey extensively on insects, digging also plays a part in food finding: it is therefore no surprise to find that these are more expert diggers than the red fox. This is particularly true of the fennec and the bat-eared fox. The dependence of the latter on termites that have to be dug up has already been mentioned and both Smithers (1966a) and Turner (1968) comment on how readily and how fast captive *Otocyon* will dig. Vogel (1962) and Gauthier-Pilters (1962) both remark on the same propensity in captive fennecs. Although analyses of fennec feeding habits have not been made it would be surprising, in view of their habitat, if they did not feed largely on insects.

Leakey (1969) records *Otocyon* using a rather unorthodox method of adding a little variety to the diet. He happened to see a falcon catch a rock pigeon which was too heavy to carry off and the bird therefore gently parachuted down with its prey in its talons to make its meal on the ground. There were, however, other more directly interested spectators: three *Otocyon* saw the falcon start its descent, ran to intercept it as it grounded and promptly attacked it. In attempting to drive them off, the falcon had to let go its prey, which was instantly seized by one of the foxes and carried off to the safety of the den, where its two companions joined it. On two other occasions exactly the same behaviour was seen and Leakey is of the opinion that it must be not uncommon for bat-eared foxes that live near cliffs to learn this trick: he believes that they then keep an eye on the doings of the falcons, so as to be ready to take advantage of this rather easy way of stealing a meal.

Many of the Canidae have the habit of burying any surplus food remaining after a meal. The technique of burying appears to be very uniform. A hollow is excavated by scratching alternately with the fore paws, the food is then thrust in with the snout and is covered up by pushing earth over it, also with the snout. A domestic dog burying a bone uses this method and I have seen a cocker spaniel 'bury' a bone on a sofa, using the whole repertoire of movements, including the 'covering', regardless of the fact that there was no earth with which to cover the bone. Vogel (1962) reports a fennec performing similar 'burying' on a bare floor and Fox (1969b) has seen the behaviour in captive red, grey and arctic foxes.

Food-burying in jackals has been mentioned. Here the storage is probably for a period of no more than twenty-four hours. In foxes, however, the habit of making caches is much more highly developed and the hoards may be used either as short-term or as long-term stores. Fisher (1951) watched red foxes hunting in stubble fields just before dark. Voles, mice and grasshoppers were being caught but instead of being eaten at once were buried - several items in each cache. Two out of three caches were gone by the time darkness had fallen and Fisher concluded that by intensive hunting during the most favourable hours, the foxes caught more prey than would have been possible had they taken time off for eating; certainly they caught more than enough to satisfy their immediate hunger. The caches made by the foxes studied by Scott (1943) were mainly longer-term winter reserves, although stores were made in every month except May, September and October. Scott was of the opinion that the foxes remembered the location of their caches and he found that *Peromyscus* which he himself buried were not discovered, although they were interred on pathways used by the foxes. Tinbergen (1965) took this kind of investigation considerably further. The foxes he studied preyed upon eggs and young birds from a nesting colony of black-backed gulls. Eggs were removed by the foxes and cached and were not utilised until later on in the season. When this happened, the tracks left in the sand showed that the fox came fast and direct to the region where he had buried the eggs and then cast about until he located them by smell. Eggs buried in the same neighbourhood by Tinbergen and his colleagues were also found and dug up, indicating that while the general area was remembered, the exact location of individual caches was not.

The red fox shows variability not only in the length of time between hiding and using the food but also in how the stores are organised. Tinbergen's foxes scatter-hoarded, burying each egg separately but many of the caches described by Scott (1943) and by Fisher (1951) contained more than one item. It is therefore possible that the location of a major hoard of this type, which must have taken several visits to accumulate, is remembered more accurately than the positions of scatter-hoarded single items.

For the arctic fox, previously accumulated stores must play a vital role in survival over the winter. According to Pedersen (1962) the caches are large and each contains a great many food items. One which he excavated contained 36 little auks, 4 snow-bunting, 2 young guillemots and a 'large number' of little auk eggs; he estimated that this was sufficient to last the fox for a month.

(*iv*) *Nyctereutes procyonoides*, the raccoon dog

Apart from the inclusion in the diet of a number of aquatic species – fish, frogs, water beetles and molluscs – the diet of the raccoon dog is rather like that of a fox. Small rodents, especially *Microtus*, form the basic food along with insects and plant food, including cultivated cereals, as they become seasonally available. Frogs are most abundant in the diet during their spring breeding period and small rodents show a peak in early spring, after the snow has melted and before the vegetation has grown sufficiently to provide effective cover (Bannikov, 1964). Novikov (1956) states that fish are taken during their upstream spawning runs and that near the coast, raccoon dogs scavenge on the shore, very much as the arctic fox does. In the more northerly part of its range *Nyctereutes* lays down abundant fat reserves in autumn and hibernates through the worst part of the winter.

B. Procyonidae

There are no exclusively predacious procyonids: vegetable food always plays a significant part in the diet and in some species is the major component.

(*i*) *Procyoninae*

The raccoon, *Procyon lotor*, is the species whose feeding habits have been most extensively investigated. Stuewer (1943), working in Michigan, recorded the food remains from stomachs and scats during the course of the year. Only for a short period during spring and early summer did animal food constitute more than half of the volume of food remains. *Microtus* occupied first place with crayfish second. As fruit and berries ripened towards the end of summer these became increasingly important and for the period of July–September they constituted 78% of the remains with insects and crayfish taking second and third places. During winter, vegetable food predominated, with grapes (which remain on the vines for a long time after ripening), corn and acorns as the three main items. In spring acorns constituted the major food with corn and invertebrates (snails and earthworms) next in importance.

The findings of other workers are in general agreement with the results of Stuewer's study. Hamilton (1951) found that from April to October plant food made up 59% by volume of the stomach contents of raccoons in New York State. Mammals contributed 15%, with *Microtus* the most frequent. Frogs and salamanders were eaten but crayfish remains were not found. Schoonover and Marshall (1951) studied scats collected from mid-June to mid-September in Rice Lake National Wildlife Refuge in Minnesota. The area includes two major types of habitat: a low-lying region along the river, where crayfish are to be found but where there are few fruit-bearing shrubs or trees and an upland area where the latter are abundant. To start with, the raccoons concentrated their activities along the river banks and crayfish remains made up the bulk of the scats. The few found in the upland area contained insects, frogs, small mammals, birds, grain and old acorns from the previous year's crop. Towards the end of July, fruit and berries started to ripen: activity shifted to the uplands and berries became the main food, although some crayfish were still taken. From mid-August to mid-September, when the berry season was over, corn and acorns appeared in increasing volume. Over the entire period, the three main foods in order of importance in the scats from the river area were crayfish, juneberries and acorns, while for the upland region they were juneberries, grasshoppers and acorns.

Cagle (1949) found that in the Mississippi Delta Wildlife Refuge, the gut contents of twenty-nine raccoons killed in May were all composed principally of crayfish and five included shells of turtle eggs. Curiously enough, although a species of fiddler crab was common in the area, it did not figure in the raccoon's diet. Tester (1953), working in Colorado, found that corn was the main constituent of the scats collected in autumn, with crayfish and grasshoppers in second and third places. In Texas, over a two-year period, Wood (1954) found that more than half of the food volume consisted of plant remains, mainly acorns: second and third places were occupied by insects and small mammals, of which cottontails were the most numerous. This study was carried out in an upland area where crayfish were scarce and were therefore correspondingly rare in the raccoons' diet, accounting for only about 3% of the total volume. By way of contrast Wood points out that in an area in Michigan where crayfish were abundant, Dearborn (1932) found that they provided more than half of the raccoons' food.

According to Gander (1966), raccoons will not accept carrion, which appears to be almost the only form of food they refuse. Omnivorous though they are apart from this, raccoons are not undiscriminating and show two very definite dietary predilections. Firstly, they are extremely fond of vegetable foods; acorns are a great favourite and cultivated grains are also relished. Indeed, Sherman (1954) reports that in the Bahamas[1] raccoons in 1952 caused serious damage to the maize and groundnut crops. Their second great love is crayfish. If these are to be had in the habitat, then large numbers will be eaten during those months of the year when they are active and easily caught.

Nasua narica, the coati, and *Bassariscus astutus*, the ringtail or cacomistle, are the only two other procyonids for which any detailed information is available. Both are omnivorous, eating invertebrates and plant food as well as vertebrate prey and both differ from the raccoon in that they will eat carrion. Their diets, however, show considerable differences in the relative importance of the various items. Kaufmann (1962) found that on Barro Colorado Island the coatis ate mainly invertebrates and fruits. The main invertebrate foods were earthworms, snails, insects, spiders and scorpions. Land crabs were taken during the wet season and the coatis also 'dredged' for crabs in the streams. Some caecilians were taken and one frog was seen captured. Neither snakes nor turtles aroused any interest but lizards were pursued. Although the coatis chased rodents Kaufmann never saw one captured and he never saw them rob a nest or catch a bird, although Wallmo and Gallizioli (1954) say that in Arizona the coati is reputed to raid the nests of both birds and squirrels. Smythe (1970b) points out that on Barro Colorado Island fallen fruit is most plentiful in May, June and July, when the young are being reared and during this period the coatis are almost exclusively frugivorous. When fruit becomes scarcer the females and juveniles live mainly on forest floor invertebrates. The males, however, who by this time have left the groups of females and their young and taken up a solitary existence (see Chapter 7), turn their attention mainly to small vertebrate prey and spend much time hunting for juvenile agoutis. Thus, during the period when food is most difficult to find, the males are not in competition with the females and juveniles for food, an arrangement which Smythe suggests may well have survival value.

The diet of *Bassariscus* differs from that of the coati mainly in the greater importance of vertebrate prey, especially mammals and birds.

[1] The Bahama population is reputed to have originated from a single pair, liberated in 1932 or 1933.

Taylor (1954) analysed the gut contents of 256 animals killed in Texas. In summer, insects and arachnids constituted 70% of the volume of the food remains, with fruits and berries second. In autumn invertebrates had fallen to 40%, fruit and berries were still in second place but the volume of both birds and mammals eaten had increased. In winter mammals were the major item at 36%, with birds second at 24% and insects third at 20%. In spring the importance of insects and plant foods started to increase again, mammals decreased slightly and birds fell to only 7%. The mammal prey included cottontails and ground squirrels as well as smaller rodents. Wood (1954), also working in Texas, found that in ten stomachs and nineteen scats small mammals were the major food, plant material second and birds third. Edwards (1955) tells how a pet ringtail that had been brought from Texas to New York escaped and was at large for a month before being recaptured. Despite the fact that it was winter, the ringtail caught enough squirrels and small birds to keep itself in good health. Brannon (1923) recorded a ringtail being shot in a farmyard in Alabama while chasing chickens and, according to older accounts (Charlesworth, 1841), in Mexico the ringtail was formerly often to be found living around farmyards in disused outbuildings and was abundant enough to be regarded as a serious predator on domestic poultry and pigeons. According to Thomson (1842) the Mexicans sometimes kept a tame ringtail in place of a domestic cat, to kill rats and mice.

I have found no information on the kinkajou, *Potos flavus*, in the wild but Poglayen-Neuwall's (1962) observations on captive animals suggest that the main foods are insects and fruit and that kinkajous are not very actively predacious. His animals refused liver and a variety of dead small mammals and, although they killed lizards, they did not eat them. Poglayen-Neuwall suggests that the kinkajou's extremely long and slender tongue may be an adaptation to licking the nectar from flowers.

The meagre data for the feeding habits of the olingo, *Bassaricyon*, are summarised by Poglayen-Neuwall and Poglayen-Neuwall (1965). These amount to little more than that olingos are very fond of fruit and are often captured when they come to feed on a tree bearing ripe fruit. The Poglayen-Neuwalls' observations on captive animals, however, suggest that olingos are considerably more predacious than kinkajous and their diet is probably more like that of *Bassariscus*. Birds up to the size of a starling, small mammals and lizards formed a regular part of the diet they gave their animals and all were quickly killed and eaten. Raw meat was also readily accepted.

Although all five genera climb well, *Procyon* and *Nasua* spend much of their time on the ground and forage there extensively and *Bassariscus* too is at home on the ground as well as in the trees. *Bassaricyon* and *Potos* are both principally arboreal, the latter almost exclusively so. As is only natural, the species that are not highly arboreal include the two that fish for crayfish, *Procyon* and *Nasua*. The importance of vertebrate prey in the diet, however, is not simply related to the degree of arboreal specialisation. *Procyon*, *Bassariscus* and *Bassaricyon* are all effective killers but *Nasua* and *Potos* are not: *Nasua*, however, concentrates on forest floor invertebrates while *Potos* is essentially frugivorous.

All procyonids have the forearm mobility characteristic of good climbers and all use their paws extensively in procuring their food. The manual dexterity of the raccoon and its expertise in finding crayfish by touch have already been mentioned. The paws are also used in turning over litter to search for insects, etc. and for digging up earthworms (Gander, 1966). Captive raccoons very often develop the habit of 'dowsing' their food: instead of at once eating the food provided, the raccoon will pick it up and put it in its water dish. It will then feel for it with its paws, pull it out and finally eat it.

The name '*lotor*' refers to this supposed washing of the food. No wild raccoon has ever been seen to dowse and the habit remained mysterious until Lyall-Watson's (1963) careful investigation. He showed that dowsing had nothing to do with removing dirt but was simply a technique the animals discovered whereby they could give themselves the opportunity to utilise their normal method of finding food in water. In captivity this behaviour pattern normally has no outlet and the animals find out how to provide themselves with one. From observations on other species that dowse, I believe the habit also has a further function. In many species, catching food has a facilitating effect on eating and even food which is not greatly relished will be eaten if it is the animal's own catch. My tame meerkats, *Suricata suricatta*, for instance, would eat a millipede captured after much digging and scraping but refuse the same species if it were simply given to them. A tame marsh mongoose too would more readily eat things she did not greatly relish if she first put them in her water dish, chased them around with her paws and finally 'captured' them. Duplaix-Hall (personal communication) has found that clawless otters will also 'wash' their food (plate 9). Captivity diets are often monotonous, and it may well be that the animals make their food more acceptable by going through the actions of catching it for themselves in the water dish.

The coati uses its paws mainly for scraping and digging for invertebrate prey, either in the ground or in rotten wood. Kaufmann (1962) says that it never roots with the snout, although prey is located mainly by smell. Foraging is carried out in trees as well as on the ground and fruit is either picked up off the ground or plucked directly from the trees. The paws are also used in killing vertebrates; the prey is held down by the forepaws and then killed by a bite on the head. Kaufmann does not give details of how 'dredging' for crayfish is done but he mentions that captive coatis used to put bits of food into their water dish in a manner rather reminiscent of raccoon dowsing.

The kinkajou, although no killer, is extremely dexterous and uses its paws to hold and manipulate its food. Gray (1865) describes a captive one in the London Zoo holding a piece of bread in one hand and with the other pulling off small bits and putting them in its mouth 'like a child eating a cake and quite as handily'. Poglayen-Neuwall's tame ones often held their heads up vertically as they ate, which he considers may be an adaptation to feeding on soft fruits without losing any of the juice. The olingo sometimes does the same but much less regularly. The food is usually picked up in the mouth and, in general, *Bassaricyon* makes less use of its paws. Both birds and mammals are killed by a bite on the head, usually in the occipital region, and the Poglayen-Neuwalls never saw their animals shake the prey.

(*ii*) *Ailurinae*, *Ailurus fulgens*, the red panda, and *Ailuropoda melanoleuca*, the giant panda

Both species of panda are virtually pure vegetarians. The red panda, according to Morris (1965), eats acorns, roots and lichens as well as bamboo shoots. Over the greater part of its range, bamboo shoots and stems form the principal or even only food of the giant panda (Sheldon, 1937) but in the Chinghai province of Tibet they feed on gentians, irises, crocuses and grasses (Morris and Morris, 1966). Since captive pandas will accept meat, it is possible that insects and carrion may form a minor part of the normal diet.

Morris and Morris, in a rather dramatised comparison of the vegetarian regime of the panda with that of true predators, say that the panda 'has waved goodbye to the nimble-minded world of helter-skelter chases, bloated blood-feasts and sprawling catnaps. Instead it has become (literally) a manual labourer, toiling endlessly at its repetitive bamboo-picking task'. The valid comparison is, of course, not with exclusively predacious species but with other procyonids. These show both the fondness for vegetable food and the tendency to use the paws in food-collecting that one would expect to find in the panda's relatives and their prey killing techniques are not sufficient to deal with anything large enough to provide the orgiastic banquets which this description conjures up. Had the panda's ancestors been expert killers of prey larger than themselves, it is extremely doubtful if they would ever have abandoned flesh and become vegetarian but there is no difficulty in visualising a gradual adoption of this type of diet if the foods and feeding techniques of living procyonids are taken as the point of departure.

C. Ursidae

The Ursidae are the least predacious family of the Carnivora and the most highly dependent on vegetable food. The brown bear, *Ursus arctos* and the American black bear, *U. americanus*, are generally regarded as typical members of the group. For both species, vegetable foods form the bulk of the diet for much of the year but both kill small mammals and will also scavenge when the opportunity offers; both will also rob bees' nests and are very fond of sweet things (Bergman, 1936; Bennet *et al*, 1943). Shadle (1941) records a case of a black bear having eaten virtually every leaf off a magnolia tree which was heavily infested with scale insects and covered with the sweet honey-dew which they excrete. Both species will take advantage of any seasonally abundant food resource, whether insect, vertebrate or vegetable. Murie (1937), for instance, found that during a season when crickets and grasshoppers were very abundant in the Yellowstone area, of 64 black bear scats collected in August, 58 contained nothing but these insects, 3 more had in addition a few berries and the remaining 3 contained grass. Similarly Chapman *et al.* (1955) found brown bears in Montana feeding on local aggregations of ladybirds (Coccinellidae) and of army cutworm moths (*Chorizagrotis auxiliaris*) and Bannikov (1967) found that the bears in the Barzuginski Reserve, on the shores of Lake Baikal, take advantage of the caddis fly larvae, which in spring become very numerous around the shores of the lake. Pocock (1939) notes that in the Himalayas and Tibet the brown bear is more actively predacious than elsewhere and will kill even buffalo calves. Extensive predation on salmon during their spawning runs provides an extra food source for the brown bear (Shuman, 1950; Trautman, 1963) but with the autumn ripening of fruit and berries both species turn vegetarian and intensive feeding allows them to lay down sufficient fat reserves to last them through their denning period during the worst part of the winter (Cottram *et al.*, 1939).

Other bears have diverged from this unspecialised omnivorous opportunistic type of feeding and have either become more nearly vegetarian or have turned secondarily more carnivorous. In the latter category comes the polar bear, which some taxonomists place in the same genus as the black and brown bears. The polar bear's food is seals, particularly the ringed seal (*Pusa hispida*) but fish, sea birds and their eggs and carrion are also eaten. Flyger and Townsend (1968) record an instance of no less than forty-two polar bears assembled to feed on the carcase of a dead whale. Although mainly carnivorous, polar bears, like their relatives, also take advantage of the autumn ripening of berries (Meyer-Holzapfel, 1957) and stomach contents and scats containing *Equisetum* and grass as well as berries are recorded by Doutt (1967).

The Himalayan black bear, *Selenarctos thibetanus*, according to Pocock (1939), has much the same omnivorous habits as the American black bear and in addition to vegetable food eats honey, insects and carrion and is also somewhat predacious. Schaller (1969b), from an examination of eighty-two scats, collected in October in the Dachigam Sanctuary in Kashmir, found that at this time of the year the bears were feeding almost exclusively on fruits and nuts of various sorts. Only one scat contained insects (wasps) and one had hair, which may have come from carrion. Although in other areas the bears are known to kill domestic stock, this was not recorded in the sanctuary. According to the forestry staff, maize and mulberries are frequently eaten in July and August and in the earlier part of the year, before fruits ripen, grass and leaves constitute the bulk of the diet.

The sun bear, *Helarctos malayanus*, and the sloth bear, *Melursus ursinus*, are both somewhat aberrant. *Helarctos* is the most lightly built and the most arboreal of bears and feeds on fruit, small vertebrates and honey, to which it is extremely partial. According to Champion (1936) sun bears also sometimes scavenge on tiger kills. *Melursus* is a specialised eater of insects, honey and soft fruits and peculiar modifications of lips, tongue and incisors in adaptation to this diet have already been described. Ninety-two scats collected by Schaller (1967) in the Kanha Park in central India contained termites, together with a few ants or fruits and seeds. Schaller concluded that, although fruit was important in the diet from April to June, termites were the staple food for much of the year.

Little is known about the food of *Tremarctos*, the only South American bear but it is said to be virtually exclusively vegetarian. The curious conformation of the extremely strong lower jaw and its possible relation to cracking hard-shelled nuts have already been mentioned.

D. Mustelidae

The Mustelidae are such a large group that it is necessary to consider the various subfamilies separately.

(*i*) *Mustelinae*

(*a*) *Mustela* The Mustelinae include the most predacious species and, in relation to their size, the weasels and stoats are killers as formidable and efficient as the larger Felidae. The food of stoats and weasels has been studied in Britain (Day, 1968), in North America (Hamilton 1933; Aldous and Manweiler, 1942; Quick, 1951) and in New Zealand (Marshall, 1963), where weasels, stoats and ferrets were introduced during the latter part of the last century in the hope that they would reduce the rabbit population.

In Britain, lagomorphs, rodents and birds are about equally important in the diet of the stoat, *Mustela erminea*. *Microtus* is the main rodent prey, and amongst the birds, game species predominate, with passerines and pigeons making up the balance. Day found that invertebrates and plant food were negligible and there was remarkably little seasonal variation. The diet of the weasel, *Mustela nivalis*, is similar but the proportion of small rodents is higher and that of lagomorphs and birds correspondingly less. *Microtus* was again found to be the main rodent prey but the birds taken were mainly passerines. Although shrews were numerous in the area Day studied, they were very rarely taken. In New Zealand the stoat lives mainly on rabbit and carrion and also takes some insects and crayfish but data for the weasel are very meagre.

In North America, in addition to the stoat, there are two species of weasel. The larger, the long-tailed weasel, *Mustela frenata*, is peculiar to America; the smaller, *M. rixosa*, is extremely closely related to *M. nivalis* and some taxonomists regard it as only subspecifically distinct. The foods of the two American species are rather similar, with small rodents, particularly *Microtus*, constituting the main food, birds very seldom taken and insects and plant food negligible. There are, however, some differences. *M. frenata* takes a higher proportion of lagomorphs and of rats and other rodents of similar size than does the smaller *M. rixosa*. The latter, however, is the only weasel that preys significantly on shrews and although other workers have not recorded invertebrate prey remains in stomachs or scats, Osgood (1936) saw a female *M. rixosa* carrying earthworms back to her nestling young.

In America, stoats appear to prey less upon lagomorphs than in Britain and their diet is rather similar to that of *M. frenata*. Rozenzweig (1966) has summarised the data for the foods of the three *Mustela* species in America and concludes that all three feed mainly on small rodents weighing not more than 50 grams. He therefore expresses some surprise that all three can coexist. This, however, seems rather naïve: he did not compare the foods taken by two or more species in the same area, as Day did for British stoats and weasels, but used all the available information on the diets. An abundance of *Microtus*, providing, for a time, food enough for all, may tend to conceal important differences and an analysis in terms of size of prey alone is bound to overlook the more subtle ones. *M. rixosa*'s habit of eating the shrews which other species avoid, for instance, may be important when *Microtus* supplies fail. Certainly both in England and in New Zealand the greater dependence of stoats on rabbits is clear enough and in both countries the stoats were much more adversely affected than the weasels by the drastic reduction in the rabbit population that followed myxomatosis in England (Day, 1968; Jefferies and Pendlebury, 1968) and an effective poisoning programme in New Zealand (Marshall, 1963).

The polecat, *Mustela putorius*, is now so rare over most of its range that quantitative data on its diet are lacking. However, since the main reason for its near extermination was its attacks on domestic poultry, one might expect birds to figure rather more largely in the diet of the polecat than in those of the stoat and weasel.

For most of the Asiatic species of *Mustela* no information is available but Pocock (1939) says that in China the Siberian weasel, *M. sibirica*, commonly frequents human habitations in search of rats and mice and that in Nepal the yellow-bellied weasel, *M. kathiath*, is kept as a domestic rodent killer.

The weasels as a whole can thus be said to be truly carnivorous: they are specialist killers of small mammals and birds, making little use of invertebrate or vegetable foods and showing little seasonal change in diet. The mink, *Mustela vison*, as might be expected from its amphibious habits, has a more varied menu; in addition to the small mammals and birds that form the food of its terrestrial congeners it takes fish, crayfish, frogs and aquatic invertebrates and there is considerable seasonal variation in relation to the changing availability of the major categories of food. The most complete study is that of Gerell (1967) in Sweden, where the mink was introduced from America towards the end of the 1920s and in the space of about thirty-five years had spread over the whole country except for the most northerly mountainous regions. Gerell's work is based on over four thousand scats from three different areas, collected over a four-year period. In the River Ronneå area crayfish were abundant and the main foods were crayfish in summer and fish in winter (figure 5.3). This seasonal pattern relates to ease of capture, since in winter the fish are sluggish and more easily caught but the crayfish remain totally inactive and inaccessible in refuges at the bottom of the river. Frogs show peaks in the diet in April and at the end of September, corresponding to the transition period between fish and crayfish and

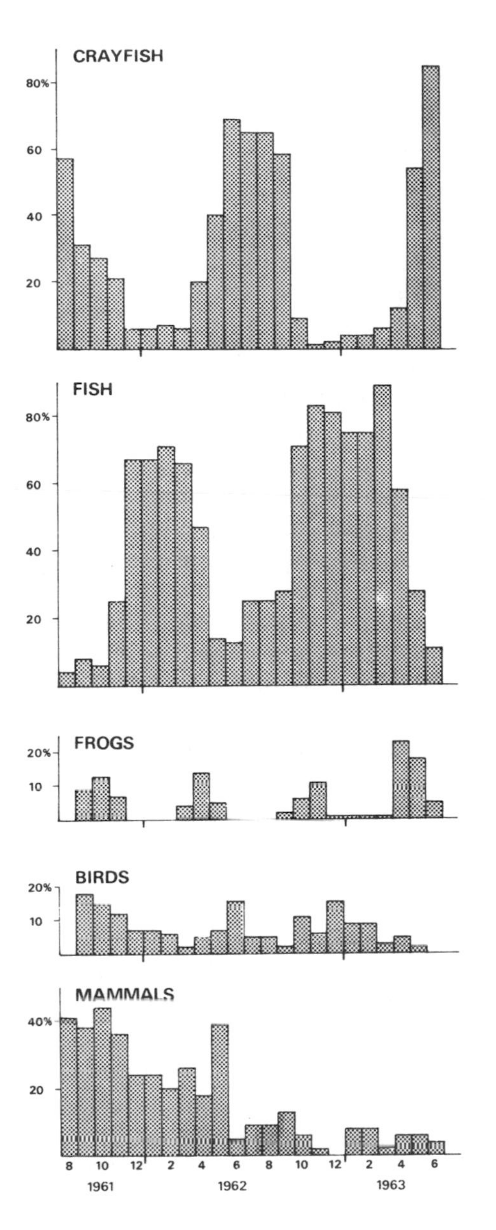

Figure 5.3. Percentage occurrence of different types of foods in the scats of mink frequenting a river in southern Sweden, from August 1961 to June 1963. [Data from Gerell, 1967]

appear to be a second-choice food, taken if nothing better is to be had. Rodents, mainly *Microtus* and *Arvicola*, were third in importance in the first year of the study but their habitat in the area was then destroyed and the mink made up for this by taking rather more fish.

Birds were of minor importance. In the other two areas crayfish did not occur and the main foods were rodents, fish and amphibia, with birds, particularly ducks, moderately common. In one area, Lake Hornborgasjön, dytiscid beetles were the main invertebrate food and, like the frogs, appeared to be taken as a second-choice food.

Erlinge (1969) studied yet another area in southern Sweden and here fish formed the main food (62%), waterfowl second (23%) and small rodents third (9%). Insects and amphibians were taken only sporadically and together accounted for only slightly over 5% of the food. Although this habitat included a trout spawning ground, trout were rarely taken and burbot was the main prey.

A considerable number of studies have been made in America (Dearborn, 1932; Hamilton, 1936a, 1940, 1959; Sealander, 1943; Guilday, 1949; Wilson, 1954; Korschgen, 1958) but many of them are based on stomach contents of animals killed during the trapping season and so cover only the winter months. In general the picture that emerges is similar: the mink takes a variety of animal foods, both aquatic and terrestrial, choosing whichever is most easily available. Wherever muskrat are common they figure largely in the diet but otherwise smaller rodents, mainly *Microtus*, take first place. Although its diet is highly varied the mink is not necessarily an indiscriminate feeder. Platt (1968) notes that during a small mammal trapping study, a mink habitually stole woodland mice (*Napaeozapus insignis*) from his traps but left *Clethrionomys*, *Peromyscus*, *Tamias* and *Blarina* untouched.

(*b*) *Martes* The martens are much more omnivorous than the weasels and both insects and fruits play a significant role in the diet, which consequently shows more seasonal variation. The two species for which there is most information are the pine marten, *Martes martes*, and the American marten, *M. americana*. For the former, Lockie (1961), working in Scotland over a two-year period, found that small rodents were important in the diet at all seasons and formed the greater part of the food except during July–October when berries, especially blueberries and mountain ash, took first place. Beetles and caterpillars were important during the summer months and small birds were taken most often in winter. Carrion and fish were subsidiary items. Hurrell (1955) comments on the fondness of a pair of captive martens for various berries and for beetles and moths.

Martes americana has been studied by a number of American workers (Cowan and MacKay, 1950; Lensink *et al.*, 1955; Quick, 1955) and the results are similar: a wide range of foods is acceptable and the marten takes advantage of local and seasonal abundance. Murie's study (1961) in Grand Teton National Park in Wyoming showed that within this area the martens in three slightly different habitats were making rather different selections of mammalian prey in relation to local abundance. In one area voles were the main prey, in another ground squirrels, while in the third, although lagomorph remains occurred only in about half as many scats as voles did, in view of their larger size they were probably the major food item. There are, however, strong indications that palatability as well as availability enters into the marten's choice of food: Murie found that shrews, *Peromyscus*, *Eutamias* and *Tamiasciurus* were very rarely taken, even where quite abundant, and both Cowan and MacKay and Lensink and his co-workers found that *Peromyscus* was not often eaten. Marshall (1946), on the other hand, found that *Tamiasciurus* was the commonest prey species in the rather small number of winter-collected scats which he examined and Lockie quotes Jurgenson (1954) as finding that in Russia the degree to which the marten preys upon squirrels shows large variations which are not explicable in terms simply of local abundance: possibly a taste for squirrel is an idiosyncrasy shown only by certain individuals or populations.

Schmidt (1943) quotes data from various Russian workers on the stone marten, *M. foina*, showing that in winter small rodents and carrion form the main food, supplemented by young hares as soon as their spring breeding season starts. The summer diet is small rodents and small birds, together with an increasing amount of fruits as autumn approaches and at the height of the berry season, 95% of the food may be vegetable. The sable, *M. zibellina*, preying mainly on small rodents and ground-living birds, also eats fruit and berries during the autumn (Bannikov, 1967).

(*c*) *Gulo* The wolverine, *Gulo gulo*, is the largest of the Mustelinae and is rather like an over-grown marten. The jaws and teeth are extremely powerful and, together with its fondness for carrion, have earned the wolverine the name of the 'hyaena of the north'. Wolverine feeding habits have been studied by Haglund (1966) and Krott (1959). The diet, like that of the marten, is varied and shows seasonal changes. For the wolverine the year can be divided into the snow-free season and the period of snow cover. Any food remaining in winter caches is finished in the early days after the melting of the snow and the main food during the snow-free months is small rodents, together with birds' eggs in spring, wasps' nests in summer and berries in autumn. During winter the main food is reindeer and carrion and, according to Haglund, wolverines will scavenge on the kills of lynxes. Hunting success in winter is largely determined by the texture of the snow: if it is not sufficiently hard to bear the wolverine's weight, then hunting is impossible; conversely, if the crust will take the weight of the wolverine but not of the reindeer, then the latter is easy prey. Haglund, however, says that most of the reindeer killed by wolverines are in poor condition as a result either of disease or of semi-starvation. Krott quotes the findings of Taplow in Russia, who found moose and reindeer equally important as winter foods but the former was almost all the result of scavenging on the remains of animals shot by hunters.

Wolverines cache any surplus food remaining after a meal and these reserves are important for their survival during periods when hunting is not possible. Their ability to locate caches is not restricted to the ones they have made themselves and wolverine are liable to raid any man-made stores.

(*d*) *Other mustelines* The mustelines so far dealt with fall into two categories: those that are virtually exclusive carnivores, preying mainly on mammals and those that are more polyphagous, eating significant amounts of invertebrate and vegetable foods and killing a wide variety of types of vertebrate prey. In the first category are most species of *Mustela* and in the second *Martes*, *Gulo* and the mink. No quantitative data are available for other genera and for most of them one can do no more than allot them, somewhat tentatively, to one or other category.

Judging mainly from observations made on captive animals, the South American tayra (*Tayra barbara*) and grisons (*Grison* spp.) both belong to the polyphagous category. Kaufmann and Kaufmann (1965) have seen wild tayras eating fruit and Brosset's (1968) tame one was very fond of honey and ate a variety of fruits and vegetables, including mushrooms. He was, however, also a formidable predator: an expert climber, he would search out birds' nests and devour eggs or nestlings and was prepared to attack any rodent, even those larger than himself. Any surplus food was cached and he accepted carrion readily. Captive grison too have been found to accept a variety of foods; fruit, insects and small vertebrate prey (Dalquest and Roberts, 1951; Kaufmann and Kaufmann, 1965; Dücker, 1968). Since grisons swim readily, it is quite possible that their normal fare includes aquatic invertebrates, frogs and maybe fish.

According to Pocock (1939) the marbled polecat, *Vormela*, preys upon small rodents, birds, lizards and insects but he makes no mention of vegetable food. The South African *Poecilogale*, to judge from observations on captive animals (Alexander and Ewer, 1959; Ansell, 1960a; Rowe-Rowe, 1969), appears to be a pure carnivore, preying principally upon small mammals. *Ictonyx* has the reputation of being a poultry raider and foods which it is known to take are small mammals, reptiles and frogs. A tame one showed itself very partial to insects and would search them out for itself and it also liked bread and milk (Mayne, 1963); all of which suggests that *Ictonyx* belongs to the more polyphagous group. There appears to be no information about the feeding habits of the Libyan weasel, *Poecilictis* or the South American *Lyncodon* and *Grammogale*, but the reduced molars and cutting carnassials of *Lyncodon* strongly suggest that it is a highly carnivorous species.

There are few records of the hunting methods of Mustelinae but tracking by scent appears to play a major role, at least in some species (Murie, 1935; Goethe, 1950). Rustling and squeaking sounds are also often important in the first alerting to the presence of prey. Once within striking distance, the typical musteline method of attacking is to leap on the prey, clutching its body with the fore limbs and to kill with a bite on the occipital region which usually crushes the back of the skull. If the prey is too large to be easily held back in this way the weasel will pull it off its feet by clasping it and throwing itself on its side, raking backwards vigorously with the hind feet as it does so. Gossow (1970) describes how a weasel managed to use the neck bite on a golden hamster that fought it off by adopting the belly-up defence posture. The weasel made a sudden lunge from the side, pushed its head under the body of the prey and, raising it slightly from the ground, seized it by the nape. His stoat used a very similar technique with rats defending themselves in the same way. With long-necked prey the killing bite is usually further back, on the neck rather than the skull and if the prey is relatively large, a few preliminary bites delivered anywhere convenient may precede the definitive attack (Allen, 1938; Progulske, 1969). The wolverine, attacking large animals such as reindeer, leaps right upon the back of the prey and rides it, clutching in typical musteline fashion and kills by bites at the neck (Haglund, 1966; Krott, 1959). Small rodents are sometimes killed by a blow with the paw but the characteristic bite at the head normally follows even if the prey is already dead (Krott, 1959).

Although in polecats smell appears to play a dominant role (Räber, 1944; Wüstehube, 1960), in most mustelines movement is one of the most important factors triggering off attack and if several moving prey animals are present, weasels will kill repeatedly before starting to eat. This characteristic, besides leading to wholesale slaughter if a weasel gets into a henhouse, is probably one of the factors that have led to the very widespread habit of storing away any surplus food for consumption later on. Food storing is recorded in *Mustela frenata* (Hamilton, 1933). *M. vison* (Yeager, 1943; Erlinge, 1969), *M.*

putorius (Räber, 1944); *M. nivalis* (Goethe, 1950); in *Martes martes* and *M. foina* (Schmidt, 1943; Räber, 1944); in *Poecilogale* and *Ictonyx* (Ansell, 1960a; Rowe-Rowe, 1969); the wolverine's extensive winter caches have already been mentioned.

(*ii*) *Mellivorinae*

Mellivora capensis, the honey badger or ratel, is the only representative of this subfamily. According to Sikes (1964), *Mellivora* becomes very omnivorous in captivity and in the wild eats honey and sweet fruits and preys principally upon small birds and mammals. Smithers (1966b) adds that it is extremely partial to insects, scorpions and spiders. Cases of honey badgers killing larger mammals are also reported, and one in the Kruger National Park was seen to kill prey as large as a ten-foot python (Anon., 1964): carrion is also sometimes taken and scavenging on the kills of tigers has been reported (Champion, 1936). According to Sikes, *Mellivora* is very dexterous with its paws and searches for small prey by turning over stones and logs and by tearing off loose bark. Although not a swift runner, it has great powers of endurance, and larger prey is captured by tenacious pursuit prolonged to the point of exhaustion. The honey badger's most curious food finding technique, however, is based on its association with the honey guide, *Indicator indicator*. The traditional story is that when the bird has located a bees' nest it gives a particular call. The honey badger responds by approaching and following the bird's cries and is thus led to the nest, which it then rips apart with its strong claws. The bird is permitted to share the feast and picks out grubs and wax, while the ratel consumes both grubs and honey. Friedmann's (1955) studies, however, suggest that this is not exactly what occurs. There is no evidence that the bird first locates a bees' nest and then seeks an assistant. It seems rather that the excited calling and short flights from tree to tree which characterise the guiding behaviour are direct responses to the ratel, which cease only when a bees' nest is located: thus the finding of the nest does not precede but follows the encounter with the carnivore. This all suggests that guiding behaviour may have evolved from a simple mobbing response to a potential predator. It is not known whether ratels respond appropriately to the bird's calls without previous experience or whether the association between the bird and the finding of honey is learnt.

(*iii*) *Melinae*

Although the badgers may possibly be a rather heterogeneous group, they share a somewhat catholic appetite and extensive powers of digging. The only two species for which detailed information is available are the European and American badgers, *Meles meles* and *Taxidea taxus*.

Neal (1948) studied the food of English badgers from stomach contents and scats and by direct observation and also summarised earlier findings. His conclusions are that vegetable food is always the major component of the badger's diet: in late summer and autumn blackberries, acorns and beech nuts; in winter, principally grass with the addition of some roots and foliage in spring. Invertebrates are second on the list, with earthworms very much favoured, together with insects, principally beetles. Slugs and less frequently snails are taken and in summer both bees' and wasps' nests are torn open and the grubs and honey devoured. The main vertebrate prey is small mammals; birds are taken only very rarely. Small rodents – rats, mice and voles – are preyed upon and badgers also appear to relish hedgehogs but in the areas studied by Neal the staple vertebrate food was young rabbits, whose remains were, naturally, most abundant in the scats during the spring and summer months. Presumably since myxomatosis the proportion of rabbit in the diet has declined and that of small rodents has increased but no data on this point are available. Captive badgers refuse carrion and badgers have never been found to store food.

The badger is too slow to catch adult rabbits and has an unusual technique for capturing the young. The brood nest is located (by scent according to Neal but possibly hearing is also involved) and the badger then digs vertically downwards to it. The technique for dealing with hedgehogs is to eat them out from below and leave the spiny dorsal skin untouched – badgers are very skilful at avoiding the spines and Middleton (1935) records that the stomach of one badger that had eaten four adult hedgehogs contained only four spines. Insect larvae are dug up and cow dung is turned over with the paws in search of dung beetles.

Milner (1967) describes how in the Conway valley in north Wales large patches of grass were found torn up and left in piles, which killed the herbage beneath. The authors of this peculiar type of damage to pastures were not seen at work but there was very strong circumstantial evidence that the culprits were badgers. It occurred only in late summer and early autumn and Milner thought that the badgers were probably searching for insect larvae living in the top soil amongst the grass roots.

The American badger, in contrast to *Meles*, is virtually an exclusive carnivore and plant foods do not figure to a significant extent in the diet. Errington (1937b) and Snead and Hendrickson (1942) both found that ground squirrels of the genus *Citellus* were the major prey item to be found in the scats of badgers from Iowa. Mice, mainly *Microtus*, came second and insects third. The latter are most abundant in the scats in late summer and autumn and bees and wasps are the favourite species. Birds are taken now and then and badgers have been seen to kill rattlesnakes (Jackley, 1938).

Balph (1961) describes *Taxidea* lying in wait inside a burrow entrance in order to pounce upon ground squirrels. Knopf and Balph (1969) record them first blocking up the accessory entrances of the ground squirrels' burrows and then digging down to the nest via the main entrance. This technique was seen only in one area in Cache National Forest, Utah, and there was no evidence of it being used in other places where a study of the ground squirrels was being carried out. It may therefore have been a learnt method, practised by a single family group of badgers. Unlike *Meles*, *Taxidea* does make food caches (Snead and Hendrickson, 1942) and presumably, therefore, is not averse to carrion.

No studies of the Asiatic badgers, *Arctonyx*, *Melogale*, *Mydaus* and *Suillotaxus*, have been made. The ferret badgers (*Melogale*) are good climbers and Pocock (1939) says that they feed on insects, lizards, birds' eggs and fruit. Everts (1968) adds to the list birds and small mammals. Since *M. personata* has larger teeth than *M. moschata* he suggests that it may take a higher proportion of vertebrate prey.

Pohle's (1967) observations on tame *M. personata* support the view that this species is highly carnivorous. Small mammals were the most favoured prey and were expertly killed with an accurate neck bite. Fish were also seized by the head and caught without difficulty in a dish containing 10 cm depth of water. The animals were less adept at dealing with birds but they did kill and eat them; they also relished lizards, insects and earthworms. At first every type of fruit and vegetable offered was refused and only with difficulty were the animals finally induced to accept a food mixture including some vegetables.

(*iv*) *Lutrinae*

(*a*) *Fresh-water otters* The otters whose foods have been extensively studied are the fish otters, belonging to the genus *Lutra* and the sea otter, *Enhydra lutris*. A great deal of information relating to these two genera is given by Harris (1968).

Studies on *Lutra* have been largely concerned with providing the basis for assessing the effects of otter predation on game and food fishes. Erlinge's (1967b, 1969) investigations on *Lutra lutra* in southern Sweden are the most complete. His first study (1967b) was based on the analysis of 14,615 scats, covering the entire year for twelve localities, including lake, stream and marsh habitats. His later paper (1969) deals with 350 scats from an area which included a trout spawning ground. Stephens (1957) gives some less detailed data for the same species in Great Britain and a number of American workers have studied the closely related *Lutra canadensis* in a variety of habitats (Lagler and Ostenson, 1942; Greer, 1955; Ryder, 1955; Hamilton, 1961; Sheldon and Toll, 1964; Knudsen and Hale, 1968). Of these latter studies the most extensive is the work of Greer, based on 1,374 scats and covering the whole year for two localities.

The results of all these investigations give a concordant and coherent picture. The main food of both otters is fish, which Erlinge estimates never to constitute less than 60% of the diet. Crayfish and amphibia (the latter in America including *Necturus* as well as frogs) are important and show marked seasonal fluctuations, reflecting the way in which ease of capture is related to prey behaviour. Crayfish are preyed upon during their active period when the water is warm but are inaccessible in their winter refuges. Amphibia, on the other hand, are often easily caught in their winter retreats for they are not hidden away under stones but simply sit on the mud at the bottom of ponds and ditches; there may also be a peak of frog catching during the spring spawning period.

Small mammals and waterfowl figure regularly in the otters' diets but are minor components. Reptiles and molluscs are taken only occasionally and are of negligible importance. Aquatic insects are also eaten in varying amounts and appear to bulk larger in the diet of *L. canadensis* than of *L. lutra*.

American and European observations are also in agreement that coarse fish form the major part of the food and that very few trout are taken. In Erlinge's study (1969) burbot was the favourite fish and occurred in 39·5% scats while trout remains were found in only 5·3%. Lagler and Ostenson (1942) and Knudsen and Hale (1968) also found that trout constituted only a small proportion of the fish consumed. Sheldon and Toll (1964) found no evidence of trout predation: although three trout species were present in the reservoir they studied, the otters neglected these and concentrated on the species frequenting the shallows. One curious finding not reported by other workers was that ripe blueberries were eaten in late summer.

Lutra lutra, although essentially a fresh-water species, is also to be found living on the coasts of Great Britain. In this habitat crabs and other Crustacea provide an extra food source (Elmhirst, 1938). Rowbottom (1969) records a case of a coastal otter in Scotland caching the remains of a large fish, apparently as a protection against sea gulls.

One species of *Lutra*, *L. felina*, the sea cat, is entirely marine and occurs along the west coast of South America. Its feeding habits have not been described but one would expect that, like coastal *L. lutra*, it would take invertebrates as well as fish.

Procter (1963) gives some information on the food of *Lutra maculicollis* in Lake Victoria. The main food again is fish, chiefly *Haplochromis*. Other foods recorded are molluscs, crabs, nestling waterfowl and eggs and also mice. Procter once saw an otter eating the sedge, *Cyperus longus* but no other vegetable food is known to be eaten in the natural habitat. The favourite foods of a tame specimen were fish and raw meat but neither was acceptable unless perfectly fresh. Insects and frogs were taken but crabs were only partly consumed. Such delicacies as cake and butter were relished and the otter used to make her own way to the vegetable garden and help herself to carrots, beans, peas and potatoes (Mortimer, 1963).

Detailed studies on the other genera of fresh-water otters have not been made but *Aonyx*, although it does include fish in its diet, is generally believed to feed very largely on crabs and certainly the scats commonly found consist mainly of crab shells (Carter, 1956). *Aonyx* will travel considerable distances over land and thefts of poultry at some distance from water are recorded. A tame one caught fish, molluscs and frogs for himself and was very fond of small rodents (mole rats), but neither fish nor flesh was accepted unless it was quite fresh. His liking for such things as bread and butter, cakes and sweets suggest that some vegetable foods may occasionally be taken (Eyre, 1963).

Zeller (1960) says that fish was the favourite food of a giant Brazilian otter, *Pteronura*, in the Cologne Zoo but according to Cabrera and Yepes (1960) this species also eats small mammals and waterfowl.

Erlinge's (1968b) observations on captive European otters showed that they hunted mainly by sight. A pursuit might take two to three minutes, the otter surfacing briefly for breath about every thirty seconds and finally seizing the prey with a quick, snatching bite. Erlinge found that if the otters were hungry the selectivity shown was in relation to ease of capture: whatever was most easily caught was eaten first and the order of 'preference' shown for different fishes could be altered by mutilating the tail fins of the faster species to slow them down. If the otters had already fed then, although they might chase further prey, they often did not eat them. Here they showed distinct preferences, with trout least often rejected. The fact that in natural conditions otters eat so few trout is thus because they are harder to catch than the other species available, rather than because they are not relished. Erlinge also found that, if offered an egg, his animals would treat it as a plaything and even when it got broken would not eat the contents. This agrees with his failure to find any evidence of eggs being eaten in the wild.

In view of the difficulties of observing them it is not surprising that we know little about the hunting techniques used by otters in natural situations. Sheldon and Toll (1964) saw an adult and a juvenile *L. canadensis*, probably mother and young, apparently cooperating in driving fish into the shallow water of a small cove, where they were easily caught. Both dived repeatedly and usually surfaced with a fish. When hunting independently the adult otter swam in a leisurely manner at the surface, with the head pointed downwards so that the eyes were under water. Every now and then a sudden dive was made and the otter would presently surface again, sometimes with a fish in its jaws but the success rate was much lower than when the two were hunting in the shallows. The otters were also seen searching along the shoreline under tree stumps, presumably for crayfish.

According to Hershkovitz (1969) the giant otter, *Pteronura*, uses the cooperative driving technique very effectively – a whole group of them will unite in driving fish into a shallow backwater, where they are easy prey.

Everyone who has had experience of the less exclusively piscivorous otters has remarked on their manual dexterity. Pocock (1921b) writes of an *Amblonyx* 'manipulating and playing with a marble in a manner recalling that of a conjurer juggling with a cricket-ball'. Eyre's tame *Aonyx* was an expert opener of tins and bottles and Maxwell (1960) provides a series of excellent photographs to illustrate his account of the manual skill of Edal, his tame *Aonyx*. This remarkable facility must surely be related to catching crabs and crayfish by feeling for them with the paws, much as a raccoon does.

(*b*) *Enhydra lutris*, the sea otter Investigations on the food of the sea otter fall into two groups; those concerned with the northern population of the Aleutian and Commander Islands and those dealing with the population of the Californian coast at the southern end of the range. In both areas the sea otters have been found to live very largely on invertebrates with fish often forming only a minor item in the diet. However, since there are differences in the invertebrate faunas of the two areas, the otters obviously cannot find exactly the same food.

The most extensive investigation of the northern otters is that of

Barabash-Nikiforov (1935, 1962) in the Commander Islands, based on an analysis of 1,480 scats. He found that sea urchins were the most important item and their remains formed 59% of the total scat volume; molluscs were second at 23·3% and crabs third at 10%. Although the otters were seen to eat octopus, the beaks very rarely appeared in the scats and Barabash-Nikiforov was of the opinion that the heads were not eaten. Octopus may therefore be more important in the diet than appears from the scats. The value of 6·7% which he found for fish must also give an under-estimate of their importance, since the proportion of digestible material in a fish is higher than in a sea urchin. From the 480 scats that were fresh when collected, Barabash-Nikiforov was able to make some assessment of seasonal variation and found that, while sea urchins and molluscs were year-round staple foods, crabs and fish were much more frequently eaten in summer than in winter.

The animals in the Aleutian group of islands have been studied by Williams (1938), O.J.Murie (1940) and Jones (1951), who all analysed relatively small numbers of scats, and by Wilke (1957), who records the stomach contents of five animals. The results agree with those of Barabash-Nikiforov: sea urchins appear to be the major food, with molluscs second, Crustacea third and fish a minor item. Kenyon (1969), however, has made a much more extensive study and, by comparing the results of analyses of stomach contents and of scats, has shown that, as Barabash-Nikiforov suspected, the latter do give a serious under-estimate of the proportion of fish in the diet. For the animals at Amchitka Island, stomach contents showed that 50% of the food volume was made up of fish, while molluscs and echinoderms accounted for 37% and 11% respectively. Scat analyses from the same area, expressed as frequency of occurrence, gave echinoderms first at 95%, molluscs second at 46%, while fish were found in only 15% of scats. The comparison of percentages by volume and by frequency of occurrence is not altogether easy but the scat analyses certainly do suggest that echinoderms are the main food and fish a minor item, whereas the stomach contents show quite clearly that the latter is the most important type of food.

The results of studies on the Californian sea otters are more diverse. The principal food species involved are the abalone (*Haliotis rufescens*), which is not present in the northern habitats; the red and purple sea urchins (*Strongylocentrotus franciscanus* and *S. purpuratus*) and a mussel (*Mytilus californianus*). The earliest work is that of Fisher (1939), who found that abalone was the main food with red sea urchins and crabs also common in the diet. Limbaugh in 1961 also found the abalone dominant and the red sea urchin occupying second place. In 1964 and 1965, however, Hall and Schaller and

Boolootian (quoted by Ebert, 1968) reported a very different state of affairs: the abalone had sunk to fourth place and represented slightly less than 10% of the remains while the dominant foods were mussel and sea urchin. The sea urchins, however, were different species: *S. purpuratus* in Hall and Schaller's investigation and *S. franciscanus* in Boolootian's. The most recent study is that of Ebert (1968), made in the same area but using the technique of watching the otters feeding and recording what they bring up at each dive. He, like the earlier workers, found abalone once more the main food, crabs second and mussels and sea urchins only minor items.

In Ebert's estimation the otters themselves had, at the time of his investigation, all but obliterated the once-abundant sea urchin stocks and were making serious inroads on the abalones. Since then the position has again altered and the two sea urchins have recently increased in numbers so much as to constitute a veritable plague (North and Pearse, 1970). The earlier observations suggest that this is not an entirely new phenomenon but has resulted from a normal oscillation having gone beyond its usual limits. The sea otter population is clearly below the level required for effective control of the urchins but the reasons for this are not fully clear. Two factors may be involved. Sea urchins are amongst the animals capable of absorbing organic material directly from the water. The increasing pollution of the sea would therefore be expected to favour their increase, while commercial abalone fishing may have made life more difficult for the otters. In the past abalone and sea urchin have alternated as the otters' main food source: when one became scarce they relied upon the other. Commercial fishing may have reduced the abalone stocks so that they no longer sufficed to maintain the otter population, once the sea urchins had been eaten down. If this is the case the otters must have declined in numbers and with reduced predation the sea urchins would be expected to begin to increase again. Since the otters breed very slowly, the natural result might well be that the sea urchins would reach plague dimensions long before the otters could re-establish their numbers sufficiently to keep them in check. It is also obvious that the number of otters required for control may now be considerably higher than it was in the days when the sea was clean. Protection of abalones as well as of otters may therefore be the only way of re-establishing a stable system. It is worth noting here that in 1928 the licensed commercial abalone fishermen in the area numbered 11: by 1963 this had increased to 505.

Sea otters are not animals of the open sea but are essentially littoral. Two of their main foods, molluscs and sea urchins, are sedentary creatures and the species of fish they capture are rather sluggish, bottom-living species (Kenyon, 1969). The otter's food-finding problem is therefore not one of capture by swift pursuit but rather of how to deal with prey that is protected by a shell and may also be difficult to dislodge from its hold on the rocks. The sea otter solves both these problems by using a stone as a tool in two different ways. In breaking open sea urchins and mussels a flat stone is used as an anvil. The otter brings up a suitable stone along with its prey, turns upon its back and places the stone on its chest: then, clutching it in both paws it bangs the shell down upon the anvil until it breaks. Hall and Schaller (1964) describe this behaviour in detail. In getting an abalone off a rock the stone is used as a hammer and, according to Ebert (1968), the mollusc is pounded until the shell is smashed. Presumably, if taken unawares a single blow will serve to dislodge it and Fisher (1939) saw abalones brought up with only a single piece broken out of the side of the shell. Kenyon (1959) tells how a captive animal used a stone to break a metal bolt holding down a drain cover and, if there was nothing else upon which to use her talents, she would pound the concrete edge of her pool.

One rather surprising thing about the use of an anvil stone is that although both the Californian and the Aleutian otters use this technique (Murie, 1940), Barabash-Nikiforov (1962) never saw the Commander Island animals do so. The sea urchins, a species with short spines, were simply turned about in the paws, the spines were brushed off and they were then bitten open. It seems very improbable that there is a genuine behavioural difference between the otters of the islands at the western and eastern ends of the Bering sea; most likely Barabash-Nikiforov's animals were living on prey soft enough to be crushed in the jaws and stones were not used simply because they were not required. This explanation is supported by the observations of Hall and Schaller (1964), who found that the use of anvil stones varied greatly from place to place in relation to the hardness of the common prey species. They also quote Lensink as saying that at Amchitka Island young otters often use stones but adults do not, which he interpreted as being because the adults can crush most of the food items but the youngsters' jaws are not yet strong enough and they therefore require the assistance of the stone. Kenyon (1969) did not see anvil stones in use at Amchitka Island at all and judged that this was because at the time of his observations the mussels were not growing large enough to require the use of a stone – a view confirmed by the behaviour of Susie, a female captured in this area and kept at Woodland Park Zoo. Although she came from a non-tool-using area she used a stone as soon as she was given molluscs too large to be simply crushed in the jaws.

Since the type of prey taken does not require long pursuit it is not necessary for the otters to make very prolonged dives. Kenyon (1969) found that when foraging at depths of 10–25 m an adult female usually stays down for about a minute, an adult male for about a minute and a half. Although sight is undoubtedly used in prey capture, molluscs and echinoderms hidden in crevices must often be located by touch and some tests made with Susie at Woodland Park Zoo showed that she possessed remarkable powers of tactile discrimination. She preferred blue mussels to crabs and when given a bucket full of turbid water containing four mussels, two hundred small crabs and some pebbles of various sizes she made no attempt to search visually but, putting in both paws, felt about and within a few seconds had pulled out the four mussels. She had been blind in one eye when captured and was blind in both eyes for a full year before she ultimately died but she never had any difficulty in finding food or in selecting out the things she liked best.

If food is abundant the otters will gather several urchins or mussels on a single dive and carry them to the surface held in the fold of loose skin under the armpit. Kenyon (1969) found that they did not use right and left sides at random but almost always used the right paw to put the food under the left arm. Duplaix-Hall (personal communication) has recently found that captive *Aonyx* and *Amblonyx* also show some degree of handedness. When carrying objects in their characteristic manner, walking on three legs and holding the object pressed against the body with one fore paw, it is predominantly the left paw that is so used (figure 8.5, p. 343). These observations are of some interest since they are the only known case of any species other than man consistently showing definite handedness.

(*v*) *Mephitinae*, skunks

Skunks are omnivorous: in addition to small vertebrate prey, including nestling birds and eggs, they eat insects and various sorts of vegetable food and often feed on garbage round human habitations. Cuyler (1924) adds to this list the information that striped skunks, *Mephitis mephitis*, dig for fresh-water clams and hunt along streams for frogs and crayfish and will even 'slosh minnows out of the water with their paws' – a habit not mentioned by any later worker. Detailed studies, however, are not very numerous. Kelker (1937) was concerned only with insect foods and found that in summer and early autumn, insects are a food of choice for *Mephitis* and are commonly the major constituent of the diet. The most extensive study of the striped skunk is that of Hamilton (1936b) in New York State, based on stomach contents and extending over the whole year but analysed only into two periods – October to March and April to September. In the summer months insects were the major food with fruits second and small rodents, principally *Microtus*, third. In winter, vegetable foods came first, rodents second and carrion third. In both, birds made up less than 2% of the total bulk. Table 5.2 gives the three types of food found to be most important by other workers at different seasons and in different areas. The main foods are clearly arthropods, small mammals and vegetable foods. Reptiles and Amphibia are also recorded but are a minor item.

Table 5.2
Foods of striped skunks in different seasons

Season	Locality	1st place	2nd place	3rd place	Author
autumn	Iowa	arthropods	mammals	birds	Selko (1937)
autumn	Maryland	arthropods	mammals	plant	Llewellyn and Uhler (1952)
winter	Maryland	arthropods	mammals	plant	Llewellyn and Uhler (1952)
all year	California	arthropods	plant	mammals	Dixon (1925)
(? summer)	Texas	insects	birds	mammals	Wood (1954)

For the spotted skunk, *Spilogale putorius*, information is even more scanty. Gander (1965) used to put out food for a spotted skunk that had developed the habit of visiting his house each evening. It relished the dog biscuits and raisins that he supplied and he also saw it kill mice and eat fallen mulberries. The investigations of Crabb (1941, 1948), however, appear to be the only detailed study. His work was based on scats collected in a farming area in Iowa over a period of a year and analysed into four periods.

Winter (*December to February*). Mammals were the most important food and along with *Microtus* and farmyard rats, which the skunks killed for themselves, included a good deal of carrion from the corpses of cottontails shot by farmers. Plant food was second in importance and consisted mainly of maize.

Spring (*March to May*). Mammals were still the major food but arthropods, mainly insects and millipedes, had increased in number and were present in almost half the scats.

Summer (*June to August*). Arthropods now occupied first place and plant food had increased slightly: mammal prey showed a corresponding decline.

Autumn (*September to November*). Arthropods had declined slightly but were still the dominant item. Plant food, especially fruits, was important in early autumn but fell off later, while mammals increased slightly from the summer value but were still in second place.

This study suggests that the food of *Spilogale* is not qualitatively very different from that of *Mephitis*. The main vertebrate prey of both is small rodents and both take insect and plant food as available. There are, however, quantitative differences. For *Spilogale* mammals appear to be relatively more important and arthropods dug from the soil correspondingly less so than for *Mephitis*. This accords with the characteristics of the dentitions and feet in the two genera: *Spilogale* has more cutting carnassials and shorter claws than *Mephitis* (van Gelder, 1959).

No study appears to have been made of the South American skunks belonging to the genus *Conepatus*. The rather large, naked rhinarium, to which they owe their common name of hog-nosed skunks, together with the small carnassials and large molars and the long claws suggest that invertebrate prey rooted out of the ground may be the most important type of food and Cabrera and Yepes (1960) say that beetles, spiders and millipedes are the favourite foods.

Little seems to have been written about the hunting methods of skunks. Stebler (1938) says that a tame *Mephitis* killed *Peromyscus* by a bite on the back of the skull and he also describes its method of dealing with an egg. When first given an egg the skunk tried to bite it open but failed. He then walked round it, hitting it with his fore paw, with equally little result. Finally he straddled it, grasped it in his paws and threw it backwards between his hind legs. This was not at once successful but after a few attempts the egg finally broke and he licked up the contents. Thereafter he always threw an egg in this manner without hesitation. Van Gelder (1953) describes a rather similar series of events with a captive *Spilogale*. When first given an egg the animal spent nearly five minutes in vain attempts to bite it open, then suddenly switched to a different technique. The egg was thrust back with the paws and given a quick kick backwards with one hind foot to speed it on its way. At the fifth attempt, the egg broke and she ate it. This differs slightly from the throwing described by Stebler in *Mephitis*, in which no kick was involved and the animal's later behaviour was also a little different: when given eggs subsequently she did not at once use the throw-kick method but always first tried to bite them open.

E. Viverridae

The viverrids are not, as a group, highly specialised killers. Typically they take relatively small vertebrate prey along with both invertebrate and plant foods. A certain amount is known about the foods of many of the African species but even for them detailed information is scanty and there have been few quantitative studies. The Asiatic and Madagascar species are even more poorly known, and for many of them one can do no more than judge from the dentition whether vertebrate, invertebrate or plant foods are likely to constitute the major part of the diet.

(*i*) *Cryptoproctinae*

Cryptoprocta ferox, the fossa, is not only the largest of the viverrids but, to judge by its dentition, one of the most actively predacious. Unfortunately there is virtually no information about its feeding habits beyond the fact that it has been known to steal poultry. Grassé (1955) declares that it preys upon lemurs, which seems extremely likely. A captive female described by Albignac (1969a) ate raw zebu flesh, birds and rats but no further details about its food are given. Bennett's (1835) original description of *Cryptoprocta* quotes the owner of the animal as declaring that it was 'the most savage creature of its size I ever met with: its motions and power were those of a tiger: and it had the same appetite for blood and destruction of animal life'. On the other hand, a letter from Mr M.W.Bojer (1834) describes this very same animal as being 'quite domesticated and fond of playing with children' and says that it 'ran about the house and yard free and sprightly, eating everything'. The latter remark suggests that although a predator, *Cryptoprocta* is not necessarily an exclusive flesh-eater. In justice to Bennett it is only fair to add that Bojer does remark that although gentle at other times, the animal was ferocious when given flesh.

(*ii*) *Viverrinae*

(*a*) The genets: *Genetta* The genets, with their reduced molars and cutting carnassials, would be expected to be amongst the more predacious of the viverrids, with the forest species taking a higher proportion of roosting birds, nestlings and eggs and the savanna ones more dependent on small rodents. This appears to be substantially correct, although other foods such as lizards and insects are also taken. Poultry killing is very frequently reported. Verschuren (1958) found that in Garamba National Park the stomach contents of *G. tigrina* (mainly a savanna species) were principally small rodents and bird remains were rare. Ingles (1965) found that three stomachs of *G. rubiginosa*, the 'bush genet', included by Coetzee (1967) in

G. tigrina, had seeds, insects, one frog and some fish remains – the latter probably not its own kill but stolen from a drying rack. Azzaroli and Simonetta (1966) record rodents, grasshoppers and scorpions from the stomachs of four *G. genetta* and two bats of the genus *Tadarida* from a *G. tigrina* stomach. The bats need not necessarily have been caught in a daytime roost, since *Tadarida* not infrequently forage on the ground where, if taken by surprise, they would be very easy prey.

(*b*) *Poiana*, the oyan Walker (1964) quotes a letter from Dr H.K. Kuhn to the effect that in Liberia *Poiana* eats insects, young birds and plant material but there appears to be no other information.

(*c*) *Osbornictis*, the water genet The dentition of *Osbornictis* is clearly adapted to deal with slippery vertebrate prey such as fish and frogs. Verheyen's (1962) rather surprising statement that on the Lugula river, Crustacea are said to be the main food has already been mentioned.

(*d*) The civets: *Civettictis*, *Viverra*, *Viverricula* The civets are terrestrial and omnivorous, feeding on small vertebrates and insects and taking a considerable amount of vegetable food. The three main foods in the stomachs of two *Civettictis* from the Kruger National Park were carrion, guinea fowl and insects: minor items were scraps of kudu skin (probably scavenged from the kill of a larger predator), grass and leaves, a small portion of *Thryonomys* and some guinea-fowl egg shell (Bothma, 1965). The latter suggests that the guinea-fowl killed were probably taken on their nests while incubating. Astley-Maberley (1955) records small mammals, birds, reptiles, amphibia, eggs, carrion, insects and vegetable foods and says that poultry is sometimes taken but that insects and vegetable foods form the greater part of the diet. Dr J.M.Lock informs me that in Queen Elizabeth National Park in Uganda, civet scats often contain the remains of large millipedes and Leakey (1969) reports that, as judged from the scats, the main foods in East Africa are insects and centipedes, frogs, toads and lizards: he also mentions seeing one searching for crabs on the seashore. A captive pair of mine ate grass as well as fruit and were extremely partial to insects. Birds, eggs and small rodents were much relished: lizards, snakes and fish were accepted and live fish or frogs were caught at once and without any hesitation when first provided in their water dish. The large Indian civet, *Viverra zibetha*, is recorded as fishing (Walker, 1964) and the behaviour of these captive animals suggests that *Civettictis* also takes aquatic prey.

In Madagascar three stomachs of the introduced small Indian civet, *Viverricula indica*, were found to contain insects together with remains of a snake, a rodent and a bird (Rand, 1935).

(*iii*) *Paradoxurinae*, *Hemigalinae* and *Galidiinae*

Nandinia binotata, the African two-spotted palm civet, is probably the best-known member of the Paradoxurinae. *Nandinia*'s natural food consists of fruit, together with whatever vertebrates and insects are available. Bates (1905) reported that in the Cameroons, *Nandinia* was trapped using fruit as a bait and also that the animals came and scavenged round the villages and ate the flesh from a chimpanzee skeleton he was preparing. Walker (1964) reports the finding, in stomach contents, of rodents, both arboreal and terrestrial and he also says that *Nandinia* will kill pottos. A tame male belonging to a friend of mine killed a potto and when he encountered a half-grown mona monkey asleep he attacked it at once and would certainly have killed it had it not been rescued in time. Attacks on such large prey suggest that *Nandinia*, despite a liking for fruit, is also extremely predacious: if sleeping monkeys are attacked one would expect that any roosting bird encountered would suffer the same fate.

A liking for fruit appears to be characteristic also of the Asiatic paradoxurines. Martin (1929) had his tomatoes raided by the masked palm civet, *Paguma larvata*, and trapped the pair of animals responsible with a bait of the same fruit. *P. hermaphroditus* owes its common name of 'toddy cat' to its partiality for the juice tapped for palm wine. Jayakar and Spurway (1968) found that a captive one throve best on a diet including a considerable amount of fruit and was very fond of cakes and sweets. While it took eggs and small vertebrate prey, it refused most forms of butcher's meat. Since the establishment of cocoa plantations in Malaya the small-toothed palm civet, *Arctogalidia*, has developed a liking for the young pods and may develop into a pest. The binturong, *Arctictis*, rather slow moving and with flat, crushing teeth, is probably the most frugivorous of all but is also known to take carrion.

Very little is known of the feeding habits of the Hemigalinae and Galidiinae but they do not appear to be highly specialised killers and rely largely on insects and small vertebrate prey. Rand (1935) records three stomachs of *Fossa fossa* as containing insects and one of them had a lizard as well, while four stomachs of *Galidia elegans* contained insects, together with lizard and bird remains. That small mammals are also preyed upon to some extent is indicated by the observations of Eisenberg and Gould (1970). They found that captive *Fossa* and *Galidia* would kill and eat the insectivore *Hemicentetes semispinosus* and obtained evidence of *Galidia* killing this species in the wild. Neither, however, would kill the more densely spined *Setifer setosus*.

Chrotogale and *Eupleres* both have rather peculiar dentitions with small teeth and weak jaws, suggesting a greater reliance on invertebrate prey and reduced importance of vertebrate and plant foods. Walker (1964) mentions two *Chrotogale* stomachs having been found to contain earthworms. The otter civet, *Cynogale*, has already been mentioned as having a dual-purpose dentition, adapted to cope both with slippery fish and frogs and with shellfish such as Crustacea and possibly molluscs.

(*iv*) *Herpestinae*: mongooses

The mongooses conform to the typical viverrid feeding regime of small vertebrates, arthropods and plant food. The relative importance of the three categories varies from one species to another but accurate information is not available for more than about a quarter of them.

(*a*) *Herpestes sanguineus*, the slender or red mongoose This is one of the most predacious of the mongooses, the main food being small mammals and birds. Bates (1905) reported it as a poultry thief in the Cameroons and it has the same reputation throughout its range. Verschuren (1958) recorded eight stomachs containing rodents only and one with bird remains and Ansell (1964) found one with a rat and a small lizard. At the same time, insects are not entirely neglected; Ansell (1965) saw one catching grasshoppers along the edge of a bush fire. This animal subsequently proved to have the remains of a squirrel in its stomach.

(*b*) *Herpestes auropunctatus*, the small Indian mongoose This species has been most studied not in its original home but in the West Indies and Hawaiian islands, where it was introduced in the last century to act as a rodent controller. The various investigations that have been made in these areas have shown that, although preying extensively on small mammals when these are abundant, *H. auropunctatus* is also highly versatile and will take other small vertebrates, arthropods and also plant food, in varied proportions, according to local availability.

The earliest investigation is that of Williams (1918), carried out in Trinidad. From an examination of stomach contents he found that insects, especially grasshoppers, bulked very large in the diet and that spiders, scorpions and centipedes, as well as an occasional crab, were also eaten. Birds (including some domestic poultry) and rodents were eaten in about equal numbers: lizards, snakes, frogs and toads were also significant food items.

In Hawaii, Baldwin *et al.* (1952), in an analysis of eighty-six scats, found that almost 40% of the food volume was rodents, with insects and vegetable foods forming the two other main items. Bird remains accounted for only 4% of the total. Mice were the main rodent prey and prickly pear the major vegetable food. Pemberton's (1925) earlier study was concentrated on the sugar fields, where rodents are more abundant, and here they constituted a higher proportion of the total food; indeed, more than half the scats contained nothing but rodent remains. The fact that in neither study were reptiles significant as prey is simply a reflection of their scarcity on Hawaii.

La Rivers (1948) investigated a rather peculiar habitat: a small peninsula jutting into the estuary near Pearl Harbor on Oahu Island. Here the mongooses foraged along the shore line and even into shallow water, turning over stones in search of crabs. Higher up the beach they also dug out ghost crabs from their burrows. Lizards and frogs were other major prey items; birds and insects were of minor importance. The mongooses also took carrion and made a nuisance of themselves by continually stealing the rotten meat which La Rivers was attempting to use as a bait for insects.

As far as rodent control goes, *Herpestes auropunctatus* cannot be said to have been an unqualified success. It deals effectively with terrestrial species that breed in burrows, such as the brown rat, *Rattus norvegicus* but it is hardly to be expected that a diurnal terrestrial predator should effectively control a nocturnal species nesting for preference above ground level in trees or roofs, like the black rat, *R. rattus*. In Puerto Rico, the result of mongoose predation was that while the brown rat decreased, the black became more common and the mongooses then fed mainly on insects and lizards (Pimental, 1955). In Hawaii, Barnum (1930) found that on mongoose-free Kauai Island, brown rats outnumbered black but on Maui and Oahu, where mongooses were introduced, the opposite was the case, presumably as the result of differential predation. In the Virgin Islands the story is the same (Seaman and Randall, 1962). The principal rat species being the black rat, control was not achieved and the mongooses, although living mainly on insects, have made themselves unpopular by making up for the lack of accessible rodent prey by raiding poultry.

Helogale parvula, *Herpestes ichneumon* and *Cynictis penicillata* are mixed feeders, taking a fair proportion of vertebrate prey. Zumpt (1968a, b), from an examination of fifty-eight stomachs of the latter species containing identifiable remains, found that the principal foods were rodents or insects, depending on relative seasonal abundance. Bird and reptile remains were recorded only once but bird-hunting was observed in the field and the animals were also seen scavenging on carrion from carcases and eating a succulent plant.

Mungos mungo, the banded mongoose, is largely insectivorous, but lizards, snakes, toads and small mammals, as well as both eggs and nestling young of ground-living birds, are also taken whenever the chance offers. Wild fruit is eaten now and then but probably mainly as a source of fluid, since tame animals evince no great enthusiasm for this form of food (Simpson, 1966). The groups which Neal (1970a, b) studied in Queen Elizabeth National Park in Uganda proved to be living almost entirely on arthropods. Abundant elephant and buffalo droppings provided the mongooses with an easily exploited food source and during the period of study (January to March) the major part of their diet consisted of dung beetles and millipedes which they dug out of the droppings. No remains of vertebrate prey were to be found in the scats, nor was any significant amount of plant food consumed.

Crossarchus obscurus, the kusimanse, is primarily a forest floor forager and its main prey is soil invertebrates, together with occasional small vertebrate prey. Hayman (in Sanderson, 1940) found only crabs and insects in the stomachs he examined; Booth (1960) records insects, earthworms, mice and lizards in the stomachs of shot specimens and I have seen lizard eggs dug up and eaten. Although the kusimanse is a good climber, Booth doubts if it ever hunts arboreally: I have, however, seen a tame one chase a lizard up a tree and also search about among the branches so extensively as to suggest that birds' eggs and nestlings may be taken now and then. Two specimens of the rather similar *Dologale dybowski* examined by Verschuren (1958) proved to have nothing but insects in their stomachs.

Ichneumia albicauda, the white-tailed mongoose, is nocturnal and largely insectivorous. It frequently scavenges around human habitations, where insects attracted by lights are one of the food sources exploited, and it also searches the roads for the remains of traffic casualties.

The broad flat molars of *Rhyncogale melleri* suggest some non-predacious dietary specialisation but there is little information about its feeding habits. Thomas (1894) quotes a collector, Mr Whyte, as saying that 'wild fruits are always found inside the stomach of this mungoose' and Ansell (1960c) records it as eating termites. The teeth would be well suited to deal with both these types of food. *Bdeogale* also has crushing molars and Ansell (1965) captured this species in a trap baited with pumpkin seeds; he records termites and centipedes in the stomach of another specimen.

Suricata suricatta, the slender-tailed meerkat, although it will readily kill small vertebrates and also takes plant food, is primarily insectivorous. Zumpt (1968b) found thirteen out of eighteen stomachs with arthropod remains and the only vertebrate prey was one frog and some scavenged meat. Plant material and seeds were found in three of the stomachs. In captivity *Suricata* becomes extremely omnivorous and takes a wide variety of fruits and vegetables (Ewer, 1963b).

Atilax paludinosus, the marsh mongoose, lives wherever there is reasonable cover close to streams and marshy ground, and frogs and Crustacea constitute major foods. It can, however, also make a living in relatively dry areas and in Ghana, Booth (1959) records its presence in the Shai Hills, where there is little water and Crustacea are not available.

Amongst the Asiatic mongooses, *Herpestes urva*, the crab-eating mongoose, also feeds largely on similar aquatic prey. Ghose (1965) has described what he considers to be a new species, *Herpestes palustris*, from swamplands to the east of Calcutta. It feeds mainly on fish and the aquatic snail, *Pila globosa* and, like *H. urva*, can emit an obnoxious odour.

Nothing is known of the hunting methods of the viverrids other than the Herpestinae but Leyhausen (1956, 1965b) has described the killing techniques of *Genetta*, *Viverricula*, *Viverra* and *Nandinia*. None shows an accurately oriented killing bite on the neck or occiput of the prey and all usually kill with repeated bites. This may take the form of what Leyhausen calls '*Nachbeissen*', which I refer to as iterant biting. A series of bites is made very swiftly, without ever withdrawing the teeth fully, so that although there may be no more than four punctures in the skin, where the canines originally entered, the area beneath is severely mauled. Iterant snapping, exactly similar except that the teeth are fully withdrawn between successive closures of the jaws, may also be used to shift the grip forwards towards the head from an initial not very favourable location. Wemmer (personal communication), working with *Genetta tigrina*, has found the method of killing to be very similar to what Leyhausen described in *G. felina* (? = *G. genetta*). Wemmer's animals were young and he found that when given a very large or particularly belligerent rat they made extensive use of the hind feet in subduing it. The genet first seizes the prey with fore paws and mouth and pulls it backwards under its own body: it then rolls over on to one side, at the same time clutching the rat with one hind foot on either side of the body (plate 8). The genet has sufficient lateral mobility at the ankle joint to permit it to hold the prey effectively in this way, literally grasping it between the feet, with a dexterity which one normally associates only with the fore paws. Wemmer has found that *Paguma larvata* also uses its feet in this manner.

Iterant biting is clearly a means of increasing the effectiveness of a bite not so oriented that a single jaw closure is likely to be quickly lethal. It is not often seen in herpestines but *Suricata* will use this technique if trying to bite something so large or resistant that the jaws cannot be closed in a single bite. The action is so fast that a friend subjected to this form of attack by one of my tame animals described it as 'like being bitten by a sewing machine'.

In the civets the absence of an accurately oriented neck bite is correlated with a very strong tendency to shake the prey. In *Civettictis* the shaking may be violent enough to smash the vertebral column of an adult rat. The civets will also often bite and then, either instantly or after a brief shake, throw the prey violently to a considerable distance and attack it anew. If they are somewhat nervous the civets will approach cautiously, deliver a quick bite and leap back again and only gradually work up to a genuine attack. In my captive African civets I find that according to the intimidatory characteristics of the prey, there is a complete series from the highly nervous 'run-away bite', through bite-and-throw; bite-shake-a-little-throw; bite-and-shake-vigorously (without throwing); to the bite without any shaking characteristic of a fully confident attack on prey from which no retaliation is anticipated. From watching their modes of attacking different types of prey I have come to the conclusion that their extremely violent shaking is a technique primarily adapted to dealing with snakes. The snake is powerless to strike back while being shaken and once its vertebral column has been broken in several places, it is virtually helpless. It is noticeable that in dealing with a snake, once the stage of bite-and-throw is reached, the civet will leap in the air with all four feet clear of the ground as it releases its hold; a procedure which I assume is likely to take it clear of any quick return stroke the snake might make.

Shaking is performed only in a vestigial form by genets and neither Leyhausen nor I have ever seen *Nandinia* shake its prey. Shaking and throwing of prey are obviously techniques unsuitable for an arboreal predator: to throw the prey away will probably result in losing it – almost certainly if it has not been seriously injured in the first attack – and violent shaking is not very easy if your own foothold is not entirely secure. The absence of shaking in *Nandinia*'s prey-catching repertoire and its vestigial appearance in genets are therefore explicable not as reflecting a primitive condition but as secondary adaptations to an arboreal habit.

The hunting techniques of the largely insectivorous species of mongoose are relatively simple. *Mungos, Helogale, Crossarchus* and

Suricata all forage in groups, turning over small stones and bits of wood, scraping or rooting at every likely-smelling crevice and maintaining continual vocal contact with each other. Arthropods located in their refuges are dug or scraped out but any prey, whether it be a grasshopper or a cricket, a lizard or a mouse, that is incidentally flushed is leapt upon with surprising speed and agility. Although the principal prey is arthropod, both *Crossarchus* and *Suricata* will kill a small mammal with a single bite, accurately oriented at the back of the skull, and Zannier (1965) says that *Helogale* also kills with a neck bite. When dealing with lizards or snakes, which have not got a clearly-defined neck, there is a very definite tendency to bite at whatever part moves most rapidly. Small prey is usually shaken as it is picked up, an action whose function appears to be to prevent insects, centipedes and the like from reaching back and clutching at the face with some of their numerous appendages. I have never seen either *Suricata* or *Crossarchus* shake a mouse: the killing bite at the back of the head is lethal and shaking would only interfere with its correct orientation. Curiously enough, a scorpion is treated with very little respect: it is picked up more or less anyhow and no special measures are taken to disable the sting. Fischer (1921) describes *Herpestes edwardsi* treating a scorpion in exactly the same casual manner.

I have only once seen a *Herpestes sanguineus* foraging. Its behaviour, as befits a true predator, was very different from that of the insectivorous species. It moved swiftly and silently, with a cat-like elegance, taking every advantage of cover, sniffing and searching visually but without pausing to scrape and scratch at crevices.

There are several sorts of food, such as pill millipedes, eggs, snails and the cells of mason wasps, which are protected by a hard outer covering. For an animal with short jaws, these are not easily bitten open unless they are very small, and it is therefore not surprising to find that a number of mongooses have special techniques for dealing with them. The commonest is the same as the method used by the striped skunk: the object is held in the fore paws and thrown backwards between the hind legs and the process repeated until it breaks. Although the action itself is not a learnt one most mongooses will learn to take up their position in front of some suitable hard vertical object. This way of breaking an egg was first described by a naval surgeon, Dr T.R.H. Thomson, in a tame animal said to be *Herpestes fasciatus*, which presumably was the banded mongoose, *Mungos mungo* (see Gray, 1848). In 1896 Lyddeker reported the same trick in *Helogale* and Zannier (1965) has confirmed the observation. Brownlow (1940), Haines (personal communication) and I have seen the same technique used by *Herpestes urva*, *H. ichneumon* and *H. pulverulentus* respectively. Hinton and Dunn (1967) quote Stern-

Figure 5.4. Banded mongoose, *Mungos mungo*, smashing a pill millipede by throwing it against a rock. [From Eisner and Davis, 1967]

dale (1884) as finding that *Herpestes edwardsi* opens eggs by holding them between the paws and biting a hole in the little end. Other species will open small eggs in this way, and *H. edwardsi* is a relatively big mongoose: it therefore seems possible that, given a bigger egg, *H. edwardsi*, like its congeners, might also break it by throwing it backwards between the hind legs.

Throwing of all sorts of inedible substitute objects, as well as of eggs and millipedes, has repeatedly been seen in *Mungos mungo* and Eisner and Davis (1967) filmed the process and give drawings from their film (figure 5.4). Most of the descriptions relate to captive animals but Eisner (1968) points out that in 1946 Wager described having seen wild mongooses in Natal smashing a pill millipede open by throwing it against a tree. *Suricata* shows the same throwing pattern but in a very incomplete form. My tame animals were unable to open an egg and, after pulling it about with their fore paws for some time, would lose interest and abandon it. Although their manipulation of the egg included pulling it backwards under their bodies, they never actually threw it between the hind legs but I have seen this done by a pair of animals in the Brookfield Zoo. They finally succeeded in breaking the egg but their actions lacked the decision and deftness characteristic of *Mungos mungo*. Possibly *Suricata* might throw a smaller object than a hen's egg with greater skill but I am inclined to the view that in this species the throwing pattern is vestigial and is probably of no real significance in normal life.

The problem of dealing with eggs too big for throwing is one that cannot arise very often but *Mungos mungo* nevertheless appears to be capable of solving it. When van Lawick-Goodall and van Lawick (1966) discovered that Egyptian vultures could smash open an ostrich egg by hurling stones at it, it occurred to them to find out whether any other egg-eating animals could do the same. Two tame *Mungos mungo*, when presented with an ostrich egg, began by trying to throw it backwards between their legs in the usual way but when they found this impossible, both resorted to throwing stones at the egg, using the normal technique (van Lawick-Goodall and van Lawick, 1970).

The marsh mongoose, *Atilax*, uses a slightly different method: grasping the egg, or whatever it may be, in both paws and rearing up vertically to its full height, it hurls the object down on the ground with great violence. Steinbacher's (1939, 1951) marsh mongoose could do this with sufficient force to smash a nut. Hinton and Dunn are mistaken in saying that Cansdale's (1960) marsh mongoose ever threw things backwards in the *Mungos* manner.

Dr Thomson, as well as being the discoverer of *Mungos mungo*'s egg-breaking technique, is apparently the only person to have described how one of the Madagascar 'mongooses' deals with the problem. His animal was a *Mungoictis lineata* that lived as a ship's pet. It had a passion for eggs and if given one 'would roll it towards a projecting timber or gunslide; then, lying down on its side, the little creature would grasp the egg with all its feet and throw it by a sudden jerk' (Gray, 1848).

The evolution of these three different throwing patterns within the Viverridae is not altogether surprising. Such special techniques must have evolved by elaboration of the animal's normal ways of using teeth and paws on a recalcitrant object. The natural use of the paws in the scrapers-out-of-insects is a backwardly directed scratch, tending to drive the object posteriorly beneath the body. That this should lead to the evolution of backwards throwing is only natural. *Atilax*, more dexterous, with longer fingers and less prone to scraping, may have begun by holding its egg in its paws and attempting to bite into it. To rear up in frustration and finally to drop it may have been stages on the way to actual throwing of the recalcitrant object downwards. *Mungoictis*'s technique of lying on its side to throw would be easily derived from a weasel-like method of attacking prey. Clutching with all four limbs, combined with throwing the body on the side, is a technique some of the felids as well as the genets also use against relatively large prey and it is taking no great risk to predict that, when the prey-catching behaviour of the Galidiinae is investigated, this form of attack will prove to be included in their repertoire.

For arboreal species, throwing is not a very suitable technique and the linsang, *Prionodon linsang*, has a way of breaking an egg in its paws. Walker (1964) quotes a description of the method which I cannot follow, beyond the fact that the egg was not thrown but broken by the paws.

For species with slightly longer jaws, biting an egg open is less of a problem and the toddy cat does this very deftly, holding it between its two paws the while (Jayakar and Spurway, 1968); genets too will bite an egg open and lick it out very cleanly (Dücker, 1965a; Rowe-Rowe, 1971).

The viverrids do not, of course, have a monopoly of neat egg handling. The red fox will hold an egg between its two paws and, after an initial puncturing with the canines, nibble away the top and lick out the contents without spilling a drop. The Peruvian desert fox will do the same (Birdseye, 1956), and Cott (1953) quotes Harrison Matthews' description of a tame coati using a very similar method with equal skill. One may, however, doubt the truth of the story that the aardwolf will smash ostrich eggs by banging one against another.

A further point of interest emerges from a study of mongoose prey-killing techniques. The speed and accuracy of the killing bite used on vertebrate prey is quite astounding, even in the least predacious species. The orientation owes nothing to practice or learning. I have seen a young, hand-reared *Suricata* make her first mouse kill with a single, perfectly placed bite; similarly, when a shrew unexpectedly ran out of a burrow in front of a portly, middle-aged kusimanse, kept as a pet since early youth and with very little opportunity for mammal killing, the shrew was dead faster than my eye could follow the mongoose's single-bite strike. Such expertise is the trade mark of a true predator and one can hardly imagine its being evolved by animals that made their livings as *Suricata* and *Crossarchus* do today, mainly as scratchers up of unconsidered trifles. Its presence in these two strongly suggests that the mongooses must have been originally more predacious and that it is the solitary hunters, like *Herpestes sanguineus*, that most closely approach the ancestral type. Diversification of feeding habits to exploit the variety of food sources available must have occurred later and with it all the behavioural complexity involved in the lives of the social species. Adaptations for insect eating led to changes in the dentition away from the strictly carnivorous pattern but the killing bite was not lost. It still pays off sufficiently often to warrant its retention in the behavioural repertoire.

This interpretation of the behavioural evidence leads to the same conclusions as were reached by Gregory and Hellman (1939) on the basis of comparative anatomical studies. Moreover, only a history of this type could account adequately for the evolution of so many species within so closely knit a group as the Herpestinae.

F. Hyaenidae

(*i*) *Hyaeninae*: the hyaenas

Heavy hammer-like teeth and extremely strong jaws and jaw muscles make it possible for hyaenas to crack larger bones than other carnivores can cope with and their highly efficient cutting carnassials can deal with tough hides and tendons. Hyaenas can therefore make a meal from the remains of the kills of even the largest fully predacious species. These obvious scavenging adaptations have sometimes tended to obscure in the popular imagination the fact that, although they often find it convenient to scavenge, hyaenas are themselves effective primary predators, capable of making their own kills. Moreover *Crocuta*, at least, hunts by night but often scavenges during the daytime. Daytime observations therefore tend to give an exaggerated impression of the importance of scavenging. In fact, in a study of the spotted hyaenas in Ngorongoro and Serengeti, Kruuk (1966) found that in more than a thousand observations of the animals feeding, the kill was their own in 82% of cases; in only 11% had the prey definitely been killed by other species and the remaining 7% were doubtful. Indeed, in Ngorongoro the traditional relationship between lions and hyaenas was reversed and the lions got most of their food from kills made by the crocutas. Different hunting groups showed evidence of individual prey preferences but, nevertheless, as with most predators, availability played a large part in determining the type of prey taken. The percentages of scats which contained hair of the three major prey species in the two habitats were as follows:

	Ngorongoro	*Serengeti*
Wildebeest	83	54
Zebra	46	30
Thomson's gazelle	16	53

Pienaar (1969) records that in the Kruger National Park the crocutas not only kill most of the large ungulate species but will also attack young or aged individuals of the major carnivores, including lion and in Serengeti Schaller and Lowther (1969) record the killing of lion and jackal. Attacks on man are reported now and then; these usually relate to people sleeping in the open or to small children left unattended (Balestra, 1962). Small prey and vegetable foods are also taken and in the stomachs of five animals Balestra found a variety of small mammal and bird remains as well as crabs and snails, cassava and maize. Deane (1962) found a considerable amount of grass in two stomachs he examined. Van Lawick-Goodall and van Lawick (1970) report that *Crocuta* cubs frequently eat the fresh droppings of large ungulates. This may well be an important source of vitamins for them but it is much more difficult to understand the significance of the observation that adult crocutas appear to regard the fresh droppings of *Lycaon* as a great delicacy.

Crocutas will scavenge on domestic garbage when the opportunity offers and in Ethiopia it is not uncommon for them to search for food round human habitations at night. Kruuk (1968) describes one case in which this activity has developed to a point where the crocutas are the official street cleaners and garbage disposal squad. In the town of Harar they are encouraged, given a small but regular food ration and permitted to roam the streets at night unmolested. In return, they remove all edible rubbish and play an acknowledged role in the city's sanitary system.

In the coastal regions of South West Africa the brown hyaena, *Hyaena brunnea*, is well known as a seashore scavenger, a habit which has earned it the name of strandwolf. It has the general reputation of being less of a killer than *Crocuta* but according to Pienaar (1969) the brown hyaenas in the Kruger Park are more aggressive and predacious than the crocutas; they attack a wider range of prey species and kill larger individuals. Kudu is the favourite prey and even adult males can be overpowered. Small prey, including ground birds, eggs, tortoises, fish and insects, is also taken and wild fruits are eaten. Pocock (1939) reports that, although striped hyaenas, *Hyaena hyaena*, scavenge extensively, they also kill a variety of domestic animals - goats, sheep, calves and even small dogs.

(*ii*) *Protelinae*

The aardwolf, *Proteles cristatus*, has a very reduced dentition and feeds for choice on harvester termites. Carrion and occasionally small vertebrates, as well as vegetable foods, are also taken. Of five stomachs whose contents are listed by Bothma (1965), four contained only harvester termites and the fifth had mainly carrion along with one rodent, one tortoise, a few insects and some fresh grass. Ansell (1964), Azzaroli and Simonetta (1966) and Wells (1968) record six more stomachs containing termites only.

The spotted hyaena is the only species whose hunting behaviour has been studied. Kruuk (1966) found that the crocutas might hunt alone, in small groups or in packs of up to thirty. Solitary hunting is usually restricted to small prey and Kruuk recorded a success rate of four out of twenty-one attempts. Pack hunting is almost entirely a nocturnal occupation and, as in *Lycaon*, when a herd is attacked there is no organised pursuit of a single individual - each crocuta chases after its own prey. Bites are made at flanks and legs and as soon as one animal begins to flag it is soon surrounded by the rest of the pack and killed by bites mainly directed at loins and hindquarters. Kruuk says that small prey is often shaken but gives no further details of the attack.

A few crocutas will sometimes associate themselves with a *Lycaon* pack and wait about nearby until the dogs rouse from rest and set out to hunt. The crocutas are, of course, slower, but if the chase is relatively short they catch up in time to take advantage of any remains. They do not necessarily wait for the *Lycaon* to finish eating but dodge about on the outskirts of the feeding group, watching for an opportunity to dart in and snatch away the remains of the prey, very much like a scrum half in a game of rugby trying to predict where the ball is likely to appear.

Proteles, although so dependent on termites, does not appear to be able to dig effectively (von Ketelhodt, 1965, 1966) and cannot break into termite mounds. This accounts for the fact that it is the harvester termite, *Trinervitermes*, which normally constitutes the food, since this species can be collected on the surface and does not have to be dug out.

G. Felidae

There are no largely vegetarian or even omnivorous felids: alone amongst the families of the Carnivora the Felidae are all essentially predators. They are the killers *par excellence* and it is not surprising that the greater cats should always have excited admiration as well as awe, nor that the 'King of Beasts' should be a felid. Indeed, even the domestic puss has inspired more poets than have ever seen fit to sing the praises of the dog. Apart, however, from its intrinsic interest, an understanding of the feeding habits of the felids, particularly the larger species, is essential for the proper management of game parks as well as for an intelligent policy towards carnivores in farming areas and there is by now a considerable body of knowledge about the majority of the large species.

(*i*) *Panthera*

(*a*) *Panthera leo*, the lion Detailed studies of the prey of lions made in a number of different localities have shown that the preferred prey is medium-sized ungulates, somewhat above the lion's own body weight. Considerably larger species, however – giraffe, buffalo and eland – are also taken, and Pienaar (1969) records attacks on young elephant. Although not important as a source of food, other carnivores are sometimes attacked: Pienaar records killing of crocuta, leopard, cheetah, jackal, civet-cat, ratel and caracal; Schaller and Lowther (1969) of crocuta, leopard, cheetah and jackal and Edmond-Blanc (1957) describes an unsuccessful attack on a leopard. Cannibalism also occurs, and the corpse of any conspecific killed in a fight is often treated as food: Schaller (1969c) saw a lioness eating one of her own cubs, killed by a marauding male belonging to another pride. The lion has no aversion to carrion and will stay with a kill until it is finished, even if by then it is far from fresh. Swanepol (1962), for instance, describes a pride remaining to feed on a giraffe kill for five nights, by which time it was stinking and fly-blown. The lions' habit of appropriating the kills of crocutas has already been mentioned and they will sometimes dispossess a leopard. Lions will also now and then take small prey such as rodents and tortoises and, as rivers dry up, will hook out fish trapped in shallow pools: they will also eat termites when a flight makes them easily available in large numbers and grass and various fruits are eaten now and then (Guggisberg, 1960). Nevertheless, despite the existence of all these minor food sources, it is the medium-sized ungulates that constitute the lion's staple prey: the major prey species in various localities where detailed studies have been made are shown in table 5.3.

Table 5.3
Percentages of total lion kills recorded constituted by the more important prey species in different localities

East Africa					
Nairobi National Park 1954–66 (Foster and Kearney, 1967)	%	Serengeti, Nairobi and Tsavo National Parks (Wright, 1960)	%	Serengeti (Kruuk and Turner, 1967)	%
Wildebeest	48	Wildebeest	49	Wildebeest	49
Zebra	21	Zebra	15	Zebra	26
Kongoni	11	Thomson's gazelle	10	Buffalo	8
Warthog	10	Buffalo	5	Grant's gazelle	5
				Thomson's gazelle	5
				Ostrich	5
South Africa					
Kruger National Park (Pienaar, 1969)	1933–46 %	1954–66 %	Timbavati Private Reserve (Transvaal lowveld) (Hirst, 1969)	%	
Wildebeest	33	24	Wildbeest	53	
Impala	16	20	Giraffe	19	
Zebra	18	16	Zebra	8	
Waterbuck	16	10	Kudu	7	
Kudu	9	11	Waterbuck	6	
			Impala	6	
Kafue and Manyara					
Kafue National Park (Mitchell *et al.*, 1965)	%	Manyara National Park (Makacha and Schaller, 1969)	%		
Buffalo	30	Buffalo	62		
Hartebeest	16	Zebra	18		
Warthog	10	Impala	12		
Zebra	7	Baboon	6		

Where wildebeest are abundant, these constitute the principal prey: failing this, if buffalo are numerous they take first place. Zebra make a significant contribution to the diet in all six areas. The percentage of the total represented by the commonest prey is highly variable, depending on what other food sources are available. Thus in Kafue, where 30% of the kills were buffalo, nineteen prey species were recorded but in Manyara, where the proportion of buffalo was twice as much, only five species were represented in the kills recorded. Similarly in the Kruger Park, where wildebeest made up less than a third of the prey, a total of thirty-five species were preyed upon but in Timbavati, where they accounted for more than half of the prey, only six prey species were recorded. The unusually high predation on giraffe in the latter reserve was largely made up of animals killed towards the end of the dry season, when they were in poor condition.

In the Kruger Park, Pienaar calculated preference indices, as already described and found that waterbuck, kudu and wildebeest were all taken in proportions well above their relative abundances. Impala, despite the fact that they occupied second place, had a low preference rating and the proportion killed was well below the relative abundance. Lion therefore do not effectively control impala and, in the southern part of the park where *Lycaon* were originally shot out, they are now being encouraged, in the hope that they may exert a check on the over-abundant impala.

(*b*) *Panthera tigris*, the tiger The only detailed study of tiger predation in India that has been made is that of Schaller (1967), who also summarises all the scattered observations of earlier writers. The picture that emerges is broadly comparable with what has been recorded for the lion. The main prey is ungulate: various species of deer, antelope, the gaur (Indian bison) and pig. Other carnivores, including bears, wolves, lynx and badger are sometimes killed and, as with the lion, cannibalism also occurs. Small prey taken now and then includes lizards, snakes, turtles, frogs, fish, crabs, locusts and termites. Carrion is accepted and eating of grass and various fruits is also recorded.

Schaller's own observations, carried out in a central region of the Kanha National Park, constitute the only quantitative data. Prey remains were identified in 335 scats and in 100 kills made within the study area. The results are shown in table 5.4 as percentage occurrence in scats and as percentage of total kills.

For comparison, Schaller quotes data from a Russian study of tiger foods in two areas from the far east of the USSR. The numbers of kills recorded were rather small but suffice to show that in one area wild pig was the principal prey, with wapiti and musk deer in second and third places: in the other, wapiti outnumbered pig but the two together accounted for thirty-two of the forty kills recorded. Deer of various species made up 49% and 65% of the prey in the two areas and in both, brown bear kills were recorded more than once.

Table 5.4
Tiger prey species in Kanha National Park

	Scats (% occurrence)		% of total kills	
Chital (*Axis axis*)	52·2	71·2%	38	82%
Sambar (*Cervus unicolor*)	10·4		18	
Barasingha (*C. duvanceli*)	8·6		26	
Gaur (*Bos gaurus*)	8·3		11	
Domestic cow and buffalo	7·6		0	
Langur	6·2		1	
Porcupine	2·6		4	
Grass	2·3		—	
Soil[1]	3·8		—	

[1] Droppings consisting entirely of fine black micaceous soil were found seasonally in November and December.

In Schaller's study area wildlife was sufficiently abundant to provide the tigers with all the food they required and little domestic stock was available. In the more peripheral regions of the park the opposite was the case and there the tigers subsisted mainly on domestic cattle and buffalo, with wild game taking a secondary position.

In the past, man-eating tigers appear to have been relatively common in certain areas but today tigers are not sufficiently numerous to cause serious loss of life. According to Schaller, the swamp forest of Sunderbans in West Bengal is the only area where attacks on man are still significant, and he quotes a Government Report as recording twenty people killed by tigers in this region in 1958–9. Sankhala (personal communication) does not regard this as implying that the Sunderbans tigers are particularly dangerous but as a product of the common dependence of man and tiger on fishing for a livelihood. In an area of rather dense cover, unexpected confrontations at very close range are bound to occur and that the tiger should sometimes attack is hardly surprising.

According to Orians and Pfeiffer (1970), the tigers in Vietnam 'have learned to associate the sounds of gunfire with the presence of dead and wounded human beings in the vicinity. As a result tigers rapidly move towards gunfire and apparently consume large numbers of battle casualties.'

(*c*) *Panthera pardus*, the leopard The leopard hunts for choice in good cover and only occasionally makes a kill in the open: its favoured prey must therefore be the species that frequent the former type of habitat rather than the plains game. Within this limitation, however, the leopard is extremely adaptable and its diet probably the most varied of any of the large fully predacious species. In the Kruger Park thirty-one prey species, excluding small animals such as rodents, hyraxes and reptiles, are recorded (Pienaar, 1969); in Kafue twenty-two (Mitchell *et al.*, 1965); and in Serengeti twenty (Kruuk and Turner, 1967). Fey (1964) reports seeing a leopard in the Aberdare mountains in Kenya strolling along, turning over buffalo droppings and picking up the dung beetles with its claws, for all the world like a cat catching flies on a window pane. He also relates how during the Kariba rescue operation, a leopard remained fit and well for a fortnight on a small island with no visible means of support. It proved to be living on fish, mainly *Tilapia*, which it scooped out of the water with its paw during the hot part of the day, when the fish 'bask' near the surface.

Although it can and does kill game up to the size of an adult wildebeest and can also make a living on small game, the staple prey of the leopard is medium-sized ungulates; Kruuk and Turner found that 77% of the Serengeti leopards' kills fell into this category. The percentages of the total kills constituted by the major prey species in various areas are given in table 5.5. Wright (1960) had data for only

Table 5.5
Principal prey species of leopard in different areas

Serengeti (Kruuk and Turner, 1967)		Kafue National Park (Mitchell *et al.*, 1965)		Kruger National Park (Pienaar, 1969) 1933–46		1954–66	
Thomson's gazelle	27	Reedbuck	21	Impala	64	Impala	77
Impala	16	Puku	16	Bushbuck	13	Waterbuck	4
Reedbuck	11	Duiker	11	Reedbuck	5	Bushbuck	4
Zebra (juvenile)	4	Hartebeest	9	Duiker	4	Kudu	3
		Impala	8	Warthog	4	Reedbuck	2
		Bushbuck	4	Waterbuck	4	Warthog	1
		Grysbuck	4				

fifteen kills in East African National Parks but these are in agreement with the findings of Kruuk and Turner: seven were Thomson's gazelle, two impala and two wildebeest, with four other species represented once only. Hirst (1969) found that in Timbavati, as in Kruger Park, impala was the main prey and Lamprey (1963) found the same in the Tarangire Game Reserve in Tanzania.

In Schaller's study area in the Kanha National Park, leopards were not common but, as far as could be judged from the small sample available, the main prey seemed to be chital and langurs: twenty-two scats analysed showed that porcupine, sambar and barasingha were also taken. In Ceylon, Eisenberg and Lockhart (1972) found that twelve out of twenty-four kills were chital, with wild pig and langur second and third in importance. Verschuren (1958) records one leopard stomach from Garamba National Park containing only grasses and one with *Thryonomys* alone.

It is common knowledge that leopards are very prone to attack domestic dogs and appear to have a definite liking for canine flesh. In this connection some observations made by Kruuk and Turner (1967) are of interest. In Ngorongoro, they found that one leopard seemed to have a special liking for jackal meat and in a single month its kills were eleven jackals and two Grant's gazelle. Since jackals were not particularly numerous it must have been preying on them very selectively. This suggests that the taste for canine flesh may be individually acquired and obviously the lesser vigilance of the domestic dog would make a taste for dog rather easily come by in any area where many dogs are kept. A small minority of animals who had become confirmed dog killers could suffice to account for the leopard's reputation as a dog killer. In Serengeti, in addition to jackals, other carnivores killed by leopards include lion cubs and cheetah (Schaller and Lowther, 1969).

Leopards attack man fairly readily and Turnbull-Kemp (1967) found that of twenty-eight leopards shot as man-eaters, the majority were adult animals, uninjured and in good health. They had thus not taken to attacking man because, through age or infirmity, they were no longer capable of killing more agile prey. Schaller (1967) recorded eight killings and two unsuccessful attacks on man in the neighbourhood of Kanha Park from 1961 to 1965. He concludes rather laconically that although man is eaten now and then he 'does not on the whole contribute significantly to the leopard's diet'. This may be true today, when man has provided himself with superior weapons and an adult can be killed only if taken unawares or unarmed but there is no reason to suppose that it has always been true. Brain (1970), from a study of the Australopithecine fossils at Swartkrans and a comparison with the residues left from the kills of the living large carnivores, has come to the conclusion that these early hominid remains actually represent leopard kills.

(*d*) *Panthera onca*, the jaguar No detailed study has yet been made of jaguar predatory behaviour and only qualitative statements about the types of prey taken are to hand. If there is enough cover nearby to provide adequate daytime refuges, the jaguar may hunt in relatively open terrain but its favourite habitat is dense forest, particularly in well watered areas: the foods reported as the main prey are therefore those characteristic of this type of habitat. According to Cabrera and Yepes (1960) capybara and alligator are the foods of choice and the jaguar is also an adept at hooking fish out of the water with its paw. Aliev (1966) says that the coypu is heavily preyed upon and Hershkovitz (1969) gives peccary as the favourite food. Deer and tapirs are also taken and domestic stock may be attacked. Cabrera and Yepes record that the jaguar shares with the leopard a predilection for killing domestic dogs. The only record I have found of jaguars eating vegetable food is van der Pijl's (1969) statement that they are extremely partial to the fallen fruits of the avocado.

(*ii*) *Acinonyx jubatus*, the cheetah

The main food of the cheetah is small antelopes or the young of larger species. Smaller mammals such as antbears, porcupines and hares are also taken, as well as ground birds like bustard and guineafowl (Pienaar, 1969) but it is the smaller ungulate species that the cheetah is specifically adapted to hunt.

In the Kruger Park, Pienaar found that impala made up roughly three-quarters of the kills, with kudu, waterbuck, reedbuck and zebra, the latter mainly as juveniles, accounting for most of the remainder. In Serengeti, Thomson's gazelle is the staple prey. Kruuk and Turner found twelve out of twenty-three kills (56%) were Thomson's gazelle, with wildebeest (five juveniles and one adult) second in importance. In a later study in the same area, Schaller (1968a) recorded 136 cheetah kills of which 89% were Thomson's gazelle. The largest kill, made by a pair of cheetahs together, was a yearling topi, which would weigh about 90 lb (41 kg). Schaller is of the opinion that prey above 60 lb (26 kg) in weight is rarely killed and considers an adult wildebeest, a kongoni and a yearling zebra in Kruuk and Turner's list as rather unusually large prey. Impala did not figure in the kills recorded in Serengeti but Schaller regards this as being due to the fact that observations were made only on the plains or along the edge of the woodlands. He is of the opinion that studies in the wooded areas would certainly show that impala are also being preyed upon. In Nairobi National Park, Eaton (1970b) found that out of thirty kills, the commonest prey species were impala, Grant's gazelle, waterbuck and kongoni. McLaughlin (personal communication), who examined 183 kills in the same area, also found impala and Grant's gazelle the most frequent prey but Thomson's gazelle and hartebeest came next in order. McLaughlin was also able to confirm Schaller's estimate that most of the animals killed weighed less than 60 lb.

Cheetahs do not normally remain with a kill once they are fully fed, nor return to it for a second meal; taking of carrion is very rare, although Pienaar (1969) reports that this has been known to occur.

(*iii*) *Puma concolor*, the puma

Ranging from British Columbia in the north to Patagonia in the south, the puma covers a wider range of latitude than any other felid and the prey species taken in different localities show corresponding diversity. Hibben (1939) carried out an investigation into the feeding habits of pumas in North America which included the analysis of over three thousand scats. Deer remains (*Odocoileus* spp.) made up 82% of the scat volume with porcupines and lagomorphs second in importance at approximately 6% each. Domestic species, despite the puma's reputation as a stock raider, made up only 0·5% of the remains. Other mammals killed now and then included several species of carnivore - badger, skunk, grey fox and coyote - and there are also cases of a puma killing a smaller individual of its own species and eating it (Lesowski, 1963).

Hibben also found evidence of individual food idiosyncrasies: one animal fed only on porcupines, although deer were available and another fed for some weeks only on lagomorphs. He also mentions reports of individuals having become stock thieves and it seems likely that, as in the case of the coyote, individual 'rogue' animals have given a false impression of the habits of the species as a whole.

Later investigations (Robinette *et al.*, 1959) have, in general, confirmed Hibben's findings. Deer of one species or another constitute the puma's staple prey over most of its range, with larger rodents and lagomorphs forming secondary but significant items and other food sources negligible. Beyond the southern limit of the deer the two large prey species mentioned by Cabrera and Yepes (1960) are guanicoes and rheas. Wright (1934) quotes an instance of a puma in California caching prey and returning to eat it within a day or two.

(*iv*) *Lynxes*

(*a*) *Lynx lynx*, the northern lynx The food of the lynx has been investigated both in North America and in Sweden. In Newfoundland, Saunders (1963a) carried out a year-round analysis of stomach contents and scats. Snowshoe hare was the most important prey and occurred in 73% of the total samples; birds were second at 21%; moose and caribou, largely as carrion, third and small rodents, mainly *Microtus*, fourth. Grasses and other vegetable matter were found in approximately 10% of scats but insects were never a significant item. Dependence on snowshoe hare was greatest in winter (December to March) and rather less in summer when more birds and voles were taken. Nellis and Keith (1968), working in Alberta, found snowshoe hare, ruffed grouse and carrion were the main foods. Taking into account the weights of the prey, they calculated that 69% of the meat eaten was snowshoe hare, 17% carrion and 11% ruffed grouse. Cannibalism has also been reported (Elsey, 1954).

The dominant role played by snowshoe hare in the diet of North American lynxes seems to be the factor responsible for periodic fluctuations in abundance of lynxes (Elton and Nicholson, 1942). As long as the hares are plentiful there is food for a large lynx population but when the hares decline in numbers this is no longer true; alternative food sources are not sufficient to maintain the predator populations at their previous level and a reduction in their numbers therefore follows.

In Sweden (Haglund, 1966) the picture is different. During the winter period the main prey is deer: reindeer in the north and roe deer in the south. Lagomorphs are second in importance and birds third, with small rodents a very minor item. It is probably in adaptation to taking larger prey that the Swedish lynxes have become somewhat bigger than the North American representatives of the species. Haglund, however, regards the lynx as primarily adapted to dealing with the smaller types of prey and notes that a reindeer kill is often only partially utilised because the lynx, unlike the wolverine, cannot cope with the carcase once it has become frozen.

(*b*) *Lynx rufus*, the bobcat According to Young (1958), lagomorphs form the bobcat's main prey. Bobcats are not normally able to kill a healthy adult deer, but carcases of animals that died of starvation are often utilised during hard winters (Petraborg and Gunvalson, 1962). Matson (1948) has found that in Pennsylvania the bobcat has shown itself capable of changing its habits to cope with a changed situation. In earlier times, when small game was plentiful but deer relatively scarce, the bobcat lived mainly on small prey. At the time of Matson's study the position was reversed: deer had increased and small game become scarcer. The bobcat had increased its predation on deer, taking mainly fawns and juveniles, which it can overcome without much difficulty. Although shrews are rarely a favoured prey, Rollings (1945) considers that they may have some importance in the bobcat's diet during hard winters.

In America, *Lynx rufus*, although commonly known as the bobcat, is also often called the wildcat. The expression 'wildcat scheme', meaning an ill-founded venture, unlikely to achieve success, is not, however, a reflection on the character or capabilities of *Lynx rufus*: it is said to date back to the notes, bearing a picture of a bobcat, which were issued without financial backing by one of the early banks in a mid-western state (Young, 1958).

Amongst the larger cats, two species on which no detailed studies have yet been made are the snow leopard, *Panthera uncia*, and the clouded leopard, *Neofelis nebulosa*. The former is generally supposed to differ from the ordinary leopard in its habitat rather than its feeding habits and to live mainly on medium-sized ungulates, together with smaller mammals such as marmot and pikas. In the case of *Neofelis*, however, the extremely long and blade-like canines (plate 3) suggest some specialisation in feeding habits and are possibly an adaptation to killing birds.

Two rather smaller species for which information is also lacking are the Asiatic and African golden cats, *Profelis temmincki* and *P. aurata*. Observations made by Leyhausen (1965b) on captive animals suggest that *P. temmincki* hunts entirely on the ground but *P. aurata* is considerably more arboreal. The stomach of a single specimen of the latter shot in Senegal by Gaillard (1969) proved to contain the remains of an unidentified bird.

(*v*) *Felis*

Very few studies have been made on any of the smaller felids and it is therefore simplest to consider first what is known of the members of the genus *Felis*, as defined by Pocock (1951) and to use this as a basis for comparison with other forms.

In all species of *Felis* that have been investigated the staple prey is small rodents but killing of larger mammals up to the size of hare or marmot or of birds as large as a hen is by no means unusual. Birds figure regularly in the diet but are generally a subsidiary item. Insects and lizards are commonly taken when they are abundant and grass or other vegetable food is eaten now and then. This description applies to the European wildcat, *F. silvestris* (Lindemann, 1953; Southern, 1964); the African wildcat, *F. libyca* (Smithers, 1968a); the jungle cat, *F. chaus* (Ishunin, 1965); the sand cat, *F. margarita* (Dementiev, 1956, Lay *et al.*, 1970) and also to feral domestic cats (Errington, 1936; McMurry and Sperry, 1941; Llewellyn and Uhler, 1952; Eberhard, 1954; Heidemann and Vauk, 1970), although if it is accessible the latter will also consume considerable amounts of garbage. In Uzbekistan, Ishunin found both fruit and cereals figuring in the winter foods of the jungle cat and Leyhausen and Tonkin (1966) note that in captivity the black-footed cat, *F. nigripes*, eats grass more readily than most cats and does not thrive unless regularly provided with fresh grass. The stomach contents of a single *F. nigripes* examined by Bothma (1965) included some green grass.

Most shrews are somewhat distasteful and *Crocidura* is particularly nauseous. I have found that domestic cats will normally kill *Crocidura*, eat their first kill and vomit it up again shortly after. Thereafter they will still kill shrews but refrain from eating them. Gossow (1970) describes a stoat showing exactly the same behaviour towards *Crocidura*. *Blarina*, however, appears to be much less distasteful, and Nader and Martin (1962) record a cat with eight *Blarina* in its stomach. They also summarise other information on shrew eating but unfortunately without distinguishing genera: the general indication, however, is that domestic cats are rather less inclined to reject shrews than the majority of predators. Southern (1964) quotes Haltenorth as finding that *F. silvestris* commonly eats shrews, presumably *Sorex*.

(*vi*) *Other small cats*

Apart from the genus *Felis* there have been radiations of small cats both in southern Asia and in South America. Amongst these, the basic diet of small rodents, eked out with whatever other small vertebrate prey is locally available, has been adapted to different habitats largely by change in emphasis. Increased predation on birds, as might be anticipated, is characteristic of a number of the more arboreal species such as the ocelots. Some of the savanna species are also skilled at catching ground-nesting birds, for example the caracal, the pampas cat (*Lynchailurus colocolo*) and the jaguarondi (*Herpailurus yagouaroundi*). The latter is also said to prey upon deer (Hershkovitz, 1969); Cabrera and Yepes (1960) believe jaguarondis are capable of killing even adult deer but the evidence is circumstantial and it seems more probable that, like the slightly larger bobcat, they take fawns and juveniles. The caracal is capable of killing small ungulates: Azzaroli and Simonetta (1966) saw the kill of a Kirk's dik-dik (*Rhynchotragus kirki*) and their records of stomach contents of three caracals include the remains of another species of dik-dik, *Madoqua phillipsi*, as well as guinea-fowl and ground squirrel. In Garamba National Park, Verschuren (1958) found that serval cats took a surprisingly high proportion of plant food: four out of seven stomachs examined contained mainly vegetable material.

Although domestic cats show some skill in hooking fish out of shallow water with a paw, fish does not normally figure to a significant extent in the diet of any member of the genus *Felis* (*sensu stricto*). *Prionailurus viverrinus*, however, living in marshy or well watered areas, is reputed to depend on fish to a considerable extent; hence its common name of fishing cat. Leyhausen's observations on captive animals suggest that the flat-headed cat, *Ictailurus planiceps*, is more aquatic and may in fact be the felid most deserving of that title (personal communication). Lyddeker (1896), judging from the well developed anterior premolars of these two species, suggested that both might prove to be fish eaters - a supposition borne out by the report of Muul and Lim (1970) that the stomach of a flat-headed cat shot in Selangor contained only fish remains. Lyddeker also says that *Prionailurus viverrinus* eats snails of the genus *Ampullaria*, although 'we have no information as to how the succulent morsels are extracted from their somewhat solid shells'. The fondness for playing in water which Birkenmeier and Birkenmeier (1971) report in their two young Bengal cats, *Prionailurus bengalensis*, suggests that this species too may sometimes take aquatic prey.

Although no felid is known to rely to any great extent on vegetable foods, fondness for fruit has been noted in a few species. Morris (1965) says that wild jaguarondis have been seen eating fruit and that in captivity they relish grapes and bananas. Bothma (1965) reports finding grapes in the stomach of a caracal in South Africa and Lyddeker (1896) says that in Borneo *Ictailurus planiceps* is reputed to be very fond of fruit and is even said to dig up and eat sweet potatoes.

It is only the larger Felidae that are able, single handed, to kill prey significantly larger than themselves and this is certainly an ability that must have been evolved in the history of the group after initial adaptations for dealing effectively with smaller prey had been acquired. It is therefore most logical to consider first what is known of the hunting and killing techniques of the small cats and then use this as a basis for comparison with the larger species.

As we have already seen, the smaller cats are predominantly predators on small rodents and the typical habitat of the group is woodland. The domestic cat is the only small felid whose prey-catching behaviour has been studied in detail and, fortunately, it is a perfectly satisfactory example of 'typical cat' and its predatory technique may therefore serve as a standard of reference with which to compare other species.

Once alerted to the presence of prey the cat approaches cautiously, taking advantage of whatever cover is available. Leyhausen (1956) has described in detail how the first 'slink run', in which the cat moves forward relatively swiftly with the body flattened close to the ground, is followed by a pause while she 'ambushes' and watches the prey intently. A second slink run and ambush may follow, or the cat may be near enough to start the final stalk, moving forward more slowly to the last piece of suitable cover. Here she again pauses and ambushes once more as she prepares to launch the definitive attack. Her hind feet start to make alternate treading movements and the tip of her tail twitches as her eyes follow the prey's every movement; at last she breaks cover in the final sprint, with body flat to the ground and, when within reach, launches herself forwards on the prey, raising her forequarters but keeping her hind feet on the ground to give stability in the struggle that may follow. The killing bite is aimed accurately at the constriction just behind the head and death is normally the result of a canine tooth passing between two vertebrae, forcing them apart and breaking the spinal cord. One of Leyhausen's (1965b) blackfooted cats was such an adept at this technique that when it was necessary to slaughter a number of guinea pigs to store in the deep freeze for future use, the simplest and most humane way of killing them was to enlist his help. He was capable of killing thirteen in rapid succession.

Depending on the size of the prey and on how it defends itself, the paws may be used to assist in pinning it down or pulling it into a position where the death bite can be conveniently made. If rather large prey is not at once overpowered and the cat is slightly intimidated, she will release it again and make a new attack. If, however, she is fully confident she will hold her grip and, throwing herself on her side, rake backwards with her hind feet and bite home until there is no further struggling.

Leyhausen (1965b) has described the variations on this theme shown by a number of species of small felid. The variations are, in the main, quantitative rather than qualitative and relate both to the type of prey most commonly killed and to the environment in which it has to be caught. The tendency to hold on to struggling prey rather than to release and attack again is very marked in some species, for instance the Bengal cat, the ocelot and the fishing cat. This may be an adaptation both to bird killing and to fishing; for either bird or fish, once released, may be gone for ever. Another clear adaptation to bird catching shown by the ocelot is a reduced tendency to ambush. As Leyhausen points out, a mouse should be given time to get well away from its burrow before an attack is made but a small bird does not usually stay on the ground for very long and he who hesitates is liable to lose his dinner.

There are also differences in the use of the paws. Most cats will use a paw to angle out a mouse that has retreated into a crevice and may also attempt to push down a rodent that has reared up defensively. The serval is unusually adept at using the paws in both these ways. It will slap down a belligerent hamster with considerable violence and in its natural habitat will use this method to kill a snake (Brain, personal communication). Although the serval's limbs are long, they are not specialised for running and Leyhausen (1965b) suggests that the mobile paws, with their long but loosely-knit metapodials, are adapted for hooking an unwary rodent out of its burrow. In this he is certainly correct, for the serval is a past master of this technique. Dominis and Edey (1968) give a superb series of photographs showing a serval waiting with paw upraised in readiness beside the burrow of a mole rat, then hooking it out and flinging it away with a single movement, so that it can be pounced upon before it can recover. The same authors illustrate a second serval technique in which the long legs are of value. When hunting through rather long grass the serval quarters the area in a series of high, almost springbok-like leaps. Any small mammal that breaks cover is seen at once and pounced upon. One may wonder whether amongst the Canidae the long-legged *Chrysocyon* may not independently have evolved a similar technique.

The serval shares with the caracal another bird catcher's technique. Most of us have seen a domestic cat rear up to catch a butterfly with its fore paws. The serval and caracal are particularly adept at this tactic and their rearing movement carries them high from the ground with the body extended almost vertically. Dominis and Edey show a caracal fielding a dove with one paw and carrying it down to the mouth in a single movement, where it is then clasped between the two paws and bitten before the bipedal posture is abandoned.

Most cats will pluck at least some of the feathers off a bird before they start to eat, provided the bird is over a certain threshold size. They are not, however, very skilled at this and one gets the impression that the cat is eager to start eating but is frustrated by the feathers, which therefore have to be removed. The ocelot, on the other hand, plucks even very small birds so neatly that one feels it is 'really plucking' as a genuine pattern in its own right – and clearly one very desirable in the repertoire of a bird specialist (Leyhausen, 1956).

Detailed descriptions of fish-catching in *Prionailurus viverrinus* and *Ictailurus planiceps* are not yet available but unpublished observations by Leyhausen on adult animals, and those of Muul and Lim (1970) on a captive kitten of the latter species, suggest that their repertoires include something more than an elaboration of the domestic cat's paw-hooking technique and that, as true fishers, they have lost the cat's disinclination to put its face in the water. The kitten played in the large water dish with which he was provided and in 12 cm depth of water would pick up pieces of fish from the bottom, seizing them in his jaws and completely submerging his head to do so. Leyhausen's animals use exactly the same technique to catch a mouse thrown into their water dish (personal communication). The jaguar's fishing, however, is said to be simply a skilled version of the normal paw scoop but it is locally believed that the jaguar has a special way of attracting fish to the surface of the water. A number of the local fishes come to the surface if the water is slightly disturbed – a response supposedly related to their habit of feeding on fallen fruit. When Alfred Russell Wallace visited the Amazon he related that 'the jaguar, say the Indians, is the most cunning animal in the forest; he fishes in the rivers, lashing the water with his tail to imitate falling fruit, and when the fish approach hooks them up with his claws'. Gudger (1946) has summarised a variety of accounts of this story and points out that it is quite possible that, as the jaguar crouches in ambush, his twitching tail may sometimes strike the surface of the water. If the fish do indeed respond to this stimulus, then possibly the rest of the story is simply over-interpretation on the part of human observers. Gudger, however, also expresses the opinion that it is not beyond the bounds of possibility that the jaguar learns to associate success with striking the water with his tail and that the story might conceivably be true. I would add that it is also not beyond the bounds of possibility for selection to have operated to increase the jaguar's tendency to tail twitching, so that the story might conceivably be true without learning being involved. However, until some detailed investigation is made it is not possible to decide between these various possibilities.

Predation by the greater cats on animals larger than themselves presents some new problems that cannot be solved simply by using elements from the basic small cat repertoire. As we shall presently see, it is in the actual killing that the difficulties arise: the approach to the prey involves little in the way of new techniques. As with smaller game, what is required is to get as near as possible without being detected. Lion, tiger, leopard and cheetah are all good stalkers. Schaller (1967) describes how skilfully a tiger uses cover, and he also found that if the deer it was stalking detected it at some distance and gave an alarm call, an experienced tiger simply walked away and did not even try to make an attack which it obviously knew would be fruitless. The lion's approach to its prey is exactly like that of the domestic cat, including typical ambushing before the short final rush. Hunting, as they so often do, in more open country than tigers, they may have greater difficulty in remaining undetected. Schaller (1969c) found that in long grass one out of every three attacks on a Thomson's gazelle was successful but in short grass the success rate fell to one in six. There is, however, a factor which compensates for the difficulty of hunting in relatively open terrain – the fact that lions frequently do not hunt singly but as a group. Schaller and Lowther (1969) found that in one area, when two or more lions cooperated the success rate was 52%, as compared with

29% for lone attacks. They describe three different ways in which a group may cooperate. In the simplest case the members of a group simply fan out, so that prey flushed by one animal may be caught by another. Secondly, one or two individuals may circle round and approach the prey from the far side, so that if they are detected the prey will flee towards the other members of the group lying in wait. A third technique involves using familiarity with the local topography to drive the prey into a cul-de-sac, such as a narrowing tongue of land where two rivers join.

The leopard usually hunts alone but in good cover and is expert both at stalking and at lying in wait. Eisenberg and Lockhart (1972) found that in Ceylon, leopards caught langurs during the dry season by lying in wait where the monkeys had to cross a treeless area to reach a drinking pool. The leopards also caught deer by lying concealed and allowing the deer to work their way closer as they fed until they were near enough to be taken by a single very short rush. Kruuk and Turner (1967) saw the Serengeti leopards using the same technique. Alikhan (1938) reports an unusual incident in which two leopards apparently cooperated in an attack on a group of langurs in a tree. One climbed the neighbouring tree and the monkeys fled down their tree to the ground, where one of them was promptly caught by the second leopard, lying concealed and waiting.

The leopard is often said to leap down on its prey from a vantage point in a tree but neither Eisenberg and Lockhart nor Kruuk and Turner saw this form of attack. The latter authors twice saw a leopard resting in a tree alerted to the presence of a reedbuck: in each case the leopard came down from its vantage point and started a slow, careful stalk on the ground. This agrees with Leyhausen's finding that neither a cat nor a puma will ever attempt to jump on to prey from above, even when the prey is deliberately placed to make this easily possible: instead, they will jump down close beside it and attack only when they themselves have an assured footing on terra firma.

The cheetah is exceptional: hunting often alone and often in very open terrain, it is not dependent on the unseen approach and brief final rush but is prepared to chase over a longer distance. If cover is available a cheetah will use it, in typical cat manner (Varaday, 1964; Schaller, 1968a; Eaton, 1970d) and will freeze into the ambush position if the prey happens to look up. If, however, there is no cover the cheetah will make a direct approach, walking up in full view, much as *Lycaon* does, and then suddenly breaking into a run. Kruuk and Turner (1967) record one running at a herd of wildebeest from a distance of 200 metres and Schaller (1969a) found that the cheetah started its run at distances of approximately 100 to 200 metres. Although the chase is continued for some distance the cheetah does not indulge in the long pursuits characteristic of the Canidae: it usually gives up after a few hundred metres if it has not succeeded in coming to grips with the prey. This, although no great distance, is considerably more than the 50 to 100 metres that is the lion's usual limit (Kruuk and Turner, 1967).

Since Thomson's gazelle has a maximum speed of about 45 mph (72 km/hr) and cheetahs may reach 60 mph (96 km/hr), the result of a chase would seem to be a foregone conclusion: the cheetah should always be able to catch up within about a third of a mile (600 m). This, however, is not the case; the gazelle may escape by dodging, for if the cheetah loses too much ground by failing to follow a quick turn it frequently abandons the pursuit.

According to Haglund (1966) the lynx follows the usual cat method of a careful stalk followed by a short final rush on the prey which, if not caught within about 20 metres, is usually lost. The technique of lying in wait and allowing the prey to work its own way within reach is also used. Haglund also describes two of the same types of cooperation as are shown by lions. A family group of a female and her young used the fanning-out technique and, from the tracks left in the snow, he twice found evidence which he interpreted as showing two adults working together, with one driving the prey towards where the other lay in wait. Barash (1971) witnessed cooperation of the latter type in the Canadian lynx.

Once the large felid has come up with its prey, it still has two problems to overcome. If the prey stands as high as its predator, or higher, it must be brought down before it can be killed. The lion and the tiger usually bring down their prey by a modification of the domestic cat's technique. The forward lunge is given an upward direction and the paws grasp the prey by shoulders or hindquarters and pull it down. The predator's hind feet, like those of the cat, remain on the ground and the prey is not knocked over by the impact but dragged down. Although a tiger can make a jump of 20 feet, Schaller (1967) never saw one take a flying leap at its prey. The cheetah also uses its paws but in a different manner. Once within reach it makes a swift blow with one paw at leg, flank or rump, which usually fells the prey. The fact that the large dew claw commonly leaves a long gash in the flesh (Schaller, 1968a) suggests that the blow is a pull rather than a strike. According to McLaughlin (personal communication), the cheetah also occasionally uses both fore paws to pull down the prey.

The normal felid *coup-de-grâce* is the neck bite and all the larger species will use this technique on relatively small prey. A cheetah or a puma, for instance, will kill a hen with a perfectly orthodox neck bite and Schaller (1967) records a tiger killing two calves in this way.

Really large ungulates, however, have extremely thick and strong neck muscles and many have horns capable of striking down and back at a predator attacking in the usual manner. In these circumstances, if the neck bite cannot be made with ease and safety, the orientation of the killing bite is shifted from the nape to the throat. Here it may not at once be lethal but the predator can hold on, keeping its body clear of horns or thrashing hooves, and maintain the grip as long as is required. Schaller (1967) records the tiger killing in this way and found that, of nine cows and buffaloes killed by tigers, all had lacerated throats and two had also been bitten on the nape. Kruuk and Turner (1967) and Schaller (1969c) describe the lion killing with a throat bite but dorsally placed bites may also be made on shoulder or back (Leyhausen, 1965b).[1] In leopard also, Kruuk and Turner found that prey was commonly bitten in the throat but some kills were found to have been bitten in the neck. Both the lynx (Haglund, 1966) and the bobcat (Matson, 1948) use the throat bite when killing deer. The cheetah appears to use the throat bite quite regularly, even with prey as small as Thomson's gazelle (Kruuk and Turner, 1967; Schaller, 1968a, 1969a; Eaton, 1970b, d).

When a kill is made by a throat bite the immediate cause of death is not always clear. Schaller believes that the predator's grip occludes the windpipe and death results from strangulation. With very large prey this may well be true but it is certainly not always the case. I have seen film of cheetah kills where the bite was made in such a way that the teeth must have entered the floor of the braincase and death occurred far too quickly to have been the result of strangulation: it therefore seems much more likely to have been caused by damage to the central nervous system. Leyhausen is also of the opinion that with relatively small prey it is possible that the neck vertebrae may sometimes be dislocated, as in the normal neck bite but from below instead of from above. It is therefore undesirable to refer to killing 'by strangulation' if the cause of death has not been definitely determined and all that is actually known is that the lethal attack was directed at the throat region.

It thus seems clear that the normal felid neck bite is included in the repertoire of the large cats but that in addition, they can abandon it in favour of the throat grip when dealing with large or dangerously head-weaponed prey. Moreover, it is surely no mere chance that the cheetah, whose predatory technique in other respects departs farthest from the felid norm, should also be the species in which the throat bite is most regularly used: just as its limb structure reflects its

[1] Eloff (1964) describes a rather unusual technique practised by lion in the Kalahari: the long- and sharp horned gemsbok are killed by a bite near the sacrum combined with an upward pull which breaks the back.

unusual prey-capturing behaviour, so too do the cheetah's relatively short canine teeth reflect its having replaced the neck bite by the throat attack for all except the smallest prey.

A number of points of interest emerge from this survey of prey and predatory techniques. Firstly, one cannot but be struck by the fact that there are so few Carnivora for whom vegetable food does not figure to any significant extent in the diet: apart from the Felidae this appears to be true only of some of the Mustelidae and a few of the larger Canidae. There is an obvious correlation between the size of prey taken and the type of diet. The teeth that are highly adapted for predation are singularly poor at crushing up vegetable food and vice versa. Those species that depend largely on plant foods are therefore normally limited to the killing of relatively small prey and the killers of large prey do not make a great deal of use of other types of food. The truly predacious species, nonetheless, show a great deal of adaptability and are able to kill a wide variety of prey species. Moreover, although each species may have its own preferred size of prey, which it is most fully adapted to deal with, the majority can not only make do with something smaller, if need be, but will also successfully attack larger game now and then. This must have facilitated evolutionary change in either direction. A carnivore adapting to larger or to smaller prey does not suddenly have to acquire the habit of attacking species it previously neglected: it has only to shift its emphasis and kill more of the species than were previously attacked less frequently.

A second form of adaptability shown by the large predators relates to the sources of the food. The majority of large carnivores are prepared to take any food they can get and very few are finicky about whether the meat is fresh or not. Most of them will feed on a carcase, whether it be of an animal that died naturally or the remains of some other predator's kill, and, if superiority in size or in numbers permits, most are prepared to drive the rightful owners off a kill and appropriate it. The cheetah does not take carrion and *Lycaon* only rarely does so but amongst the other large carnivores there is no clear and simple division into true predators and obligate scavengers: the traditional 'scavengers' also kill for themselves and the predators scavenge off each other's kills and even appropriate those of the 'scavengers', whenever the chance offers. There is, however, a considerable difference in the maximum size of bones that can be chewed up by the different species. The hyaenas' jaws and teeth are specifically adapted for bone crushing and a hyaena, if allowed to finish a meal, leaves no residue of use to any other carnivore. At the other end of the scale comes the cheetah, unable to cope with large bones so that only the smallest prey can be completely disposed of. The hyaenas are therefore potentially the greatest scavengers and the cheetah is the most self-supporting of the large predators. In the opinion of Schaller and Lowther (1969), however, predators in general do not voluntarily abandon enough of a kill for it to be possible for any species to make a living solely by scavenging and they conclude that 'there is no ecological room for a total scavenger'.

If one thinks only in terms of the present situation, it is not easy to understand why the hyaenas ever evolved their scavenging dentitions. There is, however, a major difference between the felids of the past and the present: the sabre-tooths are gone. The sabre-tooths, in addition to their enlarged canine teeth (which I believe to have been adapted to killing by a rather special type of attack on the throat) had the most specialised slicing carnassial teeth in the whole of carnivore history but their anterior premolars were extremely reduced. They must have been able to cut through flesh with the greatest efficiency but could not have crushed the bones of anything much larger than a chicken. They must, therefore, have left splendid pickings for any species that could crack large bones and it is surely no mere coincidence that the hyaenas evolved their special adaptations in the days when the dominant cats were sabre-tooths. If one puts a hyaena's and a sabre-tooth's jaw side by side, the contrast between the heavy, hammer-like premolars of the one and the vestigial character of the same teeth in the other is so striking – the hyaena is so clearly fitted to do exactly what the sabre-tooth cannot – that it is difficult to believe that there is no functional link between them. This is not to imply that the hyaenas ever lost the ability to make their own kills but merely to suggest that they became experts in bone crushing at a time when bones to scavenge were more easily come by than they are today.

Two further consequences follow from the widespread tendencies to scavenging and kill-thieving of the modern carnivores. Firstly, since the prey often suffices for more than one meal, many carnivores have developed methods of safeguarding what is not devoured at the first sitting. Lions, dominant to all their neighbours unless grossly outnumbered, merely stay near their kill and drive off any would-be marauders. Tigers also usually rest near the kill but also frequently cache it or cover it with leaves and debris, especially if they intend to move away from its immediate vicinity (Schaller, 1967). The leopard, as befits the best climber, carries its kill up a tree (plate 6) but does not bother to do so if there are no potential food thieves about. Brain (personal communication) found that in an area in South West Africa devoid of hyaenas, kills were not taken up a tree,

1a. Fishing genet, *Osbornictis piscivora* (from Allen, 1924).

b. Tree ocelot, *Leopardus wiedi*.

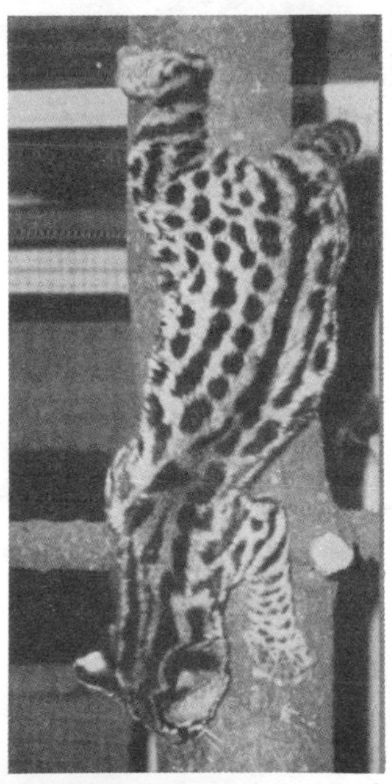

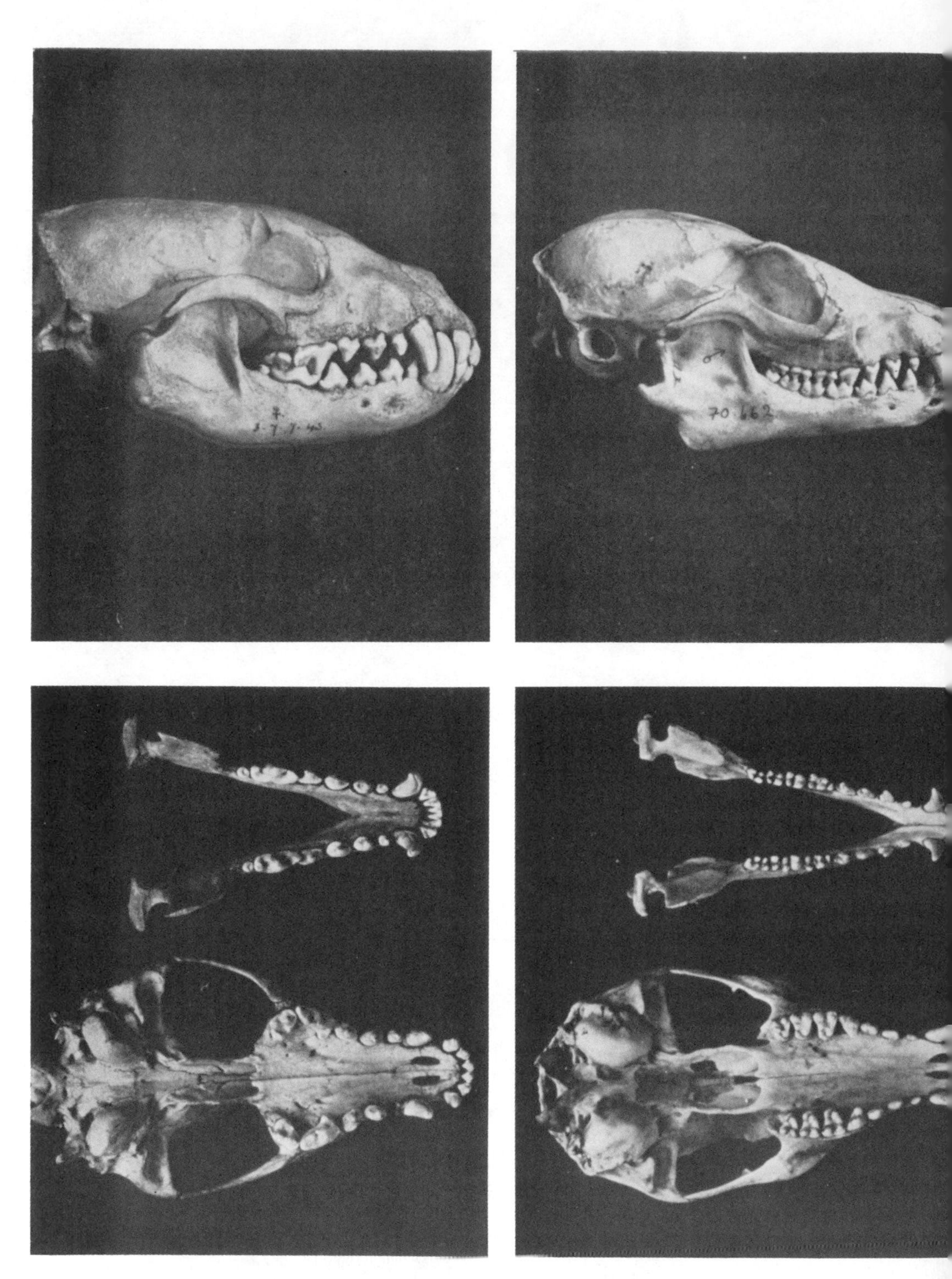

2. Skulls of bush dog, *Speothos* (*left*) and bat-eared fox, *Otocyon*.

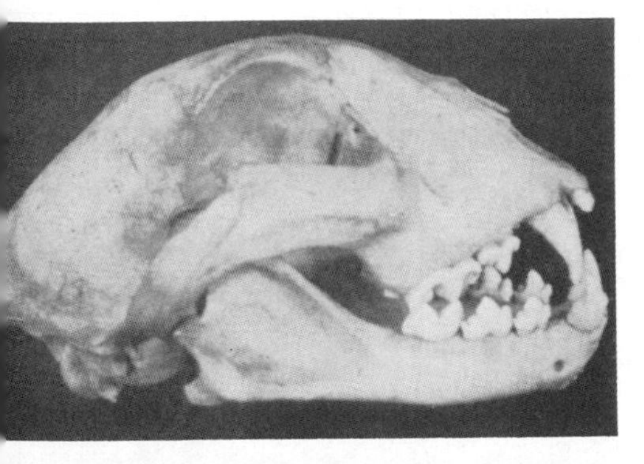

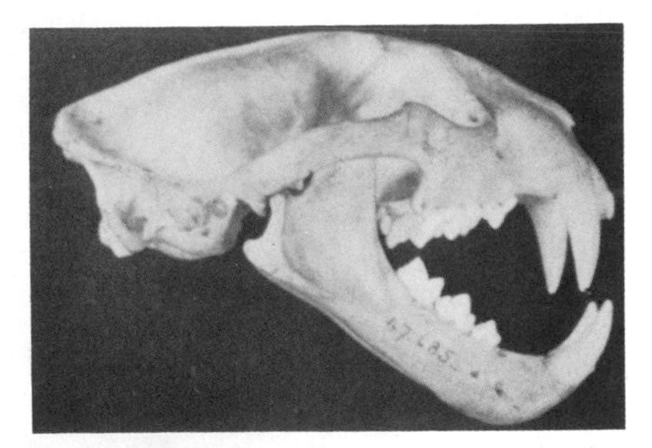

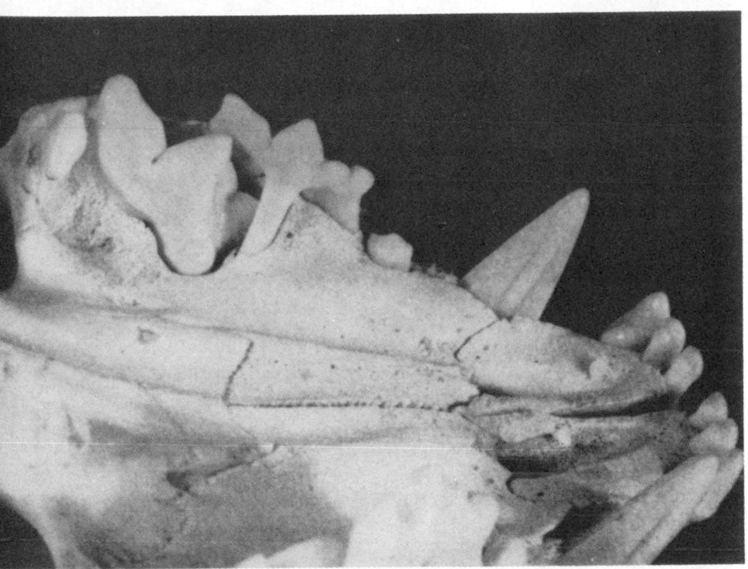

3a. Felid skulls: contrast in shape and in size of canine teeth. *Top left* cheetah, *top right* clouded leopard.

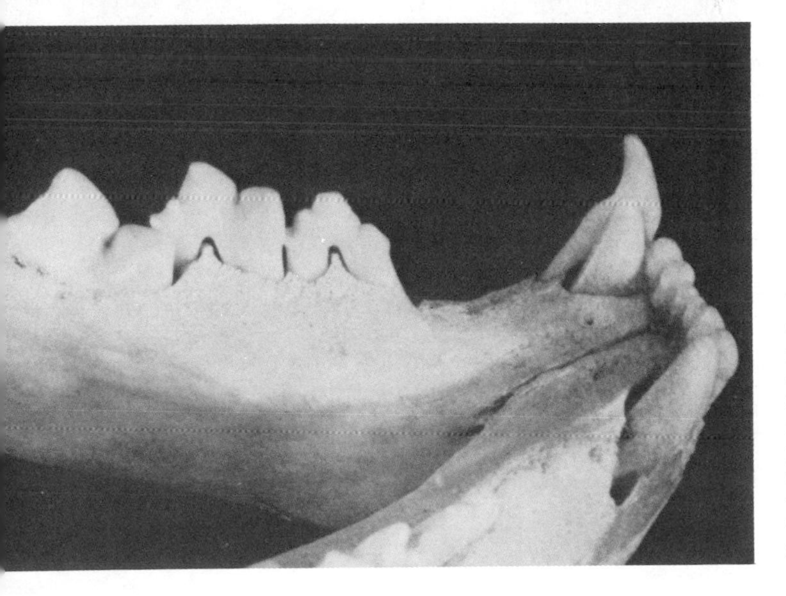

b. Upper and lower teeth of young leopard. The permanent carnassials are erupting but the milk carnassials are still in place. In the lower jaw the milk canine is still present just outside the erupting permanent tooth: in the upper jaw it has already been shed.

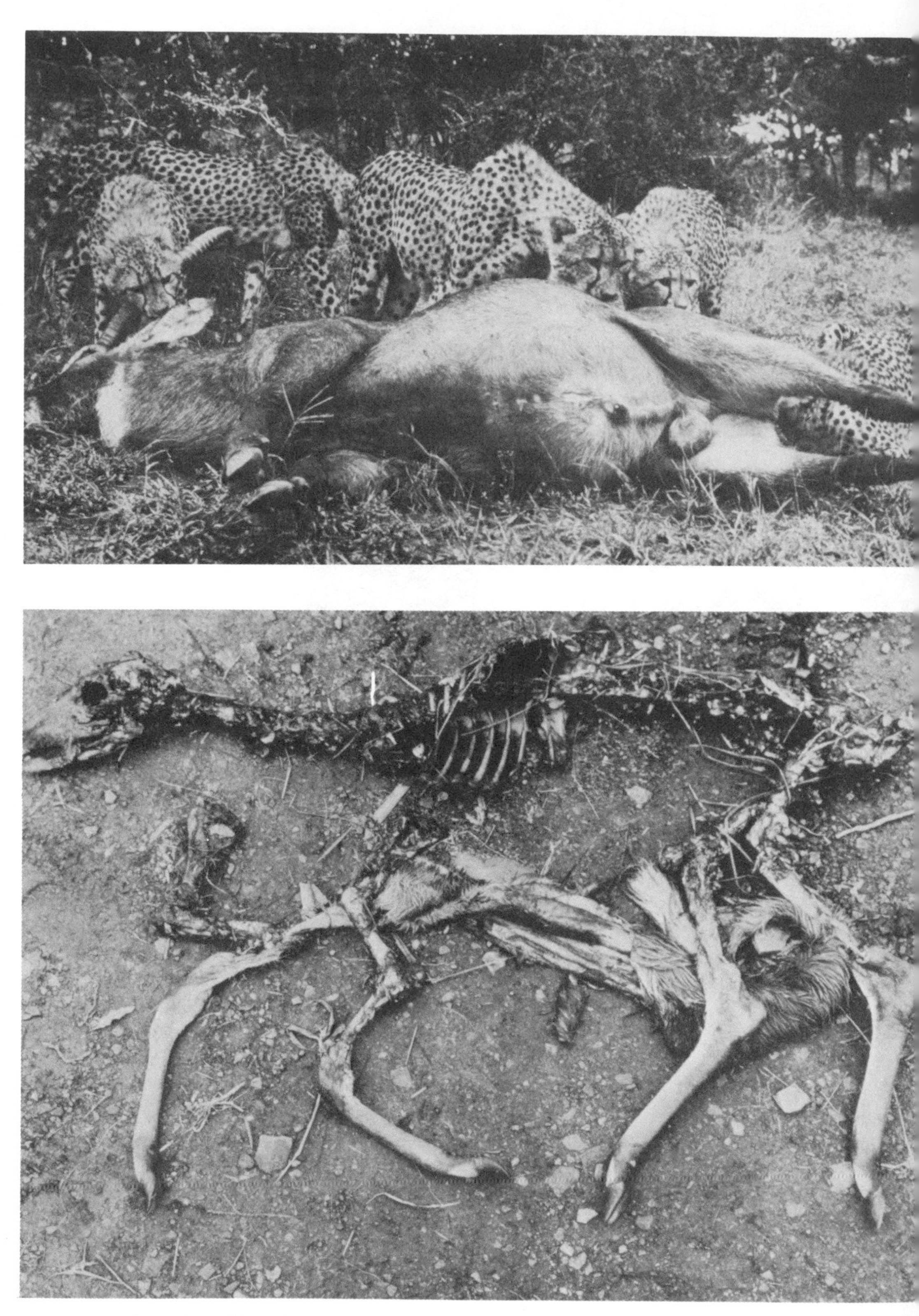

4. Cheetahs on kill and remains left after they have fed.

5a. Tigers in natural habitat.
b. Kinkajou, *Potos flavus*, manual dexterity in feeding.

6a. Leopards with kill (reedbuck) cached up tree.

b. Hunting dog, *Lycaon pictus*.

7a. Bush dog, *Speothos*.

7b. Maned wolf, *Chrysocyon*.

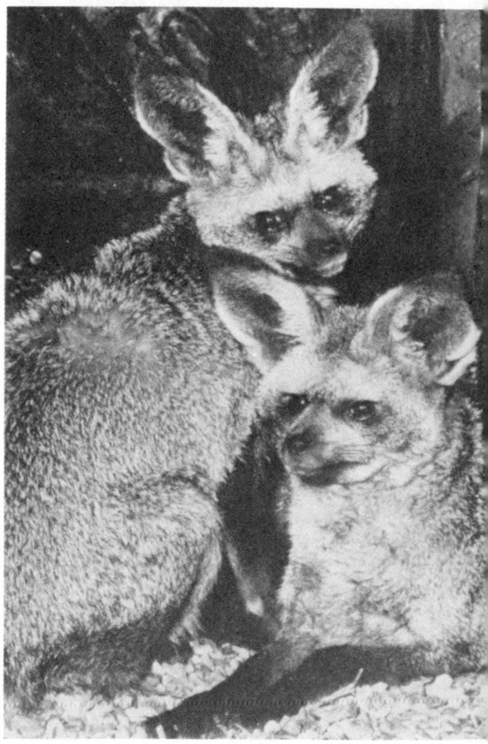

8. Heads of foxes: arctic (*top left*), fennec (*top right*), red (*left*) and bat-eared (*above*), showing relative size of ears.

9a. Clawless otter, *Aonyx capensis*, dowsing food (*top left*).
b. Genet using hind paws combined with neck bite to overpower rat (*top right*).
c. Palm civet, *Nandinia binotata*, eating a lizard.

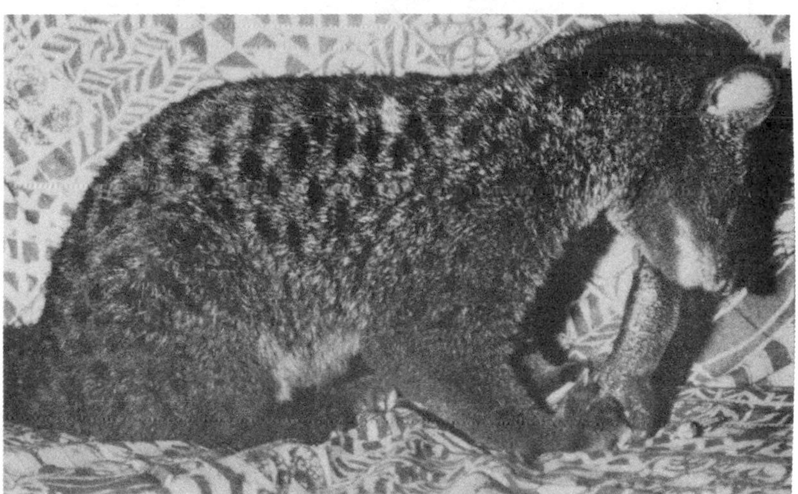

10a. Fanaloka, *Fossa fossa*, showing highly digitigrade stance.

b. Serval cat at ease.

11. Serval cat showing two stages of defensive threat.

12a. Amicable head rubbing by lions.

b. Lions mating.

13a. Abnormally patterned cheetah, originally described as a distinct species (from Pocock, 1927b).

b. Male cheetah urine-sprays against tree trunk

14a. African civet, *Civettictis civetta,* male scent-rubs on food and female bites at his neck stripe (*top*). b. Kusimanse, *Crossarchus obscurus,* scent-marking (*left*). c. Weasels, *Mustela nivalis,* mating (*right*).

15a. Fox carrying food to young (*right*).
b. Maternal care by fennec: toilet licking (*top*), carrying of cub (*above*) and protection (*below*).

16. Restorations of *Pseudocynodictis*, an early fossil canid (*top*); and *Hoplophoneus*, an early sabre tooth, both from Oligocene deposits of North America (from Scott, 1937).

whereas in the Kruger Park the only two leopard kills which he saw not protected in this way were promptly taken by the crocutas. Similarly in Ceylon, Eisenberg and Lockhart found that the same difference in the treatment of the kills in two localities was correlated with the scarcity or abundance of jackals. The jackals' own habit of caching food has already been mentioned and, according to Schaller and Lowther (1969), crocutas sometimes protect a kill by putting it in water - an observation which may explain why zoo animals frequently put their food in their water dish. *Lycaon* and the cheetah are unusual in having no method of protecting the remains of a kill; with the *Lycaon* pack there are rarely any significant remains from a meal and since the cheetah never seems to have developed a taste for carrion it is not surprising that any remains are simply abandoned.

The habit of food thieving, especially by direct driving away of the owners from their kill, has had its effects on the interspecific relationships of the various predators. Such behaviour cannot but be conducive to interspecific hostility and, as we have seen, there is a very marked tendency for carnivore to kill carnivore when the opportunity presents itself. Schaller and Lowther (1969) point out that, apart from leopards killing jackals, this interspecific enmity is rarely a matter of killing for food: it is rather the aggression against a competitor for food which one might normally expect to see aroused by an alien conspecific intruding into the preserves of a territory holder. He comments on the fact that a lion chasing a leopard or a crocuta does not behave in the same way as when attacking prey, but uses the same vocalisations and facial expressions as in aggressive behaviour towards a conspecific.

Within the Carnivora, each family has quite a varied repertoire of killing techniques and many of the actions used are common to several families: there are, nevertheless, certain family characteristics, of which those distinguishing the Canidae and the Felidae are particularly clear. It is therefore of some interest to inquire into their functional significance and possible evolutionary origin.

Since the modern carnivore families are all believed to have originated towards the end of the Eocene from the now extinct Miacidae, it is necessary to begin by considering what is known of the characteristics of the latter. Most of the known miacids were rather small and their limb structure suggests that many were good climbers. It is impossible to believe that any carnivore could have evolved a dentition of miacid type without some corresponding behavioural adaptations, the most obvious of which is not to attack at random but to direct the bite to the anterior region of the prey and aim at head or neck. This has the double advantage of giving a high probability of a quick kill and of making a retaliatory bite difficult;

moreover, aiming of the bite at head or neck is found in at least some members of every extant family. One would not expect any very precise orientation of the bite in the early stages of evolution and the miacids probably increased the efficacy of a somewhat labile orientation in the two ways adopted by living viverrids - by iterant biting and snapping and by shaking. From an ancestral killing technique of this essentially viverrid type it is possible to see how the characteristic contrasting techniques of the Canidae and Felidae could have evolved.

The Felidae must take their origin from a miacid ancestor living in good cover, where a stalk is likely to be more rewarding than a chase. Their killing technique has three characteristic features: elaborate stalking, the use of the paws in prey capture and a very precisely oriented neck bite. All are correlated, for the felid killing bite could have been perfected only with the predator in full control of the situation and choosing his time to launch an accurately aimed attack; while the adaptations of the paws as auxiliary lethal weapons would not have been possible had cursorial specialisation been required in order to get within reach of the prey. It is the use of the paws to pull down the prey into a position where a lethal hold can be obtained with the jaws that has made it possible for the larger Felidae, single-handed, to overpower prey larger than themselves. We have already seen how the oriented neck bite, which first made the felids such efficient killers, came to be supplemented but not replaced by the throat bite for the largest sorts of prey.

The ancestors of the essentially terrestrial and cursorial Canidae, on the other hand, must have been amongst the least arboreal of the miacids, hunting in poorer cover than the cat ancestors. Once the technique of chasing has been adopted and the limbs have become cursorially specialised, they are no longer free to become adapted in other ways. This means that the jaws must now be used not only to kill the prey but also to bring it down. The success rate of a lone hunter using this technique against prey larger than itself would inevitably be low and the advantages of group pursuit are obvious. One individual may lose ground after a leap and bite at the prey, but the next is ready to repeat the attack until the prey is first slowed down and finally brought to earth and, in fact, all the extant canid species that kill large prey do hunt in groups.

The cheetah, the only cursorial felid, includes the three basic felid characteristics in its repertoire of hunting techniques: stalking if cover is available; an oriented neck bite or a throat bite according to prey type and the use of the paw, albeit in a highly distinctive manner, to fell the prey. The chase is short and the cheetah abandons any pursuit which might leave it exhausted when it finally came up with its quarry. All this suggests that the cheetah's technique is not something independently evolved but simply a modification of the usual felid pattern as an adaptation to hunting in relatively open terrain. One can take the argument one step further. The cheetah possesses very distinctive anatomical adaptations to its particular method of hunting. These must have taken quite some time to evolve and it is clear that the cheetah is no evolutionary upstart but has been at the game for a considerable time. Lions cooperate in hunting but there is often quarrelling over the proceeds of the hunt and sharing of the food is much less equitable than in *Lycaon*, the wolf or *Crocuta*. Schaller and Lowther (1969) regard this as a reflection of incomplete adaptation to social life. This, in turn, suggests that, as far as hunting on the open plains goes, the lion may be a newcomer to what was once the domain in which the cheetah had no felid competitor.

However this may be, the cheetah's existence is at present extremely precarious. Although maintaining itself when not molested by man, the facts suggest that it does so with only a very narrow margin of safety, even in the comparative protection afforded by game parks. In such circumstances even relatively minor predation by man may be sufficient to swing the balance against the cheetah. If the cheetah is to survive it is essential that everything possible should be done to protect it. 'Genocide' is a word much bandied about with reference to human intraspecific conflicts, often without real justification. It is, however, literally true to say that any woman who wears a cheetah fur, or any man who embellishes his home with a cheetah skin rug, is assisting in and setting the seal of his approval on a genocide that can never be made good by any substitute. The cheetah is unique: there is no source from which its genotype could ever be reconstituted.

Before leaving the subject of food and feeding it is necessary to say a word about water relations. A meat diet has both a high water content and a high protein content. The latter results in a high rate of urea production, which requires a considerable volume of water for its excretion. No carnivore is known to be able to concentrate urea to the levels characteristic of desert rodents but this does not signify very much, since desert-living species have not yet been investigated. A cat, however, can achieve a much higher urea concentration than the human kidney is capable of producing: Schmidt-Nielsen (1964) quotes values of over 2000 mM/l of urea for the cat, as compared with less than 800 mM for man. Even at such concentrations, the volume of urine required to deal with urea clearance is so large that electrolyte excretion presents no problems for the cat.

Schmidt-Nielsen has calculated that, making due allowance for evaporative loss from the lungs and for the water lost in the faeces, a dog should be able to live on rabbit without requiring any additional water and cats have been shown to be able to survive perfectly well without drinking on a diet of fish and beef. If, however, heat stress is added and water is required for evaporative cooling, then this is no longer true and the water contained in the food becomes insufficient for the animal's needs. In carnivores sweating is of negligible importance and cooling is by panting, with evaporation taking place from the surface of the nasal passages, the tongue and the mouth. Panting is remarkably efficient as a means of heat dissipation and a dog with access to drinking water can maintain its normal body temperature for at least seven hours at an external temperature of 43°C (≏ 109°F), provided the humidity is not very high. Watching a dog pant, one is inclined to feel that it must be an inefficient process because the very rapid respiratory movements, which reach frequencies of over 200 per minute, must require so much energy expenditure. This, however, is erroneous. Crawford (1962) has shown that the energy is much reduced by the fact that the animal pants at the resonant frequency of its respiratory system. This is why panting is such an all-or-none phenomenon: a dog either pants at virtually its maximal rate or it does not pant; there is no gradual increase in frequency. The same is true of the cat, which pants at a rate of approximately 250 per minute: presumably here too the frequency is determined by the natural resonance of the system and the energy expenditure is therefore minimal.

Schmidt-Nielsen *et al.* (1970) have recently shown that, in the dog, the efficiency of heat dissipation is still further increased by the fact that during panting the air is taken in mainly through the nose but passed out by the mouth. There is thus a through-flow arrangement, which increases the rate at which heat can be dissipated, since there is no exchange of heat and moisture between incoming and outgoing air. This arrangement means that much of the evaporation takes place from the surface of the turbinal bones, rather than from the mouth and tongue. The dog possesses a large nasal gland, producing a serous secretion, which presumably has the function of supplying the moisture required for evaporative cooling from the surface of the nasal cavity. Schmidt-Nielsen points out that the nasal gland is thus functionally analogous to the sweat glands in man and suggests that its secretory activity may therefore show the same sort of thermoregulatory control as do sweat glands. This possibility, however, remains to be investigated.

No studies have been made on the water economy of wild carnivores but a number are known not to require water every day.

Schaller (1968b) found that on the Serengeti plains a lion fitted with a radio transmitter collar and tracked continously for three weeks drank only seven times in twenty-one days, and Eloff (in press) records a pride in the Kalahari going without a drink for nine days. According to Schmidt-Nielsen (1964) the fennec is the only desert-dwelling carnivore known to be capable of living entirely without water but I suspect that the sand cat, *Felis margarita*, may also be able to do so. Both species survive by restricting their activities to the cool night period and evade the heat of the day by resting in the burrows they have excavated and, according to Avarguès and Goudeau (1962), the fennec effects some economy of energy expenditure by allowing its body temperature to drop by some two degrees Centigrade during its daytime resting period. Schmidt-Nielsen found that in a tame fennec, even on an unrestricted diet, with water freely available, the urea concentration in the urine was higher than the maximal concentration a cat can achieve and quotes a sample value of 2620 mM/l. Presumably, if necessary, a considerably higher concentration could be reached and the fennec may actually be within the range of the desert rodents in this respect.

The sea otter must also be capable of producing a very concentrated urine. The highly lobulated and extremely large kidney suggests that they do possess this ability but no actual measurements have been made. Although sea otters normally live without access to fresh water and have been seen to drink sea water (Johnson *et al.*, 1967), they do sometimes come ashore and travel some distance to reach a fresh-water stream. Barton (1968), who saw the Amchitka Island otters do this, regarded the drink as a luxury rather than a necessity for the otters; something they enjoyed but could perfectly well do without.

As far as is known, all carnivores drink simply by lapping, as cats and dogs do. The wolverine, however, commonly makes treading movements with the fore-paws as it laps. This Krott (1959) interprets as related to the fact that the wolverine often gets its drinking water from marshy areas and the treading movement serves to push down the vegetation and squeeze out the water so that it forms a puddle from which lapping is possible. This action is so much a part of the drinking pattern that it is done even when there is no need for it and it is actually difficult for a captive animal to drink from a dish without upsetting it. The movement could easily be derived from the infantile milk tread (see p. 333) and may have begun as an accompaniment to obtaining water from marsh vegetation by sucking.

Chapter 6

Signals and social organisation

FOR animals to interact with each other and influence each other's behaviour it is not necessary for them to have actual physical contact. Throughout the animal kingdom extensive use is made of signalling systems, whereby the actions of the signal sender may, from some distance, influence the responses of the receiver. This introduces a time factor into their relations: the signal enables the receiver to respond appropriately not to something the sender actually does to him *now* but to what he is likely to be about to do or to what he has done in the past; to evade a projected attack, for instance, or to follow or avoid the path previously traversed by a conspecific. True social life is dependent on the existence of signalling systems but even solitaries commonly use signals, often as a means of avoiding close contact.

Visual, vocal and olfactory signals form a series with different time and distance characteristics: vocal and ordinary visual signals are non-persistent and vanish with the cessation of the activity producing them; olfactory signals may persist for days or even weeks. Visual signals can carry only over relatively short distances and their main use is in face-to-face encounters; vocal signals can be received over considerable distances and may be particularly useful where thick cover or darkness limit visibility; olfactory signals can be responded to when the animal that made them is far away, out of earshot as well as out of sight. There is, however, a special type of visual signal which also has lasting properties and is independent of the continued presence of its maker, namely those cases where the environment is altered in a lasting way, for example by scraping the bark off a tree to leave a visible mark.

Signals play a major role in the life of mammals and to deal fully with the signals used by carnivores would require a book in itself. It is proposed here to deal mainly with the way olfactory signals are used in organising the spatial relations between conspecifics but it is desirable first to say a little about other types of signals and their functions. Some of the special signals between sexual partners, between the young and between parents and young will be mentioned in the section on reproduction and only more general interactions will be considered here.

Most mammals are capable of a great range of expressive movements: their soft skin with erectile hair, their mobile ears and their form of limb articulation which permits an erect or a crouched carriage, make possible great variations in general appearance. In addition, the tail is not directly involved in the locomotor apparatus and is therefore free to be used as an organ of expression: indeed, in a few mammals, including, I believe, the bobcat, it has no other function. The way in which special markings accentuate parts of the body used in visual signalling has already been mentioned in Chapter 3.

In the majority of mammals the animal's mood is reflected in its carriage. When alert and confident, the bearing is erect and the ears are held up: caution and timidity are accompanied by a tendency to crouch and lay the ears back. Such simple expressive movements in fact prepare the animal to take the sort of action which its current mood is likely to dictate – to go forward boldly and investigate whatever comes or to retreat and seek cover – and they do not necessarily act as signals. More specific actions and postures which undoubtedly do act as signals have, however, been elaborated in most carnivores. Amongst these threat signals, indicating that an attack is imminent (offensive threat) or that if molested further there will be retaliation (defensive threat), are the most striking (plate 11).

Threat is based on three simple principles. The first is to display the weapons used in attack: in carnivores this consists of showing the teeth by opening the mouth and, in some species, by a specific pulling up of the lip to bare the canine teeth. The latter is familiar to us in the domestic dog but it is not universal in the Canidae. It is present in the wolf and coyote but Kleiman (1967) could see no trace of it in *Lycaon*. Seitz (1959) says that the golden jackal bares the fangs, but Kleiman never saw this in the black-backed or side-striped jackals. The aardwolf, with its very reduced dentition, normally threatens with the mouth closed. The second principle of threat is to display maximal apparent size. This is achieved by erection of the hair and sometimes by turning broadside on to the opponent. In many carnivores the specially long and erectile hair, forming a mane on the shoulders or all along the back and on the tail, accentuates the effect. Such apparent size increase is particularly characteristic of defensive threat, where the aim is to deter an opponent from making an attack and the mane of *Civettictis*, a very unaggressive and rather timid animal, is remarkably impressive. The third principle depends on the disconcerting effect of any sudden changes in the sensory input which is being received. In carnivores threat is very commonly accompanied by short, sharp vocalisations; spitting, hissing or growling. The erection of the hair and the changes of posture involved in threat, such as the familiar arching of the back in cats and also the laying-down of the ears which makes such a dramatic alteration in facial contour, are often abrupt and so contribute to the startling effect.

The cat's arched back threat is purely defensive and the holding back of the forequarters which helps to produce it signifies an unwillingness to advance towards the opponent. In mongooses, arching of the back occurs when a distinct element of aggression is also present. *Suricata*, with hair fluffed up and back arched, will move towards the object of its threat in a curious version of the gallop which I have called 'stiff-legged rock'. The limbs are more extended than in normal locomotion, the vertical component of the stride is increased and the horizontal decreased, so that the body rocks up and down but makes little forward progress at each step. *Crossarchus* has a somewhat similar pattern of behaviour which I call 'threat walking'. The hair is fully erected, the back somewhat arched and the limbs much extended; the animal moves forward at a walk, not a gallop, and shows more vacillation than *Suricata*; it may alternately move forwards towards the opponent and backwards away from him. The tail, with maximal piloerection, is held sloping downwards and at every step is swung violently from side to side so that, viewed from in front, it suddenly appears and disappears first to the left and then to the right of the almost globular body. Although to a human observer they may have a comic element, both performances are extremely impressive.

In addition to threat, many species have postures which signify something less extreme, simply the assertion of dominance without the implication that an attack is likely to be made. These are particularly characteristic of species like the wolf, which are social but in which the members of the group do not enjoy equal status. A dominant animal has priority of access to such desiderata as food, favourite resting places or females in oestrus and subordinates will yield place to him.

The main function of threat and of the less striking actions that signify dominance is to produce appropriate responses in the recipient of the signals. It would, however, be very surprising if their performance did not also have its effects on the performer, increasing the self-assurance of the animal displaying dominance, making the animal that bares its teeth more prepared to fight. Such effects are certainly present in ourselves. An experienced professional actor may be able to render on the instant any emotion you care to name but the amateur, required to appear on stage in a high fury, will usually be detected in the wings a few moments before he is due to enter, gritting his teeth, clenching his fists and helping himself into the required mood by performing the appropriate actions. Shakespeare was well aware of this effect, as witness Henry V's exhortation to his troops:

Then imitate the action of the tiger,
Stiffen the sinews, summon up the blood . . .
Now set the teeth and stretch the nostril wide,
Hold hard the breath and bend up every spirit
To his full height.

More recently Wilz (1970) has shown the same principle in action in the stickleback. The performance of parental behaviour patterns by an over-aggressive male changes his mood, so that his tendency to attack his female is reduced and he can then court her more effectively.

Threat behaviour can be of value in a variety of circumstances, both in intra- and interspecific encounters. In social species, however, signals with the opposite meaning are also essential, indicating friendliness, lack of aggression or submission. In an integrated group, where there is cooperative behaviour, something more than tolerance is required. The animals must not only be able to refrain from attacking each other; they must seek contact with each other. If dominance relations are strongly developed this may present difficulties. In his studies of the wolf, Schenkel (1967) showed how this problem is circumvented by the existence of two sorts of submission, which he distinguishes as passive and active (figure 6.1). The former is shown by an inferior when a superior approaches or appears to menace him in any way and has the effect of inhibiting attack. Active submission makes it possible for the inferior to

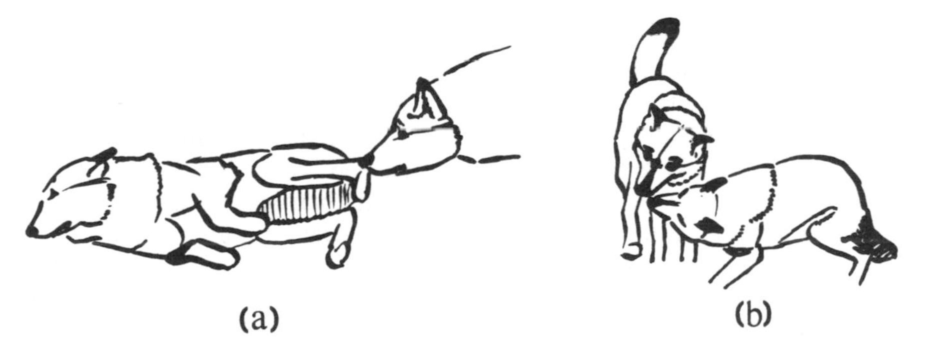

Figure 6.1. Wolf showing (a) passive and (b) active submission. [From Schenkel, 1967]

approach the superior without, so to speak, giving offence or implying any challenge to the latter's status. In both cases the attitudes and postures adopted are derived from infantile behaviour towards the parent: they thus tend to evoke parental responses rather than those which would be directed to a rival. Passive submission consists of lying down and exposing the belly by rolling over on the back, with the tail turned slightly forwards between the legs and the ears laid back, very much as the cub behaves when being cleaned by his mother. In active submission the ears are again laid back close to the head, the tail is held low and may be wagged; a slightly crouched posture is adopted as the superior is approached and attempts are made to nuzzle and lick at his snout. This behaviour Schenkel regards as derived from infantile food begging. The same ways of indicating submission are shown by dogs and may be quite widespread amongst the Canidae. Kühme (1965b) has seen both forms in *Lycaon* and in addition describes how the adult males of the group he studied would sometimes creep under the belly of the oldest female and lick her teats, an action even more clearly related to infantile behaviour. Van Lawick-Goodall and van Lawick (1970) also describe turning away of the head, thus presenting the side of the neck towards the partner, as a submissive signal both in *Lycaon* and in the golden jackal.

In the social mongooses submissive signals do not seem to play any important role, possibly because they are essentially contact animals – the members of a group normally not only show mutual friendliness but seek actual physical contact with each other. The possession of food creates virtually the only situation in which contact is avoided: a potential food thief will be driven off with a snap and a growl; he, however, merely draws back promptly and a fight does not develop. I have not seen anything that could be described as a submissive signal in *Suricata* or *Crossarchus* and Zannier (1965) does not mention any in *Helogale*. Mere withdrawal if threatened, coupled with a sense of ownership of prey on the part of the animal that has captured it, seem to suffice to avoid conflict.

Compared with birds, mammals are very silent animals. The security conferred by their powers of flight has permitted birds to utilise extensively a form of communication which inevitably draws attention to their whereabouts. By mammalian standards, however, carnivores are rather vocal and for the same reason; they are the species that can best afford to be noisy. Wolves and lions, with no larger predator to fear, may howl or roar and have nothing to lose if the message is picked up by others than those whom it directly concerns. Even small carnivores, however, make considerable use of vocal communication. I do not know of any carnivore that does not add to the impressiveness of its threat by including a vocal element of growling, hissing, spitting, snarling or screeching. In solitary species, beside the socially negative threat vocabulary, there is little need between adults for more than a single sound with the opposite significance to indicate the readiness for social contact which is a prerequisite for mating. In social species, intraspecific communication is much more complex and the vocabulary correspondingly more extensive. This is illustrated in the vocabularies of the different species of otters studied by Duplaix-Hall (personal communication). She has found that the more social the species, the larger the vocal repertoire. A comparison between the vocabularies of the solitary weasel, *Mustela rixosa* and the social meerkat, *Suricata suricatta*, shows the same thing very clearly. This comparison will also serve to illustrate the various functions of short-range vocal communication: for a more extensive review dealing with mammals as a whole Tembrock (1963) may be consulted.

Huff and Price (1968) found that the vocal repertoire of adult weasels included three agonistic sounds: a harsh chirp, a hiss and a squeal, used in threat or stress situations, sometimes separately, sometimes in combination. The only other sound the animals made was the trilling call given by both sexes when ready to mate. This call is not specific to the mating situation but is a more generalised indication of the need for contact with a conspecific, for Heidt *et al.* (1968) found that the female used the same call to summon her young. *Suricata* also has three threat sounds, a growl, an explosive spit and the harsh repetitive scolding, which the Germans expressively call '*keckern*'. In addition, however, the adult vocabulary includes seven other sounds:

(i) The almost continuous contact sound or 'conversation' that accompanies almost any activity and, like the flight calls of birds, serves to prevent straying. Social mongooses all have contact sounds of this sort, which is not surprising, in view of their small size and the ease with which they may be concealed even in low vegetation.

(ii) Satisfaction noise made when eating, which can alert a conspecific to the presence of a rich food source.

(iii) Settling-down sound, made when going to sleep. There is no evidence that conspecifics respond to this sound but since the family group all sleep together, it may well have a mood synchronising effect, making the companions more ready to rest also.

(iv) The pure fear call, which serves as a warning of the presence of an aerial predator and to which the caller's companions respond by gazing upward to locate the source of danger.

(v) The mixed fear and aggression call which serves as a warning of danger on the ground.

(vi) Alarm barking, which is contagious and alerts to some general but not clearly localised source of disturbance.

(vii) There is also a dissatisfaction call, which in my captive animals I heard made by the male only. This remained enigmatic until after I had learned something of the vocabulary of *Atilax*. Both animals made a very similar call to me, indicating dissatisfaction with something which I was expected to set right; for instance, I should adjust my position so as to make it easy for them to climb on my lap. In *Atilax*, however, there was a complete gradation from this lowest intensity call to the truly horrifying maximal intensity of the threat screech. The 'dissatisfaction' noise is therefore an extremely low-intensity threat, signifying nothing more than dissatisfaction with the current situation and thus equivalent to 'I want . . .'. In *Suricata* the sound exists only in this low-intensity form and the threat vocabulary does not include a screech. It was frequently addressed to me but I never heard it made to a conspecific, so there is no evidence as to what their response might be.

Suricata's adult vocabulary thus includes at least ten sounds which are easily distinguishable to the human ear, to eight of which appropriate responses by conspecifics are equally easily seen and of all these, only three are socially negative threat sounds. There is, however, one curious deficiency in their vocal repertoire. Although most carnivores will yelp, yowl, squeal or scream in response to sudden pain, *Suricata* appears to have no pain cry. A tail inadvertently trodden on or a paw caught in a door evokes only growling and an attempt to remove the offending object. *Nandinia* (Vosseler, 1928), *Helogale* (Zannier, 1965) and *Genetta rubiginosa* (Rowe-Rowe, 1971) also lack a pain cry but the significance of this peculiarity remains obscure. *Crossarchus*, however, does have a pain scream.

In the social Canidae, where visual signals are so important in close-range communication, the vocal repertoire is not as large as in *Suricata*. The Canidae, however, make use of loud and carrying sounds as a means of long-range communication in a way not found in smaller and more vulnerable species. The howling of wolves is the best-known example. Howling acts as a means of maintaining or re-establishing contact between members of a group who are separated and thus facilitates coordination of their activities. If one animal howls, others will answer, thus indicating their whereabouts and enabling any member who has become separated from his group to find them again: indeed, Theberge and Falls (1967) were able to utilise this response to locate the members of the group of wolves they studied in Algonquin Park: if they howled, the wolves would reply and thus disclose their positions. Howling also has another function: it is contagious and if one member of a group howls the rest join in and all will howl together for some time. This choral howling, uniting group members in a common activity, appears to have a socially bonding effect, much as community singing has in ourselves. Crisler (1959) found that she and her husband could cement the ties of friendship between them and their tame wolves by joining in such choral howling sessions.

The calls of different species are generally highly distinctive and it should therefore be perfectly simple for an animal to distinguish the cries of a conspecific from those of other species. In many cases, however, voices also have individual characteristics. Theberge and Falls found that although the howls of a single wolf showed considerable variation, there were also personal characteristics, typical of the individual. Burrows (1968) could distinguish the barking of individual foxes by differences in pitch and in the number of barks forming a sequence and Eisenberg and Lockhart (1972) have found that leopards show individual differences in the number of calls per strophe. Van Lawick-Goodall and van Lawick (1970) found they could identify crocutas by their distinctive voices and the voices of individual cats and dogs are also distinguishable. It is therefore highly probable that the animals themselves can learn to recognise each other's voices and know who it is that is calling or recognise that a voice is an unfamiliar one. This means that the same call can function in two different ways: it can facilitate the locating of companions and at the same time serve to discourage the approach of strangers. According to Schaller (1967, 1969c) the roaring of tigers and lions has both these functions and Eisenberg and Lockhart (1972) were of the opinion that the same was true of the grunting call given by the leopards they studied in Ceylon. The howling of golden jackals too may have a social bond-forming function between members of a family group and at the same time serve as a territorial warning to outsiders. Van Lawick-Goodall and van Lawick (1970) describe how the individuals in a group will howl not in unison but in succession and when they have finished the neighbouring groups will follow suit, one after the other. In this way each group declares its presence to all the rest. It is as though one group having announced 'Here we are, in *our* place', the rest replied, 'and we are here in *ours*'. The barking of foxes, on the other hand, may be essentially a socially negative sound, acting as a deterrent to intruders. Burrows (1968) found that foxes were particularly prone to bark during winter, at the time when territories were being set up, with a peak during the mating period; he also found that they kept to definite routes while they were barking. He considered that the barking was aggressive and believed that it was done by 'dog foxes barking out their territory and warning off would-be intruders'.

Although vocalisations clearly can be used in this way to warn off strangers from an occupied territory, they have the disadvantage that one cannot always be making a noise. Chemical signals have the advantage of persistence: like the GONE TO LUNCH note pinned on the door, their message can be read in the author's absence. They therefore constitute the ideal means of indicating ownership of a territory and this is one of the major functions of scent marking.

In Chapter 3 the various types of special scent-producing glands which are found in carnivores have been described. In addition, both urine and faeces can carry scent and both can be used as marking substances; saliva, too, may sometimes have an auxiliary function in marking. Faeces are normally coated with mucus secreted by the gland cells of the large intestine but they must also carry the scent of the modified skin glands around the anus and the secretion of the anal sacs could easily be added. Since sex hormones are excreted in the urine, urine may carry information about sex and reproductive condition and may be enriched by the scent of the accessory reproductive glands in the male. Schaller (1967) believes that when the tiger and lion use urine as a marking substance the secretion of the anal sacs is added to it. I take this to mean that the backwardly directed jet of urine acts as a propellant for drops of anal sac secretion which are ejected at the same time but the details have not yet been described, nor has it definitely been shown that this does occur.

The place chosen for setting a mark will obviously affect the likelihood of its being noticed by a conspecific and in many cases the behaviour used in making the mark is also adapted to make it more easily detected. The simplest way for a quadruped to urinate without soiling its coat is to spread the hind legs a little, squat slightly and direct the urine downwards: the simplest method of applying the secretion of the anal sacs is to squat down and dab or drag the anal region on the substratum as the secretion is expelled. These are the ways in which the two activities are most commonly performed but in a number of species the behaviour is modified so as to place the odour-carrying substance above ground level, at approximately nose height. The most familiar example is, of course, the way a dog cocks one leg and directs his urine sideways against some vertical object. Leg cocking is shown by the males of every species of canid that has been investigated, and in a few species the female also shows the same pattern. More usually she raises the leg less than the male and in many species does so only around the time of oestrus: at other seasons she merely squats and in some species this is the normal female behaviour at all seasons. Kleiman (1966b) summarises the information available but it is not yet possible to see whether there is any correlation between the micturitional behaviour of the female and the roles played by the two sexes in territorial and social organisation.

The felid habit of directing the urine backwards when using it for marking has already been mentioned: this, like the canid leg-cocking, has the effect of distributing the urine on a bush, tree trunk or the like, in such a way as to catch the attention of any passing conspecific nose. This method of urinating is very widespread amongst the Felidae. Fiedler (1957) has recorded it in lion, tiger, leopard, jaguar, serval, ocelot and cheetah[1] but he never saw captive pumas of either sex micturate otherwise than in a squatting posture. The retromingent habit is also found in civets: in *Civettictis* only the male urinates in this way and in *Viverricula*, according to Dücker (1965a), male urine spraying is restricted to the breeding season.

In a number of mongooses, although the anal sac secretion is applied to a horizontal surface using the simple anal drag, it may also be deposited at nose height if a suitable vertical object is available. *Suricata* achieves this by cocking one leg, very much like a dog micturating, at the same time everting the anal pouch and smearing the secretion down the object to be marked (Ewer, 1963b). The large Indian mongoose, *Herpestes edwardsi*, may do the same but will also set the mark the full body's length above ground by rearing up on the fore paws into a hand-stand position (Dücker, 1965a). *Atilax*, *Helogale* and *Crossarchus* (plate 14) also use the hand-stand technique but in *Atilax* and *Crossarchus*, at least, I have found the male much more prone to do so than the female. In the two species with which I am most familiar, the structure of the anal pouch is clearly related to the method of applying the anal sac secretion. In *Crossarchus* (figure 6.2), a series of dorsally-situated longitudinal grooves is ideally placed for leading out the secretion as the anal region is wiped downwards in the hand-stand position: in *Suricata* the opening of the pouch is horizontally elongated and the anal sacs open one on either side, in a position suited to the leg-cocking technique. Fiedler (1957) records a male genet (*G. tigrina*) using the hand-stand technique but he did not see the female do so. The use of these methods of applying the anal sac secretion, although commonest in the viverrids, is not restricted to them. Fiedler saw the male red panda mark raising one leg, as *Suricata* does, and Kleiman (1966b) describes how the female bush dog urinates in the hand-stand position but from her account I suspect that the secretion of the anal sacs is also being applied.

The civets, *Viverra*, *Viverricula* and *Civettictis*, apply the secretion of the perfume gland in a slightly different way. If a vertical surface is to be marked, the animal backs up to it, much like a tom-cat about to urine spray; the lips of the gland are then everted and it is pressed

[1] According to Eaton (1970c) only the male cheetah urine-marks in this way.

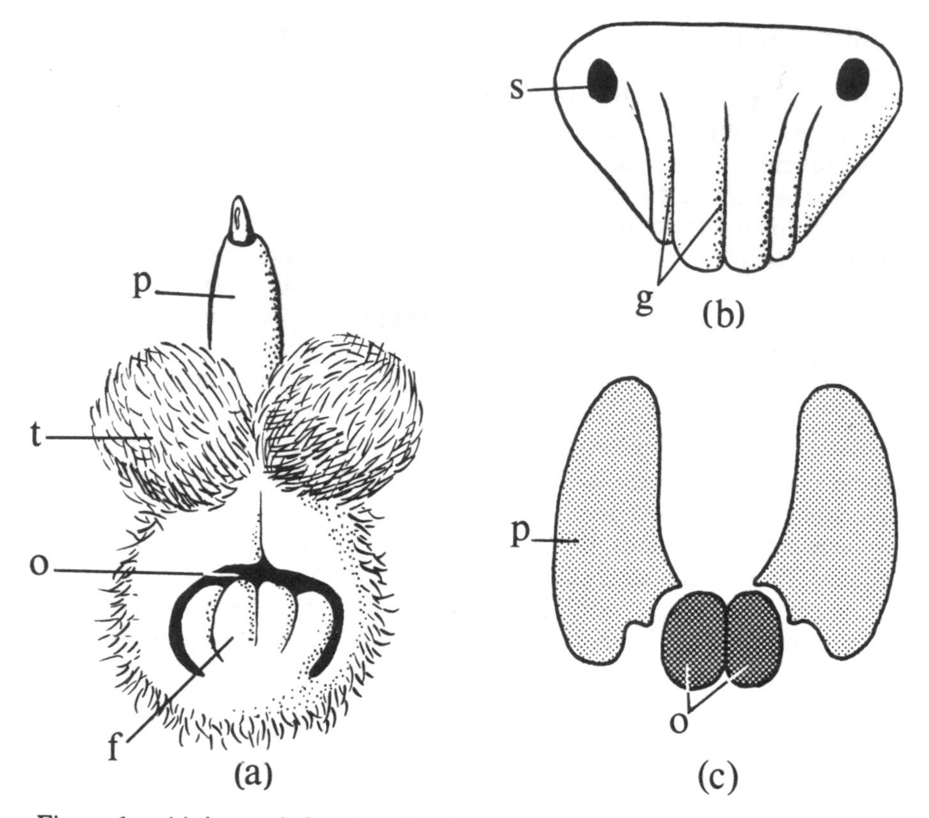

Figure 6.2. (a) Anogenital region of male *Crossarchus obscurus*. f, folded dorsal wall of anal pouch; o, opening of pouch; p, penis; t, testis. (b) Anal pouch pulled open dorsally to expose the openings of the anal sacs, s, and the grooves, g, serving to lead out the secretion. (c) Mark of adult male civet on a vertical surface. o, heavy deposit of secretion from the openings of the inner pocket; p, lighter deposit made by the everted lips of the gland.

firmly against the surface, leaving a double mark corresponding to the openings of the gland on either side (figure 6.2). On a suitable flat surface *Civettictis* will deposit a line of such marks, all on one level and each quite distinct. Burrows (1968) describes how the fox applies anal sac secretion by backing up, raising the tail and rubbing the anal region vigorously against a tree or other upstanding object and similar marking is done by the female giant panda (Morris and Morris, 1966). The little African polecat, *Poecilogale*, backs up in much the same way to deposit its faeces against a vertical surface (Alexander and Ewer, 1959). Possibly *Crocuta* may back up to apply the secretion of the anal glands to a vertical surface but the boundary that has to be marked usually has nothing more convenient than grass stems on which to deposit the secretion. Van Lawick-Goodall and van Lawick (1970) have described how this is done. Squatting slightly, the animal moves forwards slowly, so that a stem is bent down between the hind legs; the anal pouch is then everted and the secretion of the glands smeared on the stem. A whole group may successively mark the same object in this manner, the top-ranking female being the first to deposit her scent.

If trees are the objects to be marked, then it is possible to produce a visual signal by chewing off bark or branches or by scratching with the claws. These activities will incidentally transfer the scent of saliva or foot glands but further odours may be added by rubbing various parts of the body on the tree or by urinating. The use of special marking trees is probably most familiar in brown and black bears. The bark is ripped off with teeth and claws and Tschanz *et al.* (1970) have described in detail how the brown bear rubs various parts of its body, particularly the neck and shoulders, on the marking tree (figure 6.3). This rubbing is sometimes accompanied by urination. Less widely known is the use of trees as marking signs by the wolverine. According to Haglund (1966), the same tree is repeatedly chewed and bitten so that its dilapidated appearance becomes very

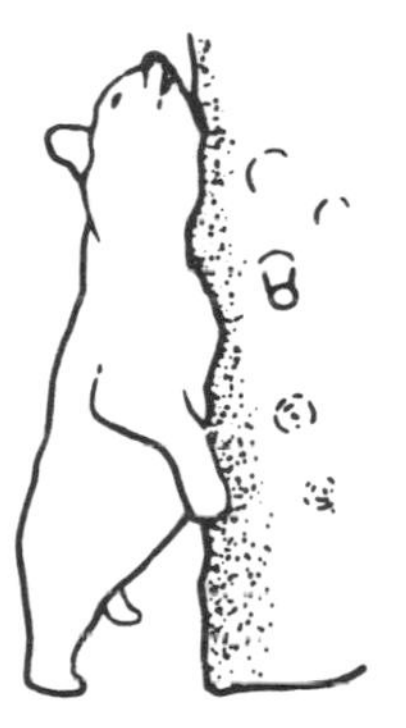

Figure 6.3. Brown bear rubbing its body on marking tree. [From Tschanz *et al.*, 1970]

obvious but he does not mention whether urine or the secretion of the ventral gland is deposited at the same time.

Tree scratching is familiar in the domestic cat and is often referred to as claw sharpening. Although it may serve to neaten up the claws by removing loose fragments which are ready to flake off, the way in which strategically situated trees are selected and repeatedly used indicates that this scratching has a definite communicatory function. Tree scratching is recorded in the African and European wildcats (Smithers, 1968a, Southern, 1964), in the bobcat (Young, 1958) and also in the leopard, where it is combined with rubbing the body and urinating (Eisenberg and Lockhart, 1972). Schaller (1967) says that although tigers do occasionally scratch trees this does not appear to be a very important method of marking and was seldom done by the tigers in his study area. The same appears to be true of the lynx (Saunders, 1963b) and puma (Hornocker, 1969). The tiger and puma, however, do use scratch marks but in a slightly different way. Schaller (1967) has described in some detail how the tiger makes scrape marks which are frequently associated with the deposition of faeces and occasionally with urine. He points out that during the dry season scrapes are not very visible on the bare dry soil but droppings remain obvious for a long time. In the wet season the opposite is true: dung beetles remove or bury the droppings within a few hours but the paw scrapes remain as conspicuous marks where the grass is torn away and the soil exposed. The scrape and the faeces therefore complement each other. In the puma similar scraping associated with urination or defaecation has been described by Hornocker (1969). Amongst the viverrids, *Crossarchus* will mark on top of or beside the urine or faeces of a number of other species and when doing so will apply the secretion of the anal sacs, urinate and also scratch intensively.

The various ventrally situated glands which are present in arboreal species are presumably simply rubbed on a branch but *Nandinia* is the only species in which I have witnessed the process. The body is lowered until the perfume gland is in contact with the branch and is then dragged forwards slightly. In watching this process it is obvious that the pre-genital position of the gland is very convenient for marking a branch. It is therefore probably in adaptation to their arboreal habits that in the Asiatic paradoxurines, too, there is a tendency for the perfume gland to be more anteriorly situated than it is in the terrestrial civet cats, where the 'backing up' method of application is used.

Before turning to consider the functions of scent marking in social organisation, it is necessary to ask what sort of information a scent mark may transmit. The most detailed investigation that has been made on the constitution of marking substances relates not to a carnivore but to the tarsal glands of the black-tailed deer. Brownlee *et al.* (1969) have shown that the male and female glands differ in the relative proportions of the components which make up the secretion. Furthermore, there is behavioural evidence that the deer can distinguish individual odours, presumably on the basis of more subtle quantitative differences. It is easiest for us to appreciate what this means in terms of a visual analogy. It is as though all females carried a flag coloured some shade of red or orange and all males had a blue or green one. Within these broad limits, each individual has his or her own characteristic hue, so that sex is at once obvious and with a little more attention one can say . . . 'but I have never seen exactly that shade before: he/she must be a newcomer.'

There is no reason to suppose that carnivore secretions are simpler, so as a working hypothesis we may take it that the scent mark carries information about sex and also that it has individual characteristics. The fact that in carnivores, sniffing at the partner's anal region is a very usual routine in any encounter between conspecifics strongly suggests that this is indeed the case. Moreover, if dogs can distinguish two people by their smells, it is hard to believe that they cannot distinguish individuals of their own species in the same way. It should therefore be possible for a mark to be read as signifying 'a male was here and it was the same fellow as is always around in these parts' – or, conversely, 'and he is a stranger to me'. Since scents gradually evaporate, the difference between a fresh mark and an old one should also be detectable and the information 'and he has only just gone' may be added. There is also the possibility that the extent of scratches or the height above ground of a mark may give some measure of the size and strength of their author. Certainly in my pair of civet cats, the male is bigger than the female and his mark on a table leg is regularly some 5 cm higher than hers. Some messages may therefore also say 'and he is bigger than I am'.

It is not, of course, either necessary or justifiable to suppose that the animals read and analyse the information in these human terms: all that is required is that they should respond appropriately to the different characteristics of the marks. Thus, in another male a male mark may arouse some combination of fear and aggression, the tendencies to retreat or to seek a fight being differentially affected by size and freshness. Eaton (1970c), for instance, found that cheetahs would avoid following the route taken by another group if marking points carried urine traces less than twenty-four hours old but if the marks were older, although they were noticed and sniffed at, the animals would continue their way along the same route. Familiarity, too, may cause responses to wane, so that the mark of a known neighbour may have much less effect than that of a stranger. A female mark, on the other hand, may be of little interest or may arouse sexual tendencies, depending on whether its author is in oestrus or not and on the condition of the male that encounters it.

Differential effects of the same scent on different species are illustrated in some unpublished work by Dr N.B.Todd, which he has kindly allowed me to quote. He tested the responses of various species to catnip and found that a number of viverrids, although clearly able to smell the catnip, showed little interest: for them it carried no message and had no particular significance. I have found the same to be true of *Civettictis* and *Crossarchus*. In the Felidae, the cat-type sexual response was widespread but not universal. Amongst the large cats, lion, leopard, jaguar and snow leopard responded sexually but adult tigers did not. The behaviour of immature tigers was interesting: they were not sexually excited but instead showed violent alarm and retreated promptly. In view of the propensity of the larger species of felids to kill each other's young and of the fact that until quite recently the ranges of tiger and lion overlapped very considerably, both the absence of sexual response in the adult tiger and the fear generated in the young may be adaptive: the former may reflect a sexual isolating mechanism; the latter may be protective.

Apart from illustrating differential responses to the same odour, these observations also demonstrate the psychological effects of odours. A piece of material impregnated with catnip is neither a mate nor an enemy but it can evoke sexual responses or flight: to do so it must act on the central nervous system so as to change the animal's mood – in short, it 'makes him sexy' or it 'makes him afraid'.

Appreciation of these psychological effects is essential to our comprehension of the role played by scent deposition in social life. The signals we ourselves use can have two different sorts of function. Some are simply informative and show us how or where to do something which we have already decided to do – for example the notice that tells us the next train is an Edgware Line, or the voice that directs passengers travelling by Flight XYZ to Rome to proceed at once to Gate No. 24. The poster which shows us a luscious female leaning against a romantic ruin, all bathed in glorious sunlight, on the other hand, is trying to make us feel that we simply must go to Rome on our next vacation. The first type of signal is purely informative and we respond to it by using our intellects: the second succeeds by working on our emotions. Of course, the two functions are often combined. The poster designed to lure us to the motor show with pictures of our dream girl driving our dream car also tells us which buses will take us there. The same is true of animal scent marking: some of the messages may be purely informative but in others the hidden persuaders may be more important, influencing behaviour not only by evoking immediate mood shifts with appropriate responses but also by causing more long-lasting physiological changes in the nervous and endocrine systems. This is particularly true of behaviour related to reproduction. Successful mating requires not only that male and female should meet but that they should be physiologically and behaviourally adjusted to each other. In very many carnivores there is a great increase in the frequency of scenting behaviour during the mating period and, while many species remain to be investigated, I know of no species in which this has been shown not to occur. In territorial species this increased activity is, no doubt, partly bound up with the fact that ownership becomes vitally important at this period and in solitary species it probably assists the partners to find each other. It very likely also has a physiological role, increasing the readiness of the partners to respond sexually to each other and synchronising their activities, so that mating finally occurs at the time when fertilisation and conception are most likely to follow. Thus such increased scenting associated with reproduction does not automatically imply that the species concerned is either territorial or solitary. Such behaviour can also have a role in the life of a social or a non-territorial species.

Scent marks are not randomly distributed. In general, they are set on much frequented pathways, particularly at crossing points, at specially important places such as the entrance to a den or beside a food cache, or else near the boundary with a neighbour's territory. We can hardly suppose that the setter of the mark is thinking out what its effects are likely to be on some other individual at some future time: the mark must surely have significance here and now for its author and making it results from the present situation acting on him in his current mood and physiological condition. The tendency to think in terms of the information content of animal signals is rather apt to make one neglect the emotional aspects involved and, in particular, to forget that both the making of a mark and the smell of it later on can have psychological effects on its author as well as on other individuals. Captive animals will usually mark if they are brought into an unfamiliar place or if they encounter a new and unfamiliar object or person in their home: if they are slightly alarmed they may also mark and a sudden scare in many species causes an apparently involuntary emission of anal sac secretion. All this suggests that marking is done in situations producing some combination of anxiety and aggression and that the familiar smell of the animal's own mark has the effect of making him more self-assured or less anxious. Some such relationship would accord very well with the places where marks tend to be set. Near a neighbour is the place where anxiety and aggression are most likely to be aroused and there must always be some tension at the moment of emergence from the safety of the den. There is, however, much more to it than this. From watching the behaviour of wild animals or of tame ones in relatively free conditions, it is clear that, once routine marking points have been established, the marks are repeatedly renewed without there necessarily being any particular tension involved and the animal may make his way to the 'right' place with every appearance of deliberate intention in just the same way as he may go to a routine feeding or drinking place. Adamson (1968) tells how one of his lions, riding home on the roof of his Land-Rover, would insist on being allowed to get down and renew the scent of every routine marking point along the way. One must conclude that marking, although it may originally have been a direct response to situations of tension, has become, in the course of evolution, an activity in its own right which the animal can and does perform without having to be scared or angry.

There are also cases where the opposite has happened: the link with situations causing fear has become closer and the anal sac secretion has become primarily a defence mechanism, used most frequently against other species, rather than a means of communication with conspecifics. Mustelids are in general rather prone to emit anal sac secretion if startled and in many of them the odour produced is distinctly unpleasant and likely to cause a predator to decide that he has not, after all, discovered a potential good dinner. This tendency has reached its highest development in the skunks, where the secretion, containing butyl mercaptan, is nauseous in the extreme and can be projected to a considerable distance. The same is true but to a lesser degree of the secretions of the African mustelines *Ictonyx* and *Poecilogale* and also of the mongooses *Herpestes urva* and *H. vitticollis*. The occurrence of black and white warning colours in these species has already been mentioned. In the skunks, however, evolution has gone still further, and there is special threat behaviour associated with the use of the anal sac defence mechanism (figure 6.4). The spotted skunk, *Spilogale*, will run at an opponent, then check abruptly and throw up the hindquarters so that it balances on its hands with its tail turned to one side and the everted openings of the anal sacs directed forward against the enemy (Howell, 1920; Johnson, 1921): in this position the skunk may walk on its hands for a distance of over a metre (Crabb, 1948). This threat, however, is largely bluff: it is very rarely followed by actual discharge of the secretion (Walker, 1930; Gander, 1965). Walker found that if the skunk was actually about to 'fire' it usually remained with all four feet on the ground but bent its body into a curve, so that both head and tail faced the enemy and in this position the discharge occurred.

Figure 6.4. (a) Striped skunk in firing position. (b) Hand-stand threat posture of spotted skunk. [Drawings, F. Bourlière]

According to Crabb (1948), this performance may be preceded by stamping with the fore feet in rapid alternation. The striped skunk, *Mephitis*, also adopts the hand-stand threat posture (Seton, 1920) but much less readily than *Spilogale* and the commonest and most easily evoked threat in this species is fore foot stamping, combined with arching of the body. The animal may then rock back and forth from hind to fore paws and may even go over into the full hand-stand position but this is rather unusual (Verts, 1967). The skunk's ability to hit its target is based on a good supply of ammunition rather than on any great accuracy of aim. According to Verts, the skunk swings its body through an arc of up to 45° as it sprays, thus greatly increasing the chance that some, at least, of the fluid will find its mark.

In species where the anal sacs have taken on a purely defensive function, one might expect either that territorial marking would not be practised, or that some other secretion would be used for the purpose. The former seems to be the case in the non-territorial skunks and possibly the development of ventral glands in some mustelines may be related to a largely defensive use of the anal sacs. The wolverine emits its very smelly anal sac secretion only when alarmed and uses the ventral gland for marking (Krott, 1959) and the same is true of various species of *Martes* (Markley and Bassett, 1942; Herter and Ohm-Kettner, 1954). Schmidt (1943) describes the pine and stone martens as using the anal sacs for marking but his photograph quite clearly illustrates the use of the abdominal gland. The pine marten also makes use of urine and faeces as marking substances (Lockie, 1966).

In other cases it is less easy to understand why the animal should require more than one scent gland. Clearly it is reasonable to suppose that if there is more than one sort of scent gland, then more than one type of message has to be transmitted. The converse, however, is not necessarily true. Where glands are histologically complex and therefore presumably capable of producing more than one type of substance,[1] the psychological state of the animal may affect the composition of the secretion and a single gland may have more than one function. Donovan (1969), for instance, has found that if a dog on the veterinary surgeon's table emits its anal sac secretion in alarm, then objects smeared with this secretion are avoided by adult animals of both sexes, a response not found to secretion from an animal that is at ease. Psychological effects of this sort may well be involved in the anal sniffing which forms a normal part of an encounter between two dogs. Animals of approximately equal status will hold the tail high and permit anal investigation; inferiors will usually keep the tail firmly down between the hind legs and attempt to avoid being sniffed. Possibly the characteristics of the anal sac secretion mirror the animal's degree of self-confidence and his status may be declared not only by his overt behaviour but also by the olfactory characteristics of the secretion he emits. Secretions may also change in relation to reproductive condition. Donovan (1967) has found that the anal sac secretion of a bitch in oestrus is sexually exciting to a male but he remains unmoved by that of a female not on heat. It is not known whether such changes in anal sac secretion are widespread but clearly a species in which the composition remained unaltered might require a separate sexual signal and, conversely, a separate signal which remains invariant might be required to carry messages which are uninfluenced by breeding. More information about the chemical composition of anal sac secretions in relation to reproductive cycles is required before we can see whether there is any correlation of

[1] Albone and Fox (1971) have recently found that the anal sac secretion of a female red fox contains at least twelve volatile components: the majority of these are saturated carboxylic acids but the base, trimethylamine, is also present.

their properties with the presence or absence of auxiliary marking methods.

One species in which different secretions have been shown to have different functions is the dwarf mongoose, *Helogale parvula*, studied by Dr A. Rasa. She has found (personal communication) that the cheek gland secretion carries a hostile message and it is possible to evoke a threat display from a dwarf mongoose by presenting it with the secretion of a conspecific smeared on a glass slide. She has also shown that the mongoose can tell the difference between the anal sac secretions of different individuals and that the secretion of an unfamiliar animal arouses interest, without signs of fear or aggression. It would therefore seem that in this species the anal sac secretion carries no specific threat but is quite simply the personal signature of the animal that made the mark: cheek gland marking, on the other hand, is an olfactory threat but it is an anonymous one; it does not itself carry a personal signature. In fact, the signature is always added, for the animals never mark with the cheek glands without also marking with the anal sacs. Exactly how one mongoose responds to another's mark will, no doubt, depend on whether the signature is familiar, how fresh it is and the detailed circumstances in which it is noticed. A fresh foreign mark near one's own burrow might well have a different effect from a rather stale one, encountered when foraging near the limits of one's range. It seems clear, however, that while there is nothing intrinsically aggression-rousing in anal sac secretion, the same is not true of the cheek gland secretion: this does have effects on the emotional state of the animals and the responses to it may therefore be simpler and less dependent on the whole complex of circumstances in which it is encountered. In the one case the animal is, so to speak, free to weigh up the situation and act accordingly: in the other, his judgement is swayed by emotion.

Another interesting point discovered by Dr Rasa is that while the anal sac mark lasts for something of the order of a fortnight, the olfactory threat is short-lived and the smell is no longer detectable by the animals after forty-eight hours. The adaptive significance of this is clear. The signature, indicating who habitually frequents the area, is lasting but the threat, which presumably relates to some specific incident, is not. The mongooses may not quite fulfil the biblical injunction 'let not the sun go down upon your wrath' – but they come very close to doing so.

In the banded mongoose, too, cheek marking appears to have threat significance, to judge from the behaviour of a captive specimen in the possession of a colleague of mine. This animal, originally friendly, presently became intractable and bit all members of the household. Biting was usually preceded by intensive marking with the cheek glands as well as the anal sacs. In *Crossarchus* I have never seen cheek gland marking unaccompanied by anal sac marking: I have not, however, been able to demonstrate that the former acts as a threat in this species, nor that the secretion evokes any signs of aggression.

In a social species, a personal signature, in addition to whatever role it plays in territorial demarcation, may also have a function in the integration of the group. Within the group, smells must be recognised and responded to as signifying group member and friend but aliens must be recognised as such and treated differently. In *Helogale*, members of the group or sexual partners mark each other with anal sac secretion and routine marking of particular objects in the environment is regularly carried out. Keeping one's smell fresh in the memory of one's fellows appears to be a matter of vital importance, as illustrated by the following incident. On one occasion two of Dr Rasa's animals were lost for three days, during which time they could obtain neither food nor water. When found they were thin and weak but on being returned to their fellows, their first action was not to eat the food which they clearly desperately needed but to go and renew their marks on the routine marking places.

In *Suricata* the problem of designating group membership appears to have been simplified. In my family group, the female rarely marked but the male did so very frequently and his behaviour was such as to ensure that his scent was carried by every member of the group. When he marked, he very often rubbed his body along the mark, thus scenting himself all over very thoroughly. Since the group all sleep together piled on top of each other in a heap, all must become impregnated with his scent. Furthermore, the male also marked the entrance of the sleeping box very assiduously, with the result that every individual frequenting it automatically brushed against his mark as they went in and out. Even in the relatively asocial cat, scents may have a role to play in indicating friendship. Prescott (in press) has recently shown that rubbing against a conspecific of a glandular area situated between the eye and the ear is a frequent accompaniment of friendly interchanges (see plate 12).

The function of the perfume gland of viverrines is still very imperfectly understood. My very friendly pair of *Civettictis* often mark on top of each other's marks and they neither mark nor emit scent when alarmed. This suggests that the main function of scent deposition may be to assist male and female to find each other. On the other hand, they will mark foreign objects, although not immediately in the way a mongoose will: it seems that they must wait until some degree of familiarity has made the object less alarming before they are ready to mark it. Possibly the perfume glands were evolved originally as pure sexual signals. They may at first have been used in conjunction with anal sac marking but have gradually come to take over the functions of territorial marking as well: anal sac marking has consequently become redundant and been abandoned. Certainly I have never seen either sex mark with the anal sacs but it is possible that their secretion may be added to the faeces.

One curious form of behaviour connected with scents remains to be mentioned. A number of carnivores will rub the chin, neck and shoulders or even roll the whole body in a variety of strongly smelling substances (Heimburger, 1959). Many canids do this on carrion and civets (plate 14) are exceedingly prone to roll in any strong-smelling or unfamiliar animal food before they eat it. This response is evoked most readily by anything which is slightly rotten and by the gut of any form of prey – bird, reptile or mammal – but I have never seen them roll on vegetable foods. Crocutas also roll on similar objects in much the same way but the thing which throws them most readily into an apparently almost ecstatic frenzy of rolling is the mass of undigested hair which they frequently vomit up after a good meal (van Lawick-Goodall and van Lawick, 1970). This behaviour is not understood: it is not even clear whether its main function is to transfer the animal's smell to the object which is rubbed or vice versa. In the civets it certainly has connections both with eating and with sexual behaviour. In the absence of any knowledge of the chemical stimuli actually involved, however, one cannot tell whether the behaviour is merely the result of some fortuitous similarity between the animals' own secretions and the decomposition products of decaying animal protein or whether it has some deeper significance. The rolling of domestic cats in catnip, however, very probably does depend on a chemical coincidence of this sort (Palen and Goddard, 1966).

One is naturally inclined to assume that the more complex scent-producing organs are and the more important the role their secretions play in the life of a species, the more highly evolved they must be. This is true only up to a point, for secondary simplification can also occur, particularly in species where individual recognition plays an important part in social life. The persistence that gives odour marks their value in certain situations is the very quality that renders them unsuitable in others. A vocal or visual signal can be varied from moment to moment, according to changing circumstances; it can be directed to one individual and not to another and the responses to it can be equally directional. The same is not true of a scent mark: it stays where it is, unaltered except for its gradual fading. The responses to it must necessarily be of a rather generalised nature: fear or aggression may increase or decrease but it is not the mark itself that must be attacked or appeased. It is therefore not surprising that in the Felidae and the larger social Canidae, which on other grounds are regarded as advanced, scent-producing organs are not particularly complex and visual and vocal signals play a relatively large role in social integration.

Chapter 7

Social organisation and living space

BEFORE starting to deal with living space, it is desirable to define the ways in which certain terms will be used. Few mammals are truly nomadic: they usually remain within a certain area at least long enough to become familiar with its main features, so that it constitutes their home at any rate for the time being. Home range may therefore be defined as the area covered by an animal in its normal day-to-day activities. The usefulness of the concept of home range is not destroyed by the fact that occasional sorties or prospecting trips may be made beyond its limits.

Although we normally think of a piece of ground as an area which we may map completely and may buy, sell or fence in as a unit, there is no reason to suppose that a home range has exactly the same sort of significance for other species. For some it may consist rather of a set of important places used in different ways: feeding and drinking places, resting or basking places, defaecation sites and marking points, etc. together with a series of pathways linking them in various ways – there may be intervening regions almost unfrequented and of little or no significance. It is therefore possible for neighbouring ranges either to overlap or to interpenetrate without effectively overlapping. The whole of the home range is not necessarily equally utilised: there may be a region, usually round the den or resting place, in which the animal spends most of the time and in which the majority of its activities are concentrated. Such a region may be designated as a core area within the home range.

A territory is defined as an area from which conspecifics are actively excluded. Clearly a territory may be congruent with a home range or may be smaller. If the core area is defended but the periphery is not, then the core area constitutes a territory within the home range. In the extreme case the only place actively defended may be the den or favourite refuge.

The definitions which have been given imply that the term 'territory' should be applied only to an area which one knows is in fact defended. 'Home range' has no such implications and may be used without prejudice as to whether the area is or is not defended: it does not necessarily imply absence of defence and it is the only term one can legitimately use if there is no information as to whether conspecifics are or are not excluded.

The concept of home range is straightforward: the animal is familiar with a certain area, knows its way about and is aware when it is venturing into *terra incognita*. Home range, in short, is a function of the reactions of the animal to its habitat and these reactions are probably very much like our own – a sense of assurance and ease in the known and familiar, of anxiety or excitement in the new and strange. Territory is more complex. Defence implies an opponent and to hold a territory implies the presence of neighbours. What we call a territory is the result of the ways in which neighbours react to each other. These reactions are very strongly influenced by the animals' responses to home or to foreign ground but they can and do alter in relation to other circumstances. We cannot therefore expect to find that territories are always as clearly defined, as permanent, or that they have such clear rules governing their use as, say, a series of suburban gardens. Moreover our view is an external one: we say that the animal behaves in certain ways because it is or is not in its own territory. We cannot expect such abstraction and objectivity from other species and it is equally legitimate to say that it is because the animals behave in particular ways that the territories become demarcated as they do.

Obviously no individual mammal can hold a personal territory the year round for all of its life. There must be tolerance at least of a mate for a brief period and the female must share her territory with her young for longer. Animals living as solitaries for part of the year may defend a joint territory as a mated pair during the breeding season, while in a truly social species the territory is shared by all the members of the group and only strangers are excluded.

The area occupied by an animal, whether defended or not, clearly must contain all the necessities of life. A sufficiency of food, water and shelter are the most obvious requirements but provision must also be made for the establishing of normal social relationships with conspecifics, including those relations involved in reproduction and care of the young. It is therefore necessary to take account not only of food requirements but also of social organisation and breeding behaviour when considering living space and its utilisation. Clearly both kinds of factor may bring about seasonal changes in requirements and to understand fully how any species uses its living space, a year-round picture is required. On *a priori* grounds a few simple correlations are to be expected. By and large, the space required should be larger the larger the animal, but a polyphagous species, since it exploits a greater number of food sources, should require less space than one with a more restricted menu and, other things being equal, a group must utilise a larger area than a single animal. Hunting technique is also a relevant factor, for the predator that runs down large prey in a long chase is likely to cover more ground than the animal that hunts by stealth. One would therefore expect that the large, highly predacious, group-hunting Canidae would be the most widely ranging of all carnivores, while some of the smaller omnivorous procyonids and mustelids might have the most restricted ranges.

The methods used to study social organisation and living space are all slow and toilsome procedures: deductions from repeated trappings, tracking with or without the aid of radio transmitters, prolonged observation of individuals recognisable by natural or artificial markings – all require not only time and patience but also a study area free from extraneous interference and disturbance. It is therefore no surprise to find that our knowledge is so far from complete that there are relatively few species for which reasonably satisfactory information has been obtained. This means that no very consistent or logical treatment within the various families is possible and one must perforce deal first with whatever species have been studied in most detail.

A. Canidae

In considering the Canidae, it is convenient to begin with the red fox, in which social organisation is relatively simple and to deal then with the group hunting species. Since foxes are highly polyphagous they can find adequate food in a variety of habitats but suitable places for earths for the rearing of the young may be harder to find. Good den sites are therefore a major determining factor in fox social organisation and movement patterns, so that when, in autumn, the vixens start to take up residence in their chosen earths they effectively set the distribution pattern for the coming season. Burrows (1968) has made an extensive study of the foxes in Gloucestershire, in the English midlands, from which the following picture emerges.

The selection of earths by the females is followed by a period of great activity amongst the males coinciding with the maturation of the gonads. This starts in November and most males are fertile by December. During this period there is much barking and the males compete for possession of an area containing one or more females. This may involve actual fighting, usually of a stylised character, screaming threat duels, barking and scent marking. Any excess males must either accept peripheral areas lacking females or travel in search of unattached females elsewhere. Huntsmen say that in winter dog foxes pursued by hounds often cover long distances: these are thought to be animals that, having moved away from their previous homes in search of a territory, attempt to make their way back when they find themselves in danger. Burrows was able to plot the size of the winter territories ultimately set up by three dog foxes in his neighbourhood. The largest was oval and about three-quarters of a mile ($1\frac{1}{4}$ km) across its greater diameter; the smallest almost circular and rather less than half a mile ($\frac{3}{4}$ km) across. The three were separated by the three arms formed where two streams met. Vincent (1958) also found three neighbouring territories between which a creek formed a major boundary.

Mating takes place during January and thereafter the males become quieter, territorial defence wanes and boundaries gradually fade. Although the den with the young still remains the focus of activity, the animals begin to wander further, so that the territorial arrangement gradually passes over into a system of widely overlapping home ranges occupied by mutually tolerant individuals. These ranges, however, are not very large, usually no more than two or three square miles in area. Murie (1936), working in Michigan, found that a family of foxes hunted over approximately 2 square miles (5 km^2); Scott (1943) in Iowa estimated the range of adults as about $1\frac{1}{2}$ square miles (4 km^2); Ables (1969) estimated winter ranges for Wisconsin foxes as 142–400 acres (about 0·5–1·5 km^2). Storm (1965), by radio tracking, found that during summer an adult male in Illinois had a home range measuring $1{\cdot}9 \times 1{\cdot}4$ miles ($3 \times 2\frac{1}{4}$ km) but that while the cubs were small he did not often go further than 440 yards (400 m) from the den. If there is food shortage, however, foxes are quite capable of making longer journeys and Englund (1965a) records them as travelling over 6 miles (10 km) from their home areas to feed on garbage during hard winters.

Foxes have often been said to be monogamous but the evidence strongly suggests that, although the single mated pair may often form the social unit, it is by no means uncommon for a male to be associated with more than one female. The three males studied by Burrows had one, two and three females in their respective territories and two litters sharing a den have often been recorded (Sheldon, 1949, 1950; Fairley, 1969a). It seems likely that mutually tolerant females living in close proximity are usually closely related animals – mother and daughters, or sisters who have remained together in their original home after the death of the mother. During the summer the earths are seldom used and the foxes usually lie up in cover above ground. By autumn the young are independent and dispersal takes place with the onset of a new annual cycle in October or November.

The data which have been collected by tagging and radio tracking studies are all in accord with this picture. Five adult females tagged by Sheldon (1950, 1953) in New York State and recaptured at least one winter later had moved distances no greater than 2 miles: adult males, on the other hand, had shifted various distances, ranging from 3 to 40 miles, between October and February. This strongly suggests that adult females remain resident but males may move significant distances during the winter pre-territorial phase. Phillips and Mech's (1970) record of an adult female taking twelve days to find her way home after being tagged and liberated 35 miles (56 km) away also supports this view: an animal on familiar ground could have returned much more speedily. Data for cubs show that, while a few may be recaptured the following season near their birth place, many disperse over long distances and all the records of major translocations refer to the movements of young animals away from home during their first winter. The two longest records are 160 miles (260 km) (Errington and Berry, 1937) and 245 miles (400 km), the latter being covered by a juvenile male between August and the following May (Ables, 1965), and a juvenile female tagged by Longley (1962) was recovered two summers later in a den with her family of 8 cubs 126 miles (200 km) from her original home. Although Sheldon's (1950) data suggest that young males start to disperse a little earlier than females, there is not yet sufficient information to tell whether the proportion of stay-at-homes is higher amongst the female cubs than amongst the males. Certainly some individuals of each sex do succeed in remaining on home ground: Ables (1965) recorded a juvenile male recaptured next summer within 300 yards (270 m) of the original tagging site; Sheldon (1953) had a similar record of a female recaptured within half a mile and Fairley's (1969b) data include records of both male and female cubs that did not move significantly during their first winter.

Data for other species of foxes are scanty. Both Egoscue (1956) and Kilgore (1969) believe that the kit fox, *Vulpes velox*, is monogamous and that the mated pair remain together the whole year round. Egoscue quotes the case of an adult female captured and kept as a pet for some time, who escaped and made her way back to her home 20 miles (30 km) away and of a male who remained in the same area for two successive seasons. Tagging studies of the grey fox, *Urocyon*, by Sheldon (1953) and Sullivan (1956) suggest a pattern of winter distribution similar to that of *Vulpes*, but it is not clear whether there is also a territorial phase during the breeding season. Chesemore (1968), working in northwest Alaska, found a tendency for seasonal movement of arctic foxes. In summer there is concentration round den sites on the tundra, followed by winter movement towards the coast and the sea ice and during this period long distances may be covered.

Wyman (1967) found that in Serengeti mated pairs of both golden and black-backed jackals defended territories of approximately 2 miles (3 km) diameter around their breeding dens but he does not make clear whether any territory is maintained once the pups have left the den. The pair of golden jackals studied by van Lawick-Goodall and van Lawick (1970) in Ngorongoro had a hunting range of something between 1 and 2 square miles (2·5–5 km^2) but on the Serengeti plains they found that ranges were larger, from 4 to 9 square miles (10–23 km^2). Within the range of the Ngorongoro pair there was a clearly defended strip of ground some 150 yards wide by half a mile long (about 11 hectares) within which the breeding den was situated. While there was some overlapping of hunting ranges, no trespassing into the territory was seen. In Serengeti, van Lawick-Goodall and van Lawick found that although most of the jackals were year-round residents, there were some amongst the black-backed species that moved with the seasonally migrating game and so were nomadic for at least part of the year.

Rausch (1967) gives the numbers in a wolf pack as ranging from two to twenty and the Isle Royale pack, which has been studied over a number of years, has varied from eleven to twenty-two (Jordan *et al.*, 1967). These figures are in accord with Murie's (1944) earlier estimate, based on his studies in the Rocky Mountains, that the unit is a family group, rarely more than ten or twelve individuals, except when two such groups unite. It seems more likely, however, that the large packs are family units which have grown to a size where they are unstable and likely to subdivide, rather than that they are formed by the joining of two independent family units. This is what appears to happen on Isle Royale, which is too small to support two packs: the animals that fail to remain with the main pack either become loners or live peripherally for a time but do not succeed in establishing a second breeding group. Nothing appears to be known about social organisation within the pack in the wild but in captive animals there are very definite dominance relations both between males and females (Woolpy, 1968).

The range of movement is limited during the summer breeding period, when the animals must remain within reach of the dens but once the cubs are old enough to run with the adults this restriction vanishes and the sedentary phase gives place to a more mobile period. The traditional picture of winter hunting activity is that the wolves work round their hunting range in a regular circuit, so that each part of it is covered in turn. The investigations that have been made, however, give no evidence for any such regularity of movement, although there are main travel routes which are used repeatedly (Thompson, 1952; Stenlund, 1955). Stenlund's studies in Minnesota suggest that each pack remains within a fairly definite home range but that boundaries are not very distinct and there is some overlapping. The size of the range must obviously depend not only on the size of the pack but also on the abundance of suitable prey and the figures quoted in table 7.1 give some idea of the variation in size of the winter hunting ranges. In addition, in the parts of Canada

Table 7.1
Ranges used by wolf packs of different sizes

Locality	No. in pack	Range (sq. miles)	Range (km^2)	Area per wolf (sq. miles)	Area per wolf (km^2)	Author
Isle Royale	11–22	210	544	10–20	25–50	Jordan *et al.* (1967)
Minnesota	7	122	315	17	45	Stenlund (1955)
Wisconsin	3–4	150	388	40–50	97–129	Thompson (1952)
Alaska	10	473	1250	47	125	Burkholder (1959)
Michigan	4	260	673	65	168	Stebler (1944)

where migratory caribou are the main food, the wolves move with them, wintering in the forest and moving to the tundra in summer, covering distances which may be as much as 500 miles (800 km) (Kelsall, 1968). For comparison, Ozoga and Harger (1966) estimated the winter ranges of coyote groups in two areas in northern Michigan to be 36–50 square miles (93–130 km^2) and there was evidence of some overlapping. The population densities were estimated as one coyote per 2 square miles (5 km^2) in one area and one per 4 square miles (10 km^2) in the other.

The wolves' summer ranges during the denning period are, of course, much smaller. In Algonquin park, Joslin (1967) found the same pack occupying the same area of 7–8 square miles (18–21 km^2) during the period from 1961 to 1963, although their numbers increased during this time from three to six adults. Occasional sorties were made beyond the main area but these probably did not take them outside a limit of 25 square miles (65 km^2). Kolenosky and Johnston (1967) studied the movements of individual animals in the same area by radio tracking. A lactating female, during three days of more or less continuous tracking, stayed within an area of approximately one square mile ($2\frac{1}{2}$ km^2). Since there were plenty of beavers to be had near the den it was not necessary for her to forage far for food. A sub-adult female moved somewhat further – up to $3\frac{1}{2}$ miles ($5\frac{1}{2}$ km) from her starting point – and a sub-adult male ranged slightly further still.

Tagging studies by Banfield (1953) give evidence that cubs may either remain with the parent pack or move off to another area once they become independent of the mother. For example, a young male tagged in his first summer was recaptured more than two years later 162 miles away.

Lycaon groups are of the same size order as wolf packs. Schaller and Lowther (1969) give 2 to 32 with an average of eleven animals as normal for Serengeti and Pienaar (1969) gives an average of eleven in the southern part of the Kruger National Park but rather fewer farther north. He also mentions that very large groups of up to fifty have been recorded. Estes and Goddard (1967) found that in Ngorongoro a pack of twenty-one left after a four-month residence and about a month later seven of the animals returned for a shorter visit. Subsequently six of these seven returned and remained for a further period in the crater. This suggests that the original group of twenty-one had split into two smaller groups and Schaller and Lowther (1969) say that some of the Serengeti packs split up and later reunited.

Lycaon are more mobile than wolves and travel longer distances; even during the more or less sedentary period while the pups are being reared their ranges are larger. Kühme (1965c) found that in Serengeti a group of eight adults, with fifteen pups to care for, at first had a hunting range of 20 square miles (42 km^2) when game was plentiful but as food became scarcer this expanded to 60–80 square miles (150–200 km^2). Four other packs occupied adjoining areas but the groups kept separate and it was not clear whether the hunting ranges were defended. There was, however, exclusion of hyaenas and jackals from the immediate vicinity of the half-grown pups. Schaller and Lowther (1969), also working in Serengeti, a few years later found five packs occupying an area of 96 square miles (250 km^2) but their hunting ranges overlapped and they regard *Lycaon* as being completely non-territorial.

Once the young are able to accompany the adults the pack becomes much more mobile. According to Schaller, they roam so widely that successive sightings of a known pack may be separated by months and one group seen at Seronera was next encountered at Olduvai, 45 miles (72 km) away. As one might expect, some packs follow the migrating herds of game to the woodlands in the dry season and out on to the plains with the onset of the rains.

Neither Kühme (1964, 1965a) nor Estes and Goddard (1967) nor Schaller and Lowther (1969) saw any evidence of a hierarchical structure within the *Lycaon* pack: no individual was seen in a peripheral relationship and fighting was not observed. In situations where one might have expected a conflict between two animals, both acted submissively and hostilities did not develop. It was therefore concluded that the *Lycaon* pack is unique in having a completely egalitarian social structure in which there is no hierarchy. Van Lawick-Goodall and van Lawick (1970), however, give a radically different picture of *Lycaon* social life. In the animals they studied, they found that very definite dominance hierarchies existed amongst the females as well as the males. Hostile interactions were not uncommon and fighting was seen, including one extremely serious fight between two females in which the loser, although not killed, was very badly mauled. Exactly what precipitated this conflict was not fully clear but its consequence was a reversal of the social positions of the combatants: the erstwhile subordinate was the winner and became dominant to her rival as a result of her victory.

The reason for the discrepancies between the conclusions reached by van Lawick-Goodall and van Lawick and those of the earlier workers is probably very simple. The latter concentrated mainly on hunting and food-sharing behaviour: these activities are normally carried out in a very amicable manner and food sharing goes on with less jealousy and more consideration for the younger or weaker members of the pack than in any other species, while in the communal greeting ceremony which precedes setting out to hunt, the general excitement becomes so great that what is virtually a 'free-for-all' develops and the normal dominance relations are obscured. Van Lawick-Goodall and van Lawick, on the other hand, began by concentrating on the behaviour of a single pack around their breeding den. The group consisted of eight adult males and four females, of whom only one had pups, a litter of 8. Their varied colour patterns made it relatively easy to recognise each individual and, studying the same known animals over the whole of the period during which they remained held to a definite home by the presence of the pups, the regularities in their relations to each other soon became clear. Moreover, *Lycaon* shows more rivalry when competing for the privilege of attending to the young than in almost any other circumstance: with only eight pups to twelve adults, dominant and subordinate roles were bound to show up with particular clarity. One may therefore conclude that the social structure of a *Lycaon* pack is not unique. The normal hierarchy is in operation but hunting and food sharing are carried out in such a way that there is little rivalry and the dominance relationships are not very obvious. Even in these activities, however, such relations are not entirely in abeyance. In van Lawick's first pack it was always the top-ranking male who decided when they would set out to hunt and what direction they would take, although this did not preclude independent chases being started by other dogs, once prey was sighted, nor did it mean that he was always the leader in a chase.

B. Mustelidae

(i) *Mustelinae*

The smaller mustelids have often been regarded as vagrants but this impression has arisen because of failure to realise that if an efficient territorial system is in operation then, unless the region is under-populated, there are bound to be homeless individuals who move over considerable distances in search of an unoccupied area where they can establish themselves. In fact, in those mustelines that have been carefully studied, the animals will, as Lockie puts it, 'stay put if they can hold a territory'. Lockie (1966) studied stoats and weasels in two areas in Scotland and found that the animals he trapped could be assigned to the three categories defined by Hawley and Newby (1957) in their study of the American marten. These are resident territory holders, temporary residents, who remain for short periods living a peripheral existence and transients, who merely pass through. In both stoats and weasels the resident males held clearly defined territories with stable boundaries the year round. In winter there was little overlap but in summer, when the animals were more active, this increased slightly - or possibly it would be better to say that there was more trespassing. No newcomers succeeded in establishing themselves except when a resident was removed or vanished, presumably killed. If no transient happened to be about at the crucial time the other residents usually extended their boundaries to take over the vacant area. Lockie found that stoat male territories were up to 50 acres (20 hectares) in extent; those of weasels considerably smaller, from 2½ to 12 acres (1-5 hectares). The latter corresponds with Polderboer's (1942) finding that least weasels (*Mustela rixosa*) moved about within an area of about two acres during three days of tracking. Lockie points out that although the food requirements of a stoat are about twice those of a weasel their territories were considerably more than twice as large, although at the time of his study both species were living largely on voles. This, he suggests, may be because the smaller weasel can pursue the voles below ground in their burrows, whereas the larger stoat is restricted to surface hunting and may therefore require a bigger territory.

The female stoats and weasels occupy smaller ranges within the territories of the males. The boundaries of female ranges are therefore not contiguous and Lockie suggests that the main defence carried out by a female is against the male in whose territory her living quarters are situated. In captivity a male is always dominant to a non-breeding female and, indeed, may be very aggressive towards her, but when she becomes pregnant the roles are gradually reversed and she remains superior until her young have been reared. Presumably in natural conditions too the same thing happens and the female's territory may expand and contract, according to her breeding condition. In the American weasels, *Mustela frenata* and *M. rixosa*, family relationships appear to be more amicable. According to Hamilton (1933), in both species, pairs are commonly captured together and he believes that the male stays with the female and young. Moreover, he saw a male *M. rixosa* carrying back food to the young in the nest.

This arrangement of mutually intolerant males holding permanent territories within which a female (or in some species more than one) is permitted to reside appears to be characteristic of mustelines in general. Hawley and Newby's (1957) investigation of the American marten in Glacier National Park, Montana, suggests an essentially similar system. Within the area they studied, males held territories usually a little less than one square mile ($2\frac{1}{2}$ km^2) in extent. The females held smaller territories, about a third of a square mile, within those of the males. Considerable movement of transients through the area was going on and when the young reared in the study area became independent, they also moved away, many of them covering considerable distances. One of the home juvenile males, for instance, was later captured 25 miles (40 km) away.

Having remained stable from 1952 to 1954, the resident marten population declined from twenty-seven to fourteen in the following year, coinciding with a decline in the numbers of small rodents in the area. It was not clear whether the losses to the marten population were the result of deaths or of emigration. In Lockie's area the weasel territories remained stable over two years despite fluctuations in the numbers of voles. He considered that the weasels held as large a territory as they could defend and that this gave sufficient food in all situations except the most extreme. A series of accidents then resulted in the deaths of six of the ten resident males and the system had not been re-established by the time the study ended.

According to Mech (1965), the male mink covers a range of up to 3 miles (5 km) in diameter, whereas females remain within a much smaller area of between 20 and 50 acres (8–20 hectares). His statement that after mating with one female, the male will search for another suggests that two or more females may sometimes live within the range of a single male. Gerell's (1970) study, made in southern Sweden and involving radio tracking as well as trapping, gives rather more information. He found that the animals concentrated their activities in the immediate vicinity of waterways and rarely went far inland. He therefore expresses their ranges not as areas but as length of stream bank or lake shore. The stretches of water utilised were 1–3 miles (2–5 km) in length but during the mating period in early spring the males sometimes made longer journeys outside their usual ranges, presumably in search of mates. In one of the areas studied, Gerell tracked two neighbouring adult males, a juvenile male and an adult female. The adult males' ranges were about 3 and 2½ miles (5 and 4 km) long and showed slight overlap: the female's range lay within the larger male area (♂1) and was contiguous with that of the other male (♂2) but did not overlap it. Gerell interprets this as indicating territoriality, with the males attempting to exclude each other and the female defending her area against ♂2 but not against ♂1, within whose territory her smaller domain was situated. A little further downstream a juvenile male occupied a stretch of water approximately the same length as the female's, separated by an unoccupied region from ♂2's territory; three juvenile females were trapped during the tracking period in the part of ♂1's area not utilised by the female. ♂1 thus apparently was prepared to allow juvenile females within his territory but by the time the observations were made (September–December) male young of the year were no longer tolerated. From other trapping data, Gerell found that the break-up of the family starts in July: the young males are the first to disperse, followed later by the females, some of whom may remain within the home area until the following spring.

Krott's (1959) studies of the wolverine show that here too the male holds a territory from which other males are excluded but within which females are permitted to live. Females are also mutually intolerant but since their territories are smaller two or even three females may establish themselves within the territory of a single male. The territories are large: one male's covered 770 square miles (2,000 km^2) while two females held areas of approximately 150 to 200 square miles (400–500 km^2). In Krott's opinion food supply is not the only factor that decides the size of the territory; availability of suitable denning sites also counts.

(*ii*) *Lutrinae*

Controversy has raged as to whether the European otter, *Lutra lutra*, is territorial or a vagrant of no fixed abode. Erlinge's (1967a, 1968a) careful studies on this species in Sweden have shown that in favourable areas there are resident territory holders but in addition, transients and temporary residents may make up about a third of the population and it is these that give the impression of vagrancy.

The social arrangements are essentially as in the Mustelinae. Adult males are solitaries and hold territories between which there is only slight overlapping. That they are in fact restricting each other's movements is shown by the fact that if one male is killed, his neighbours extend their ranges to take over the vacated area. Each of the male territories overlaps or includes one or more of the smaller territories held by a female and later on exploited by her and her young, working as family group. Erlinge found that the family group ranges had a maximum diameter of a little over 4 miles (7 km), which was about half the value found for dog otter ranges (figure 7.1).

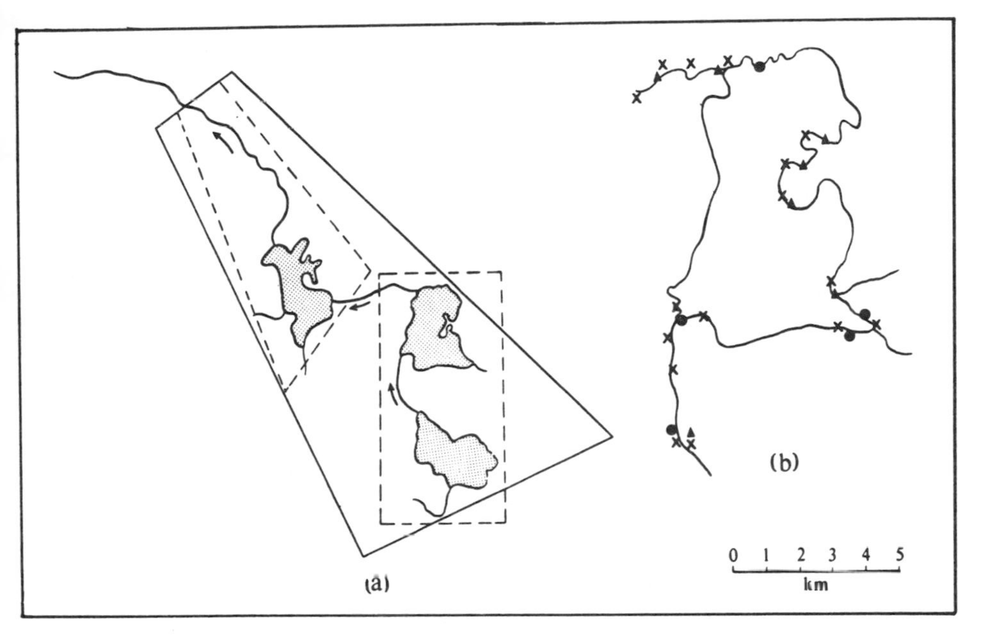

Figure 7.1. Home ranges of European otter in Sweden. (a) Ranges covered by an adult male (solid lines) and by two family groups of females and young (broken lines) round three lakes and the stream system joining them. Arrows show direction of water flow. (b) Enlarged view of central lake showing positions of resting places (triangles), rolling places (circles) and sprainting points (crosses). [After Erlinge, 1967]

The annual cycle of otter activities and the way they utilise their living space are complicated by the fact that in winter the lakes freeze over and only in flowing streams is there still open water. Spring is a season of great activity, for not only does mating occur then but the break-up of the lake ice brings back into effective use a major part of the range which was inaccessible during the winter. The whole area is repeatedly visited, there is a great deal of scent-marking activity and all the biologically important places are examined. All the favourite hunting and fishing grounds are visited and the major travel routes are used extensively, resulting in the formation of definite runways where thick vegetation has to be traversed. The otter's tendency to use regular routes is very strong; indeed, Erlinge says that in winter they cross the lake ice at the same places as they swim across in summer. Droppings, 'spraints', are usually deposited on conspicuous objects at strategic points within the territory, very often at landing places. If a rock or mound of earth is not conveniently to hand at such a sprainting point, the otter will often scrape up a mound of earth or a heap of grass and deposit the droppings on top. All the refuges too are refurbished. These include not only the complex, well-protected burrows, with entrances above and below water level, a nest chamber lined with dead leaves or grass and a lavatory chamber but also a number of much simpler resting places, often under an overhanging bank, which the otters upholster with leaves and grass and which Hewson (1969) refers to as 'couches'.

The ranges of different family groups are sufficiently separated for them not to interfere with each other and they provide a stable environment in which the young can grow up, with established feeding places, refuges and traffic routes. The dog otter, although he continues to traverse his whole territory, does so less actively than in spring and takes little notice of the family groups. In autumn he again becomes more active and deposits scent marks more often. Presumably, for the male the territory has a significance which is mainly sexual, being the area within which he is free to mate with the resident female(s), undisturbed by competitors. The autumn peak of territorial activity will serve to prevent encroachment on the part of the young males of the previous year who, as two-year-olds, are now attempting to set up territories of their own.

Finally, as winter approaches, many of the summer feeding places go out of use and activity becomes concentrated in other parts of the range, outflows from the lakes being particularly favoured winter haunts.

Lutra is the most solitary of the otters and it is likely that the social structure of other genera will prove to be somewhat different, particularly in *Pteronura*, which is distinctly gregarious. There is, however, little information for other genera, apart from some observations on the sea otter, summarised by Kenyon (1969). Sea otters live together in groups but the sexes tend to segregate and males generally occupy areas separate from those used by females and juveniles. The male groups are the larger; between 70 and 100 males may be found aggregated at a favourite hauling-out place, whereas it is not usual to find more than 10 to 15 females in close proximity. Tagging studies have shown that individuals stay within a home range extending over 5 to 10 miles (8–16 km) of coastline. Soviet workers have found similar values of 9 to 10½ miles (15–17 km). Although the otters remain within definite home ranges there is no evidence of territorial defence and adult males, when ready to mate, periodically patrol a neighbouring female area, seeking out any receptive female. The mated pair may remain together for a few days but after that, they part and care of the young is entirely the female's responsibility.

(*iii*) *Mephitinae*

Verts (1967) carried out a series of radio-tracking studies on striped skunks in a farming area in Illinois. These were directed mainly to finding out the distances covered on individual journeys rather than to delimiting ranges or studying interactions with neighbours. Distances moved between den and feeding areas are considerable: for instance, during the autumn a five-month-old female was three-quarters of a mile from her den at the furthest point recorded and a male in April went a little over a mile and a half. Winter tracking in the snow shows that the skunk may travel more than two miles in a night's activity. Allen and Shapton (1942) got very similar results from tracking skunks in Michigan: distances of three-quarters to over two miles were recorded.

Verts estimated that the population of skunks in his study area increased from nine to thirty-seven per square mile (3–14 per km^2) between 1959 and 1961. Since his tracking studies indicate that journeys of up to a mile long are often made, it is obvious that the animals cannot be holding individual territories: either they have widely overlapping home ranges or some form of group organisation exists. Since, apart from a female with young, skunks are normally seen singly, the former is more likely but the possibility remains that while female ranges overlap males may hold territories.

Apart from these studies on range of movement, most of our information about striped skunks relates to winter denning habits. It is well known that in winter skunks den up, often communally and, although they do not hibernate, they remain inactive for considerable periods and emerge only now and then to forage. Cuyler (1924) says that several pairs of skunks, together with their young, will den up together: other workers, however, are unanimous in saying that to find more than one male in a den is extremely unusual, although a single male and several females are common and several females without a male are by no means rare (Hamilton, 1937; Allen, 1939; Allen and Shapton, 1942; Yeager and Woloch, 1962). The various numbers of females that are recorded in the four papers just cited as denning without a male or with only one male are as follows:

Without a male: 1, 1, 4, 5, 6, 6, 8

With one male: 1, 1, 1, 2, 7, 7, 10, 10

In addition, three solitary males and one pair of males together were found. These records produce a total of seventy females as against only thirteen males. The adult sex ratio, as determined from trapped animals, does not show any preponderance of females (Verts, 1967), so we are faced with the question of where the rest of the males spend the winter. Possibly they occupy inferior dens in the periphery of the habitat, further from human dwellings and so are more likely to escape notice. Apart from their denning with a male in winter, captive females will tolerate the close presence of a male only briefly during the mating period and will attempt to drive him off once they have become pregnant (Wight, 1931). All the data taken together suggest that the social organisation may resemble that of the mustelines in that a single male may hold a territory within which there are a number of resident females but may differ in that the females may be mutually tolerant and their ranges may overlap extensively.

The spotted skunk, *Spilogale*, has been studied by Crabb (1948) in Iowa. Cultivated farmlands are a favourite habitat and the farmyards themselves offer the double attraction of a supply of rodents as winter food and a variety of suitable den sites – although few modern farms can be quite such a skunks' paradise as one which Crabb described that had a number of dilapidated outbuildings, eight to ten woodpiles, several straw heaps, half a dozen derelict cars and many old hollow trees.

Crabb estimated the population density as thirteen per square mile (5 per km^2) or one skunk per 50 acres (20 hectares). Each animal had more than one den site and would move from one to another as convenient in relation to food abundance. He found no evidence that the skunks were territorial; the home ranges appeared to overlap and the same den might be used on different occasions by different animals. Unfortunately he made no distinction between the sexes and it remains possible that males may be less tolerant of each other than Crabb's account suggests.

(*iv*) *Melinae*

The European badger is social, at least to the extent that several animals, comprising more than a single pair and their offspring, make common use of a series of sets (burrows) within a communal home range. The woodland area of about 70 acres (28 hectares) studied by Neal (1948) constituted the core area of a group of nine to eleven badgers. Foraging excursions were made into the adjacent pastures but the total extent of the home range was not determined, nor was the length of these excursions. They may have been quite long journeys, for Bonnin-Laffargue and Canivenc (1961) recorded a female badger making trips of one to two kilometres from her den during winter.

During winter badgers den up, very much as skunks do but according to Neal they do so in pairs, although more than a single pair may occupy the same set. The young of the previous season usually den separately in one of the smaller sets included in the range and when the new cubs are born, about February, the male probably moves to a different part of the burrow system.

Neal found that during late summer, when mating occurs, there was a period when the whole group occupied the same set and a considerable amount of chasing, scuffling and emission of scent went on. This suggests some sort of competition for females but no details are known, nor is it clear whether a male mates only with a single female. There is, however, no disruption of the group and after the period of sexual excitement the badgers become less active as winter approaches. Neal does not discuss the problem of the dispersal of the young nor the relations between animals belonging to different communal groups but Burness (1970) was able to follow the history of a male cub, easily identified because he was an albino. He moved away from his home set when the succeeding year's litter first appeared above ground in June, by which time he would have been approximately sixteen months old. He then joined company with a young female and a very old male living some distance away. There was no evidence of any hostility being shown to him and he subsequently mated with the female.

I have not found an account of social organisation in other badgers but one would expect the much more carnivorous *Taxidea* to be less social and the males may well be territorial.

C. Procyonidae

The only procyonids on which any detailed observations have been made are the raccoon, *Procyon lotor*, and the coati, *Nasua narica*.

Stuewer (1943) made a detailed study, extending over more than two years, of the raccoons in an area of just under 4 square miles (10 km^2) in Michigan. During the period of study the area, which had previously been open, was closed to hunters and the population grew from an initial fifty-four to seventy-four animals, i.e. from about thirteen to eighteen per square mile (5-7 per km^2). The raccoons' annual cycle includes a relatively inactive phase when they den up for the winter. Apart from this period, they are active by night and retire to a den during the day. Adults den singly but pairs of yearlings, presumably litter mates, are sometimes found sleeping together. The availability of suitable den sites is an important factor in determining distribution. Stuewer found that dens twenty to forty feet up in hollow trees were the most favoured and those near water were preferentially used. Berner and Gysel (1967), on the other hand, did not find any relation to distance from water or from special food sources: the dens most used were those giving the best protection against wind and rain.

In Stuewer's study there was no evidence of territoriality and home ranges overlapped considerably. Those of males were about a mile in diameter (1·6 km), those of females a little smaller. In the Cedar Creek Natural History Area in east-central Minnesota, Schneider *et al.* (1971) studied the activity patterns of females fitted with radio-transmitter collars. The ranges occupied by four females were distinctly larger than those reported by Stuewer - roughly three kilometres in diameter - and there was considerable overlapping. In the rather atypical habitat of a marsh managed for waterfowl on western Lake Erie, Urban (1970) found the remarkably high density of forty-five raccoons per square mile (17 per km^2). Nine animals were radio-tracked and the mean sizes of home ranges were found to be from about 200 acres (80 hectares) for adults to 40 acres (19 hectares) for juveniles. Even with these rather small ranges the high population density implies that there must have been a great deal of overlapping. Sunquist *et al.* (1969) found that the range covered by an adult male, during a three-month period of radio-tracking, was an area approximately 3×1 miles ($4{\cdot}8 \times 1{\cdot}6$ km). A second adult male animal, which was completely blind, was also radio-tracked and proved to be covering a rather larger area of 4×2 miles ($6{\cdot}4 \times 3{\cdot}2$ km). Although he appeared to get about without great difficulty and could find his way back to a den, presumably he had more trouble than a normal animal in finding food and therefore had to cover slightly more ground.

Bider *et al.* (1968), working in Montreal, found that the raccoons' nightly activity followed a fairly regular programme. On emergence from the den the raccoon works its way gradually towards whatever is the favoured feeding place at the time, where it then spends a few hours. Around midnight it moves off, foraging more widely here and there and visiting minor food sources before finally starting to work its way back to a den - not necessarily always the one in which it spent the previous night (Mech *et al.*, 1966). If in the course of the foraging period a rich source of food is discovered the animal will return to it earlier the next night and so the focus of feeding activity shifts around in relation to changes in the location of the best food sources. Schoonover and Marshall (1951) also found a shifting of the main concentration of activity from stream to uplands in relation to the seasonal ripening of fruit.

Gander (1966) used to put out food each night at his home in California and one of the animals that took advantage of this was an adult female raccoon. She used to make her appearance after midnight which, according to the time schedule given by Bider *et al.*, corresponds to the beginning of the wide foraging period – in other words, Gander's feeding place counted as a minor food source. This female would not tolerate the presence of another female at the feeding place so possibly, although not territorial, the animals normally maintain a minimum individual distance both when feeding and – since two adults do not sleep together – in the daytime dens. The observations of Mech *et al.* (1966) also suggest that a minimum distance is maintained and two animals did not normally come within less than 400 feet of each other. Schneider *et al.* (1971) also found that females with overlapping ranges tended to show mutual avoidance. Despite overlapping, the integrity of the individual ranges was thus maintained without any detectable overt hostility.

In winter denning these restrictions do not apply and communal winter denning is well known. Usually the number of animals found together is not very large: Mech *et al.* (1966) recorded a maximum of 8 but subsequently Mech and Turkowski (1966) found the quite exceptional number of 23 animals in a single den – 3 adult males, 5 adult females and 15 young, of which 7 were male, 7 female and the sex of the last was not recorded. Mating occurs early in the year, in February or early March, shortly after the end of the winter denning period. According to Whitney (1952), males and females separate after mating and one male may mate successively with two or three females.

Stuewer's tagging experiments showed that adults remained as residents and during the three years of his study no adults emigrated. Some of the young also established themselves within the area but others dispersed and tagged juveniles were later recovered at various distances, the maximum for a male being 27 miles (42 km) and for a female, 16 miles (26 km). Priewert (1961) has a record of a male recaptured 165 miles (265 km) from his original home three years after being tagged as a juvenile, and Lynch (1967) reports a movement of approximately the same distance by a young male tagged in May and shot in November of the same year. Of the juveniles that joined the resident population in Stuewer's area, one female provided what is surely a longevity record for a raccoon in the wild. She was tagged as a juvenile in October 1940 and shot within half a mile of the tagging place just over twelve years later, in November 1952 (Hangen, 1954).

The social organisation of the coatis living on Barro Colorado Island has been studied by Kaufmann (1962). It has little in common with what has been described in the rather unsociable raccoon but shows some similarities to that of the mustelids in that a solitary male inhabits an area overlapping the range of several females. It differs, however, in that the females are gregarious and are dominant to the male at all times, not only when they are pregnant. The basic social unit is the female with her young of up to two years old. Several such family groups commonly unite to form a rather loosely integrated band, which may also include adult females that have not got young. When about to give birth, the females leave the band but rejoin it with their young as soon as these are old enough to leave the nest. Each band occupies a home range within which it is possible to distinguish a core area in which about 80% of the time is spent, while the outlying parts of the range are visited only sporadically. Ranges overlap but core areas do not and if two bands do happen to meet they show mutual avoidance with very little sign of active hostility.

At the age of two years the males leave the group in which they grew up and live as solitaries in home ranges which overlap those of their groups. The females are hostile to adult males for the greater part of the year but the one whose range they share and with whom they have some familiarity is permitted to join them during the mating season, which lasts for about a month. During this period he moves and forages with the band, grooms the females and mates with them but remains subordinate to them. The group male is hostile towards any other male who attempts to approach the band and the females also refuse to tolerate the presence of a foreign male.

Presumably the female bands originate as extended family groups but it is not clear how new groups are formed nor what becomes of the males who fail to become attached to any group nor yet how a new male establishes relations with a group, as must happen when the attached male dies or becomes too old to drive off a rival. The bands studied by Kaufmann had a maximum of twenty and a minimum of two members. Core areas were between 35 and 50 acres (15–20 hectares) in extent, home ranges rather more than twice as large and the distance covered in a day's foraging was of the order of a mile ($1\frac{1}{2}$–2 km).

Virtually nothing is known of the organisation of the other procyonids. Apart from a female with her young, kinkajous are normally seen singly but several may be found feeding together on a tree bearing ripe fruit. Poglayen-Neuwall (1962) regards this simply as an aggregation of mutually tolerant animals at a rich food source and not as reflecting any permanent group structure. It does, however, suggest that kinkajous are non-territorial and that their ranges must overlap considerably. In captivity they are mutually tolerant and a hierarchy in relation to access to food is developed. The lesser pandas, which are also mutually tolerant in captivity, may have a similar form of social organisation but nothing seems to be known of their behaviour in the wild. For the giant panda, the only observations are those of Sheldon (1937) made during a collecting trip. He judged that the pandas were solitary and covered considerable distances but probably each remained within a single main valley. He found that an animal was not to be seen for two consecutive days in the same place and concluded that they must work their way gradually round their feeding range: since they spend ten to twelve hours a day feeding, their day's journey carries them well away from their starting point.

D. Ursidae

The only bears for which any information is available are the closely related brown bear, *Ursus arctos*, black bear, *U. americanus* and polar bear, *Thalarctos maritimus*. There is general agreement that bears are typically solitaries but considerable disagreement as to further details. At one extreme, Krott (1961, 1963) declares that the brown bear is completely non-territorial; indeed, that the animals pay no attention to each other and are 'socially neutral': at the other, Meyer-Holzapfel (1957) regards bears as straightforward, typically territorial animals. The known facts are not entirely in accord with either view but suggest a slightly modified form of territoriality. Quite apart from the fact that it is *a priori* difficult to believe that any mammal is socially neutral, Krott's view provides no explanation of the bears' habit of marking trees by clawing and chewing off the bark and urinating and rubbing upon them;[1] not to mention the fact that fighting and threat duels do occur. On the other hand, Meyer-Holzapfel's view does not explain why marking trees are most frequently located not at boundaries but at much frequented places such as points where pathways cross or near favourite feeding places. Furthermore, it does not accord with the fact that many bears may aggregate where there is a particularly rich food source, for example, brown bears beside a river during the spawning run of salmon (Trautman, 1963; Erickson and Miller, 1963) or polar bears around a whale's carcase (Flyger and Townsend, 1968).

To reconcile the various apparently conflicting observations it is necessary to appreciate a very elementary but very important point made by Leyhausen (1965a): in natural conditions there is no such

[1] The tests made by Frei (1968) of the responses of captive bears to the marks of a wild conspecific are not valid, since the test material included pine needles but the control did not and bears are known to roll and rub upon clean pine needles.

thing as a truly solitary mammal. The owner of a territory holds it by virtue of interactions with his neighbours and in time he becomes familiar with them and, probably equally important, with their smells. A social order gradually develops, in which known neighbours respect each other's rights and can therefore treat each other with a tolerance which would not be possible between total strangers. In such a system two animals may use the same pathway, provided they do not do so at the same time and provided there is some 'convention' which governs priorities. One of the functions of marking points near cross roads may well be to indicate continuing 'ownership', despite permission to neighbours to use a right of way: a very fresh mark may also carry the information that its author cannot be far away and that by waiting for a little while an encounter may be avoided. A territorial system including this tolerance within definite limits of known neighbours would accommodate most of the observations made on brown and black bears, with one further complication: an abundant but transitory source of food is not individual property; it is not defended and may be shared with neighbours. However, even though a number of animals may gather together to exploit the food, they normally take care not to approach each other very closely (Trautman, 1963).

Although technically solitary, the female is in the company of her cubs for the whole of their first season. Born during the winter denning period, they remain with their mother until the following winter and, indeed, may sometimes den with her again and not leave her until they emerge the following spring, a little over a year old.

Polar bears have to cover such long distances in search of food that any defence of the range covered is out of the question and they cannot be regarded as territorial. Normally, however, they avoid each other's company and are not seen together except when some major food source causes aggregation (Flyger and Townsend, 1968). According to Harington (1968), movements of the pack ice tend to bring the bears together in the main denning areas in autumn and, although they den separately, two dens may be quite close together. Presumably in relation to the vagaries of ice movement, there are big variations from year to year in the numbers of animals denning in a particular area. Harington also mentions that adult males do not always den up in winter: there are some that remain active the year round.

In bears the winter denning period lasts for longer and the animals make fewer excursions out of the den than is the case with any other species of carnivore. In the far north they may spend as much as six months in their dens (Bergman, 1936; Rausch, 1961; Harington, 1968) and a female black bear studied by Matson (1954) made only three trips out of her den between 5 December and the middle of the following March, when she finally emerged.

Whether one should regard the bears' winter lethargy as hibernation or not is very much a matter of definition. Body temperature does not fall very much: Folk (1967) recorded a maximum depression of 4°C in a black bear with an implanted temperature-recording capsule which gave readings every ten minutes through an entire winter. This is in agreement with Rausch's (1961) reading of 33°C (normal body temperature 38°C) for the rectal temperature of an animal shot in its winter den. There are, however, some striking metabolic changes. Hock (1960) recorded values for oxygen consumption as low as half the normal rate and Folk (1967) found that the heart rate could fall as low as ten beats per minute as compared with forty per minute during ordinary sleep in summer. Folk's most interesting finding, however, was that, even during winter sleep, there was a circadian rhythm in heart rate: although the rate remained low for most of the time there was, on most days, a period of at least thirty minutes around midday when the heart rate rose to something approaching the normal summer sleeping rate.

Classical hibernation, with body temperature and metabolic rate remaining low for long periods, is characteristic of small herbivores, particularly rodents, living in climates where there is a hard winter. It is easy to see why this should be so. Winter is a time when plant growth ceases and vegetable food becomes exceedingly scarce but it is preceded by an autumn fruiting period when food is available in abundance. A vegetarian can therefore take advantage of the way the plants' annual cycles are geared to the climate and use the autumn abundance to lay up stores enough for the winter, either by hoarding or by laying down fat reserves, or a combination of the two. Hibernation has the advantage that the store accumulated need be sufficient only to maintain a greatly reduced metabolism.

For a flesh-eater the winter presents a different problem: it is not preceded by a period of plenty when food is to be had simply for the picking. It is therefore no surprise to find that the carnivores that do den up for winter and remain inactive without feeding for long periods, even if they do not truly hibernate, are all more or less omnivorous species. They can and do take advantage of the rich autumn fruit crop to fatten up before retiring to winter quarters. Even the otherwise exclusively flesh-eating polar bear gorges on berries in autumn.

E. Viverridae

It seems strange that no detailed study of home range and movements or of social organisation appears to have been made of any viverrid species in its natural home. We know little more than that, apart from the mongooses, some of which are gregarious, they almost all appear either to be solitary or to stay together as pairs and the only estimates of range are those of Taylor (1970). On the basis of trapping results in East Africa he concludes that mongooses have ranges of from 250 acres (1 km^2) for *Herpestes sanguineus* to 3 square miles (8 km^2) for *Ichneumia albicauda*. There have, however, been some studies of the small Indian mongoose, *Herpestes auropunctatus*, in Puerto Rico and in Hawaii. From the results of trapping, Pimental (1955) concluded that in Puerto Rico the home ranges of males were considerably larger than those of females, but neither sex covered great distances. His estimates for the diameter of the home ranges were 137 yards (125 m) for males and 89 yards (81 m) for females.

Tomich (1969), working in Hawaii, made a more extensive study over a three-year period. He found that the animals were relatively sedentary and the longest movements he recorded were two males that moved 5 miles (8 km) and just under 3 miles (4·8 km) in the course of twelve and a half and thirty-one months. The diameter of the maximum usual range was estimated as about a mile for a male and a little under half a mile for a female but the distance covered in any one day would be about a quarter as much. This is in better accord with Taylor's findings than the surprisingly small ranges given by Pimental.

Although no systematic studies have been made a certain amount is known about the habits of many of the African mongooses (see Roberts, 1951; Booth, 1960; Ansell, 1960c; Smithers, 1966b), much of which is summarised in Walker (1964). A number of species are fully social, both foraging in a group and denning together. This is true of *Mungos mungo*, *M. gambianus*, *Crossarchus obscurus* (and probably the other species of this genus), *Helogale parvula* and *Suricata suricatta*. The size of the groups is rather variable: usually they appear to be composed of two or three family units and number some ten to fifteen. In both species of *Mungos*, however, the groups are often larger; up to twenty-five are recorded for *M. gambianus* (Booth, 1960), according to Smithers (1966b), packs of thirty to forty *M. mungo* are not uncommon in western Rhodesia and on the Accra plains a group of over twenty *Crossarchus obscurus* has been seen. *Herpestes ichneumon* is often seen singly or in pairs but Booth also records seeing small groups and Roberts reports groups of up to fourteen.

In the Cape grey mongoose, *Herpestes pulverulentus*, although family parties den together, foraging is usually an individual activity. *Cynictis penicillata* also normally forage individually but den communally, often sharing with *Suricata* and the ground squirrel, *Geosciurus inauris*, a large burrow system, which is almost entirely the work of the squirrels. According to Roberts (1951), there may be up to fifty *Cynictis* in such a colony but this is exceptional. *Liberiictis kuhni* is a curious species, probably related to *Crossarchus* and known only from a few specimens collected in northeastern Liberia. According to the local villagers, it is somewhat social and family parties of three to five are normally seen together (Walker, 1964).

The development of social life amongst the mongooses seems to be related to a number of different factors. Firstly, the type of food taken is relevant: the more predacious species are solitary since, with the killers of small prey, one hunter tends to interfere with another. The social species are the mixed foragers in whose diet invertebrates play a major part. Diurnal habits and an open habitat, both of which facilitate keeping contact, also favour group formation: all the social species are predominantly diurnal and no nocturnal species is social. The use of burrows as refuges is also a character of social species: a large communal burrow system gives better protection than a series of small individual burrows.

Although their ranges have not been determined, a number of the social mongooses are known to be much less sedentary than Tomich (1969) found *Herpestes auropunctatus* to be. Nothing, however, is known about inter-group relationships or about the formation of new groups, nor is it clear whether the solitary species are territorial.

Mongooses are often compared with the smaller mustelids and certainly there are some ecological parallels between the two groups. It is, however, clear that viverrid social structure has nothing in common with the typical mustelid arrangements. The frequency with which even the 'solitary' species are seen foraging in pairs makes it clear that the relations between male and female are very much closer and more amicable than in the mustelids. Social life in the mongooses is based on a family group which includes the male as well as the female and her young and it is easy to see how the close relationship between the mated pair in the solitary species could have facilitated the development of larger and more stable associations. Another point of some relevance here is the relative sizes of the two sexes. In the small mustelids, where the male dominates the female, often quite aggressively, there is usually a marked size disparity, the male being considerably the larger. In mongooses there is no such obvious male superiority in size and, indeed, it is possible that in some species the female is actually slightly larger than the male. I have never made series of measurements but in both *Suricata* and *Crossarchus*, which I have kept as pets, my biggest animals have been females and they have been dominant to the male in relation to access to food. According to Dr A. Rasa, females are also dominant in her captive groups of *Helogale*.

Before leaving the viverrids, the case of *Poiana* must be mentioned. As far as is known, this species lives much in the same way as the genets and one would expect it, like them, to be a solitary. Walker (1964), however, quotes Kuhn as saying that small groups of *Poiana* live together. They are said to construct a nest of leaves, which they use communally for a few days before moving on. Both the construction of the nest and the social grouping are surprising. It seems possible that the interpretation given of the latter is based on observation of a female and her young moving her living quarters when disturbed and whether the nest is actually made by *Poiana* or taken over from some other species remains to be definitively established.

F. Hyaenidae

The spotted hyaena, *Crocuta crocuta*, is the only hyaenid whose social life has been investigated. Kruuk (1966, 1968) found that in the enclosed environment of Ngorongoro the crocutas were social and territorial. There were 420 animals living within the 100 square miles (259 km^2) of the crater, organised into eight groups, or 'clans', with numbers ranging roughly from ten to a hundred. Each clan had its own territory, constituting the hunting range and within it a denning area. When a chase carried a group outside their own territory and the kill was made on ground belonging to a neighbouring clan, the latter usually attacked the intruders, drove them off and appropriated the kill.

In Serengeti Kruuk found a slightly more complex state of affairs. There were resident clans, as in Ngorongoro but in addition some clans did not have a fixed abode but followed the migratory wildebeest. There was also a third category, the 'commuters', who had permanent dens to which they regularly returned but they often made long excursions away from the dens to wherever the main concentrations of game were at the time. These trips could cover distances of up to 50 miles and the animals might be away from base for several days at a time. Kruuk does not discuss what happens during the period when cubs are being reared but, presumably, both the migrants and the commuters must have a more sedentary phase during this period.

One curious feature of *Crocuta* social organisation, which may possibly reflect the close relationship of hyaenids with viverrids, is that the females are rather larger than the males and are dominant over them. Kruuk (1968) found that males might occasionally move from one clan to another but he did not see such shifting on the part of a female. The curious penis-like structure of the clitoris of the female is presumably in some way related to her social dominance. Mutual genital sniffing, preceded by a display of the penis or clitoris to the partner, is one of the forms of greeting that occurs between clan members (Wickler, 1964) and it may be that a female would find it more difficult to maintain her status relative to the males were she not equally well endowed with the optical signal involved in this exchange.

No detailed studies have been made of the brown or striped hyaenas. They do not, however, appear to have such a complex social organisation as *Crocuta*. Although a number of animals may assemble at a carcase, they are usually encountered alone or in pairs. *Proteles* is also usually seen singly or in pairs but, according to Roberts (1951), groups of five or six are also reported now and then.

G. Felidae

The popular concept of any complex phenomenon must needs be a simplification. When we deal with something which is not only complex but also imperfectly understood, simplification tends to become distortion: what we do not know is necessarily omitted and the pieces we do know, naturally, expand to fill up the vacant areas in our picture. The popular notion of the typical felid is such a distortion. Apart from the lion, which we all know to be 'social' without necessarily having any very clear idea of what the term implies, the cat is a solitary and a ruthless killer: proud and aloof, he lives in isolation, caring for none but himself. In the breeding season the male, the by-word for promiscuity, searches out any and every available female with whom to satisfy his sexual urges and then departs once more into his splendid solitude.

It is *a priori* rather improbable that any mammal with so little socially positive behaviour would be able to survive. Moreover, there are plenty of indications, if we are prepared to see them, that cats are not so self-sufficient and asocial. The strong bonds of affection that can be formed between a cat and a human companion suggest a capacity for some form of friendship. This ability to form a bond with man is not an artificial product of domestication, for Smithers (1968a) found that *Felis libyca* showed the same sort of personal affection. Those of us who have kept male and female domestic cats together know that it is not uncommon for there to be a genuine and personal affection between a male and a female. Such facts suggest that the social life of the Felidae may have a more positive side and that there may be more contact with conspecifics than is allowed for in the traditional 'cat that walks by himself' picture.

(*i*) *Domestic cat*

Although we know remarkably little about the social life of domestic cats and although domestication may have brought about some modifications in behaviour, there is a good reason for dealing first with this species, for it was his studies of *Felis catus* that first made it clear to Leyhausen that even the most solitary species do have some form of communal organisation; they are not simply a set of mutually intolerant individuals – a collection of Ishmaels, with their hand against every man. Observations in an urban environment suggest that, on the contrary, neighbours know each other and have an established rank order and that the males in particular accord each other a considerable degree of tolerance which is not shown to total strangers. A strange male, imported into an established community, is normally attacked and there may be a series of fights before he stabilises his position in relation to the previously resident males. In one such case a young male was persistently attacked by one particular large adult male, presumably the local top ranker. These fights were extremely violent and the established male actually carried the battle into our house and even upstairs: this is not usual – generally a pursuit is abandoned when the loser retreats into 'his' house.

Leyhausen (1965a) discovered that familiarity with neighbours is not merely the result of chance encounters or odd trespassings that are detected. There are also what he describes as 'social gatherings', when both males and females come together at a meeting place, usually adjacent to or on the fringes of, rather than within, any territory. There they sit, often within a few yards of each other, without hostilities. Leyhausen describes this happening by night in a small square on the outskirts of Paris. I have seen exactly the same sort of gathering in the middle of the morning on a bombed site in north London. There can be little doubt that such meetings do have some important role in cat communal organisation and similar assemblies have been reported in a few wild species, often but not invariably associated with mating. Schaller (1967) found that jungle cats congregated together, apparently while mating and reports of aggregations of lynx and tiger are mentioned below.

Since an urban cat's world places restrictions on free movement of both animal and observer and provides neither the need nor the normal opportunities for hunting, a study of territory and its utilisation in such circumstances would be neither very easy nor very meaningful. Leyhausen and Wolff (1959) therefore selected for such a study a farm, surrounded by meadows and woods, where a cat could range at will and do much of its own hunting. There were two resident female cats at the farm: one was entirely subordinate and rarely left the immediate vicinity of the farmhouse; the other ranged freely in the surrounding country and was the subject of their study. The two most widely separated points she was seen to visit lay approximately 800 m apart, one to the northeast, the other to the southwest of the homestead. Northwest and southeast, the distance between the farthest points was about 300 m. To say that the cat therefore had a home range of 800 m × 300 m (24 hectares, or about 64 acres), however, would be extremely misleading. The cat used and frequently traversed the whole of the area of the homestead itself but this was not true of the rest of her range. There were a number of favourite hunting, sunning and resting places which she visited regularly, each of which might be reached by a number of different pathways. Large intervening stretches of ground were rarely or never traversed.

When a strange female appeared near the farmhouse she drove it off and chased it some 150 m towards the neighbouring farm. Two males, however, frequently visited the farm without being attacked by either of the home females and a completely unknown male, brought to the farm by one of the investigators, was also accepted without trouble. This suggests that in wild relatives, male and female might occupy largely overlapping ranges but the female intolerance shown indicates that female ranges are likely to be effectively separate. In view of what has been said about the use of the living space, it would nevertheless be possible for the ranges to interpenetrate considerably without mutual interference.

One would expect a history of domestication to have involved selection for tolerance and wild relatives might show greater hostility to strangers. This appears to be true of *F. libyca*, for Smithers (1968a) found it impossible to get his three animals (two females, one male) to share the house: one always drove the others away. The rather clear demarcation of the much frequented core area from the rest of the range of Leyhausen and Wolff's cat is probably an artificial product of the unnatural concentration of places of importance provided by the farmhouse.

One further point of practical importance emerges from this study. The responses of one cat to another are not simply a matter of size, sex, age and the situation in which they meet. Past history and personal relationships matter and the full complexities of cat sociology will therefore not be fully comprehensible until studies extending at least over a cat's whole lifetime have been made. The same applies with equal force to wild felids. Although quite extensive studies have been made on a number of species and the major features in their social organisation are now clear, the finer nuances will be deciphered only when it becomes possible to carry on studies over generations in a single area, where every individual is known and his history has been recorded. This, however, remains a dream of the future, and for the meantime, we must be content with less extensive studies.

(ii) Lynxes

The northern lynx has been studied in Sweden (Haglund, 1966) and in North America (Saunders, 1963b; Nellis and Keith, 1968). Haglund found that a male had a territory of about 115 square miles (300 km^2), round which there was a marginal area into which he made occasional trips and where contacts with neighbours were sometimes made. Reciprocal visits by outsiders into the male's area were also recorded now and then. Saunders also recorded a male, during the mating season, making a trip 5 miles beyond his normal range. Haglund found that a female with two juveniles lived in the same area as the male but her range was smaller. During most of the year they moved independently but in March, around the time of mating, the pair travelled together. Berrie (in press), working in Alaska, found that two males held overlapping territories which also overlapped with those of two adult females. The female ranges, however, did not overlap each other and, while the males showed some degree of mutual tolerance, adult females were intolerant of each other. According to Haglund, the lynx lies up during most of the day and usually leaves its day bed before darkness falls. The night's travel may be broken up by one or more short rests in temporary beds and the animals do not necessarily return to the same day bed each morning.

The size of the hunting range must be largely governed by availability of prey, and he found that ranges were bigger in the north of Sweden than in the south. The estimate of a maximum population density of the order of five per 100 km^2 (thirteen per 100 sq miles) given by Schauenberg (1969) for lynx in eastern Europe and the Soviet Union suggests that at such densities the ranges are rather smaller than Haglund's 300 km^2 for four animals. The ranges reported by the American workers are also smaller: three ranges plotted by Saunders (1963b), from his tracking studies, were 6, 7 and 8 square miles (16–21 km^2) in extent and in Alaska, Berrie (in press) found ranges of from 6·5 to 16 square miles, with an average of 10·7 square miles (27 km^2).

Estimates for an average night's travel (table 7.2) show that in the most favourable conditions the lynxes in Alberta covered only about a quarter as much ground as the Swedish animals. According to Nellis and Keith (1968) a lynx does not kill every night but every second or third night. These workers found that the distance travelled per kill was about 8 miles in two good years and 20 miles in a winter when poor snow increased the predators' difficulties. Saunders' data give a value of about 9 miles (16 km) per kill.

Table 7.2
Estimates of the average distance travelled by *Lynx lynx* in a night's activity period

Worker	Locality	Distance miles	km
Haglund (1966)	Sweden	12·4	19
Schauenberg (1969)	Czechoslovakia	11·2	18
Saunders (1963 a, b)	Newfoundland	5	8
Nellis and Keith (1968)	Alberta, 1964–6	3–4	5–6
Nellis and Keith (1968)	Alberta, 1966–7	7–10	11–16

Although long movements are sometimes recorded, lynxes tend to be sedentary, provided conditions remain favourable. Nellis and Wetmore (1969) trapped an adult male repeatedly in the same area over a period of three years. On the last occasion one foot was frozen as a result of the trapping. The paw was amputated and the animal kept in captivity until the wound had healed and he was then released. In the meantime, however, another lynx had appeared in his previous home. Whether because, maimed as he was, he could not now displace the newcomer or because of the unpleasant consequences of his last trapping, he left the area and was shot some five months later raiding poultry 102 miles (163 km) away.

Although both Haglund and Saunders found that the lynxes sometimes made trips beyond their normal range, they do not report anything resembling the domestic cats' social gatherings. Lindemann (1955), however, says that lynxes have definite meeting places, which are not inside any individual territory and are visited by all the animals in the region, particularly in the mating season.

These fragmentary data all suggest that lynxes are territorial but that hostility is limited to members of the same sex, so that a female's territory may overlap or lie within that of a male. The size of the territory appears to vary within wide limits and Haglund's observations suggest that in a large territory there is no single fixed home refuge. Although details are not clear, there are definite suggestions of some form of social contact other than what is incidental to territorial defence.

The hunting range of the bobcat, according to Young (1958) is also very variable in size. It may be as much as 25 to 30 miles (40–50 km) in diameter but is often much smaller. In the Savannah River Plant area, for instance, where small rodents provide abundant food, Provost (in press) has found that home ranges average less than 2 square miles (5 km^2) and Marston (1942) found that three places at which an animal was located within a month were within an area about 7½ miles (12 km) in diameter. Rollings (1945) found that in Minnesota the night's travel was usually from 3 to 7 miles (5–11 km) and in Massachusetts, Pollack (1951) estimated it as 2 to 5 miles (3–8 km) while Provost's animals moved only about 2 miles (3¼ km). Provost found that droppings were deposited at special places, particularly round the periphery of the range, which suggests territoriality; he also noted that, as in the lynx, adult females showed greater mutual intolerance than did males. However, apart from the fact that they use definite crossing places when moving over a mountain ridge from one valley to another, we have no data on how the bobcat uses its range and social gatherings have not been reported.

(*iii*) *Puma*

The most detailed study of the puma is that of Hornocker (1969), made in a region of central Idaho where prey (elk and mule deer) was abundant and there was very little human interference. The investigation extended over four years and combined winter tracking with trapping and marking. The adult residents in the area numbered from five to eight in different years and in addition a number of transients passed through. The males occupied non-overlapping ranges of up to 25 square miles (65 km^2): female ranges were smaller, from 5 to 20 square miles (13–52 km^2) and a male overlapped with two or three females. The size of a female's range changed from one winter to another in relation to how many young she had to provide for. The females appeared to make the necessary readjustments peacefully and female ranges sometimes showed some overlap. The animals seemed to avoid each other by mutual consent; although those whose ranges overlapped might use the same pathways, they refrained from doing so at the same time. Transients did not attempt to stay in the already occupied areas and were permitted to pass through without molestation. Clearly the system can be maintained for long periods without recourse to fighting and although there are, no doubt, circumstances in which fights do occur, Hornocker came to the conclusion that they are very rare.

Hornocker does not discuss what happens in the mating season but clearly the pattern of mutual avoidance must be modified. According to Young and Goldman (1946), several males will follow a female who is on heat.

(*iv*) *Leopard*

Eisenberg and Lockhart (1972) have made a study of the leopard in Ceylon where, in the absence of any larger predator, they are less retiring and also more terrestrial than is usual in regions where tiger or lion are also present. The study area was mainly forest but contained a number of water holes and small lakes surrounded by clearings. The leopards used definite pathways between these clearings, which were their main hunting grounds. Home ranges were small, the largest covering 3·8 square miles (10 km^2). A male and a female could have overlapping ranges round one of these feeding areas but there was no overlap between animals of the same sex and sub-adult males occupied inferior peripheral regions. Ownership was maintained largely by indirect signalling behaviour – vocalisations and marking – rather than by actual contests or even direct confrontation. Eisenberg and Lockhart were dealing with a resident population and did not record any long-range movements. Schaller (1967), however, found that, although there were no resident leopards in the area where he worked in Kanha Park, sporadic visits were made by transients throughout the year.

(*v*) *Tiger*

In the area studied by Schaller (1967) there were four resident adults, three females and one male, of whom the male and one female were regularly present within the main study area. The female's total range during a year of observation covered 25 square miles (65 km^2), that of the male 30 square miles (78 km^2). The two ranges were completely overlapping but the animals had separate core areas. The two other resident females had their centres of activity peripheral to Schaller's main study area and were therefore seen less frequently. In addition to the residents, seven transients were seen; they remained in the area for periods varying from a day or two to a month before passing on. Schaller summarises previous estimates for the size of tiger ranges. These vary from 1,500 to 80 square miles (3,900 to 207 km^2). In view of Schaller's findings it seems likely that the higher estimates do not relate to established resident animals but to transients that were on the move. Since many of the records refer to man-eaters, they were probably subject to hunting and were, in any case, not typical.

Schaller's observations showed that the ranges were not occupied by a single owner to the exclusion of all others. The transients that remained for a time and then passed on were using the same ground as the residents, and the three resident females used the same range.

At the same time there was never more than a single resident adult male and only one of the transients was a male. Schaller concluded that the male actually held a territory which was shared freely with females but within which other males were not permitted to remain. The home females appeared to be mutually tolerant but there was nevertheless some limitation, for the number of residents did not increase. Moreover, although the two would sometimes share a kill, there was often obvious tension between them, more so than when the male and one of the females were together at a kill.

When in oestrus, a tigress may wander more widely than usual: she may thus attract the attentions of several males and fighting between the latter may occur. Such fights, although often bloody, are very rarely lethal, according to accounts quoted by Schaller. Large aggregations of tigers have also been reported now and then, not apparently associated with the presence of an oestrous female. Nothing is known about the frequency or function of these aggregations but they are reminiscent of the social gatherings of domestic cats.

(*vi*) *Lion*

Lions, as is well known, are the most social of the Felidae and in this they contrast with the closely related but more solitary tiger. This difference reflects the fact that lions typically live in more open terrain than tigers do. The tiger is hunting relatively scattered prey, in good cover, where a careful solitary stalk is likely to be the most successful technique and where cooperative hunting, depending on visual contact, would not be easy. The lion operates where prey is more concentrated, cover poorer and the relative advantages of cooperative versus solitary hunting are reversed. The size of the social group depends both on the availability of prey and the density of cover: the largest prides are found where the terrain is most open and prey most plentiful (Guggisberg, 1960).

The most complete studies that have been made of lion social structure are those of Schaller (1968b, 1969c) and Schaller and Lowther (1969). According to them the members of a pride are rarely all together at one time. In one pride studied over three years, for instance, the eleven females which it included were never all together at once. The members of the pride separate and rejoin; males are particularly prone to wander and were not infrequently seen as much as 15 kilometres (10 miles) from the rest of their group. This means that simple records of numbers of lions seen together are likely to be under-estimates of the full numbers of the prides to which they belong and records of prides lacking a male may not be correct. This is almost certainly true of Adamson's (1964) report of

6 out of 12 prides in Serengeti being composed only of females and young. The total numbers in the groups which he counted ranged from 7 to 23. According to Schaller and Lowther (1969) the Serengeti prides normally have 1 to 4 adult males, together with several females and their young. One, which they cite as typical, comprised 2 males, 11 females and 12 cubs under two years old - a total of 25 animals, of which 13 were adult. The composition of the pride studied by Schenkel (1966) in Nairobi National Park was 2 males, 3 females (one very old) and 6 cubs - 11 in all, of whom 5 were adult. In the harsher environment of the Kalahari, Eloff (in press) found that prides were small and the largest number of individuals he saw together was 13.

Four prides in Serengeti, studied in some detail by Schaller, occupied ranges which overlapped considerably (figure 7.2). Each pride, however, had its core area in which most activities were concentrated and there was little or no overlapping between core areas. The two largest ranges were about 150 square miles (400 km^2) in area, the smallest about 80 square miles (210 km^2). In Lake Manyara

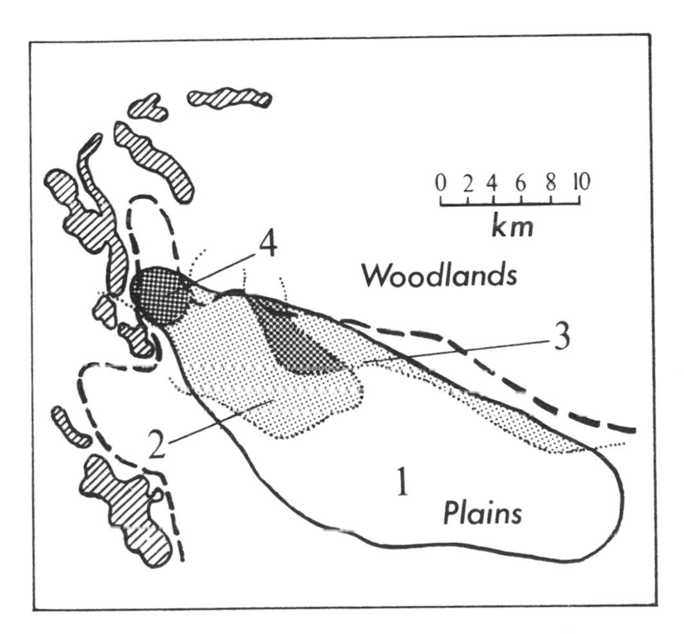

Figure 7.2. Lion home range in Serengeti. The solid line shows the range used by pride 1; dotted lines enclosing stippled areas show zones of overlap with three other prides, 2, 3 and 4. High rocky ground shaded; heavy broken line, boundary between woodland and open plain. [After Schaller and Lowther, 1969]

National Park ranges were smaller; one pride which was studied there (Makacha and Schaller, 1969) had a range of about 10 square miles (25 km^2) and Schenkel estimated ranges in Nairobi National Park to be 10 to 20 square miles (25–50 km^2).

The population of lions which an area can support depends on the prey available: Schaller and Lowther (1969) give values ranging from one lion per square mile (1 per 2·6 km^2) in Lake Manyara Park where prey biomass is highest, to one per $6\frac{1}{2}$ square miles (17 km^2) in Kruger Park, where it is lowest. Although the number of lions in a pride is not constant there is also a rough inverse correlation of range size with prey abundance.

With ranges as large as those of the lion any strict maintaining of boundaries is not possible but encounters in zones of overlap are usually avoided. If prides do meet, chasing is common but actual fighting is rare: indeed, Schaller (1968b) had the impression that in a chase the pursuer was taking good care not to catch up with his quarry – he was 'seeing him off' rather than trying to catch him. Fights do, however, occur and Guggisberg (1960) describes a number of cases where lions killed each other. Schaller (1969c) records a case where a male was killed by the three males of a neighbouring pride. The strangers then invaded his territory found three cubs and killed them.

Strangers are usually excluded by members of their own sex but Schaller regards the males as the main defenders. Females alone might keep out foreign females but would be unable to maintain their area in face of an invasion led by foreign males, and Schaller also notes that breeding success was very low in prides that had few males. Since prides do defend their ranges, even if boundaries are not inviolate and overlapping does occur, the area they utilise can be regarded as a territory. A territorial system of the sort outlined is the typical arrangement but anomalies are occasionally found. In Lake Manyara Park, for instance, two groups of females and their young inhabited neighbouring but overlapping ranges in the usual way. However, instead of each pride including one or more males, there were two males who covered the whole area making up the ranges of the two female groups (Makacha and Schaller, 1969); Guggisberg (1960) has described a similar case.

In addition to the resident territorial prides Schaller (1968b, 1969c) found that there were also nomadic lions who followed the migrations of the game and did not have permanent territories. Although there might be a temporary delimitation of ranges on the plains, once the game moved on with the coming of the dry season these had to be abandoned. Many of the nomads were males but there were also some females; possibly the latter had left their original home because it no longer provided sufficient food. The preponderance of males is not surprising, since young males are driven from the pride at three years of age and must wander unless they can find an area of their own or take over where there is a vacancy.

Lions are not great travellers. One of the nomads, who was fitted with a radio transmitter and tracked for twenty-one days during his stay in the plains, covered an average of 5 miles (8 km) in twenty-four hours and his longest journey was 13⅓ miles (22 km) (Schaller, 1968b, 1969c). In the Kalahari, where food is hard to find, Eloff (in press) found the distances covered daily ranged from 2 to 21 miles (3–34 km). It is thus not surprising that, although there are nomads as well as resident lions, there are no lion 'commuters' as there are in the much more cursorial crocutas.

(*vii*) *Cheetah*

Studies on cheetahs have been made by Eaton (1969a, b; 1970a–d) and Schaller (1968a, 1969a, 1970) but their social organisation is still far from clear. The difficulty is that while cheetahs are certainly not highly social animals they are also not exactly solitaries. Apart from association of a female with her young, they are usually but by no means invariably seen singly. Both Eaton and Schaller encountered groups of various kinds but we have as yet very little understanding of the significance of these associations. It is therefore probably best to begin with Adamson's (1969) description of the history of a female cheetah released in the Meru reserve in northern Kenya and of her 3 surviving female cubs out of a litter of 4. When she came on heat, the female wandered farther afield than at other times and the male with whom she mated remained in company with her for some time. He was not, however, seen with her after the birth of the young. Break-up of the family took place rather abruptly when the cubs were eighteen months old: two of them remained together, the third stayed apart on her own. Although ranges were not accurately determined, the mother and her daughters had divided between them an area which was about 13½ × 8 miles, i.e. just over 100 square miles (260 km^2). The mother apparently had the largest area, more than a third of the total but no details of other cheetahs present in the reserve are given and these ranges were therefore not necessarily the animals' own exclusive territories.

In the Serengeti most of the cheetahs follow the seasonal movements of their main prey, Thomson's gazelle (Schaller, 1970). During the rains, in the early months of the year, the game is on the plains and so are the cheetahs. Schaller (1968a) found that a female with two cubs stayed within an area of about 4 square miles (10 km^2)

for over three weeks while prey was plentiful. Later on in July, when the grass dries up, however, the cheetahs move over distances of up to 25 miles, from the open grasslands to the edge of the thornveld. Schaller found that individuals tended to return to the same area in successive seasons and that they occupied ranges of 20 to 25 square miles. There was often overlap of ranges but neighbours seemed to avoid each other by mutual consent. Eaton (1969b, 1970c) also found that ranges overlapped but mutual avoidance prevented hostile encounters. According to McLaughlin (personal communication), two family groups in Nairobi National Park had ranges of about 80 km^2 but a pair of males occupied a slightly larger area, approximately 100 km^2.

The data quoted so far all suggest that cheetahs are essentially territorial but that, like the domestic cats, they tolerate known neighbours within certain limits and active defence of the territory is rarely resorted to. This, however, cannot be the whole story, for groups other than family parties of a female and her cubs are seen quite frequently. Schaller (1968a) saw a kill made by two males and groups of more than one family may gather at a kill. Both Eaton and Schaller saw groups of adult males or adults of both sexes but associations between adult females only were not seen; although Varaday (1964) says that family parties including a male are encountered, Eaton and Schaller did not see this. Schaller (1970) is inclined to regard most groups of adults simply as litter-mates who have not yet parted company. This, however, cannot apply to all cases. Groups far too big to represent a single litter are reported (Graham, 1966) and I have myself seen one such group, including both full-grown and sub-adult animals, simply sitting together, with nothing to suggest why they had congregated. It seems much more likely that although the family does break up the individuals may not completely lose contact with each other but may meet now and then to renew personal acquaintance with each other, even after they have themselves produced families.

It should also be borne in mind that the population density in most of the areas that have been studied is extremely low – one cheetah per 33 square miles in Serengeti and one per 27 square miles in Kruger National Park (Schaller, 1970) and it may be that if the animals were more numerous a less casual social organisation might be discernible. Certainly in Nairobi National Park, where the density is much higher and there is one cheetah per 2 square miles, Eaton (1970c) found more tendency for the animals to form groups. He also found that in groups including males, one male always acted as leader: he determined the direction of movement and decided when and where to hunt and he was also the most vigilant member of the group.

It seems, then, that although cheetahs do possess a social organisation, it is distinctly looser and more facultative than that of lions. Depending on the circumstances cheetahs may exist as solitaries, or family ties may be maintained long enough to give rise to groups with considerable stability. Such a system, neither fully social nor yet truly solitary, would fit in very well with some of the characteristics of personality and temperament of the animal that Adamson describes as 'my lovable but aloof friend who, for me, would always remain a spotted sphinx'.

Carnivores must always exist at lower densities than herbivores and they are, in general, better equipped for self defence. It is therefore not surprising that only in special cases do they form stable groups beyond the limits of the single family unit of parent(s) with dependent young. As already indicated, social carnivores belong to two sharply contrasting categories. Highly predacious species that kill large prey by running it down, rather than by stalking, are commonly social, the advantage they gain being the greater success achieved by group pursuit. The large Canidae and the spotted hyaena exemplify this situation. The somewhat anomalous case of the lion, non-cursorial but social, appears also to be related to the greater efficiency of cooperative hunting in relatively open terrain, even for a stalker. The other social carnivores are not highly predacious but foragers after small prey, depending very largely on invertebrates, and most of them are small. The variety and types of food they take make it possible for a group of animals to find sufficient for their needs within a day's travel, and the advantages gained are largely those of group vigilance and defence. The mongooses and the coati provide examples of social species of this type.

Most carnivores, however, are not gregarious and the social species are in a minority. It would, nevertheless, be misleading simply to label the rest as 'solitaries'. For a predator, rearing a family is an exacting business: the young cannot be independent until they can make their own kills and in the more predacious species caring for the young until they have reached this stage would be an onerous task for a completely unaided female. In fact, the social organisation even of the 'solitary' species is such as to produce some sharing of the burden. In some species, as we shall see later, the male participates actively and directly in providing food for his offspring but in many cases his role is more subtle and less obvious. Amongst the 'solitary' species the commonest form of social organisation is one in which males maintain large territories within which the female (or sometimes more than one) occupies a smaller area and there rears her family. She may herself do all the hunting, she may now and then chase out an intruding female but what she does not have to do is to defend her territory against encroachment by foreign males. This is her own male's responsibility and she has no need to dissipate energies in defending an area large enough to provide for the future needs of her family: she need only go about the day-to-day business of hunting for them *now* in the part of her range where food is most readily obtained *today*. Erlinge's (1968a) description of the family relations of the otter shows clearly how this arrangement works. This pattern of males holding large ranges within which one or more females rear their young is typical of the Mustelinae but we meet it again in the Canidae (the fox) and it is the commonest arrangement in such Felidae as have been adequately studied. Even in rather omnivorous species, like skunks and raccoons, where ranges overlap widely, a fuller investigation may show that males are in fact playing a role in regulating the land tenure system.

Possibly the most striking feature of carnivore social organisation is the rarity with which serious fighting appears to be required in 'maintaining law and order'. In many species territories are so large that boundaries cannot possibly be constantly supervised. If direct attack were the only means of preserving ownership, one would therefore expect that trespass would be frequent and that, when detected, it would lead to instant attack by the owner. Such an arrangement would be both inefficient and costly, and we know that the situation is in fact very different. A territory owner does not himself have to be visible to make his presence known. This is done both by vocal signals and by olfactory ones which, in some species, are accompanied by visible markings to make them more obvious. By and large, the signals of ownership are respected: a transient does not attempt to take over from an established owner and in return is given free passage. This does not, of course, imply that fighting never occurs but only that its function and occurrence are restricted and that it is not the normal routine means of preserving the social order.

Chapter 8

Reproduction

A. General biology

WHILE the reproductive processes of placental mammals are all broadly similar, it is natural to expect differences in detail from order to order and from species to species, in relation to habits and habitat. Several aspects of the predatory way of life are likely to have effects on reproductive biology and it is therefore desirable to consider these factors before dealing with the reproductive biology of the various families. This may incidentally serve to throw some light on the significance of two physiological peculiarities, induced ovulation and delayed implantation, which, although not restricted to carnivores, are found more commonly in them and in the closely allied Pinnipedia than in other orders.

One such consideration is the fact that since many predators are solitary and range over a wide area, the preliminary stages in the reproductive cycle may have to be prolonged. It may take some time for male and female to locate each other and still longer before they can come together and permit physical contact with a partner who may have been previously unknown or even hostile. Since carnivores are well armed these preliminaries cannot be hurried and in most species the female's oestrous period is long: she is commonly receptive not merely for a period of hours but over several days.

During the actual process of mating both vigilance and ability to take quick avoiding action are reduced and the pair are more than usually vulnerable to attack. It may therefore be advantageous for coupling to be of brief duration. This consideration, however, weighs much less heavily on predators than on prey species and one may therefore expect to find that carnivore matings may be relatively prolonged as compared with those of the species that constitute their prey. This indeed appears to be the case (see table 8.1). In a number of species, particularly amongst the mustelids, copulation is very prolonged. In others, such as the majority of viverrids and the felids, although individual mountings are relatively brief, there is repeated copulation over a long period. The domestic cat will mate repeatedly for up to three or four days and Sankhala (1967) finds

Table 8.1
Longest copulations recorded in various species

Species	Duration (minutes)	Author
Mustela nivalis	90	East and Lockie (1965)
M. rixosa	60	Heidt *et al.* (1968)
M. furo	*c.* 180	Hammond and Marshall (1930)
Martes americana	90	Markley and Bassett (1942)
M. martes } *M. foina* } *M. zibellina* }	>60	Schmidt (1934, 1943) Siefke (1960)
Lutra canadensis	24	Liers (1951)
Enhydra lutris	>14	Kenyon (1969)
Mephitis mephitis	20	Wight (1931)
Ursus arctos	26	Mundy and Flook (1964)
Canis familiaris	>60	Beach (1968)
C. lupus	120	Gensch (1968)
C. latrans	15	Kleiman (1968)
Lycaon pictus	5	van Lawick-Goodall and van Lawick (1970)
Cuon alpinus	20	Sosnovskii (1967)
Speothos venaticus	15	Kleiman (1968)
Chrysocyon brachyurus	12	Da Silveira (1968)
Vulpes vulpes	40	Tembrock (1957b)
Fennecus zerda	75	Petter (1957)
Nyctereutes procyonoides	10	Kleiman (1968)
Galidia elegans	13	Albignac (1969b)
Fossa fossa	5	Albignac (1970)
Cryptoprocta ferox	165	Vosseler (1929b)
Genetta spp.	5	Dücker (1965a)
Civettictis civetta	<1	(own observations)
Herpestes ichneumon	5	Dücker (1960)
Helogale undulata	5	Zannier (1965)
Mungos mungo	>10	Neal (1970)
Cynictis penicillata	9	Jacobsen (pers comm)
Crocuta crocuta	12	Schneider (1926: quoted by Matthews, 1939)
Panthera tigris	rarely>3	Sankhala (1967)
P. onca	<1	Stehlik (1971)
Caracal caracal	10	Gowda (1967)

that a tigress may mate three to twenty-three times a day for anything between three and twenty-one days. In this connection some observations on the black-footed cat made by Leyhausen and Tonkin (1966) are of interest. In this species the female's period of heat is unusually short: she accepts the male for only five to ten hours and there are rarely more than a dozen copulations in all. Leyhausen and Tonkin regard this curtailment of mating as a protective adaptation against predation in a species which is not only very small but also lives in areas where little cover is available.

The phenomenon of induced ovulation may be correlated with the prolonged oestrus which is characteristic of the majority of carnivores. In most mammals ovulation is spontaneous and takes place at the appropriate stage of the female's oestrous cycle whether she mates or not. In induced ovulators this is not the case and eggs are shed from the ovary only in response to the stimuli provided by copulation. Since egg and sperm do not remain for very long capable of interacting to give a fertile zygote, it is clear that correct timing of ovulation in relation to mating is essential. For ovulation to occur as a response to mating seems to be such an obvious solution that one might expect to find it universal. However, if the female's receptive period is short there is no need for this: an egg shed at the end of the receptive period is bound to be correctly timed. If the receptive period is both prolonged and rather variable, however, triggering of ovulation by copulation may be the simplest solution, both in terms of ensuring correct timing and of endocrine control. Even such triggering is by no means simple, since it requires the translation of a message delivered in terms of nerve impulses into the endocrine language to which the ovary is responsive – a task which is performed by the hypothalamus. Relayed thither, the nerve impulses produced by mechanical stimulation of the cervix cause the liberation of a hormonal 'releasing factor'. This activates the pituitary, causing the secretion of luteinising hormone which, in turn, acts on the ovary to cause ovulation.

Asdell (1966) says that 'most carnivores are induced ovulators', which may well be true but in fact induced ovulation has been demonstrated only in a few species: the cat, several species of *Mustela* and *Martes*, the raccoon and the American river otter. Spontaneous ovulation has been definitely established in even fewer: the dog, the fox, the stoat and the spotted skunk. Obviously much more information is required before any firm conclusions can be reached but it is tempting to suppose that the common occurrence in carnivores of a large baculum and the penile spines that are present in many species are connected with ensuring that the stimulation provided during copulation is sufficient to cause ovulation and hence that induced ovulation is indeed of wide occurrence. Why it should apparently be absent in some species remains obscure. It must, however, be borne in mind that even in spontaneous ovulators it has been shown that some release of luteinising hormone in response to cervical stimulation can occur (Zarrow and Clark, 1968). The difference between spontaneous and induced ovulation may therefore be one of degree rather than of kind and the fact that some carnivores do ultimately ovulate without copulation does not preclude the existence of some normal triggering effect of the stimuli resulting from mating.

Most carnivores are able to prepare a refuge of some sort for the reception of their young although only a few species, such as foxes and some of the smaller mustelids and viverrids, can actually dig a burrow. One would therefore expect that the young would in general be born neither as altricial as those of burrow-dwelling rodents, nor yet as precocial as those of ungulates, born without the protection of any refuge. Broadly speaking, this is true: the young of most carnivores are born furred but the eyes are not yet open and although they are capable of some crawling, locomotory powers are very limited. Departures from this intermediate semi-altricial condition are not very numerous. Where the young are particularly well protected, there is a tendency for them to be born at an unusually early stage of development. This is true of some of the smaller mustelids and viverrids, whose young are naked or virtually naked at birth. It is also true of the bears, whose young are extremely small at birth and in some species, including the polar bear, very poorly haired. This is correlated with the fact that they are born during the winter denning period and their first few weeks of life are spent in the den, in constant contact with the mother.

At the other end of the scale come the sea otter, the spotted hyaena (Grimpe, 1923; Schneider, 1926; Deane, 1962; Pournelle, 1965) and the fanaloka (Wemmer, 1971), all of which have unusually precocial young. In the case of the almost completely aquatic sea otter the reason for this is obvious. The female cares for her single youngster very assiduously: he need not be able to swim long distances, for at first she carries him on her chest most of the time but he must be able to look after himself when she dives for food (Fisher, 1940; Kenyon, 1959). Obviously if he is to survive at all he must be born considerably more advanced than, say, a kitten or a puppy. The case of the spotted hyaena is not so simple. The young are born in dens and are no less protected than those of other hyaenids: their advanced state may therefore be a matter of ensuring that they do not tie their rather mobile parents to one spot for too long a period. In the absence of any information about the conditions in which the young of the fanaloka are reared, it is not possible to offer any suggestions about the significance of their precociousness.

With young typically born in the intermediate semi-altricial state, one might expect that gestation periods would not be particularly prolonged and that breeding would be moderately rapid. There is, however, another factor which complicates the issue. In many herbivores the young become independent of the mother directly after weaning but in true predators this is not possible. Eating meat is one thing, catching it another. With highly skilled predators there is a significant time interval between weaning and independence, during which the parent(s) must kill for the young until the latter are both strong enough and experienced enough to fend for themselves. This period of juvenile dependence has profound effects on the breeding cycles of carnivores. A rat may have overlapping cycles and become pregnant on a post-partum oestrus with the next litter, while lactating for the current one, but no pregnant lioness could expect to make sufficient kills to support herself, a suckling litter and another set of half grown cubs. If a suckling litter is lost, the female will come on heat again almost at once, presumably in response to the sudden cessation of the suckling stimulus. If the litter is successfully reared, however, this does not happen. Although lion cubs are weaned at about eight months (Adamson, 1969), the female continues to look after them until well on in their second year of life and during this time she does not come into oestrus again. As a result she normally breeds approximately once in two years. The same is true of the tiger (Schaller, 1967), the puma (Robinette *et al.*, 1961), the cheetah (Adamson, 1969) and very probably of all the larger felids. Adamson's cheetah ceased to lactate when her cubs were in their sixth month of life but did not mate again until ten months later, when they finally became independent. There is thus an inhibition of oestrus while caring for weaned but dependent young: the physiological mechanism underlying this, however, has not been investigated, nor is it clear what are the relevant stimuli provided by the youngsters. Schaller (1969c) remarks on the fact that, within a pride, lionesses tend to synchronise their breeding and come on heat at more or less the same time. He suggests that the presence of one oestrous female may act as a stimulus to the others but it is also possible that, as long as there is a sufficient number of dependent youngsters in the pride, no female will come on heat and that all are released from this inhibition at approximately the same time as the cubs become independent.

Apart from the larger Felidae, prolonged post-weaning maternal care accompanied by breeding no more frequently than every other year is found in the wolverine, the spotted hyaena, the sea otter and the majority of bears.

In regions where there are major seasonal changes and part of the year is a lean time for predators, a new complication arises. Litters must not only be well spaced but they must be born at the right time of year. The 'right time' may mean in very early spring, giving the young as long as possible to grow before they have to face their first winter. This would normally mean pushing the mating season one gestation period's length back into the winter. This itself may be undesirable, for mating is a time of considerable activity and a winter mating time may place too much strain on the adults, or in some cases it may be ruled out because they have denned up. The phenomenon of delayed implantation may represent the solution to this dilemma, since it appears to be basically an adaptation which permits the timing of mating and of parturition to be independently adjusted to the environmental situation.

Once an egg has been fertilised it starts its development as it travels down the oviduct and very shortly after reaching the uterus implantation normally takes place; the formation of the placenta is initiated and gestation proceeds without interruption. In delayed implantation the initial stages of cell division occur and the egg reaches the uterus in the blastocyst stage. It then becomes inactive; the metabolic rate falls, cell division ceases and implantation does not take place. The free blastocyst remains thus quiescent for a period which may be as long as several months; it then resumes activity; implantation and formation of the placenta take place and gestation proceeds in the normal way.

The physiological basis of delayed implantation is not yet fully understood in any species. It is not clear what causes the blastocyst to become inactive, nor yet what triggers it off to resume its development at the end of the period of delay. There is, however, sufficient information to suggest that the mechanisms are not identical in all species that show delay. The control of implantation is likely to be complex, for we are not dealing with the timing of a single event; the changes taking place in the uterus in readiness for gestation must be correctly synchronised with the activities of the blastocyst and this is likely to involve a network of interactions in which there is ample room for variation from species to species.

Amongst the carnivores, delayed implantation has been found to occur in a number of the Mustelidae and Ursidae. The details are summarised in table 8.2 (p. 306) and figure 8.1. In the brown and black bears the adaptive significance of the delay is clear enough. Mating occurs during the summer when life is relatively easy but takes place early enough not to interfere with the period of heavy feeding during the autumn, when fat reserves for the coming winter are being accumulated. Parturition takes place in the winter den and

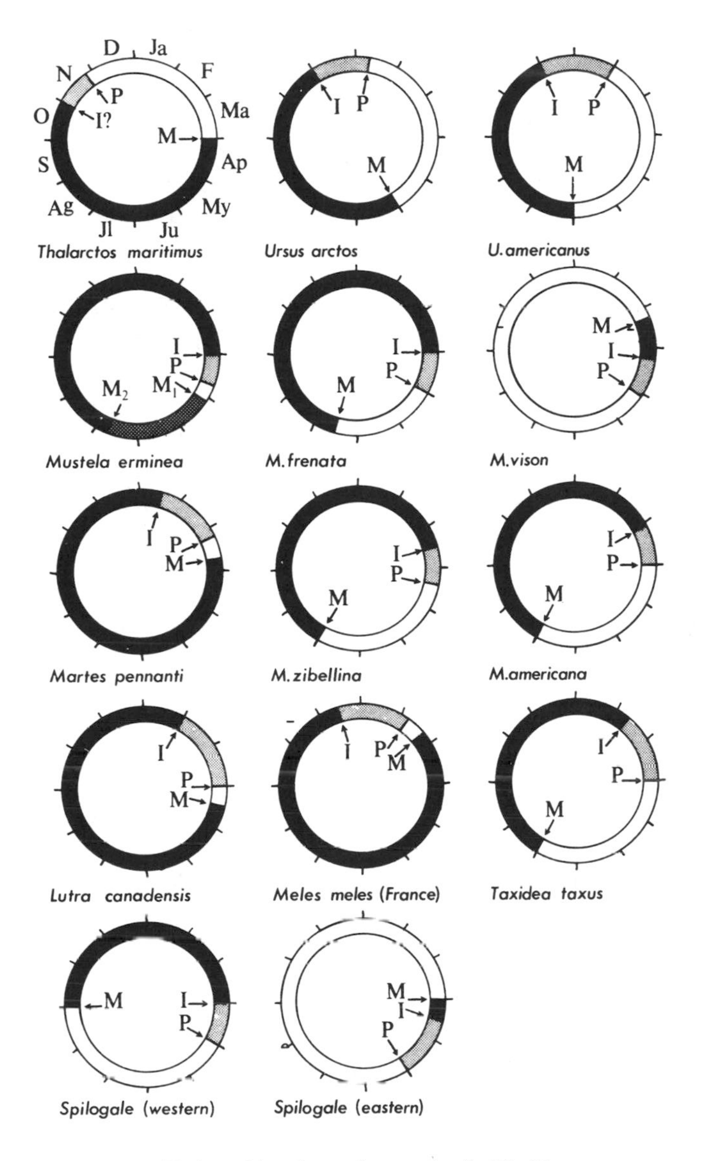

Figure 8.1. Delayed implantation cycles in Ursidae (top line) and Mustelidae. The months of the year are read clockwise, as shown in the first figure. Arrows M, I, P, show dates of mating, implantation and parturition. Period of delay shown in black; active gestation stippled

the cubs are old enough to accompany their mother when she emerges in spring. In the polar bear the cycle is similar but slightly advanced in relation to time of year. Since the animals are widely dispersed during the summer but concentrated in the denning areas in winter, it is obviously advantageous for them to mate shortly after the spring emergence, when finding a partner is simpler than it would be later on in the season. Apart from these three species, the only other bear in which delayed implantation is definitely known to occur is the Himalayan black bear, *Selenarctos thibetanus*. It is not possible to consider the significance of the delay in this case, since the only information available is that a captive female that died in October had an unimplanted blastocyst (Dittrich and Kronberger, 1963).

From figure 8.1 one can see that amongst the Mustelidae delayed implantation in general results in the birth of litters early in the year and, at the same time, permits mating to occur during the favourable season. There are, however, several problems. The genus *Mustela* includes two species with a long delay, *M. erminea* and *M. frenata*, but *M. nivalis* and *M. rixosa* breed successfully without a delay although the latter's range extends as far north as the Arctic Circle. The function of a delay as brief as that of the mink is obscure: why should they not mate at the end of March instead of the beginning or, conversely, have their young in April instead of early May? It is tempting to suggest that this genus does not represent a set of stable adaptations to present-day conditions but a series of incomplete adjustments to the climatic changes that have been going on since the last glaciation. Certainly the curious situation in the spotted skunk, *Spilogale*, is one in which recent history must be taken into account. *Spilogale* is basically a Central American animal but has been extending its range northwards since the Wisconsin glaciation. The route followed has diverged into two main branches, to the east and the west of the Rocky Mountains (van Gelder, 1959). No investigation of the breeding cycle has yet been made in Central America, and without a knowledge of the situation in the ancestral home it is pointless to speculate about how or why the cycles of the eastern and western subspecies have come to differ.

The case of the sea otter also presents some difficulties. Earlier workers observed young pups at all times of the year and concluded that there was no defined breeding season. If this were the case, it would be difficult to understand what possible value delayed implantation could have. Kenyon's (1969) studies, however, show that although some young may be born in any month of the year, in the harsh conditions of the Aleutian Islands, there is a very marked peak of births in late summer and a peak of mating activity a little later.

The delay therefore has the usual function of ensuring that neither mating nor parturition need occur in the more inclement months. In the California populations breeding activities occur considerably later, with the peak of births in December to February (Sandegren *et al.*, in press). Possibly this is similarly related to the strong north and northwest winds characteristic of late spring and early summer in these lower latitudes.

We must now turn to a more detailed consideration of what is known about reproduction in the different families. Where sufficient information exists it has been tabulated (pp. 306–313), but I have refrained from including data from compilations such as Asdell (1946), where the details given are based on sources of highly variable reliability. Where mating and parturition take place over long periods but with a distinct peak the peak month is italicised in the tables. Brackets denote a time of mating or parturition not given by the author quoted but calculated from the known gestation period. Data for litter numbers include, where possible, both the limits and the most unusual numbers. Some authors give only the mean, a figure less useful than the mode, for no animal gives birth to, say, 6·48 young and it would be more helpful to know that the commonest number is 6, or maybe 7. Gestation periods are often known only within rather wide limits, given by the time between parturition and the first and last of a series of matings which have gone on over a number of days.

Data from captive animals must be treated with some caution. Breeding cycles may be altered if there has been a major change in latitude and numbers of young may be either higher or lower than is normal in the wild, depending on how favourable or unfavourable the conditions of captivity were.

It is convenient to deal first with the residual members of the Ursidae and Mustelidae which have not been covered in table 8.2.

(*i*) *Ursidae*

Apart from the species showing delayed implantation, which have already been dealt with, very little is known. The Malayan sun bear, *Helarctos malayanus*, in Berlin Zoo did not breed seasonally but, although cubs were born at all seasons, there were peaks in April and in September to October and one female produced two litters in a year. In all cases only a single cub was born and the gestation period was 95–6 days (Dathe, 1963, 1966, 1970). Records for the breeding in captivity of other species, given by Crandall (1964), are as follows:

The spectacled bear, *Tremarctos ornatus*: litters of 1–3 cubs. In Buenos Aires Zoo young were born from June to September after a gestation period of 8–8½ months.

The sloth bear, *Melursus ursinus*: litters of 2 and 1 cubs born in November and December in San Diego Zoo. According to Norris (1969), although in the wild breeding is not strictly seasonal in this species, there is a peak of births in August to September.

The long gestation period of *Tremarctos* strongly suggests that there is delayed implantation but the 95–6 days of *Helarctos* is more difficult to interpret. Post-implantation gestation lasts 8 or 10 weeks in *Ursus*, so 13½ weeks could be interpreted either as including a relatively short delay or merely as representing a rather slow direct development.

Apart from the fact that the polar bear has two pairs, the only data about teat number that I have been able to find are the statements by Pocock (1939) that *Selenarctos* has three pairs and by Davis (1964) that most bears have this number.

(*ii*) *Mustelidae* (table 8.3, p. 307)

Those mustelids that do not have delayed implantation show a variety of breeding patterns, from strictly seasonal with a single annual litter, like the striped skunk, to completely non-seasonal, like the otter in Britain. This group also includes the species with the shortest gestation periods so far recorded in any carnivore – *Mustela nivalis* and *M rixosa*, with periods of approximately 5 weeks.

(*iii*) *Canidae* (table 8.4, pp. 308–9)

The Canidae are typically seasonal breeders and have one litter a year. In species inhabiting temperate or northern regions the breeding season is strictly defined and the pups are born in early spring. Day length would appear to be the determining factor, since reversal with translocation across the equator has been shown both in the fox imported into Australia and in the maned wolf, *Chrysocyon*, in northern hemisphere zoos. In tropical species the breeding season is less sharply defined than at higher latitudes but no species is known to show complete absence of seasonality. This suggests that the group did not evolve in the tropics but in a region where seasonal changes were significant, which accords with the postulate made on quite independent grounds of a centre of evolution in the northern temperate zone.

Gestation periods range from just over 7 weeks in the smaller species to about 10 weeks in the largest. Litter numbers are frequently high and include the largest known in any of the Carnivora. There is no strict correlation of large litters with communal organisation and we do not at present know enough about canid behaviour to be certain whether large litters in non-social species are found only where the male assists the female in rearing the young, but this seems very likely to be the case.

(iv) *Procyonidae*

Breeding data for procyonids are scanty and the raccoon is the only species which is reasonably well known. Apart from the kinkajou, the species listed below are all seasonal breeders.

(a) *Procyon lotor*, the raccoon The following data are taken from Stuewer (1943) and refer to observations made on animals in Michigan.

Mating: early February to early March
Parturition: April to May
Gestation: 63 days
Litters: 3 to 7 young with mode 4

Gander (1928) gives 65 days as the gestation period. Litter numbers given by Sanderson (1950), Biggers and Creed (1962) and Montgomery *et al.* (1970) for other areas in the United States fall within the limits given by Stuewer.

(b) *Nasua nasua* and *N. narica*, the coatis Gander (1928) records mating in March and parturition in mid-May, after a gestation period of 71–3 days for *Nasua* sp. and Brown (1936) gives a gestation period of 77 days for *N. nasua*. Kaufmann and Kaufmann (1963) say that on Barro Colorado Island the mating season of *N. narica* lasts about a month and that the young emerge from the nest at the age of 5 weeks. Smythe (1970b) points out that in this area the reproductive cycle is so organised that when the young are born, ripening fruit provides an easily available source of food for the mother. By the time the young leave the nest, fallen fruit is so abundant that they can easily find all the food they require during the early stages of their post-weaning development. Wallmo and Gallizioli (1954) found that in Arizona the young of this species were first seen in June, which would correspond with birth in late April or May and mating in early February, in agreement with Gander's observations. In Santa Barbara Zoo two litters, each of 5 young, were born during the last week of April and the first week of May (McToldridge, 1969). In

London Zoo six litters of *N. nasua*, each of 2 young, were spread from April to November (Zuckerman, 1953).

(*c*) *Bassariscus astutus*, the ringtail The young, numbering 2–4, are born in May to June and the female has 2 pairs of teats (Richardson, 1942; Taylor, 1954). Crandall (1964) reports one litter of 3, born in June in New York Zoo.

(*d*) *Potos flavus*, the kinkajou Breeding is not seasonal and young may be found at any time of year. Gestation lasts 98–115 days; there is usually a single offspring but occasionally twins are born (Poglayen-Neuwall, 1962).

(*e*) *Ailurus fulgens*, the red panda Crandall (1964) reports the birth of 2 young in June in Darjeeling Zoo to a wild-caught female who was pregnant when captured. In San Diego and London Zoos five litters, all of 1 or 2 young, were also all born in June. Pocock (1939) says that the gestation period is approximately 3 months and the normal litter number is 2. His statement on page 254 that there are 4 pairs of teats is presumably an error and should read '4 teats', since ten pages later he says that there are 2 pairs.

(*f*) *Ailuropoda melanoleuca*, the giant panda Morris and Morris (1966) summarise the breeding information available from Pekin Zoo. There are probably two annual breeding peaks, in March and October, but gestation lasts 5 months and, unless a litter is lost, a female will not breed more than once in a year. The litters are of 1 or 2 and the female has 2 pairs of teats.

(*v*) *Viverridae* (table 8.5, p. 310)

The Herpestinae are the only viverrids whose breeding habits are reasonably well known. They are typically polyoestrous and have two or more litters per year. *Cynictis* and *Suricata* appear to be exceptional in having only one moderately extended breeding season. The observations on which this statement is based were all made in the southernmost part of the animals' ranges and it is possible that further north, where the winter is milder, they may have more. Up to 6 young in a litter are recorded but 4 is the most usual number and in most genera the female has 3 pairs of teats. According to Roberts (1951) *Ichneumia* and *Rhyncogale* have only 2 pairs but I have found no information about their litter numbers. Roberts gives 3 pairs of teats for *Atilax*, Booth (1960) reports only 1 pair and my tame female had 2 pairs, so clearly there must be considerable variation in this species.

Civets and genets possibly have two annual breeding peaks, while the scanty information that exists for the Paradoxurinae suggests that they may be non-seasonal. This is quite reasonable, in view of their essentially tropical distribution.

(*vi*) *Hyaenidae*

Crocuta crocuta: the spotted hyaena is the only species whose breeding biology is adequately known. In Tanzania, in Kruger National Park and in London Zoo breeding is non-seasonal and cubs are born at any time of year (Matthews, 1939; Fairall, 1968; Zuckerman, 1953). Further south, in Zululand, Deane (1962) found that births occurred at any time except from June to August, the coldest months of the year.

Matthews (1939) quotes from Schneider a gestation period of 110 days. There are 1–3 cubs, most frequently 2, which, as previously noted, are unusually precocial. There is normally only a single pair of teats but supernumeraries are not uncommon. Matthews estimated that a complete breeding cycle occupies approximately 12 months, with a lactation period of about 6 months, followed by 3–4 weeks anoestrus. If a heat does not result in pregnancy, cycles follow each other at intervals of about 14 days. Van Lawick-Goodall, however, found that in the clan she studied in Ngorongoro the complete cycle took more than a year and the cubs were not weaned until well on in their second year of life. This long period of dependence on the mother's milk she attributes to the fact that the young are not consistently provided with meat before they are old enough to join their elders at the kill. Even when they are old enough to do this, they are not accorded any special consideration and are often crowded out by the adults, so that at first they get only a very meagre share of the spoils (van Lawick-Goodall and van Lawick, 1970).

The brown hyaena, *Hyaena brunnea*: litters of 1–2 born in January and February in Pretoria Zoo (Brand, 1963) and 2–3 born in March and July in London Zoo (Zuckerman, 1953).

The aardwolf, *Proteles cristatus*: the only information available is Shortridge's (1934) statement that there are usually 2–4 young but 6 have been recorded, and that births are in November and December, at least in the southern part of the range.

Table 8.2
Breeding data for species with delayed implantation

Species	Mating	Implantation	Parturition	Delay (months)	True gestation	Total gestation (months)	Pairs of teats	No. of young Range	No. of young Usual	Author
URSIDAE										
Ursus americanus	late June–early July	early Dec.	late Jan.–early Feb.	*c.* 5	*c.* 8 wks	*c.* 7		1–4		Wimsatt (1963)
U. arctos (Leipzig Zoo)	April–mid-July	late Nov.–early Dec.	Dec.–*Jan.*–March	4½–7	8–10 wks	6½–8		1–4	2,3	Dittrich and Kronberger (1963)
U. arctos (Alaska)	May–mid-July							1–4	2,3	Hensel *et al.* (1969)
Thalarctos maritimus (Prague Zoo)	mid-Feb.–mid-May		late Nov.			6–9	2	1–4		Volf (1963a,b)
MUSTELIDAE										
Mustela erminea (Britain)	March–April or June–July	Feb.–March	March–April	9–11	4–6 wks	10–12	4–6	6–13f	9	Deansley (1943)
M. erminea (N. America)	April–May or July	March	April–May	9–11	*c.* 4 wks	10–12	4	4–10	6,7	Hamilton (1933); Wright (1963)
M. frenata	July	March–April	April–May	*c.* 9½	23–24 days	*c.* 10	4		6,7	Wright (1963)
M. vison	Feb.–*March*–April	early April	early May	13–50 days	28–30 days	40–80 days		4–8c		Enders (1952); Enders and Enders (1963)
Martes martes and *M. foina*	late June–Aug.	?Feb.–March	March–April	5½–6½	*c.* 30 days	8–9	2	1–5	3	Schmidt (1934, 1943)
M. americana	July–Aug.	Feb.–March	March–April	*c.* 8	25–28 days	8½–9		1–6	3	Markley and Bassett (1942); Wright (1963)
M. pennanti	March	*c.* Jan.	Feb.–*March*–April	*c.* 10		up to 358 days				Pearson and Enders (1944); Wright and Coulter (1967)
M. zibellina	July–Aug.	March	April	7–7½	*c.* 4 wks	8–8½		1–5c	3	Schmidt (1934, 1943); Baevsky (1963)
Gulo gulo	mid-summer	Jan.–Feb.	March			*c.* 7		1–4	3	Wright and Rausch (1955); Wright (1963)
Lutra canadensis	March–April	Jan.–Feb.	March–April	8–10	*c.* 8 wks	10–12	2	max. 5c	2,3	Hamilton and Eadie (1964)
Enhydra lutris	peak early autumn (Alaska)	—	peak late summer (Alaska) Dec Jan (California)	*c.* 9	—	>9	1	1–2	1	Sinha *et al.* (1966); Kenyon (1969); Sandegren *et al.* (in press)
Taxidea taxus	July–Aug	*c.* Feb	March–April	5½–6½	*c.* 6 wks	*c.* 7			2,3	Wright (1963, 1966); Canivenc and Bonnin-Laffargue (1963)
Meles meles (France)	Feb.	Dec.	early Feb.	*c.* 10½	6 wks	*c.* 12				Canivenc (1966)
M. meles (Britain)	April–May or July–Aug.	*c.* Dec.	Jan.–*Feb.*–March	*c.* 4½ or *c.* 7½	6 wks	*c.* 6 or *c.* 9		1–4c	2,3	Harrison (1963)
Spilogale putorius (western)	Sept.–Oct.	late March	April–May	*c.* 5½–6	>3 and <7 wks	7–8		3–5c	4	Mead (1968a)
S. putorius (eastern)	March–April	*c.* 2 wks after mating	May–June	13–16 days	5–7 wks	7–9 wks		4–9ps	5,6	Mead (1968b)

Table 8.3
Breeding data for Mustelids without delayed implantation

Species	Characteristics of cycle	Mating	Parturition	Gestation (days)	Teats (pairs)	Litter nos. Range	Usual	Author
Mustela nivalis (Britain c)	Usually 1 litter/year; sometimes 2	April	May–June	35–36			6	Deansley (1944); East and Lockie (1965)
M. nivalis (New Zealand c)	1 or 2 litters/year		Oct–April	35–37		3–6		Hartman (1964)
M. rixosa (N. America c)	1 or 2 litters/year	starts Feb.	March onwards	35–38		4–6		Heidt *et al.* (1968)
M. putorius (Britain)	not known whether there is more than 1 litter/year	late March–early April	April–May	41–42		8–10		Southern (1964)
M. putorius (Germany c)				77		7		Goethe (1940)
Poecilictis libyca c				37 or 77 (uncertain)		2		Petter (1959)
Mellivora capensis (Kruger Park)	probably not seasonal	Feb., June, Dec.	March (1 litter)		2			Roberts (1951); Fairall (1968)
M. capensis (Pretoria Zoo)			Feb. (1 litter)				2	Brand (1963)
Melogale personata (Germany)					2	1–3		Pohle (1967)
Lutra lutra (Britain)	not seasonal	any month	any month	61–63	3		2	Harris (1968)
L. lutra (Sweden)	♀ may not breed every year	late winter–early spring	late spring			1–5	1–3	Erlinge (1968a)
L. maculicollis		July	September		2			Roberts (1951); Procter (1963)
Lutrogale perspicillata (Jaipur Zoo)		June?	October	63?		1–2		Yadav (1967)[1]
Amblonyx cinerea (British Zoos)	not seasonal			60–64		1–6		Leslie (1971); Timmis (1971)
Mephitis mephitis	seasonal: 1 litter/year	late Feb.–early March	early May	62–66	6 (variable)	5–9f	6, 7	Wight, (1931); Verts (1967)

[1] Yadav states that parturition occurred on 28 October after a gestation period of 63 days, measured from the last copulation on 26 June. Since these two dates are separated by an interval of 124 days, it is clear that there is an error somewhere.

c = observations made in captive animals, f = counts of foetuses.

Table 8.4
Breeding data for Canidae

Species	Characteristics of cycle	Mating	Parturition	Gestation (days)	Teats (pairs)	Litter nos. Range	Litter nos. Usual	Author
Canis lupus, timber wolf (Alaska)	seasonal: 1 litter/year	Feb.–*March*–April	late May–June	*c.* 63	5	3–11f	6·5 (mean)	Hildebrand (1952); Rausch (1967)
C. lupus, timber wolf (N.Z.)		Feb.–March	Feb., *March*, *April*, May	63		1–8c	4,5	Brown (1936); Zuckerman (1953); Crandall (1964)
C. lupus, Indian wolf (Jaipur Zoo)		Sept.	Nov.–Jan.			3–5		Yadav (1968)
C. lupus, Indian wolf (London Zoo)			March–May			3–5		Zuckerman (1953)
C. latrans (N.Z.)	seasonal: 1 litter/year	early March	March, *April*, *May*, June	58–61	4	1–11	3,4,5	Hildebrand (1952); Zuckerman (1953); Crandall (1964); Kleiman (1968)
C. mesomelas (Serengeti)	long season: 1 litter/year		July–late Sept.			2–7	3,4	Wyman (1967)
C. mesomelas (Kruger Park and Pretoria Zoo)			Aug., *Sept.*, *Oct.*, Nov.			1–8	4	Brand (1963); Fairall (1968)
C. mesomelas (London Zoo)			March–April			2–4		Zuckerman (1953)
C. aureus (Serengeti and Ngorongoro)			Dec.–Feb.			2–6	2	Wyman (1967); van Lawick-Goodall and van Lawick (1970)
C. aureus (London Zoo)			March–April			2–4		Zuckerman (1953)
C. adustus (Pretoria Zoo)			Sept. (2 litters)		2	5,7		Brand (1963); Roberts (1951)
C. adustus (London Zoo)			March, April (2 litters)			3,1		Zuckerman (1953)
Lycaon pictus (Kruger Park)	1 litter/year: in Kruger Park the season is long but in Kafue it is much shorter and the pups are born early in the dry season (Mitchell *et al.*, 1965)		*May*, *June*, ?*Oct.*					Fairall (1968)
L. pictus (Ngorongoro)			Jan.–April			up to 16		van Lawick-Goodall and van Lawick (1970)
L. pictus (Pretoria and Nairobi Zoos)		Feb.	*March–June* and *Oct.–Nov.*	72–73	6–7	2–12	7	Bigalkie (1961); Brand and Cullen (1967); Cade (1967)
L. pictus (Dublin Zoo)			Nov. and Jan. (3 litters)	78–80		4,5,12		Cunningham (1905)
Cuon alpinus (N.Z.)			March, April: Oct, Nov.	60–70	6–7	2–5		Weber (1927); Pocock (1939); Zuckerman (1953); Sosnovskii (1967)

Speothos venaticus (New York Zoo)	may have 2 heats in a year		Oct. (1 litter)	65	4	6		Flower (1880); Crandall (1964); Kleiman and Eisenberg (in press)
Vulpes vulpes (Alaska)		Feb.						Vincent (1958)
V. vulpes (Scotland)		late Feb.	April		3			Tetley (1941); Douglas (1965)
V. vulpes (England and Ireland)		Jan.–early Feb.	March				4,5 (mean)	Burrows (1968); Fairley (1969a, 1971)
V. vulpes (New York)	strictly seasonal; 1 litter/year, with slightly later breeding farther north	late Jan–early Feb.			4	1–9	5·4 (mean)	Schoonmaker (1938); Sheldon (1949)
V. vulpes (Illinois, Michigan			March			1–11f	4·9 (mean)	Storm (1965); Switzenberg (1950)
V. vulpes (Germany)		Dec., *Jan.*, *Feb.*	late March–early May	51–52		5–6		Tembrock (1957b)
V. vulpes (Canberra)		July	early Sept.			1–8	4	McIntosh (1963b)
V. vulpes (England c)		late Jan.	March–April					Rowlands and Parkes (1935)
V. vulpes (N.Z.)			April–May			2–6		Zuckerman (1953); Crandall (1964)
V. velox (Oklahoma and Utah)		Jan.	March–April		4			Hildebrand (1952); Sullivan (1956)
V. chama[1] (Pretoria Zoo)			late Sept.–early Oct.			1–4		Brand (1963)
Urocyon cinereoargenteus (Florida)	strictly seasonal: 1 litter/year	late Jan.–early Feb.				2–7 f & ps	4,5	Wood (1958)
U. cinereoargenteus (New York)	breeds slightly later and season more prolonged than in *V. vulpes*	mid-Jan.–*March*–May	*May*		3	1–7	3·6 (mean)	Sheldon (1949); Hildebrand (1952)
U. cinereoargenteus (Illinois)		Jan.–Feb.	March–May			2–6 f & ps	3·8 (mean)	Layne (1958)
U. cinereoargenteus (Alabama)		*Feb.*				1–5	4	Sullivan (1956)
Alopex lagopus (Greenland and Commander Islands)		Feb.–April	May–June	52		4–10	7	Barabash-Nikiforov (1938); Pedersen (1962)
A. lagopus (Alaska-farmed)			April–June	51–52	6 or 7	1–14		Ashbrook (1925); Hildebrand (1952)
Fennecus zerda	normally 1 litter/year but if a litter is lost ♀ comes on heat again	Jan.–Feb.	March–April	50–52		1–5		Petter (1957); Volf (1957); Saint Girons (1962); Gauthier-Pilters (1966); Koenig (1970)
Otocyon megalotis (N.Z.)				60–75		2–5		Crandall (1964); Rosenberg (1971)
Chrysocyon brachyurus (S.Z.)			June–Aug.		2	1–3		da Silveira (1968); Langguth (1969)
C. brachyurus (N.Z.)		Oct.–Nov.	Dec.–Feb.	60–65		3		Crandall (1964); Faust and Scherpner (1967)
Nyctereutes procyonoides (N.Z.)			April–June			2–8		Zuckerman (1953); Crandall (1964)

[1] The gestation period of 51–52 days, often incorrectly quoted as given by Shortridge (1934) for this species, actually refers to *V. vulpes*.

c = observations made in captive animals; f = counts of foetuses; ps = counts of placental scars.

Table 8.5
Breeding data for Viverridae

Species	Characteristics of cycle	Mating	Parturition	Gestation (days)	Teats (pairs)	Litter nos. Range	Litter nos. Usual	Author
Civettictis civetta (Kruger Park and Pretoria Zoo)	possibly 2 litters/year		Aug.–Dec.				3	Brand (1963); Fairall (1968)
C. civetta (Ghana and Cameroun)	births in any month, peak in dry season	any season	dry season	*c.* 60 or *c.* 80	2	2–3	2	Bates (1905); own data[1]
C. civetta (Tanzania)			Oct. and March			2–5		Vanderput (1937)
C. civetta (N.Z.)	births at any time during summer		*July, Aug.*	45–60		1–4		Zuckerman (1953); Mallinson (1969)
Viverra zibetha (Ahmedabad Zoo)			April and June (2 litters)				3	David (1967)
Genetta genetta (Kenya)		pregnant ♀♀ captured	Sept.–Dec.					Taylor (1969)
G. genetta (Prague Zoo)	2 litters/year		*April* and *Aug.–Sept.*	70–77		1–3		Volf (1965)
G tigrina (Kenya)		pregnant ♀♀ captured	Sept.–Dec.		2			Roberts (1951); Taylor (1969)
G. tigrina (Kruger Park)			kittens seen Feb.					Fairall (1968)
G. tigrina (N.Z.)			July–Aug. (2 litters)		2	1–3		Zuckerman (1953); Wemmer (personal communication)
Poiana richardsoni (Cameroun)			lactating ♀ taken Oct.			2?		Bates (1905)
Nandinia binotata (N.Z.)			April, July (2 litters)		2	1–2		Vosseler (1928); Zuckerman (1953); Crandall (1964)
Paradoxurus hermaphroditus (New York Zoo)			Sept. (1 litter)		3		3	Pocock (1934); Crandall (1964)
Paguma larvata (N.Z.)	not obviously seasonal		*April–June*		2	2–3	2	Pocock (1934); Zuckerman (1953); Crandall (1964)
Arctictis binturong (N.Z.)	not obviously seasonal		Feb.–April; July; Nov.	90–92		2–3		Zuckerman (1953); Gensch (1962, 1966); Crandall (1964)
Galidia elegans	seasonal: 1 litter/year	Aug.–Oct.	Oct.–Dec.	*c.* 75	1	—	1	Albignac (1969b)
Fossa fossa	seasonal: 1 litter/year	Aug.–Sept.	Nov.–Dec.		1	—	1	Albignac (1970)
Cryptoprocta ferox (wild-caught)			Dec.	>50	1	2		Albignac (1969)
C. ferox (Germany c)		April–May		>42			2f	Vosseler (1929b)

Herpestes auropunctatus (India, Hawaii and Puerto Rico)	2 litters/year	Feb.–*March* and June–July	May and Aug.–Sept.	42–43		2–4	2	Powell (1913); Pearson and Baldwin (1953); Pimental (1965)
H. auropunctatus (London Zoo)			July–Aug.			1–3		Zuckerman (1953)
H. edwardsi (India)	2 or 3 litters/year		May–June and Oct.–Dec.	68 or 92[2]		2–4		Frere (1928)
H. palustris (India)			lactating ♀ July		3			Ghose (1965)
H. ichneumon (Germany c)		April	July–Aug.	≯104, probably less	3	2–4		Roberts (1951); Dücker (1960)
H. sanguineus (Kenya)	probably 2 litters/year	pregnant ♀♀ found	March–April and Nov.					Taylor (1969)
Helogale parvula c	breed repeatedly; up to 4 litters/year	any season	any season	50–54	3	2–6	4	Roberts (1951); Taylor and Webb (1955); Zannier (1965)
Mungos mungo	non-seasonal	any season	any season	60–62	3	3–5		Roberts (1951); Simpson (1966); Jacobsen (personal communication)
Crossarchus obscurus (Ghana)	Probably non-seasonal: repeated heats in absence of pregnancy		young taken March and Aug.		3	4 c		Own data
Atilax paludinosus (London Zoo)			June (1 litter)		1–3		2	Zuckerman (1953)
Cynictis penicillata (Cape Province)		Sept.–Dec.	Oct.–Jan.	45–47 or 52–57[3]	3	1–4	2	Brand (1963); Zumpt (personal communication); Jacobsen (personal communication
C. penicillata (London Zoo)			July; Sept.			1, 2		Zuckerman (1953)
Suricata suricatta (Cape Province c)	normally 1 litter/year	Sept.–Oct.	Nov.–Dec.	≯77, maybe less	3	2–5	4	Own data
S. suricatta (Pretoria Zoo)			Sept.–Jan.			2–5		Brand (1963)
S. suricatta (London Zoo)			Feb; May; Oct. (3 litters)			2, 4 and 7		Zuckerman (1953)

[1] Two uncomplicated gestations were *c.* 65 days but two conceptions a fortnight after a miscarriage and a litter which was not reared were followed by gestations of 79 and 80 days.
[2] Frere's animal mated on 1–4 April and again on 28 April: the young were born on 5 July. He believed the first mating to have been the effective one, but erroneously calculated the gestation period as 60–65 days.
[3] The lower values are Zumpt's from captive animals: the figures of 52 and 57 days are Jacobsen's, obtained in circumstances where there was no possibility of unobserved mating.

c = observations made in captive animals; f = counts of foetuses

Table 8.6
Breeding data for Felidae

Species	Characteristics of cycle	Mating	Parturition	Gestation (days)	Teats (pairs)	Litter nos. Range	Litter nos. Usual	Author
Panthera leo (Kenya Tanzania, Kruger Park)	non-seasonal: polyoestrous	any season	any season	105–112	2 (rarely 3)	1–6		Fairall (1968); Schaller (1968b); Adamson (1969)
P. leo (Pretoria and London Zoos)	non-seasonal: polyoestrous	any season	any season	108		1–5	3,4	Zuckerman (1953); Brand) (1963)
P. tigris (India)	non-seasonal: tendency for a peak period varying with latitude	any season	any season	95–109		1–7	2,3	Schaller (1967); Sankhala (1967)
P. tigris (N.Z. and S.Z.)	non-seasonal	any season	any season	93–114		1–4	2,3	Zuckerman (1953); Brand (1963); Sankhala (1967); v. Bemmel (1968)
P. pardus (Africa, Ceylon)	non-seasonal	any season: in Ceylon most frequent Aug.–Nov.	any season: in Ceylon most frequent Nov.–Feb.	90–95	2		2,3	Turnbull Kemp (1967); Fairall (1968); Adamson (1969); Eisenberg and Lockhart (1972)
P. pardus (N.Z. and S.Z.)	non-seasonal	any season	any season	93–112		1–4	2	Zuckerman (1953); Brand (1963); Dobroruka (1968); Robinson (1969)
P. uncia (N.Z.)	births mostly in spring–early summer	*Jan.*–May	*April*–June	98–103		1–4		Marma and Yunchis (1968); Frueh (1968); Calvin (1969)
P. onca (N.Z. and S.Z.)	non-seasonal: polyoestrous	any season	any season	98–109		1–4	2	Zuckerman (1953); Brand (1963); Crandall (1964)
Neofelis nebulosa (N.Z.)	births in spring		*March–April–June*	85–90		1–4	2	Fellner (1968)
Acinonyx jubatus (Kenya and Kruger Park)	peak of births in autumn		March–May	90–93	6	3–4	4	Fairall (1968); Adamson (1969); Eaton (1970a)
A. jubatus (N.Z.)			Jan.–Dec.	91–95		1–4	3	Florio and Spinelli (1967); van der Werken (1968); Manton (1970)
Puma concolor (Utah and Nevada)	non-seasonal but peak of births in summer		*July*			1–5	3	Robinette *et al.* (1961)
P. concolor (N.Z. and S.Z.)	non-seasonal	any season	any season	*c.* 90		1–4	3	Brown (1936); Zuckerman (1953); Brand (1963)

Lynx lynx (N.Z. and c)	seasonal: 1 litter/year	March–April	late May–June	60–75		1–4		Oeming (1962); Crandall (1964); Kunc (1970)
L. rufus (U.S.A.)	usually 1 litter/year but sometimes 2		*March–June*	50–60		1–6	4	Young (1958); Gashwiler *et al.* (1961)
Felis catus (South Africa)	normally 2 litters/year but ♀ comes on heat again promptly after loss of a litter	Oct.–*Dec.*–Jan. and *July*–Sept.	Dec.–*Feb.*–March and *Sept.*–Nov.	65	4	1–8	4	Own data
F. silvestris (Europe c)	normally 1, rarely 2 litters/year: loss of litter followed by heat		*April–May*	63–69		1–8	4,5	Meyer-Holzapfel (1968); Volf (1968); Condé and Schauenberg (1969)
F. libyca (Rhodesia and Kruger Park)	not truly seasonal but peak of births in summer		*Dec.–Feb.*		4	1–5	3	Fairall (1968); Smithers (1968a)
F. nigripes (Germany c)	1 litter/year in spring	*Feb.–March*	*April–May*	63–68	3	1–2	2	Leyhausen and Tonkin (1966)
Prionailurus bengalensis (Korea)	1 litter/year but loss of litter followed by heat	Feb.–March	April–May	53–57		max. 4		Anon (1966)
P. bengalensis (E. Berlin Zoo)			March–June				3	Dathe (1968)
P. viverrinus (Philadelphia Zoo)				63		2–3	2	Ulmer (1968)
Ictailurus planiceps					4			Tonkin (personal communication)
Profelis temmincki (Wassenar Zoo)			March, July (2 litters)		2		1	Louman and van Oyen (1968)
Caracal caracal (Kruger Park)			kittens seen Nov.–May					Fairall (1968)
C. caracal (S.Z.)	births any month except June–Oct.	*Nov.*	*Jan.*	77–78		1–3		Brand (1963); Cade (1968)
C. caracal (N.Z.)	may have 2 litters/year	Feb.–April and Aug.–Nov.	April–July and Nov.–Jan.	69		1–6		Zuckerman (1953); Gowda (1967); Kralik (1967)
Leptailurus serval (Kruger Park)			kittens seen June–Aug.					Fairall (1968)
L. serval (Basel Zoo)	non-seasonal: sometimes 2 litters/year	any season	any season	66–77	2	1–4	3	Jones (1952); Wackernagel (1968)
Leopardus tigrinus (Europe c)		Dec.–June	Feb.–Aug.	74–76		1–2		Leyhausen and Falkena (1966)
L. pardalis (London Zoo)			Dec., April, June			1–2		Zuckerman (1953)

c = observations made in captive animals

(vii) Felidae (table 8.6, pp. 312–13)

The tropical Felidae are polyoestrous and breed at any season but there may be a tendency for a peak period. At higher latitudes breeding becomes seasonal but no species is known to be monoestrous. In many of the smaller species a female that reproduces early in the year may have a second litter and it is usual for the female to come on heat again if a litter is lost.

The gestation period of 50–60 days given for the bobcat by Young (1958) is surprisingly short in comparison with other species and I am inclined to question its accuracy.

The numbers of young per litter are generally rather small, 2, 3 or 4 being the commonest, even in species which occasionally produce up to 7 or 8.

It is surprisingly difficult to find any information on the number of teats. In the domestic cat 4 pairs is the usual number and one might expect to find fewer in species where the maximum number of young is low. This appears to be the case in the black-footed cat, the caracal and the African golden cat, all with relatively small litters: I am indebted to Miss B. Tonkin for the information that the former has 3 pairs of teats and the latter two species each have 2 pairs; she also reports that a female tree ocelot had only a single pair of teats and was never known to produce more than one kitten at a birth. The cheetah most commonly has 4 young but may have more. Eaton (1970c) mentions five litters with a mean of 5 but he does not give the maximum. Even if there are sometimes as many as 6 or 7 cubs, it is a little surprising that Adamson's cheetah had no less than 13 teats, which I assume represents a normal count of 6 pairs. One would, however, like to have data for a number of animals in order to find out whether there is significant individual variation. In some species this can be considerable: the case of *Atilax*, with 1 to 3 pairs has already been mentioned; *Mephitis* with a norm of 6 pairs ranges from 11 to 15 and in the red fox, although 3 pairs is usual, Tetley (1941) records a vixen with 9 teats.

B. Courtship and mating

Most of the special characteristics of carnivore courtship and mating can be related to two major factors. Firstly, as already mentioned, there is the need for well-armed partners to come together, although they may previously have lived as solitaries and been unknown or even hostile to each other. Secondly, a carnivore pairing is not always a mere temporary affair, in which sexually motivated individuals meet, copulate and part: in some cases it is the prelude to a partnership in which the male as well as the female will have parental responsibilities and which will last until the young become self-supporting. It therefore may involve the formation of a more or less enduring personal bond between the partners. What individual characteristics are involved is not yet clear but any female and any male will not necessarily accept each other and very definite individual preferences may be shown on the part of the female as well as the male. We may therefore expect to find the most prolonged preliminaries to mating in those species where the prospective partners are not already familiar with each other and in which the pair will not part directly after mating – i.e. in species which are not truly social but in which the male's responsibilities to his family include something more than ensuring their conception.

Fighting-play is one of the commonest types of game indulged in by young carnivores and fighting of a more or less playful nature is a relatively common element in carnivore courtships. It is usually initiated by whichever animal is the more eager to mate and appears to have a stimulating effect on the partner, generating a high level of excitement which then switches over into sexual behaviour. Here, however, lies danger: we are no longer dealing with the play of youngsters who may inadvertently give rather too hard a nip to ear or paw, but with adults who are accomplished killers. If play goes over into earnest, the consequences could be lethal. Two factors operate to prevent this and, although there are cases of a mate being killed, this is extremely rare. The first is the existence of highly ritualised forms of fighting in which attack is specifically directed to particular parts of the body, where even a bite that is too hard will not have serious consequences; the second is an inhibition which operates in the male against biting a female. The inhibition is not reciprocal and a smaller and weaker female may bite quite viciously to defend herself against the importunities of a male whom she is not yet prepared to accept, while the rules that govern his behaviour prevent him from retaliating. To some extent the two protective mechanisms are complementary and the inhibition against biting a female is most marked where highly ritualised courtship fighting is absent.

(*i*) *Canidae*

Amongst the Canidae the approach of the breeding season is usually heralded by an increase in territorial behaviour which in captive animals manifests itself most obviously by an increase in the frequency of urine marking. Previously tolerant animals may become increasingly hostile to members of their own sex and, if kept together, a definite rank order commonly develops. In species where a den is excavated for the young there is usually also an increase in the time spent in digging (Naaktgeboren, 1968).

The full sequence of courtship and mating has been described in very few wild canids and the red fox is the species about which most is known. Fox reproduction is characterised by a preliminary stage of pairing up, during which an attachment is formed between the partners, before the actual courtship and mating begin. This is correlated with the fact that the male will remain in attendance on the female and assist in the rearing of the family, so that a companionship between the two which is not related solely to mating is required. To begin with, the female is hostile to the male, attacks him and drives him off if he approaches. As the reproductive period draws near she gradually becomes more tolerant and he, presumably in response to changes in her smell, becomes more eager to establish contact with her and to smell her genital region. At first, however, most of the advances are made by the female and finally the stage is reached where the two are almost always together; they groom each other, rest together or near each other and indulge in vigorous play together. Tembrock (1957b) has described this sequence in captive foxes and Vincent (1958) has seen the same behaviour in the wild, progressing from initial hostility with daily fights, through tolerance to constant companionship. Although the partnership dissolves with the break-up of the family in autumn, there is a strong tendency for the same pair to choose each other in the subsequent season and when this happens the preliminaries are curtailed and the stage of companionship is reached sooner than is the case with first pairings. Commercial fox farmers have found that, unless male and female are allowed to be together early enough in the season to give time for the normal bond to be formed, mating often does not occur; they have also found, as Tembrock did, that a male will often refuse to mate with more than one female (Rowlands and Parkes, 1935).

Other species have not been studied in as much detail as the fox but it is likely that preliminary pair formation is characteristic of a number of other species. According to Bannikov (1964) raccoon dogs pair up in autumn, although mating does not take place until the next spring and Wyman (1967) speaks of a strong pair bond in both the golden and the black-backed jackals. The fennecs studied by Koenig (1970) were already familiar with each other and on friendly terms before the female first came on heat. The fact that there was no prolonged preliminary pairing stage may therefore possibly not be typical of what happens in the wild. The female made the first advances by showing submissive behaviour, lying down in front of the male and allowing him to smell her genital region and the two sniffed each other's faces. The male became extremely aggressive during the female's heat and pregnancy: he defended her and refused to permit anyone to approach her.

In canid mating the male mounts the female, clasping her round the lumbar region with his paws. In most species the female remains standing but the bush dog is exceptional in that she adopts a more cat-like pose, lowering her forequarters and belly but keeping her hindquarters somewhat raised and turning her tail to one side. Once intromission has occurred, the *bulbus glandis* near the proximal end of the penis enlarges and makes withdrawal impossible. The male then dismounts or is dislodged by the female and the two stand, either side by side or facing in opposite directions. They remain thus for a period of ten minutes or more (see table 8.1) before detumescence of the *bulbus* permits the penis to be withdrawn. A copulatory tie of this type is peculiar to the Canidae and has been reported in every species investigated, including *Cuon* and *Speothos*. *Lycaon* was formerly believed to be exceptional and both Cunningham (1905) and Cade (1967) reported that mating took place without any tie. More recently van Lawick-Goodall and van Lawick (1970) have found that the tie does occur in *Lycaon* but is of very brief duration: it usually lasts no longer than one to three minutes and five minutes was the longest witnessed.

While the tie lasts any defensive or evasive action is much hampered and the pair must be very vulnerable to attack by any larger predator. Presumably therefore, there must be some considerable compensating gain to outweigh this disadvantage. Obviously the copulating male effectively excludes any rival from mating while the tie lasts but why an exclusion of some fifteen to twenty minutes should have any particular value is not at once apparent. Recent work by Adler and Zoloth (1970) on rats, however, suggests at least a partial answer. Mating in the rat consists of a series of intromissions, leading finally to ejaculation. Adler and Zoloth have found that if the female receives further genital stimulation shortly after a mating has been completed, sperm transport is inhibited: very few sperm reach the uterus and the fertilisation rate is therefore very low. By allowing two males with different coat colours to mate successively with a female and noting the colour of the resulting progeny, they were able to show that if the second mating takes place within fifteen minutes of the first, it is the second male who sires most of the pups: if the interval is longer, the first male to mate sires most young. If in the Canidae there is a similar effect, then the existence of the tie obviously gives advantage to the male who first succeeds in mating and it is of little genetical consequence if the female subsequently mates with others. Such a system would have advantages in a social species, since the exclusion of rivals could occur without necessitating fighting within the group.

We are still, however, left without an explanation as to why the tie should occur in species like foxes and jackals that pair long before they mate and do not, apparently, show promiscuity, or what is the significance of its brevity in *Lycaon*. The condition in foxes could be explained away as a legacy from an earlier stage when mating was promiscuous and in *Lycaon* as a protective adaptation to life in a habitat where there is little cover and where there are larger predators about. Before accepting any such conclusions, however, it is obviously necessary to have information both about the relationship between mating interval and subsequent siring success when multiple matings occur and also to know much more about what normally happens in the wild in a variety of species, in terms of rivalry and promiscuity. Information of both sorts would also very likely serve to increase our understanding of the details of mating behaviour in induced ovulators.

(*ii*) *Ursidae*

Very little appears to be known about courtship and mating in bears. Meyer-Holzapfel (1957) gives little information beyond the fact that in brown bears, as the female begins to come into oestrus the male shows an increasing tendency to smell at her hindquarters and may make incomplete mounting attempts. When full copulation occurs, the male clasps the female in the pelvic region and in the mating pair described by Mundy and Flook (1964) he rested his outstretched head on her back but did not apparently grip her with his jaws. A photograph of mating polar bears in Meyer-Holzapfel's review, however, shows the male gripping the female's neck, much as a cat does. Mounting lasts some twenty to thirty minutes and copulation may occur on several successive days. Dathe (1963) describes a male *Helarctos* apparently exhibiting flehmen during mating attempts, but no other author mentions having seen this.

(*iii*) *Mustelidae*

Very few studies of courtship and mating have been made in mustelids and our knowledge is restricted to martens, weasels, skunks and otters. Mustelids are difficult to study in the wild and most of the work has been done on captive animals. This has the disadvantage that close confinement makes it impossible to observe the preliminary stages of the reproductive cycle, when the future partners start to make their first contacts with each other.

The martens are the species about which most is known. Schmidt (1934, 1943) has studied the European stone and pine martens and the sable; Markley and Bassett (1942), the American marten: their observations are in very general agreement. Mating does not take place until the current year's spring litter has been reared and if pairs are put together earlier than this they may play together if young enough, ignore each other or fight. The approach of the mating period is heralded by a great increase in scenting behaviour, both by urine deposition and by marking with the abdominal gland and the animals also become much more vocal. Schmidt describes a special piping call made by both sexes at this time and Markley and Bassett a 'clucking' sound made by the female which is not heard at other seasons. Presumably in the wild these signals serve both to indicate readiness for social contact and to facilitate meetings between the sexes; they may also have physiologically stimulating effects. Since in martens the female normally has two or three periods of heat, separated by quiescent intervals of a few days, the pair may remain together for some time. They remain in contact, sleep together, groom each other and indulge in play and chasing which may go over into fighting. There is, however, considerable individual variation and Schmidt (1943) describes some couples as being very tender and loving, others much rougher. In mating the male grips the female by the neck and if she does not at once assume the mating posture, with her hindquarters raised and tail turned aside, he will drag her about until she does so. Mating most frequently takes place on the ground but in the very arboreal pine marten it has also been seen in the branches of a tree (Siefke, 1960).

Although mating is often interspersed with fighting, this is commonly initiated by the female and does not apparently lead to injury: presumably the male is under the influence of the usual 'bite-inhibition' and the fighting is not serious but has the effect of generating an excitation which facilitates mating. Despite the manifestations of affection shown by some pairs, a genuine individual bonding does not seem to be formed. According to Schmidt (1943), a male will accept several females and a female will accept a different male for successive periods of heat.

In weasels the course of events is very similar but, since the male and female are hostile outside the mating season, animals kept in small cages are usually separated until the female has indicated her readiness for contact with the male and it is therefore likely that some of the preliminary stages have not been described. Heidt *et al.* (1968) worked with *Mustela rixosa* and have given more attention than other workers to the behaviour preceding actual mating. Pairs were kept in adjoining cages until they showed readiness for mating by giving the trilling call which Huff and Price (1968) regard as signifying the desire for social contact and the male also started to dig his way into the female's cage. East and Lockie (1964, 1965), dealing with *M. nivalis*, adopted the simple device of linking the cages by a small opening, through which the female could pass but the larger male could not. She was therefore free to seek out or to avoid his company.

When the partition separating Heidt's animals was removed the female *M. rixosa* at once entered the male's cage and started leaping playfully around him. He responded very promptly by seizing her by the scruff of the neck and mounting, clasping her in the lumbar region. She struggled and broke loose from his clasp several times but, maintaining the neck grip, he dragged her about until she adopted the right position and prolonged copulation followed. Alternating bouts of fighting and mating continued for some hours but the fighting must have been semi-playful, for no injuries were inflicted and when the male finally went to sleep the female tried to rouse him to further activities by leaping round him, as she had done in her initial approach.

East and Lockie's animals also mated almost directly after the female entered the male's cage. The matings were of the usual prolonged type and the animals lay on their sides for much of the time (plate 14).

In the striped skunk mating normally takes place in the shelter of the den. Wight (1931) describes the behaviour of a captive pair that were removed to a large open box when they were first heard making mating calls. Four days later the first mating took place but unfortunately Wight does not describe the animals' responses to each other during the intervening period. The male always took the initiative but at first his mating attempts were not successful. These were prefaced by a bout of fighting which looked quite ferocious but was in fact very ritualised; bites were inhibited and not carried home and the whole performance token rather than genuine. When mounting, the male gripped the female by the neck in his teetn and clasped her with his paws in the usual way, but he then performed an action which has not been described in any other species. Raising one hind foot, he made repeated rapid scratching movements at the female's vulva. This behaviour was shown so regularly that Wight considered it to be a titillating action, stimulating the female to assume the copulatory position. Verts (1967) confirms the observation and comes to the same conclusion. The male does not succeed in achieving intromission until the female responds appropriately by raising her hindquarters and turning her tail aside. The female remains receptive for two or three days but then becomes increasingly hostile to the male, at first merely repelling his approaches but finally going over to the attack, by which stage the two would have parted company in natural circumstances.

Verts found that if several females were present the male, after mating with one female, would immediately mount another and did not show any further interest in the first until several hours had elapsed. As Verts points out, since a single male often dens up with several females, such behaviour ensures that they all become pregnant. It also explains Wight's observation that nine out of his twelve captive females gave birth to their litters on the same night.

In the Canadian river otter both Liers (1951) and Harris (1968) find that during the breeding season the males tend to become very aggressive; Liers also reports that they tend to range over a wider area than usual at this time. Although otters sometimes mate on land, they very commonly do so in the water. This must present some problems in the way of achieving intromission and the wide heavy tail must add to the difficulties. In *Lutra canadensis*, Liers describes the male gripping the female in the usual way by the scruff of the neck and contriving to bend his body round the base of her tail so as to reach her vulva. One would imagine that a ventro-ventral orientation would be much simpler and Yadav (1967) says that captive *Lutrogale perspicillata* mate in this way. O.J.Murie (1940) also believed that he saw the ventro-ventral position adopted by sea otters in the wild. Kenyon (1969), however, states that the female floats at the surface of the water with her back extended so as to be slightly concave, thus lowering her vulva towards the male, who lies below her or slightly to one side, his belly against her back. The male clasps the female round the chest with his paws and also grips her with his teeth in a modified neck grip, by whatever part he can most easily reach, upper or lower jaw or simply the loose skin of her face. The grip is often ferocious enough to produce wounds on the female's nose (Foott, 1970).

(*iv*) *Procyonidae*

Almost nothing is known of the mating behaviour of procyonids. Increased scent marking in the breeding season has been noted in the giant panda (Morris and Morris, 1966) and in the coati, *Nasua narica* (Kaufmann, 1962) and increased vocalisation in the raccoon (Gander, 1966). Mating of the kinkajou, described by Poglayen-Neuwall (1962), shows a number of interesting features. When approaching the oestrous female, the male, before attempting to mount, may sniff and bite at the scent glands on her lower jaw and throat, which appears to have a sexually stimulating effect on him. He clasps the female with his hands turned so as to bring the enlarged and pointed radial sesamoid bone into contact with her body and rubs her flanks but he does not use the usual carnivore neck bite.

Unfortunately, it is not known whether any of these features are found in the mating of any other procyonids.

Poglayen-Neuwall and Poglayen-Neuwall (1965) give some information about the mating of the olingo but do not go into detail. Both sexes make a special mating call and the male first grips the female by the neck and may drag her about before mounting. He releases his hold, however, during copulation and if the female attempts to raise her head he will push her down again with his snout but does not renew the grip.

(v) *Viverridae*

Although reproductive behaviour has been described in relatively few viverrids and the observations have almost all been made on captive animals, they are yet sufficient to show that there is considerable diversity within the family. As in other groups, the approach of mating is commonly characterised by increased vocalisation and increased scenting behaviour. In the genets the former is more noticeable; in the civets, the latter.

Dücker (1957, 1965a) has described the mating of captive genets (*Genetta tigrina*). Cat-like in their general appearance, their mating behaviour also shows similarities to that of cats. The male makes the first approaches, following the female and making a repeated low call. At first she avoids him but presently she starts to answer his call and permits him to make contact with her. They sniff noses and rub heads together: she permits him to smell her genital region and may sniff at his. After an hour or so of such preliminaries, the female invites copulation by assuming the mating posture, with her shoulders held low but her hindquarters slightly raised and her tail turned aside. The male mounts very much as a cat does but grips her by the neck only after he has done so and Dücker does not mention any treading movements of the hind limbs. During the copulation, which rarely lasts more than five minutes, both partners make a cat-like miaouing call and, on separation, while the male licks his genitalia, the female goes through a series of excited actions, drags her anal region on the ground, circles about, rolls as a cat does and finally also licks her genitalia.

In the African civet cat too, mating shows many cat-like features. It is heralded by increased scent-marking and locomotor activity, particularly on the part of the male. In my captive pair the male would go to and fro in his enclosure at a fast walk or even a trot with his mouth held slightly open, in a manner not seen at other times. This activity suggests that normally he ranges more widely than usual during the breeding season, thus increasing his chances of finding a mate. Both sexes tend to deposit a scent mark on top of the partner's mark, a habit which would also help in locating a partner. As her heat approached the female also began to show a similar pattern of walking to and fro. At first when approached by the male she would drive him off, darting her head at him and snapping but not actually biting. He normally made no attempt to retaliate but either simply drew back or lay down on his side in the appeasement posture. When the female was on heat he also sometimes uttered the juvenile contact call when in close proximity to her but I could never detect any overt response on her part. Finally the female usually broke into a run and incited the male to follow by dashing past close beside him. On some occasions this led to wild chasing but at other times the pursuit was merely a token affair and the female almost at once adopted the mating posture. Sinking down slowly, she lies extended at full length and the male promptly mounts. The position adopted is very similar to that of a tom cat. The fore paws, placed on either side of the female's shoulders, support most of the weight, the hind feet make treading movements and pelvic thrusts at a rate of about two per second are in evidence. After a few moments the female gives a low miaou, which presumably indicates intromission. In the two copulations seen the male did not grip the female's neck during these initial stages but merely laid his head down on her neck. In the longer of the two, the female after some time gave a second miaou and crawled forwards slightly. The male remained mounted and moved with her, gripping her fur with his teeth just in front of her shoulder blades as he did so. When the male dismounted, both partners licked their genitalia. In this mating a rather short copulation, lasting only a few seconds after the female's miaou, was followed by a second, lasting forty seconds. After this the animals retired to rest together and did not at once attempt to mate again.

If sexual motivation is low, as may happen with animals that are not yet fully mature and also towards the end of a mature female's heat, then incomplete matings are often seen. The female may crouch briefly but she does not maintain the mating posture and the male attempts to mount, clasping her round the lumbar region with his paws, much as a dog does. I never saw intromission following a mounting of this sort. Usually the male is not very persistent, the female turns round and snaps at him and the mounting attempt goes over into a bout of stylised fighting.

In contrast to the civet's use of the neck grip only to restrain the female if she attempts to move away, Vosseler (1929b) found that in *Cryptoprocta* the male gripped the female by the neck at once and retained his hold during the whole of the extremely prolonged copulation. It was presumably to the long duration of pairing that Pocock's

(1921a) local informants were referring when they told him that *Cryptoprocta* mate like dogs: certainly Vosseler's observations do not suggest any other similarity. Like the genets, the fossas both miaoued during copulation. Mating in *Nandinia* has not been described but a tame male, when eager to mate, made a very soft repeated calling, possibly corresponding with the male civet's use of the juvenile contact call, and would attempt to mount one's arm or leg, clasping with his paws and biting as though trying to establish the neck grip.

Among the Herpestinae, mating has been described in several species: *Herpestes ichneumon* (Dücker, 1960), *Suricata suricatta* (Ewer, 1963b), *Helogale parvula* (Zannier, 1965) and *Mungos mungo* (Neal, 1970a) and I have also seen it in *Crossarchus obscurus*. All these are social species, so the partners normally know each other in advance and preliminaries leading to establishing contact are not required. It is therefore not surprising that when I introduced a stranger male to a female *Suricata* who was on heat, he simply grabbed her by the neck without further ado and dragged her about until she ceased resisting and permitted him to mount. In more normal circumstances, although the preliminaries are not prolonged in any of the species so far described, the beginning of the female's heat may first become obvious by the male's following her more closely and marking more frequently than usual. My male *Crossarchus* at this period behaves as though attached very literally to his partner: where she goes, he goes and where she rests, he does. He also marks very frequently and becomes rather aggressive, especially towards anyone handling the female. Increased marking is also recorded in *Helogale*.

In all the species studied, copulation is preceded by some form of chasing or fighting, during which excitation gradually works up until the point is reached when the female is ready to accept the male and he is ready to mount. In *Helogale* the two first sniff and lick each other's genitalia: the female then moves off, followed by the male. She stops periodically and he attempts to mount but at first she wriggles loose and moves off again and it is some time before she is prepared to stand for him. In *Herpestes ichneumon* the female acts in much the same way: she runs away, stops and crouches for a moment, only to run and repeat the performance as soon as the male catches up. In *Mungos* there is chasing and the two may rear up as though about to fight. When my pair of *Suricata* had come to know each other, mating was usually preceded by bouts of more or less playful fighting, in the course of which one would often grip the other's nose but without closing the jaws in a true bite. If the female was the more eager, she would incite the male to fight by nipping at him, directing her bites principally at his cheeks. In *Crossarchus* there is highly ritualised fighting with the two biting each other and delivering a brief shake, always gripping the thickly furred and more lightly coloured region at the side of the neck just behind the cheek. In *Herpestes* and *Helogale* characteristic repetitive short vocalisations accompany these activities. The first mountings are usually brief and incomplete: excitation is still rising until finally the stage is reached where a full and effective copulation, lasting some minutes, is achieved. The male clasps the female in the usual way but a true neck grip during the copulation does not occur: in *Herpestes* and *Mungos* the male thrusts his head against the female's neck with his mouth held open but he does not grip her and in *Helogale* he may bite at the side of her neck.

Neal's (1970) observations on *Mungos mungo* were made in the wild on a natural group, including a number of adults of both sexes: the rest all refer to captive animals with only a single pair of adults present. Neal saw no fighting over the female. While one male mated with her, another stood by and waited until the first had left her. He then mounted in turn and was accepted by the female. Jacobsen (personal communication) also found that in his captive animals a female mated successively with two males. This absence of rivalry in *Mungos* is of some interest and reflects a high degree of social integration.

In the mating described by Neal, the second male's first mounting resulted in a long copulation, in contrast to the series of brief mountings that characterises the beginning of a female's mating. This could either be because the female had by then become fully receptive or because the male's sexual excitement had been increased by watching the previous mating, so that he did not require a number of intromissions to achieve ejaculation.

(*vi*) *Hyaenidae*

Amongst the hyaenids the spotted hyaena is the only species in which mating has been described. Since they are social, no long period of contact making is required and Deane (1962) found that in the wild the oestrous female attracts the attentions of a number of males. There is rivalry between them and the larger animals drive the others off. Mating is prefaced by a preliminary stage in which the pair circle about, sniffing each other's anal regions. The tails are held erect and the anal pouches everted, thus exposing the secretion of the anal glands to the partner's olfactory investigation and there is much excited cackling vocalisation ('laughing'). The most detailed description of actual mating is that of Schneider (1926, summarised by Matthews, 1939). During the preliminary sniffing and circling both the penis and the clitoris undergo partial erection, and both at this stage are backwardly directed. There is then a change: the penis undergoes further erection and assumes a forward curvature while, according to Schneider, the clitoris loses erection. The male then mounts, clasping in the usual manner but adopting a more or less sitting posture: the penis now attains maximal erection and intromission follows. As we have already seen (Chapter 3), the urinogenital opening of the mature female is a ventrally elongated aperture, lying beneath the clitoris and it is thus possible for intromission to occur after the clitoris has lost its erection. The male then stands up, still clasping, and during the copulation, which lasts some ten minutes, he leans heavily on the female, resting most of his weight on her hindquarters.

The details of mating in *Crocuta* are clearly adapted to the unusual form of the female genitalia (see Chapter 3) and it is unfortunate that there is no description of mating in any of the other hyaenids which lack such peculiarities.

(*vii*) *Felidae*

Courtship and mating patterns amongst the Felidae are rather uniform and, as far as they are known, conform in general with those of the domestic cat. The special characteristics of felid courtship are correlated with the fact that the female becomes attractive to the male well in advance of the time when she is prepared to copulate – up to two or three days beforehand in the domestic cat. Two things follow from this. Firstly, more than one male may locate her, both odours and, in some species, special mating calls playing their part in this. Schaller (1967) gives references to an oestrous female being followed by several males in the tiger, jaguar and puma; van der Werken (1968) also quotes Schaller as saying that the only time he saw a cheetah on heat in Serengeti she was followed by three males. In such circumstances fighting amongst the males is common: it may be quite violent and severe wounds may be inflicted, although killings appear to be rare. Adamson (1968) found that in the presence of a female on heat his male lions became pugnacious not only to each other; they treated him also as a rival and would not tolerate his presence near their mate.

The second point that follows from the long time that elapses before the female becomes receptive is that, even when he has driven his rivals off for the time being, the victorious male is not yet free to mate: he still has to be accepted by the female. His initial advances are evaded or repulsed, the female spitting and striking out at him with her claws. The male does not retaliate: he merely draws back a little, only to renew his advances a moment later, prefaced by the typical 'entreaty' vocalisation. Adamson (1969) gives a series of photographs showing a lioness treating her much larger suitor in this way, exactly as a domestic cat will repulse a tom who is half as big again as herself. Gradually, however, her behaviour changes. She may approach the male, purr, roll invitingly on the ground in front of him and pat at him playfully but she will still flounce off with an indignant spit and a 'never-so-insulted-in-my-life' expression if he attempts to establish contact. Before she has reached the stage of acceptance, the male has been put through a severe test of persistence in his relations with the female and often of fighting prowess, too, in his contests with rivals of his own sex: faint heart assuredly never wins a feline fair lady. It seems to me quite possible that the length of the period during which the female is attractive but not yet receptive may therefore have a second function, besides providing ample time for finding a mate: it may also serve to ensure that there is competition between the males in which the qualities of strength, pugnacity and determination decide the outcome – qualities which a predator cannot afford to dispense with. The fact that the female may ultimately mate with another male besides the victorious one does not rule out this suggestion. We do not know what the relations between siring success and mating interval are in cats and, in any case, a competition does not have to be like a race, in which there is only one winner: it can also be an examination in which anyone who reaches a certain standard is awarded a 'first'.

In the domestic cat, once the female permits the male to make contact with her he mounts, gripping the fur of her neck in his teeth. The female meanwhile does not stand but lies on her belly and the male does not clasp her in the usual way with his fore limbs but makes treading movements with his hind feet against her hindquarters. If she has not already done so, this stimulates her to adopt the mating posture with rump slightly raised and tail turned aside. There may be a few unsuccessful mounting attempts before intromission and ejaculation are achieved, usually signalised by a loud screeching cry from the female and often a growl from the male. The female then writhes free and may turn and strike at the male, who leaps back to avoid the blow. When the pair know each other well, however, the female usually does not make this attack but goes over directly into a series of post-copulatory actions, eloquent of high excitement. She rolls on the ground, rubs her head on any convenient object, licks her genitalia and, after a short interval, a further copulation follows. With repeated copulations, the initiative gradually passes from male to female. Her eagerness to mate increases while his declines and presently it is she who makes the advances: rubbing against him, purring and taking up the mating position in front of him, she invites mounting and it is quite common for her to mate successively with more than one male before her heat is at an end.

In the larger cats mating is very similar and the description of mating tigers given by Sankhala (1967), for instance, is in general agreement with the account just given. There is, however, one difference. The male usually does not grip the female's neck as he mounts but bites only at the climax of the copulation. Antonius (1943) found this almost vestigial use of the neck bite to be characteristic of lion, tiger and jaguar and he also saw it in the only North American lynx mating that he observed. In the puma this behaviour, although common, is not invariable. There may also be more individual variation in other species than Antonius supposed, for, while Sankhala's observations agree with his, Gowda (1968) saw tiger matings in which the neck grip was used during the whole of the mounting and Leyhausen (1956) notes that in the domestic cat, at the stage when the female is inviting a male who is no longer very eager, he may mount without a neck grip.

The significance of the felid mating posture, with the female lying down and the male standing over her without a pelvic clasp, is not at once apparent. It may, however, be a function of the fact that of all the Carnivora, the Felidae are the most accomplished killers and are the only ones in which the fore paws as well as the teeth are potentially lethal weapons. The whole of their courtship seems to be adapted to ensure that the male does not become involved in any type of fight with the female, thus avoiding the risk of his accidentally switching over into a genuine attack. There is no playful or ritualised fighting and the male's inhibition against biting a female is extremely powerful; he does not normally establish contact with her until she permits it and although attempts at rape do occur, they are very rare. Furthermore, it may be that the normal mounting position, with pelvic clasp and neck grip, is dangerously like an attack on prey. In the prone posture, however, the female's resemblance to prey is less and she cannot easily move forwards away from the male, as an escaping victim does. Moreover, if the male supports his weight mainly on his fore limbs, as he must while treading, his whole balance is exactly the opposite of what is required in making an attack. The delayed neck bite of the larger cats is obviously interpretable in the same terms and the entire performance seems admirably suited to ensuring that there is as little danger as possible of mating accidentally turning into attack. Since the mating posture of *Genetta tigrina*, as described by Dücker (1957), is so similar to that of the Felidae, a detailed comparative study of mating behaviour and prey-killing patterns in different species of this genus would clearly be extremely interesting and might serve to throw some further light on the significance of the mating behaviour patterns. The question of whether mating usually takes place on the ground or in a tree may also be relevant; clearly in an arboreal mating it is desirable for the female to crouch and cling firmly to the supporting branch.

C. Development and rearing of the young

As already indicated, the rearing of a young carnivore to the stage where he can make an independent living as a predator involves the parent in greater responsibilities than fall to the lot of a herbivore. Parental care must be adjusted to the changing needs of the developing young and parental behaviour must thus alter in a coordinated manner with the maturation of the offspring. It is therefore necessary to say a little about the physical and behavioural development of the young before dealing with the behaviour of the parents towards them.

Although the development of the young is a continuous process, it is convenient to recognise three stages:

(i) the early nestling period, when the only nourishment the young receive is their mother's milk;

(ii) the mixed nutrition period, during which the young begin to eat solid food but still continue to take milk as well;

(iii) the period of post-weaning dependence on the parent, before the young finally become capable of finding their own food and become fully independent.

In view of the typically high protein diets of the adults, one would expect that the milk provided for young carnivores would also have a relatively high protein content. In general this appears to be the case and in the majority of species investigated the milk contains some 7 to 12% of protein (see table 8.7). Unfortunately no data are available for very small species, in which the initial growth rate of the young is very high and which therefore might be expected to have milks with higher protein contents.

Broadly speaking, the fat content of milk is related to the metabolic energy the young have to produce in order to keep warm and the high fat content of polar bear milk is therefore not surprising. The fat contents of the two husky dog samples, although high, are within the range covered by other breeds of dog and wolf milk is not very different. The general similarity between the milks of wolf, husky dog and other species of *Canis* suggests that their milks are typical for large carnivores from cold climates and supports the idea that the genus *Canis* is of northern origin.

A number of the values found by Ben Shaul (1962) are rather unexpected and some confirmation is desirable before they can be accepted as typical for the species in question: these have been

Table 8.7
Composition of milks of various carnivores.
(Values for cow and man have been added for comparison.)

Species	Total solids	% fat	% protein	% lactose	Author
Dogs (various breeds)	20·4–29·2	8·5–13·5	5·4–10·6	2·5–4·2	Luick *et al.* (1960)
Husky dogs (2)	24·5, 28·3	11·2, 13·4	8·2, 8·3	2·9, —	Lauer *et al.* (1969)
Wolf (arctic)	23·5	6·6	12·4	3·0	Lauer *et al.* (1969)
Wolf	23·1	9·6	9·2	3·4	Ben Shaul (1962)
Coyote	22·1	10·7	9·9	3·0	Ben Shaul (1962)
Golden jackal	22·0	10·5	10·0	3·0	Ben Shaul (1962)
Lycaon	23·1	9·5	9·3	3·5	Ben Shaul (1962)
Fox	18·1	6·3	6·4	4·6	Hock and Larson (1966)
Black bear	23·5	10·5	7·3	1·5	Hock and Larson (1966)
Brown bear*	11·0	3·2	3·6	4·0	Ben Shaul (1962)
Polar bear	43·5	30·6	10·1	0·5	Baker *et al.* (1963)
Raccoon*	12·0	3·9	4·0	4·7	Ben Shaul (1962)
Otter (species not stated)	38·0	24·0	11·0	0·1	Ben Shaul (1962)
Domestic cat	18·4	6·3	10·1	4·4	Spector (1956)
European lynx	18·5	6·2	10·2	4·5	Ben Shaul (1962)
Puma*	35·0	18·6	12·0	3·9	Ben Shaul (1962)
Leopard	19·4	6·5	11·1	4·2	Ben Shaul (1962)
Lion*	36·1	18·9	12·5	2·7	Ben Shaul (1962)
Cheetah	23·2	9·5	9·4	3·5	Ben Shaul (1962)
Cow	12·3	3·6	3·2	4·7	Hock and Larson (1966)
Man	12·0	3·7	1·2	6·8	Hock and Larson (1966)

*Denotes questionable value.

marked with an asterisk in table 8.7. It is, for instance, difficult to see why the brown bear should have milk so much more dilute than the black bear; the raccoon milk also has surprisingly low fat and protein levels. Amongst the Felidae, the very high fat contents of puma and lion milks appear discordant. Ben Shaul attempts to explain this by pointing out that, if the mother has to leave her young alone for long periods while she hunts, they may have a problem in keeping warm, even if the climate is not particularly cold and may therefore require extra fat as an energy source. While this may be true it is difficult to believe that the puma leaves her young for vastly longer periods than the lynx, or the lion than the leopard. Clearly some factual information on the times spent alone by the young of the various species is desirable but in the meantime it seems unwise to accept the data for puma and lion as necessarily correct.

Comparative data on the development of the young are both scattered and scanty but such quantitative data as I have succeeded in finding are summarised in table 8.8 (pp. 356–7). Most of the information is based on rather few observations and those of different workers are not always in agreement: many have been made on captive and some on hand-reared animals and the dates of first eating of solids and of weaning may therefore be a little misleading. One would expect the young to be capable of eating solids some time before they normally have the opportunity to do so; in *Suricata*, for example, they will accept solid food during their third week, although the mother does not normally start to provide them with this type of fare until about a week later: similarly, the domestic cat normally brings food to her young when they are five weeks old but they can be induced to eat about a week earlier. In a captive animal, with food provided in abundance, weaning may also occur earlier than it would in the natural situation and observations in captivity rarely make clear the point at which the young normally become fully self-supporting.

Despite these shortcomings, it is clear that patterns of development show some variation from family to family. The Canidae appear to be very uniform, with the eyes opening at about two weeks old, solids first taken at about a month and weaning after a further month. The viverrids and the smaller felids, although rather more variable, are roughly similar. As one might expect with the killers of large prey, the period of mixed nutrition is prolonged and in the larger species there is a very obvious period of post-weaning dependence.

The early opening of the eyes in the large felids may relate to the rather unprotected situations in which many of them are born. Baudy's (1971) observations suggest that the lack of uniformity reported in tigers (see table 8.8) may be, at least in part, a reflection of subspecific variation, which itself may be related to the degree of protection afforded by the den in different regions. He found that, in two litters of Bengal tigers from different parents, all the cubs were born with their eyes open, whereas in his Siberian tigers the cubs were two to three weeks old before their eyes opened. Differences such as those between the ocelot and the black-footed cat are probably adaptively related to habitat but before any detailed analysis can be offered it is necessary to have much more extensive comparative data on other small felids. The most surprising departure from what appears to be the normal carnivore developmental pattern, however, is amongst the small mustelids. Here the eyes open extremely late and the young are ready to eat solid food before this has happened. One would very much like to know at what stage the young are in fact given solid food in natural conditions: possibly one should regard their peculiar developmental characteristics as an adaptation whereby the young, born in a well protected burrow, have become increasingly immature in general while the period during which the female has to support them on her milk alone has remained constant, instead of being correspondingly prolonged.

Amongst the bears, the very small size of the young is striking. As Leitch *et al.* (1959) have shown, the weight of the litter normally bears a simple relationship to maternal weight. Throughout the placental mammals, the relative weight of the litter decreases with increasing body weight and for species ranging from whales to bats the values fall on a single curve. The values for bears, however, are discordant and well below what would be predicted from the curve. This, as already indicated, is related to the fact that the cubs are born during the winter denning-up period. The fact that the young are small even in tropical species, where there is no denning up, suggests that this is a secondary habitat and the bears originated in a northern region, with big seasonal changes and hard winters. There is nothing in the data at present available to suggest any particular dislocations in the normal pattern of development: the cubs appear simply to be unusually altricial. What is less clear is why it is necessary for the young to remain under parental care for so long, when their methods of food finding do not require any great strength or skill and one cannot imagine them normally being subject to any very heavy predation. Possibly the major difficulty which faces the growing cubs relates not to their interactions with other species but rather to those with their conspecifics. They cannot make a living until they are old enough to establish a territory for themselves in the face of intra-specific competition which must have been much more drastic before human interference had reduced the populations to their present low levels.

The behavioural repertoire of the newborn young is relatively simple: they are capable of searching for a teat and responding to it correctly when they locate it. Many of them show the 'milk tread' – a series of alternating movements of the fore paws, pushing against the mother's body as they suck – which has the effect of stimulating milk flow. This type of milk tread is familiar to us in the domestic cat and is of widespread occurrence amongst the Felidae. It is also shown by members of a number of other families and has been reported in the Viverridae (*Genetta tigrina*, Dücker, 1965a), Hyaenidae (*Crocuta crocuta*, van Lawick-Goodall 1970), Canidae (red fox, Tembrock, 1957b; fennec, Koenig, 1970) and Mustelidae (pine and stone martens, Schmidt, 1943; wolverine, Krott, 1959; badger, Eibl-Eibesfeldt, 1950; European otter, Hurrell, 1963; sea otter, Kenyon, 1969). In some species the method of stimulating maternal milk flow is slightly different. In the domestic dog, for instance, although the paws are pressed against the mother's body on either side of the teat, most of the stimulation is provided by the snout. As the puppy starts to feed, the head is bobbed up and down as he alternately pushes against the mother's body and tugs at the teat. The African civet cat is of some interest in this context, since both patterns are shown. When starting a feed the kitten first stands on all four paws and alternately thrusts its head against the mother's body and pulls back, tugging at the teat, very much as a puppy does: before very long, it snuggles closer to the mother's body, supports itself on its hindquarters and starts to use the fore paws in a rapid, rather jerky milk tread. Presumably both the form of the teats, the type of mouth movements used by the young to create suction, and possibly also the form of the paws, are involved in determining the relative importance of snout and paws in providing the stimulus for milk let-down.

In many species the young purr as they suckle. This is also familiar to us in the domestic cat and is common amongst the Felidae. Amongst the viverrids too, purring appears to be common. It has been reported in *Genetta tigrina*, *Helogale undulata*, *Suricata suricatta*, *Nandinia binotata* and *Arctictis binturong* (Zannier, 1965). Purring by suckling young is also recorded in the black bear (Herrero, 1970) and, according to Deane (1962), in the spotted hyaena, although the cubs do not purr, the mother does so as she nurses them.

In the domestic cat it has been found that within a few days a system of teat ownership has developed and each kitten sucks almost exclusively from a single one of the mother's teats (Ewer, 1959, 1961). Since the kittens' claws could do serious damage if there were prolonged fighting over a teat, it seems likely that one function of teat ownership may be to prevent this, and it certainly ensures that a whole litter can settle peacefully to feed within a very short time. In puppies fighting is less likely to be damaging and Rheingold (1963) found no evidence of teat ownership in puppies, beyond the fact that in one litter the strongest pup usually managed to get possession of the best-yielding posterior teats. In *Civettictis* the claws are not sharp and the kittens are no more likely to injure each other than are young puppies. Teat ownership was, however, shown by the two kittens of the only litter I have studied, although it was established more slowly than in the domestic cat. In this species lactation is prolonged and may go on for fifteen to twenty weeks, by which time the teeth have erupted and fights between the youngsters could be serious. Teat ownership may therefore be of value towards the end of lactation rather than during the early stages. In *Suricata* I found evidence of a form of teat recognition which does not imply exclusive ownership (Ewer, 1963a). There was very little fighting over teats, and when starting a feed each kitten usually selected its own teat. After a little while, however, they would all shift their positions, and by the end of the feeding period each would have sucked from several teats and all six teats were used by litters of three or four. In the cheetah, Adamson (1969) found that fighting over teats did occur; the cubs frequently changed position during a feeding period, twelve teats were used by a litter of four and nothing in her observations suggests that there is any teat ownership.

Most young carnivores, possibly all of them, have a tendency to come to rest only when in contact with a littermate's body. This response is dependent on temperature and in very hot conditions it ceases and the young lie apart from each other. This, however, is not a common occurrence and normally the contact response reduces heat loss, thus assisting the young to keep warm; it also ensures that they do not become scattered.

At a very early stage, even before the eyes are open, both felids and viverrids show a defensive response to any disturbance, in the form of an explosive spitting or hissing, which is quite surprisingly disconcerting.

During the nestling period, locomotor abilities develop; toilet behaviour, along with the usual comfort movements of yawning and stretching, makes its appearance and before very long the young begin to play together. The opening of the eyes does not result in any very sudden or dramatic change in behaviour, since vision is at first rather inaccurate. Play will be dealt with in more detail at the end of this chapter; here it suffices to say that many of the patterns that will later be used in prey catching and in fighting make their first appearance in play. As motor coordination and general activity increase, the point is reached when the young begin to emerge from the nest. It is usually not long after this that the parent (or parents) begins to provide the young with solid food and the period of mixed nutrition begins.

Here, in passing, a curious form of behaviour may be mentioned. In a few species, at about the time when eating of solids is due to begin, the young lick at the mother's mouth and apparently drink her saliva. Hamilton (1933) has recorded this in *Mustela frenata*; Dücker (1957) saw it in a genet, and it also occurs in *Civettictis*. 'Mouth suckling' of this sort is widespread amongst rodents but has not, to date, been reported in other carnivores. The significance of this behaviour is not understood: it is not known whether the saliva contains something of physiological importance, possibly enzymes or antibodies; whether it is merely a source of fluid; or whether the smell of the food she has recently eaten remains on the mother's mouth and this plays some role in encouraging the young to start eating.

Once they emerge from the safety of the nest the young are in a very vulnerable stage. Not yet familiar with the locality, they are in danger of straying away and getting lost: still relatively small and weak, they are also in danger of attack by predators. There are, however, two types of behaviour patterns which have by now matured and serve to minimise these dangers.

(i) *Contact-keeping*. The young show a strong tendency to remain together, so that even if they do move some distance from the nest the parent on return will locate them all. In a number of viverrids, a special 'lost call' facilitates this keeping together. If one youngster becomes separated from the rest he gives this special call and the rest of the litter respond at once by running to him. I first witnessed this behaviour in *Suricata* and was greatly surprised to find that neither parent paid any attention to the lost call. In fact, I was so obsessed with the idea that the mother 'ought' to come in response to the cry that it was some time before I appreciated its true function and realised that, although neither parent reacted to it, the littermates did so at once. The African civet cat also has a similar call but here the littermate not only runs to join the caller but also calls in answer. The female of my pair of youngsters suffered from a digestive upset and was ill and lethargic for a few days. So strong, however, was her response to her brother's lost call that if he gave it, she would reply and stagger weakly out of the nest box to join him.

As the kittens matured, contact calling became gradually less and less frequent and finally ceased altogether. When her first litter was born, however, the female used the same call to summon the young and they responded exactly as they did to each other, by calling in return and running to her. She, however, did not now respond in the same way: she did not go to the kittens when they called but clearly expected them to come to her when she did so. Duplaix-Hall (personal communication) has found that the cubs of the clawless otter,

Amblonyx cinerea, also give a lost call, to which a littermate responds in exactly the same way as the civet kittens but whether the mother also uses the call as the female civet does is not clear. Wemmer (1971) has found that a hand-reared fanaloka gives a similar call, which he interprets as having the function of keeping contact between mother and offspring but which clearly might also serve to keep littermates together.

Although so far reported only in viverrids and in *Amblonyx*, it seems likely that such vocal contact-keeping behaviour within the litter will prove to be more widespread than is at present realised. At least in the three viverrids mentioned, the calls are brief, rather abrupt and repeated several times in a series, so that they are extremely easy to locate. The orientation of both my *Suricata* and my civets to the caller was extremely swift and accurate.

(ii) *Alarm behaviour*. Most young carnivores respond to a sudden alarm by some form of defensive behaviour; either by fleeing, or taking cover, or turning to face the danger with bristled hair and giving a sharp hissing or spitting vocalisation. A number of them also show specific responses to a parental alarm call. In the domestic cat, if the mother growls the response of the young is dramatic in the extreme. They may be playing together, quite oblivious of any potential danger but if there is a sudden growl from their mother they vanish instantly, as though by magic, scattering in all directions, each hiding under whatever cover is nearest and remaining immobile until the alarm is over. This response may well be widespread amongst the Felidae, for Gashwiler *et al.* (1961) have seen it in kittens of the bobcat and Leyhausen and Tonkin (1966) in the black-footed cat. In the latter case, however, the scattering of the kittens appears to be a direct response to an alarm, rather than to an alarm call by the mother and they remain in hiding until she gives a special call which acts as an 'all clear' signal.

Koenig (1970) found that in response to the mother's barking young fennecs fled to the safety of their nest box and Tembrock (1957b) mentions a similar response on the part of fox cubs. Young *Suricata* respond to a parental 'hawk warning' call by running to the mother and remaining close beside her as long as she continues to call: if she moves to cover they move with her; if she freezes where she is they do likewise (figure 8.2). The sloth bear, *Melursus*, is unique amongst carnivores in that the young, if alarmed, not only run to the mother but climb on her back and cling to a specially developed tuft of long hair while she carries them to safety (Norris, 1969). This method of protecting the young is clearly related to the sloth bear's extremely arboreal habits.

Purring has already been mentioned: its primary significance is

Figure 8.2. *Suricata*: (a) the mother, having seen a hawk, is giving the warning call and the young have run to her; (b) still calling, she moves cautiously towards cover and the young follow closely.

probably as an 'all's well' signal between the nursing mother and her young, which may help in forming the original bond between them. Later on, however, when the young are active outside the nest, purring in the domestic cat takes on a new function as a signal between littermates. When the mother returns to feed her young the first kitten to encounter her will start to suck and will also purr loudly. The rest of the litter alert to the purring and come running to join in, the purring clearly acting as a feeding summons; in this way, no kitten misses a meal. I have not seen this 'dinner gong' function of purring described in any other species but it may well be common amongst the Felidae and the purr-like call with which the cheetah summons her cubs to come to her may well be related.

Weaning is a gradual process: the proportion of solid food in the diet increases, suckling becomes less frequent and finally ceases altogether. This, however, does not mean that the bond between parent and young is ended and that the break-up of the family follows at once. Particularly in those species that kill large and dangerous prey, there may be a further period of 'apprenticeship' during which the young learn to make their own kills. Since it takes a considerable time for them to gain sufficient strength and skill to be self-supporting, they still rely largely on prey killed by the mother and in many species she actually assists in their education. Before describing what we know of this parental training, however, it is desirable to go back to the beginning and describe from its earliest stages the mother's behaviour towards her young.

The female's care of her young actually starts before they are born. Towards the end of pregnancy, in many species and possibly in all, she starts to search out a suitable place for giving birth. Many of the Canidae, as already noted, dig extensively and clear out their burrows; Adamson's cheetah sought out a concealed and well-protected lair; domestic cats become increasingly interested in any warm, dark, sheltered resting place. If the young of a previous litter are still with her, the female becomes more hostile to them and at this stage a cat will drive off her kittens with considerable violence whenever they approach. This is a necessary process, for as parturition draws near, her maternal responses are activated and her behaviour changes dramatically: she is now prepared to accept and to care for her young. If the previous kittens were there, she would mother them and, in competition with them, the new litter would have little chance of survival. In wild species the dispersal of the previous litter is normally a more gradual process, not involving violent rejection by the mother. Two litters in a year is exceptional for wild members of the genus *Felis* but has become the norm in the protected conditions of domestication: the kittens must therefore be driven away before they are in fact ready to go and the mother's violence towards them is a product of domestication.

In carnivores, as in most other mammals, once the young are born the mother licks them free of their embryonic membranes, licks their fur dry and eats the membranes and the afterbirth: there is thus very little soiling of the nest. Where litters are large, the food value of the placentas is considerable and the female therefore does not need to leave her young in order to hunt during the period immediately following parturition. Licking the young dry has the obvious function of preventing their becoming chilled, as drying the hair out permits it to take up its normal insulating air layer. In the sea otter pup this process is of particular importance. Unlike seals, sea otters do not have a blubber layer and depend entirely on the properties of their fur for insulation; the air layer must also add significantly to buoyancy. A pup who was not properly dried out would therefore very soon die and the sea otter gives her youngster an unusually prolonged initial grooming. Sandegren *et al.* (in press) record a female spending three and three-quarter hours in grooming her newborn pup, interrupting the process only once, when she turned her attention briefly to grooming herself instead of the pup.

Quite apart from providing the young with warmth, protection and food, the prolonged initial contact with her young which the female's post-parturitional cares entail may have another function: namely, to facilitate the formation of the specific bond between mother and young which must be established if the litter is to be successfully reared. Immediately after parturition, a domestic cat is ready to care for young but her responses become specifically directed to her own kittens only after she has had some time to become familiar with them. She will, in fact, accept strange offspring, even those of another species, if given them early enough. A friend of mine successfully added two baby giant rats to his cat's newborn litter and zoos not uncommonly use bitches to foster the young of other species of carnivore, if their own mothers do not seem likely to care for them adequately.

Leyhausen and Tonkin (1966) record an interesting case in which a captive black-footed cat had given birth to a litter but owing to some disturbance, her maternal responses appeared to be upset and she remained inattentive to her young. It was therefore decided that the kittens must be removed and hand-reared and first they were taken and weighed. When placed on the scale one gave a loud distress call and this was sufficient to trigger off the female's latent maternal responses. She reacted at once, ran up and retrieved the kittens one after the other, carried them back to her nest box and from then on cared for them in the normal manner.

Once the bond with her own young has been established the female will no longer behave maternally towards strange young, even of her own species, and it becomes very difficult to induce her to adopt a foreign youngster. The bond with her own young, however, is not permanent. Obviously it must ultimately be dissolved and the young must become fully independent: in fact, the process starts as soon as they become sufficiently strong and active to leave the shelter of the nest. Once this has happened the female ceases to differentiate very sharply between her own and foreign youngsters. One can easily see this happening if two domestic cats in a household have kittens of roughly the same age. Once the young are active outside the nest, it is seldom long before one finds the kittens sometimes feeding from the 'wrong' mother and being permitted to do so. Similarly, van Lawick (personal communication) succeeded in getting a female *Lycaon* whose pups had just begun to play actively outside the den to adopt an abandoned youngster of approximately the same age. Verts (1967) reports two cases of young striped skunks, aged five and seven weeks, being adopted by females with litters of their own. A comparable accidental exchange of brown bear cubs is mentioned by Erickson and Miller (1963) and in lions the development of some degree of communal care of the cubs is a normal happening. The lioness rears her young apart from the pride during their first few weeks of life but once they become active she returns with them to the pride and, if there are several litters, there is a very strong tendency for them to become the common property of all the mothers and to suckle indiscriminately from any lactating female.

If two or more females give birth in close proximity, they may each accept both their own and their neighbour's young and the litters may become joint property, the young suckling from both mothers. This may happen if two vixens share a burrow and Neal (1970) found that it is the normal arrangement in the communally living banded mongoose. What happens in the earliest stages, when the young remain within the shelter of the den, is not known but in one of the groups Neal studied, eight youngsters, old enough to emerge from the den but too young to forage with the group, were being cared for by three lactating females. One of these remained behind with the young during the day and was seen to suckle them briefly now and then. When the other two females returned in the evening from the day's foraging, all three mothers fed the youngsters: there was considerable shifting around from one female to another; the mothers were clearly used by the young as communal food sources and there was no specific attachment of any youngster to a particular female.

Most young mammals are at first incapable of micturating or defaecating except in response to external stimulation of the perineal region. The mother includes toilet licking as part of her routine care of her offspring, in the course of which she consumes the urine and faeces which the young produce in response to her stimulation (plate 15). In altricial species, independent micturition and defaecation do not take place until the young have reached the stage of moving about actively outside the nest and there is therefore no fouling of their living quarters. Apart from simple sanitary considerations, accumulation of urine and faeces might well have the further undesirable result of attracting the attention of predators. This is presumably why the requirement for perineal stimulation in the first days of life is so very widespread and why even the precocial fanaloka does not defaecate or micturate unaided until well into its second week of life (Wemmer, 1971). It would also explain why, even after their pups can defaecate on their own, many bitches will clean up the droppings from the immediate vicinity of the nest.

During their first days of life the female spends much of her time with her young. She assists them to feed by adopting a suitable nursing position, she cleans them and her presence also helps to keep them warm (plate 15). In addition she will, if necessary, remove the young to a new nest. This may be done if there is disturbance of any sort or if the nest becomes unsuitable through some accident, such as flooding and also, possibly, if it becomes flea-ridden. In some cases moves are so frequent that it is difficult to believe the old nest can have become unsuitable and it may be that the frequent shifting is protective, reducing the chance of the nest's location becoming known to potential predators in the area. Van Lawick-Goodall's and van Lawick's (1970) golden jackals, for instance, moved their pups to a new burrow five times in their first twelve weeks of life and Adamson's (1969) cheetah shifted her much more exposed youngsters twenty times during their first six weeks.

In carnivores the usual method of transport is to pick the youngster up by the scruff of the neck, to which it responds by remaining limp and passive until it is set down again (plate 15 and figure 8.3). This method is the one normally seen in the Felidae, including the cheetah (Adamson, 1969). Van Lawick-Goodall and van Lawick (1970) report the spotted hyaena carrying cubs in this way and it is a method commonly but not invariably used by the Canidae. It is also found in some of the Mustelidae, for example the polecat (Leyhausen, 1956) and the wolverine (Krott, 1959). Stegman (1937) saw a striped skunk carry her young in this way but Verts's skunk picked her young up by the middle of the back. In this connection a peculiarity in the development of the stoat, *Mustela erminea*, first described by Bishop (1923) is of some interest. Both stoats and weasels are born almost naked but their patterns of hair growth are different. In the former,

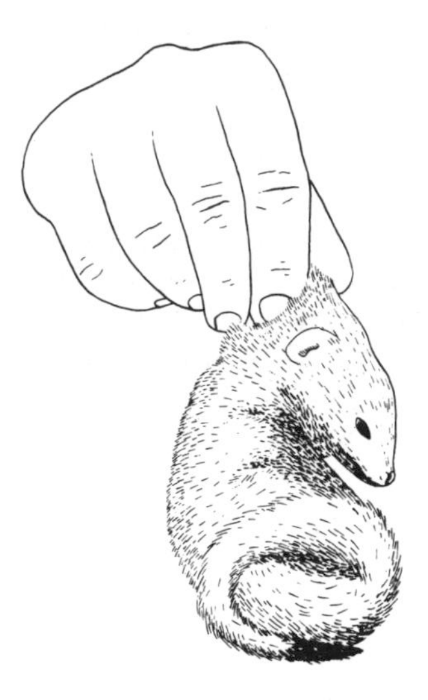

Figure 8.3. Baby red mongoose, *Herpestes sanguineus*, showing typical passive carrying posture. [Drawn from a photograph by N. Jacobsen]

Figure 8.4. Young stoat (crown-rump length approximately 105 mm.), showing development of long hair on neck. [After Bishop, 1923]

growth is most rapid on the nape of the neck and by the age of two or three weeks the stoat has developed a distinctive mane (figure 8.4) which is not found in the weasels, *M. nivalis*, *M. rixosa* or *M. frenata* (Hamilton, 1933; East and Lockie, 1965; Heidt *et al.*, 1968). It is tempting to suggest that this precocious growth of neck hair serves as a protection when being carried by the mother and it would be extremely interesting to know whether in fact the neck grip is used only by the stoat. Unfortunately, the only information I have been able to find is that the female *M. rixosa* kept by Heidt *et al.* carried her young gripped by the middle of the back and a statement by Guggisberg (1955) that the stoat grips her young by the neck mane. These data, although in accord with what has been suggested, are very scanty and clearly more information is required.

Figure 8.5. Small-clawed otter (*Amblonyx*). (Top) adult male walking bipedally while carrying a large cub, held between fore paws and chin. (Bottom) female walks on three legs while carrying a coin held between left paw and cheek. [Drawings, N. Duplaix-Hall]

Amongst the otters, both *Lutra lutra* (Stephens, 1957) and *L. canadensis* (Harris, 1968) are said to carry their young gripped by the neck in the orthodox manner. In *Amblonyx* and *Aonyx* carrying in this way is seen now and then, especially if long grass or movements of the youngster create difficulties. The normal method of carrying the young, however, is to hold them clasped against chest or chin by one or both fore paws (figure 8.5) and hobble along on three legs or even walk bipedally (Duplaix-Hall, personal communication). This is of some interest, since on other grounds the sea otter is regarded as more closely related to these species than to the fish otters. Some method of transporting the young other than the neck grip would have had to be evolved before care of the young away from land became possible and the sea otter uses the same method of clutching the young to her chest, the only difference being that she swims on her back as she does so.

In the Viverridae there appears to be some variability in how the young are carried: Dücker (1957) records a genet (*G. tigrina*) and Louwman (1970) a banded palm civet (*Hemigalus derbyanus*) picking up their kittens by the middle of the back. The same grip is sometimes used by *Civettictis* but I have seen an experienced female use both the neck and the back grip. In *Helogale*, too, according to Zannier (1965) there is variability in the orientation of the grip. Jacobsen (personal communication) finds that in *Mungos mungo* the young are usually carried by the neck but sometimes are gripped by any part of the body in an apparently random manner. In *Suricata* I found that a young female picked her young up more or less randomly, without precise orientation but another, who was older when she had her first litter, always used the neck grip. Jacobsen has also seen the neck grip used by *Herpestes sanguineus*.

Leyhausen (1956) has found that in the domestic cat the proper orientation may be lacking in a young mother and the cheetah which Schaller (1970) saw carrying her cubs by back or leg, instead of by the neck as Adamson (1969) reports, was a young animal. In the cat, at least, this appears to be a matter of maturation rather than of learning, since an older female will use the neck grip even with her first litter. The variability recorded in *Mungos*, *Helogale* and *Suricata* may also possibly reflect incompletely matured behaviour in young animals but it also seems to me possible that in these species the orientation of the grip may be rather easily disrupted by any factors which upset or disturb the animal. A detailed study of the circumstances in which random orientation occurs would clearly be of interest.

It could be argued that in moving her young in response to disturbance, etc., the female is merely shifting home away from a place which, for one reason or another, has become distasteful to her and the fact that this benefits the young is incidental. The raccoon, however, provides an example where the mother's behaviour appears to be related to the welfare of the litter and not directly initiated by her own discomfort. Dens in hollow trees are favourite places for the rearing of the young but at the stage where they first become active and start to try to crawl out of the nest, there would be considerable danger of a youngster falling and either being injured or failing to find his way back to the nest. First excursions from the nest are normally made at some time between six and nine weeks old. Montgomery *et al.* (1970) report a case where a raccoon moved her three youngsters over a distance of about half a mile from a tree nest to a ground den when they were forty-five days old and quote another case where a similar move from tree to ground was made when the young were six weeks old. Schneider *et al.* (1971) report similar moves being made by two females when their cubs were forty-seven and sixty-three days old. This behaviour on the part of the mother would therefore appear to be a response to changes in the young, rather than in the nest; one would expect their increasing activity to be the relevant factor.

Another response which is shown only when the young have reached a certain stage of development is that of bringing food to them, but whether it is their size, their activity, some combination of the two or some quite different factor which evokes this behaviour is still not known. In the domestic cat I found that a female brought food to foster young older than her own litter when they were of the correct age and Robinson (1952) found that a coyote fed a young swift fox for longer than usual; presumably because, being a smaller species, it remained the correct size for evoking the response for a longer period.

The response of feeding the young is not restricted to lactating females but may be shown by females who are not rearing young and, in some species, by males also. Leyhausen (1965b) records non-breeding females of the domestic cat and the European wildcat behaving in this way. According to Young (1958) the male bobcat as well as the female will bring food to the young and Oeming (1962) found that in captivity the male Canadian lynx stays with the family, grooms the young and plays with them. The male Siamese cat also often shows similar solicitude for his kittens. Although male parental care is uncommon in the Felidae, amongst the Canidae it is usual for the male as well as the female to assist in feeding the young. This is known for the fox (Tembrock, 1957a, b), the coyote (Robinson, 1952), the wolf (Crisler, 1959), the golden and black-backed jackals (Wyman, 1967), the hunting dog (Dunton, 1960; Kühme, 1964, 1965a, c), the raccoon dog (Bannikov, 1964; Novikov, 1956), the arctic fox (Barabash-Nikiforov, 1938) and the fennec (Gauthier-Pilters, 1967).

Amongst the viverrids it has been recorded only in *Helogale* (Zannier, 1965) but may well occur in other social mongooses.

The form in which food is first provided for the young varies from species to species in relation to the usual type of prey. If it is of suitable size, then the prey may simply be killed and carried back to the young. This is the method adopted by the smaller carnivores that feed mainly on small rodents, for example the cat and the fox. Cats will not only bring home rodents and small birds for their young but also, if they are to be had, large insects such as locusts. Captive weasels and martens also carry killed prey to their young in the nest, starting when they are aged five to six weeks. This has been recorded in *Martes martes* and *M. foina* (Schmidt, 1943), in *Mustela nivalis* (East and Lockie, 1964), *M. rixosa* (Heidt *et al.*, 1968) and in *M. frenata* (Hamilton, 1933). In the latter species the male as well as the female carried food to the young but whether this normally happens in the wild is not known. Osgood (1936) reports a rather unusual case of a stoat, observed in the wild, carrying back earthworms for her young.

In the pack-hunting Canidae the prey is normally too large to be carried back whole and the adults swallow lumps of flesh and subsequently regurgitate for the young. This is familiar in the domestic dog (Martins, 1949) and is also the method used by wolves (Crisler, 1959), coyotes (Snow, 1967), jackals (Wyman, 1967) and *Lycaon*. The latter are unusual in that the adults are prepared to share food in this way with each other as well as with their young. The animals returning from a hunt with full stomachs not only regurgitate for the pups but will also disgorge, although slightly less readily, if solicited by one of the adults that remained at home to guard the young and did not participate in the hunt (Kühme, 1964, 1965a, c). In both coyotes and wolves, once the cubs are capable of chewing large pieces of meat, the adult may simply carry a lump of meat to the young without first swallowing it and Martins (1949) also found that the domestic bitch would carry bones to her pups.

There is some dispute about whether the larger felids ever regurgitate for their young. Carr (1962) believes that the lioness will do so but other workers do not agree. Most carnivores, given the opportunity, are rather prone to bolt down too much food too quickly and then promptly sick up the excess. Even if regurgitation for the young is not a normal practice, if such incidental vomiting by chance took place in the presence of the young it could give a misleading impression. Florio and Spinelli (1967) quote a case of this happening with their captive cheetah but do not believe that this means that feeding of the young by regurgitation is a normal habit. The large felids seem rather to carry the prey whole to the young, if it is not too large, or simply to lead the young to their kill and permit them to share in feeding on it. This is true of the lion, leopard and cheetah.

Although crocutas may occasionally carry back pieces torn from a kill to their dens, this does not appear to be done very regularly and there is no very organised method of providing the young with meat. Bones broken away from a carcase by adults and subsequently chewed by juveniles have been found by Sutcliffe (1970) in some East African *Crocuta* dens. Whether the young themselves brought these back to eat in the safety of the lair, or whether adults sometimes carry food back to their young was not clear at the time. Since then, van Lawick-Goodall has found that in the clan she studied, although the adults did not regularly carry food to the young, they did now and then bring back some remains from the kill - generally bones without much in the way of meat on them. If the kill was made fairly near the den, youngsters also occasionally carried something back, usually horns or some other part for which competition was not particularly keen (van Lawick-Goodall and van Lawick, 1970).

Suricata has a different method of dealing with the problem of providing the young with solid food. The main prey is insects; these are individually small and collection of a stomach full may take a long time. Neither carrying back of individual catches nor regurgitation is therefore very suitable. In the type of foraging used to gather such prey, however, efficiency in catching is not impaired by the presence of inexperienced young. They therefore accompany the parents on their foraging trips and when the mother digs up an insect, instead of eating it herself, she holds it in her jaws, approaches the youngster and incites him to snatch the morsel from her. If he fails to do so she will in the end lay it down in front of him. In *Suricata* I never saw the male behave in this way but in *Helogale* both sexes do; indeed, Jacobsen (personal communication) found that with his animals the male was more assiduous than the female in attending to the needs of the young.

In pack hunters, such as the social Canidae, the young can make the transition to killing their own prey gradually. Once they are strong enough, they can run with the hunting group and, although they will at first be in the rear of the chase, they will catch up in time to join in eating the kill and will gradually become more adept and come to take a full share in the hunting and killing. Van Lawick-Goodall and van Lawick (1970) found that *Lycaon* pups began by following the hunt only for a short distance; they then gave up and waited for the adults, who returned presently and led them to the kill. Presumably they gradually extend their range but are fetched to the kill whenever the chase proves too long for them. For the lone hunters by stealth, however, the problem is more difficult, and here there is a period during which the young are provided, under maternal supervision, with the opportunity to make their own kills and are allowed to discover how best to utilise their repertoire of instinctive movements and so to perfect their techniques. How this is done is most familiar in the domestic cat. The female's behaviour goes through a series of stages (Leyhausen, 1956; Ewer, 1969). At first she carries home prey which she has killed and eats it herself in the presence of her kittens. A little later she leaves the prey for the young to eat and finally she brings home live prey, sets it down before the young and allows them to kill it. She does not help them to make the kill but if the prey escapes from them she will catch it again and bring it back to them. Kruuk and Turner (1967) and Schaller (1970) both saw a cheetah bring back a live Thomson's gazelle and liberate it for the cubs in much the same way and Schaller (1967) describes a tigress pulling down a buffalo and leaving it for her cubs to kill. The Canadian river otter also brings back live prey to her young and releases it for them to kill (Liers, 1951).

In none of these cases can the parent be said to teach the young how to kill. She simply presents them with prey and prevents it from escaping. In this situation the natural responses of the young automatically lead to their learning how to improve their killing technique: the behaviour of the mother is adjusted to the developing repertoire of responses of the young and gives them the opportunity to educate themselves by easy stages.

There is, however, a further problem. Prey must first be caught before it can be killed. In this connection some observations made by Schenkel (1966) in Nairobi National Park are of interest. He describes how two lionesses used to permit their cubs to join in some of their hunts from the age of about six months onward. He was of the opinion that if the mothers seriously intended to kill, they left the cubs behind but at other times would encourage them to follow and participate in the hunt. Once more, the mothers appeared to be giving the young the chance to educate themselves but here in the techniques of stalking the prey rather than of killing it. Their experience must soon teach them that if you expose yourself to the view of the prey too soon, it will take to flight and escape you and the value in cooperative hunting of lying in ambush must also become apparent. Eaton (1970d) saw very similar training of the young in the cheetah. At first the mother did not permit her cubs to accompany her and used to give a special call as she left them, to which they responded by remaining together until she returned and, with a different call, summoned them to her kill. When they were six and a half months old, however, she began to allow them to follow her and participate in the hunt. Here too the young were given the opportunity to gain experience in how to get within reach of their quarry and how to bring it down. Possibly, however, there may be a little more to it: in such hunting excursions with the mother a tendency to stay with her and do as she does may speed the process of learning. Certainly in the domestic cat it has been found that a kitten learns faster to press a lever for a food reward if it is first allowed to watch its mother doing so than if its 'teacher' is an unrelated but friendly female (Chesler, 1969). Be that as it may, it is clear that in both lion and cheetah, the mother's treatment of her cubs does put them in the way of mastering all the techniques that they will presently require in order to become full-fledged, self-supporting hunters.

The education of the sea otter pup may also involve learning based on watching the mother. The first introduction to solids presents no problem. The youngster stays close to the mother and when she dives for food he waits for her to surface again and is ready to share her catch. This must be facilitated by the fact that her chest serves both as his cradle and as her dining table. In this way the pup can become familiar with all the local delicacies, but diving and catching them for himself must at first present difficulties. The young remain with the mother for over a year, by which time they are almost full grown. During this long period of dependence Hall and Schaller (1964) regularly saw large juveniles dive after the mother when she went down. Although they were not able to stay under as long as she could, it seems very likely that by diving with the mother they learn the sort of places in which food is to be found and may also, by watching her, learn something of the methods of collecting and opening the different sorts of prey. The technique of using an anvil stone, however, is not dependent on watching the mother for, as has already been mentioned, it may be used by pups in areas where the stronger-jawed adults have no need of a stone to break open their prey. Some observations by Sandegren *et al.* (in press) suggest that this technique may be based on an innate movement pattern. According to these authors, the pup is not at first permitted to take food from the mother, although he may try to do so; she does, however, give him the empty shells from which she has extracted the food. These he mouths and manipulates in his paws, possibly using them as play objects. Sandegren and his colleagues noticed that during the course of these activities, pups 'often made the stereotyped hammering movements towards the chest, which later are used when hard food items are cracked on a rock resting on the chest'.

It is quite clear from the behaviour of females with their first litters that their assistance to their young is at first quite automatic and uncomprehended. They literally do not at first appreciate why they behave as they do, although they may subsequently come to learn what their behaviour is for. This was extremely clear with my female *Suricata*, who would attempt to 'feed' her young even when they were all eating from a dish and her efforts actually hindered rather than helped them. Robinson's coyote was a male, captured before his eyes were open and hand-reared. He had no experience of conspecifics and had never had to make his own living. Nevertheless, when at two years old a pup of his own species was put with him, he promptly regurgitated food for it. Another point which is very striking is the accuracy of timing of the parental responses. In three litters from two cats I found that food was first carried to the kittens when they were 35, 35 and 36 days old and Martins (1949) found that three bitches - two mongrels and one fox terrier - first regurgitated for their pups aged 21, 22 and 24 days.

It is easy to see that in simple nutritional terms it is desirable for the young to receive solid food as soon as they are capable of digesting it but in terms of behaviour, too, correct timing is important. The cat normally brings live prey to her kittens when they are two and a half to three months old. Their prey-killing patterns have by then matured but the kittens are still a little hesitant about making their first attacks. As Leyhausen (1965b) has shown, competition with littermates is often the factor which finally raises their eagerness to the point when the first swift attack with the fully oriented neck bite is launched and the first kill is made. There is a sensitive developmental period, during which this triggering-off of the attack occurs most easily. If no kill is made during this period then, as the kitten grows older, it becomes increasingly difficult to bring it to the point of launching an attack. A kitten taken from its mother too young and so deprived of the opportunity to make its first kill at the right time may well grow up a non-killer, showing little or no interest in its normal prey. Adamson's (1960, 1969) difficulties in inducing both her lioness and her cheetah to make their first kills very likely reflect the same thing: both had already passed the age at which first kills are most readily made and at which the mother normally provides the opportunity to do so. There is no more striking example of the detailed adaptiveness of behaviour than the way the responses of the adults and of the developing young have been shaped by natural selection to fit in with and complement each other. The evolution of the mammalian placenta may be a remarkable example of structural and physiological coadaptation, but the behavioural relations between mother and offspring are no less intricate and wonderful and there is still much for us to learn about them.

In this context one final point deserves mention. Everyone who has reared wild mammals in captivity knows that maternal responses are not always normal even though apparently healthy young may have been born. Cases in which the mother fails to rear her young are common and it seems that maternal behaviour is remarkably easily disturbed by any departure from the normal situation. One is tempted to regard this as a sort of failure and to ask why there is not a more adequate factor of safety. This sensitive dependence on 'right' conditions may, however, itself be adaptive. If a litter is lost the female may breed again successfully a little later on but if the mother dies her young will not survive and there will be no second chance. It is therefore desirable for maternal responses to fail if conditions are wrong, rather than that the female's own existence should also be jeopardised. Our first reaction may be one of horror at the apparent callousness sometimes shown but such an attitude is out of place, for this is a question not of morals or ethics but simply of survival value. In terms of natural selection, altruism is advantageous only so long as it results in more surviving offspring: the female who leaves most progeny behind is not necessarily the one whose maternal responses are most firmly fixed but may rather be the one in which they are most advantageously balanced against the behaviour making for individual survival.

Before leaving the subject of the development of the young, it is necessary to return to the question of play. Play is not an exclusively mammalian activity but mammals are more playful than any other animals and, although play is not unknown in adults, it is much more common in the young. Play is therefore most important in species with rather a long period of childhood, during which the young are cared for by the parents and do not have to make their own livings independently. It is thus only natural that carnivores should be amongst the most playful of mammals and their play has been the subject of a number of studies (Eibl-Eibesfeldt, 1950; Tembrock, 1958, 1960; Rensch and Dücker, 1959; Ludwig, 1965; Poole, 1966).

Play is not something involving distinctive patterns of behaviour not seen in other contexts; on the contrary, it is made up of actions that are also used in earnest – very commonly those concerned with prey capture, fighting and escape behaviour. The difference is that in play the motivation for the actions is different. Two kittens may chase each other in play, but the one that flees is not terrified and the pursuer is not bent on attacking his quarry or driving him away: the roles of pursuer and pursued can alternate as rapidly as in a children's game and, indeed, at the age of about seven weeks, kittens begin to play running games with such enthusiasm that what starts as a chase may end up with several of them galloping more or less abreast. Similarly, in play fights, biting and scratching are not carried out at full strength and it is quite obvious that there is no intention of wounding or killing the opponent. Moreover, since the normal goal of all these actions is not relevant in play, there is nothing to act as an end point and bring the whole to a halt, as the killing and eating of the prey or the flight of a defeated rival does in the genuine situation. In play, therefore, actions may occur over and over again and may switch back and forth from attack to defence, from fighting to prey killing, with the partner acting as a substitute for prey one moment and for a conspecific rival the next. Another feature of play is that it usually has a characteristic sort of exuberance; all the actions are carried out with just a little more energy than is strictly necessary, giving the whole performance its typical air of eagerness, excitement and enjoyment.

Since the actions which first appear in play also have their context in earnest, the play of the young, like the behaviour of the adults, shows features which are species-specific and there are characteristic differences between species. Puppies will worry and shake; civet kittens will neck-fence; kittens will ambush and stalk and throw themselves defensively on their backs, raking with the claws of their hind feet. The play of the black-footed cat is in general very like that of a kitten but my partially tame youngster would very readily chase a piece of paper on the end of a string and leap on it, turning a somersault as he did so, in a way which is much less common with domestic cat kittens. I believe this is related to the readiness with which the diminutive black-footed cat will hurl himself at relatively large prey, biting, clutching with the fore paws and raking with the hind claws all at once. This technique means that he must be prepared to roll over with his prey without losing his grip and the somersault is the play version of this form of attack. Leopard cubs also somersault very readily in much the same way. The most striking example amongst the Felidae of a species-specific type of play, however, is the way in which cheetah cubs will knock each other over with the typical paw slap that will later be used to fell their prey (Prater, 1935). In the largely insectivorous aardwolf and bat-eared fox, on the other hand, where prey catching is not a task demanding very much in the way of swift and accurate motor coordination, play related to escape behaviour predominates (von Ketelhodt, 1966). Otters, whose catching of invertebrate prey requires a great deal of manual dexterity, are very prone in captivity to invent all sorts of manipulative games (figure 8.6): Hall and Schaller (1964) quote Lensink as having seen sea otter pups in the wild 'pounding rocks and bits of coral together' in an apparently playful manner and, as already mentioned, the movements characteristic of the use of an anvil stone may appear in an apparently playful context.

Play clearly has a number of different functions. Obviously the young must become strong and active before they are ready to launch

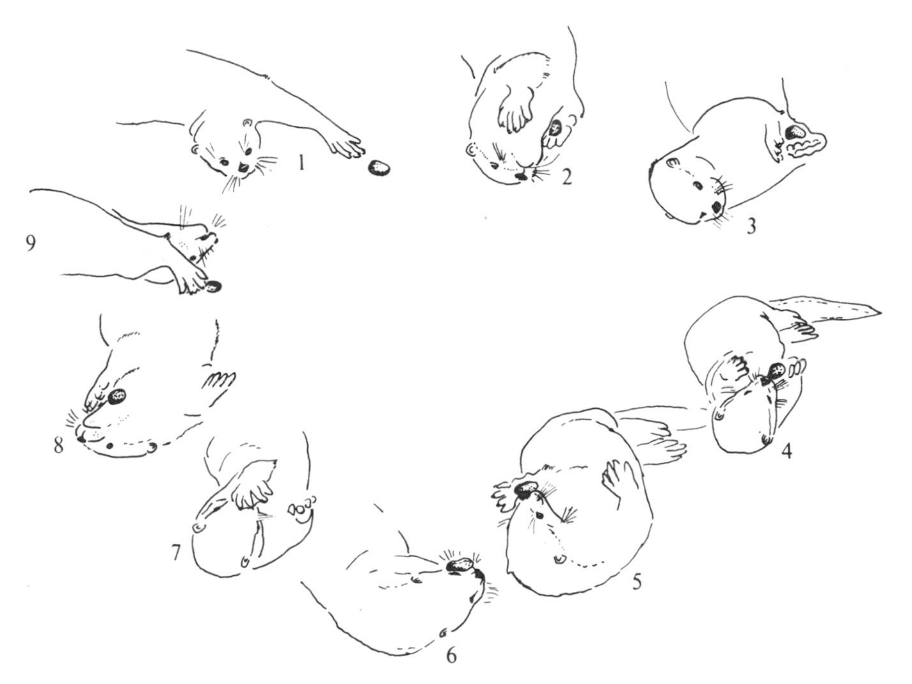

Figure 8.6. Small-clawed otter (*Amblonyx*): manipulative play with a pebble; 1–9, successive drawings from photographs of a single play sequence. [Drawings, N. Duplaix-Hall]

forth on their own and, like any other form of exercise, play will promote general physical fitness. It does a good deal more than that, however. Species-specific action patterns may mature endogenously but the animal still has to learn how to use them effectively. It is worth noting that the actions used in play are precisely those that require accurate coordination and in which initial failure in the earnest situation carries the risk of serious penalties. In its play the animal is continually experimenting with the relations between its own actions and the external world. It learns how to judge times and distances and how to aim a blow; how far you must jump to land on a moving rival; how fast you must run to intercept him; how you must time your paw stroke to hit him and so on. Every object is investigated, especially those that move. Things are smelled at, scratched, patted, bitten or shaken and thus the world of experience is continually enlarged. In animals kept as pets, this exploratory aspect of play is particularly obvious and leads to the invention of all sorts of games, utilising whatever objects are available.

Play also has a number of social functions. Young carnivores must stay together when their mother leaves them to go off hunting and in this connection the role of contact-keeping calls has already been mentioned. A tendency to be distressed if left alone is a negative factor and is only one side of the story: complementary to it is the fact that play provides the young with a communal activity and gives them a positive reason for seeking each other's company. The experience of each other gained in play may also have far-reaching effects on adult social behaviour. Scott (1962) has found that, if puppies are deprived of contact with their own species during the period from about three to ten weeks, when playfulness is at its height, their social responses will not develop normally: they will grow up asocial neurotics, failing to respond normally to their fellows. The period of intensive play may also initiate the establishment of dominance relationships within the litter and so reduce the necessity for serious rank-determining fights later on. In this connection Fox (1971b) makes some very interesting observations. According to him, in some of the less social Canidae (red fox, coyote, golden jackal) the pups do not play together until after dominance relationships have been settled by a serious fight at the age of approximately one month. This seems rather surprising and one would like to see it confirmed by more extensive observations on full-sized litters reared by the mother in conditions as near normal as possible.

Although play diminishes in intensity as the youngster matures, it does not necessarily vanish entirely. Van Lawick-Goodall and van Lawick (1970) found that in the black-backed jackal a meeting between two of the nomads who were following the migrating herds of game was very likely to lead to a play session, despite the fact that the animals were adult. They were of the opinion that this permitted the animals (neither of whom was on home ground) to determine their relative strengths and settle their social status without the need for fighting.

Adult carnivores will often cooperate when invited to play by their young and when live prey is given to captive animals they will often 'play' with it for some time before killing it. In the natural situation many hunts are unsuccessful and the animal must be prepared to hunt more often than it kills. In captivity the opportunities for killing are restricted and those for hunting are even more severely curtailed. The animal is therefore deprived of the opportunity to perform some of the most important patterns in its whole behavioural repertoire: it is ready, in fact eager, to make a kill if given the chance, but it is even more ready to carry out all the preliminary actions of catching the prey. Leyhausen (1965b) has studied this type of play-with-prey, especially in Felidae and concludes that what the animal is actually doing is making the prey provide it with a means of utilising some of its frustrated eagerness to hunt. He describes, for instance, how an adult serval cat invented the trick of picking up a live mouse very gently, carrying it across her cage and pushing it into a crevice, so that she might then be able to hook it out again with her paw; an action which, it will be recollected, the serval uses very expertly in capturing small rodents in the wild. She would repeat this several times before finally killing and eating her mouse. Lyall-Watson (1963) has given a very similar explanation for the raccoon's 'dowsing' behaviour, described in Chapter 5, by means of which the raccoon gives itself a substitute for its normal technique of fishing for crayfish and the like. I believe that such substitute hunting behaviour also makes the animal more eager to eat and that the captive animal quite literally 'gets more fun' out of its rather monotonous diet if it can devise a sort of substitute for the hunting which is the normal preliminary to a meal.

In the play of youngsters, too, there is probably an element of the same substitute effect. Their fighting and prey-catching patterns have matured but there is as yet no opportunity to use them in the genuine situation. Play provides the means of doing all the things the central nervous system is now prepared for and it does so without any of the risks inherent in the real situation. This, however, is not the only factor which makes young animals playful. Two adults kept together and never fed live prey may play together now and then but they will not do so with the same frequency and intensity as two youngsters kept in exactly the same conditions. There is some change in central nervous organisation with age, which is manifested as a decline in playfulness. My male *Suricata* showed this very plainly. At a year old he often joined in the games of his first litter of young. By the time his second family was born, however, he had become more staid and could very rarely be induced to participate in play although nothing in his living conditions had altered in the meantime.

Some workers consider that in young animals there is a quite non-specific drive towards activity which finds its outlet in play; others that there is a specific drive for play as such. I do not find it easy to see exactly how one can, in practice, distinguish between these alternatives: it therefore appears to me best to regard play as a true activity in its own right and to acknowledge that the internal motivating factors which affect it, whatever name we choose to bestow on them, do undoubtedly change with age. Readiness to show reproductive behaviour also changes with age and, in this case, we know that alterations in hormones play a major role. In the case of play, however, we are still in the dark as to the physiological or biochemical bases of the changes that characterise the normal developmental sequence from youth to maturity: I do not even know of any experiments having been done to find out whether treatment with sex hormones affects playfulness.

Table 8.8
Development of the young

Species	Birth weight (g)	Eyes open (days)	Eat solids (days)	Weaned (weeks)	Independent (months)	Author
CANIDAE						
Vulpes vulpes	110	*c.* 14	*c.* 30	*c.* 8	*c.* 4	Tembrock (1957a,b) Leitch et al. (1959)
Fennecus zerda	23–32	12–16	*c.* 25	8–10		Koenig (1970)
Canis familiaris		*c.* 12	*c.* 21	5–6		Martins (1949); Rheingold (1963)
C. lupus				6–8	*c.* 4	Tembrock (1957a,b)
C. latrans		10–17	28	*c.* 5		Snow (1967)
C. aureus			*c.* 35	*c.* 9		van Lawick-Goodall and van Lawick (1970)
Nyctereutes procyonoides		9–10		*c.* 8	4–5	Novikov (1956); Bannikov (1964)
Cuon alpinus	200–350	13–15	in 2nd month			Sosnovskii (1967)
URSIDAE						
Ursus arctos	265–625			nearly 2 yrs		Dathe (1963); Hensel *et al.* (1969)
U. americanus	310	40				Matson (1954); Leitch et al. (1959)
Thalarctos maritimus	410–840			nearly 2 yrs		Dathe (1963); Volf (1963a,b); Hess (1971)
Helarctos malayanus	325					Dathe (1963)
PROCYONIDAE						
Procyon lotor	*c.* 150	*c.* 21	63	16		Stuewer (1943); Montgomery (1969)
Potos flavus	191	19–22	54–91			Poglayen-Neuwall (1962); Clift (1967)
Bassariscus astutus		31–34	*c.* 35			Richardson (1942)
Ailuropoda melanoleuca	*c.* 150					Morris and Morris (1966)
MUSTELIDAE						
Mustela nivalis	2–4 (1 day old)	26–32	18–24			East and Lockie (1964, 1965); Hartman (1964)
M. rixosa	*c.* 1	26–29	28–35			Heidt *et al.* (1968)
M. erminea	1·7	35–41	21		1st kills 11½ wks	Hamilton (1933); East and Lockie (1964)
M. frenata	*c.* 3	36–37	21			Hamilton (1933)
M. putorius		29	23			Goethe (1940)
Martes martes and *M. foina*		34–38	36–45			Schmidt (1943)
M. americana		*c.* 28		6–7		Markley and Bassett (1942)
Mephitis mephitis	33	21–24	36–38		*c.* 3	Stegman (1937); Verts (1967)
Lutra canadensis		22–35	58	16–20		Liers (1951); Wayre (1967)
Lutrogale perspicillata		10	90	18		Yadav (1967)
Enhydra lutris	1250–2250	at birth		*c.* 52		Kenyon (1959)
VIVERRIDAE						
Viverra zibetha		10	*c.* 30		*c.* 6	David (1967)
Civettictis civetta		0–4	*c.* 30	*c.* 20		Own observations

Genetta tigrina		5–18	42–91	8		Dücker (1957); Wemmer (1971)
G. rubiginosa		10	*c.* 35		1st kills 7 months	Rowe-Rowe (1971)
G. genetta		8		25		Volf (1969)
Prionodon linsang	40	18–21				Louwman (1970)
Galidia elegans	*c.* 50	6–8	30	*c.* 8	*c.* 12	Albignac (1969b)
Fossa fossa	65–70	at birth	33	*c.* 10	*c.* 12	Albignac (1970); Wemmer (1971)
Cryptoprocta ferox	*c.* 100	12	90	16		Albignac (1969a)
Arctictis binturong			56			Gensch (1962)
Hemigalus derbyanus	125	8–12	*c.* 70			Louwman (1970)
Mungos mungo		10			2½–3	Simpson (1966)
Herpestes auropunctatus				4–5		Pearson and Baldwin (1953)
Cynictis penicillata			16	6		Zumpt (personal communication)
Helogale parvula			20			Zannier (1965)
Suricata suricatta	25–36	10–14	23–30	7–9		Dücker (1962); Ewer (1963b)
HYAENIDAE						
Crocuta crocuta	1500	at birth		*c.* 75 (wild) *c.* 25 (captive)		Matthews (1939); Pournelle (1965); van Lawick-Goodall and van Lawick (1970)
Hyaena brunnea		5				Lang (1958)
FELIDAE						
Panthera leo	1200–1400	at birth	56	33	24	de Carvalho (1968); Adamson (1969)
P. tigris	1100–1800	0–17	*c.* 90		18–24	Schaller (1967); Sankhala (1967)
P. pardus	430–567	6–9	42	12	12–18	Wilson and Child (1966); de Carvalho (1968); Adamson (1969)
P. uncia	368–708	8	*c.* 30	8–10		Jones (in press)
P. onca	680–990	1–3	70	22		Hunt (1967); Stehlik (1971)
Neofelis nebulosa			42–70	*c.* 20		Fellner (1968)
Acinonyx jubatus	250–280	10–11	30–35	12–20	15½–18	Florio and Spinelli (1968)[1]; Adamson (1969); Schaller (1970)
Lynx lynx		10–12	50			Kunc (1970)[2]
L. rufus	280–340			*c.* 8		Young (1958)
Caracal caracal		9–10	50–70	10–25		Gowda (1967); Kralik (1967)
Felis catus	85–110	9–20	32	*c.* 8	*c.* 6 months	Own data
F. nigripes	60–84 (1 day old)	6–8	*c.* 35			Leyhausen and Tonkin (1966)
Leopardus tigrinus		17	53			Leyhausen and Falkena (1966)

[1] A single cub previously born to the same female appears to have been somewhat post-mature: it weighed 300 g at birth and its eyes opened at 4 days old.
[2] The cubs described by Oeming (1970), whose birth weight was only 85–115 g and whose eyes did not open until they were 3 weeks old, were probably premature.

Chapter 9

Fossil relatives

WHEN we make a new acquaintance we usually find that before very long we are asking questions about his family background and his past: where is his home; what did his father do for a living; has he any brothers and sisters; where did he go to school; what previous jobs has he held and so on. We feel that the answers to such questions help us to know and understand him better. In much the same way, we know and understand animals better if we not only know them as they are today but also know a little about their past history, and the questions we ask are much the same - where did they originate, who were their ancestors and their relatives; have they always made their livings in the same way or not?

The mammals, although one of the last two classes of vertebrates to make their appearance, can hardly be described as newcomers: their history can be traced back as far as the upper Triassic, a span of at least 190 million years. Roughly half of this time falls within the Mesozoic but it is not until we reach the Tertiary, some 70 million years ago, that we can begin to discern and distinguish the first representatives of the modern orders of mammals.

During the whole of the mesozoic 'first act' the mammals were small and inconspicuous, playing minor roles in the evolutionary drama, while all the star parts were taken by reptiles. Fossil remains are neither abundant nor impressive and most of the early mammals are known only by teeth and jaw fragments. It must have been during this first act that the fundamental mammalian characteristics were evolved - the high and relatively constant body temperature, made possible by the double circulation, efficient pulmonary ventilation and insulating hair coat; viviparity and the specialisation of skin glands to produce milk; a kidney capable of producing highly concentrated urine by virtue of the presence of the loop of Henle. Although deductions made from the skeleton do give some indications about the evolution of these characteristics, they do not show up directly and unequivocally in fossils. Teeth, however, preserve exceedingly well and mammalian physiological efficiency could not have evolved without an efficient fuelling system, which entails an effective means of masticating the food. What we know of the teeth and jaws of the mesozoic mammals makes it clear that one of the major advances that was made during this initial stage of mammalian evolution was the development of efficient teeth and jaws. Amongst the radiation of early mammals of diverse types there was one group which was vastly superior in this respect to all the others: the Pantotheria. The pattern of tooth structure, jaw muscles, jaw and skull architecture which they evolved provided the take-off point for the new Tertiary radiation which ultimately gave us the mammals of today.

The pantotheres were small animals and almost certainly largely insectivorous. The masticatory apparatus they evolved was a dual-action system, capable first of crushing and pulping up the food, then of chopping or cutting it. The teeth of the living opossum are not very different, and Crompton and Hiiemäe (1970) have recently analysed how the opossum's masticatory system works. The details do not concern us here: what matters is only the general principle.

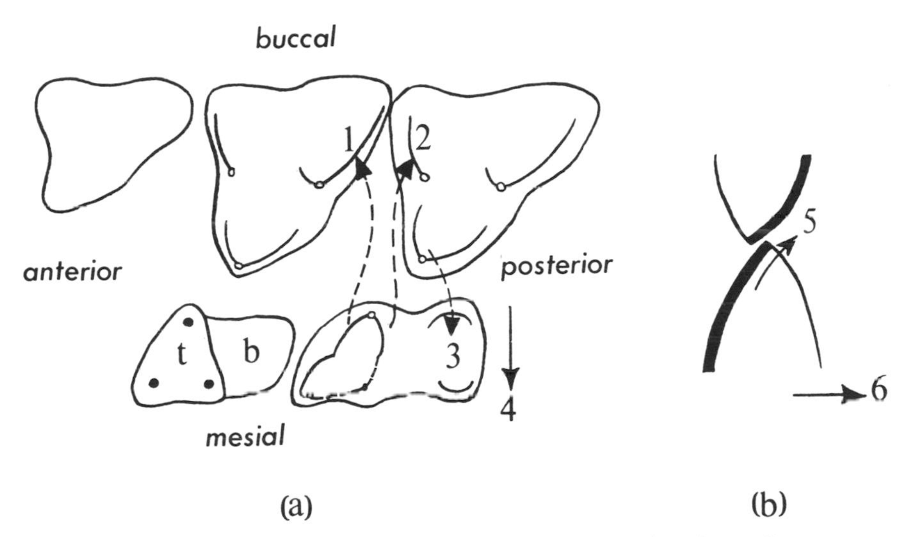

Figure 9.1. Basic mammalian tooth mechanism (a) Occlusal surfaces of upper (above) and lower (below) molar teeth. The anterior lower molar is diagrammatic and shows the anterior triangle, t, of high cusps and the posterior basin, b. As the jaws close the cutting edges shear against each other as shown by the arrows 1, 2. In the final stages of the bite the inner cusp on the upper tooth beds home in the basin on the lower, 3. For this to occur, the lower jaw must slide sideways, 4, during the bite, a movement automatically guided by the conical shape of the anterior cusps. This action is shown in figure (b), a vertical section through occluding conical cusps: as the thickened edges shear against each other, 5, the lower jaw is thrust sideways, 6.

The teeth of the lower jaw are roughly rectangular, with a triangle of high cusps on the front half and a lower basin at the back. The upper teeth are triangular, arranged with the apex facing inwards, so that there is a triangular embrasure between each pair: into this space the triangle of high cusps (with its apex facing outwards) of the lower tooth fits when the jaw closes. As the two sets of cusps shear past each other they make a scissor-like cutting device: since the triangles are not exactly isosceles, the longer edges, where the back of an upper tooth meets the front of a lower, form the main cutting blades (figure 9.1). If we visualise what happens when such teeth close on a piece of food we can see how the dual action of pulping and cutting comes about. The sharp cusps first stab into the food and it is crushed between the upper cusps and the lower basin. Preliminary softening up is thus achieved without full closure of the jaws but as they bite home the scissor action of the shearing edges comes into play and cuts the food into pieces – rather like softening a tough piece of steak by beating it with a tenderiser and then cutting it into pieces.

This type of dentition no doubt served the pantotheres admirably for dealing with insects and other small prey but from the present point of view what is much more important is the evolutionary potentialities inherent in such a system. It is modifiable in two different ways: exaggerate the cutting features and allow the crushing ones to diminish and you convert it to a specialised carnivorous dentition; do the opposite and concentrate on crushing and grinding and it becomes a herbivore's mechanism. The masticatory adaptations of the modern orders of mammals all appear to have been derived in this way from a single origin. It is therefore not surprising that at first the different lineages are difficult to distinguish, or that increasing knowledge has necessitated some revision of earlier attempts to deal with their taxonomic relationships.

It was formerly usual to group together as the suborder Creodonta, within the order Carnivora, a number of families which more recent work suggests are not in fact closely related, and few mammalian taxonomists today would defend the old arrangement. Two families, the Hyaenodontidae and Oxyaenidae, were definitely predators and paralleled the carnivores of today in having specialised carnassial teeth. This, however, happened independently in these early forms and in the modern Carnivora. In the latter it is P^4 and M_1 that have become the carnassials but in the Oxyaenidae the shearing teeth are M^1 and M_2 and in the Hyaenodontidae, M^2 and M_3. It is easy to appreciate that, while several small cutting edges, operating more or less simultaneously, may be well suited to chopping up insects, slicing through tough flesh and skin can be done more effectively by a single pair of large blades. We may therefore expect to find that some type of specialised carnassial shear is evolved in any truly predacious group and, failing any other evidence of close relationship, their possession of carnassial teeth is not a sufficient reason for including the Hyaenodontidae and Oxyaenidae in the Carnivora. Van Valen (1966) has pointed out that there is a very close resemblance between the two early flesh-eating families and some of the genera previously classified as Insectivora (or, by some workers, as marsupials) and he proposes to unite them in a single order, the Deltatheridia. However, while accepting that these animals should be placed in an order of their own, distinct from the Carnivora, I would support MacIntyre (1966) and Romer (1968) in preferring to retain the old name Creodonta for them and elevate the group to the rank of an order.

Two other families previously regarded as Carnivora are the Arctocyonidae and Mesonychidae, which flourished in the Palaeocene and Eocene epochs. The former are small animals, not very clearly differentiated from primitive insectivores. Their teeth are sharp-cusped but show no tendency for the development of a carnassial shearing mechanism and there is nothing about arctocyonid structure that links them specifically with the Carnivora. The Mesonychidae, in contrast, are remarkable for their large size, some of the later Eocene species being as big as a grizzly bear. The teeth are rather blunt-cusped and, to judge from the shape of the terminal phalanges, the digits bore small hoofs rather than claws. Modern studies suggest that both families are actually representatives of an early ungulate group, the Condylarthra. Their removal from the Carnivora leaves but a single basic stock, the family Miacidae, as representing the order in the early Tertiary period. The miacids were small, rather genet-like animals and appear to have been forest dwellers. The paws were wide and spreading, with the first digit somewhat opposable, which suggests that they were good climbers. The scaphoid and lunar bones in the wrist were still unfused and another primitive feature was the absence of an ossified auditory bulla. The ancestral relationship of the miacids to the modern carnivores, however, is indicated by the fact that, very early on in their history, P^4 and M_1 started to become specialised as carnassial teeth (figure 9.2).

Forest is an environment notoriously unfavourable for fossilisation and miacid material is somewhat scanty. Although there are sufficient specimens to give a reasonable picture of the structure of a few genera, the record is far from adequate. The Miacidae, as at present known, are classified in two subfamilies: representatives of one of these, the Viverravinae, are known from the Palaeocene but the other subfamily, the Miacinae, does not appear until much later. The

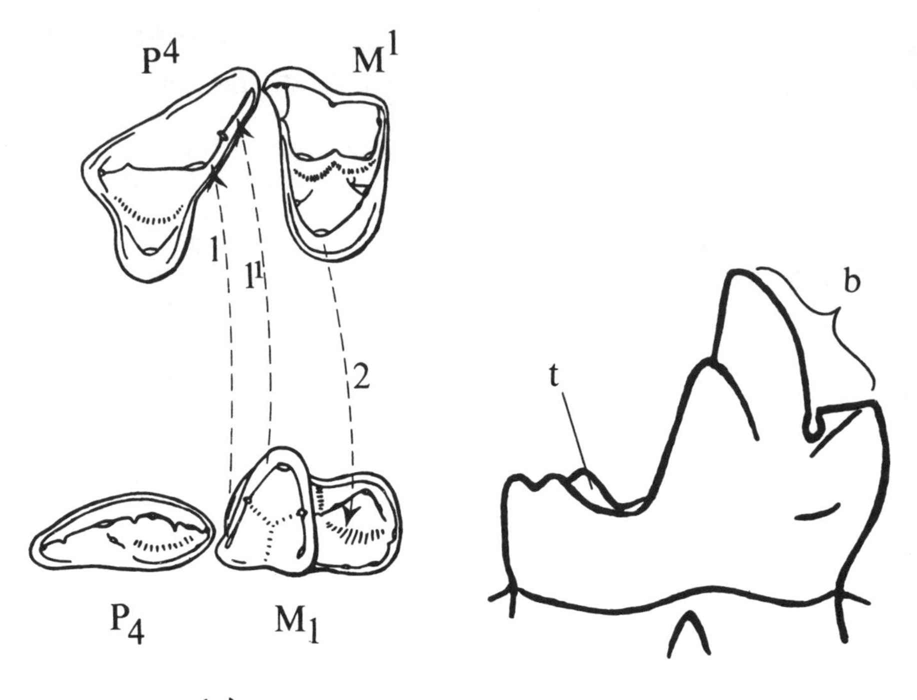

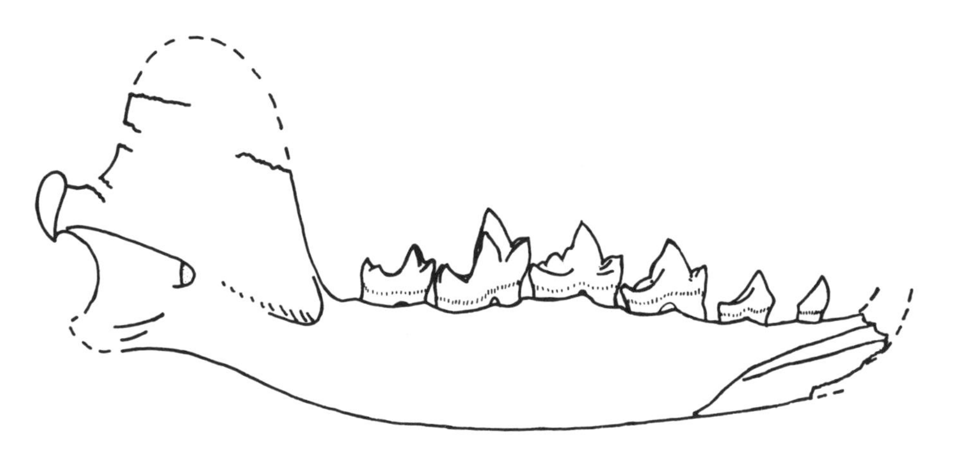

Figure 9.2. (a) Occlusal views of miacid last premolars and first molars. P^4 and M_1 are the carnassial teeth, with the shearing surfaces indicated by arrows 1, 1^1. The inner cusp on M^1 beds home in the talonid basin of M_1 (arrow 2). (b) Miacid left lower carnassial, mesial view: b, cutting edge with large carnassial notch, preventing food from slipping off the very sloping blade. t, talonid basin. (c) Miacid lower jaw, mesial view. [After McIntyre, 1966]

relationships of the two subfamilies are still obscure and the origin of the Miacidae as a whole remains shrouded in mystery. There is nothing to link them with either of the two creodont families and they may well have originated independently from the early insectivores.

The scarcity of miacid fossils also makes it impossible to trace with any certainty the details of the diversification which gave rise to all of the modern families of carnivores. This took place rather rapidly in the late Eocene and early Oligocene. The factors making for a new radiation of carnivores at this period are complex. Since carnivores occupy the terminal positions in food chains, diversity of carnivores presupposes diversity of prey; the latter in turn implies diversity of plant foods and habitats. The evolution of the flowering plants began in the Cretaceous but was still going on rapidly throughout the first half of the Tertiary and a great many modern plant families, including the grasses, make their first appearance in the Eocene and Oligocene (Chesters *et al.*, 1967). The evolution of herbivores amongst the early mammals and of fruit- and seed-eating birds must have accentuated this diversification of the flora by increasing the complexity of competitive interrelationships. Bitter substances and thorns, making leaves less palatable or harder to pluck, are defensive adaptations: seeds which pass undamaged through a digestive tract turn the apparent disaster of being eaten into a dispersive advantage and institute competition for being eaten instead of for being avoided. The hair coat of the mammals too has had its effects on plant evolution, for it provides another dispersal mechanism that has been turned to account by a myriad of plants that have developed fruiting structures with entangling devices of one sort or another. Even the grasses may have owed their initial success to the fact that they evolved a growth form in which being nipped off close to the ground by a herbivore does not have lethal consequences. Their success in turn provided new opportunities for herbivore adaptations and we know that by the Miocene, grasslands were sufficiently extensive to have made it possible for the horses to abandon browsing in favour of grazing. Seeds with hard, digestion-resistant coats provide a rich food for any small herbivore that can open the covering and reach the nutritious kernel inside and for this the incisors of the rodents are admirably suited. As Smythe (1970a) has recently shown, the plants' counter-move may be to produce synchronously so many fruits that the rodents cannot possibly eat them all: they have, however, evolved the habit of burying the surplus for future use. Since they never rediscover all their hoards, this is equivalent to the fruit tree paying the rodent to plant next year's crop for it.

Although some rodent remains are known from the Palaeocene, the major expansion of the group did not occur until the Eocene and myomorphs first appear in the Oligocene. Amongst the lagomorphs, too, the same period was one of diversification, possibly associated with the rise of the grasses, and the earliest rabbits and hares are found in the Eocene (Butler *et al.*, 1967). With these new types of prey available and with open grasslands becoming more widespread and more populated with herbivores, the stage was clearly set for a major expansion of predators and the way was open for some of them to abandon their traditional forest habitat and take to life in more open terrain: one could almost say that the evolution of the Canidae was inevitable.

The mongooses also arose from a lineage that took to life in more open country than their forebears and became fully terrestrial. Originally predators of small vertebrates, they soon radiated to produce a number of species feeding largely on insects; this in turn has given rise to some surprisingly complex interrelationships. As already mentioned, in the area studied by Neal (1970a), dung beetles provided a major food source for the banded mongoose. The beetles on which the carnivore feeds are thus living at second hand on plant food that has already been processed in the digestive tracts of the large herbivores constituting the primary consumers in the food chain. It would not immediately strike one that a buffalo or an elephant affected the life of a mongoose very closely and one would have no grounds for linking them in this way if one found them both as fossils. In view of the complex interactions of this type that do occur amongst living species, it is not surprising that when dealing with fossils we are often unable to make more than very broad general deductions about mode of life and are often at a loss to understand the full significance of the wealth of species we encounter.

This digression has carried us away from the main issue; to wit, the origin of the modern families of carnivores from their miacid ancestors. Before returning to this, however, it is worth noting that there was no comparable radiation amongst the Creodonta: on the contrary, with the rise of the modern carnivores, the Creodonta declined and ultimately became extinct. A few hyaenodontids linger on into the Pliocene but the Oligocene marks the end of the creodonts as a group of any significance. There are probably a number of reasons for the failure of the creodonts and the success of the miacids. One factor which was almost certainly relevant is the situation of the carnassial shear in the two groups. The miacid position, between P^4 and M_1, has two advantages. Firstly, as we have seen, the carnassials of the permanent dentition can come into operation before the milk carnassials are shed. It is difficult to see how so complete a functional continuity could have been achieved with the carnassials situated wholly within the permanent molar battery and quite possibly the creodont milk dentition did not include an effective shearing system. More important is the fact that the miacid arrangement has greater potentialities for adaptive differentiation. M_1 is the only molar tooth involved in the shear and as the carnassial teeth become progressively modified to form more effective cutting blades, there are still two possible pathways for the evolution of the molar tooth battery. In the species that become purely flesh-eaters, the posterior molars are progressively reduced and lost and the jaw shortened so that the force which the jaw muscles exert at the level of the carnassials is maximal. In mixed feeders, however, the molars are retained and modified in exactly the opposite way to the carnassials. The shearing edges are reduced and the crushing features emphasised, so that a dual-purpose dentition is produced, as has happened in the canids. The transition from this to a predominantly vegetarian diet is then possible: the carnassials become secondarily less blade-like and more molariform, so that they become effectively part of the crushing system.

The Miacidae thus had a dentition which held the potentialities of producing either more efficient pure predators, mixed feeders or even secondary vegetarians. The Creodonta, with their molar battery so heavily involved in the carnassial shear, lacked this flexibility and it is therefore not surprising that it was not they but the miacids that took the lead in exploiting the new resources of the Eocene and Oligocene. Behavioural factors, however, must also have been involved. No animal first evolves characters adapted, say, to chasing rabbits and then starts to chase them. There can be selective advantage in the relevant adaptations only provided attempts to catch rabbits are made moderately frequently. Behaviour must lead the way and the miacids must have been adaptable in their behaviour if they were ever to realise the adaptive potentialities of their dentition. It was therefore probably an ability to utilise a wider spectrum of prey species and a readiness to sample new foods and to move into new habitats that initially swung the balance in favour of the miacids.

As we have already seen, the modern families of carnivores fall into two assemblages: the Canoidea; comprising the Canidae, Ursidae, Procyonidae and Mustelidae, and the Feloidea; comprising the Viverridae, Hyaenidae and Felidae. Since the differences between the two groups include such features as the way in which ossification of the auditory bulla has taken place, it seems likely that the basic canoid and feloid stocks originated independently from the miacids. Possibly the two lineages resulted from rather similar happenings in two climatically distinct areas, the feloids originating in a tropical, the canoids in a more temperate region. Since passage between the Old and New Worlds via the Bering Straits land bridge was possible at various times during the Tertiary, there was no effective barrier to animals adapted to withstand a non-tropical climate. While the present distribution of the viverrids indicates an Old World origin for the feloids, it is difficult to be certain whether the canoids originated in the Old or the New World, for Eocene canids and mustelids are no less common in Eurasia than in America. This is not actually a question of great moment from the point of view of general evolutionary history, since, wherever their first beginnings may ultimately prove to have been, the canoid stock very soon populated the entire Holarctic region.

The early history of the canoids is better known than that of the feloids, not only because the American deposits have been so extensively investigated but also because the open country habitat makes for better fossilisation than does forest. Possibly because of the difficulties they at first encountered in capturing prey in the open, the Canidae did not become obligate flesh-eaters but retained their molars and developed a typical dual-purpose dentition, capable of dealing with vegetable food as well as with meat. The first dogs were relatively small and must have looked rather like a cross between a fox and a genet. The late Eocene *Pseudocynodictis*, although probably not actually ancestral to later forms, gives a good idea of what these early canids must have looked like, with a long tail, legs still relatively short, feet spreading with five digits ending in rather sharp claws (see plate 16).

Numerous undoubted canids, with the characteristic combination of cutting carnassials and crushing molars, are known from Oligocene deposits both in America and in Europe, and by the beginning of the Miocene, a major radiation was in progress (figure 9.3). Forms such as *Cynodesmus* and *Tomarctos*, ancestral to modern Canidae, have dentitions much resembling those of dogs and wolves but cursorial adaptation of the limbs is not yet very far advanced. *Temnocyon* is peculiar in possessing a lower carnassial with a single-cusped cutting talonid instead of the usual double-cusped basin. The three modern genera with a similar talonid (*Lycaon*, *Cuon* and *Speothos*) are not known earlier than the Pleistocene and whether they have independently acquired the cutting talonid or are in fact descended from *Temnocyon* remains problematical. The absence of other features linking the three modern forms, however, suggests parallelism rather than common ancestry.

In addition to these more orthodox canids, the Miocene radiation included a number of species in which the carnassials were reduced in size and the molars enlarged. Some of them, the so-called 'bear-dogs', attained gigantic size; *Amphicyon*, for instance, measured over one and a half metres, not counting its long tail. The name is misleading,

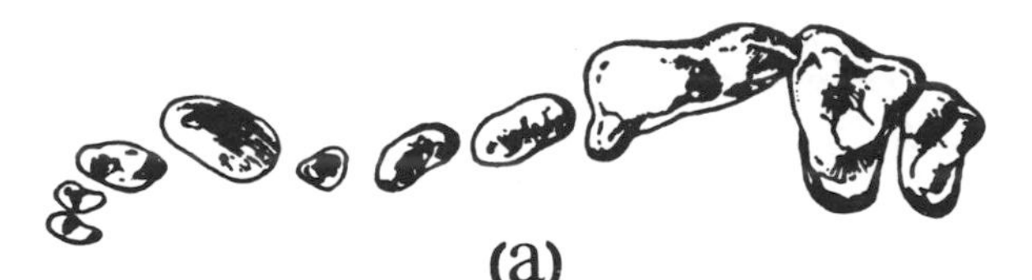

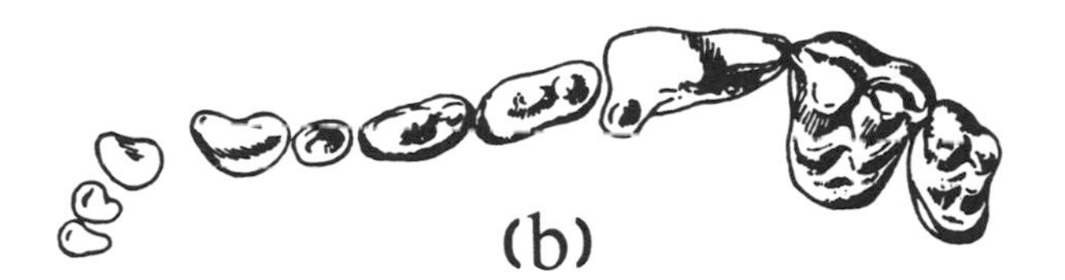

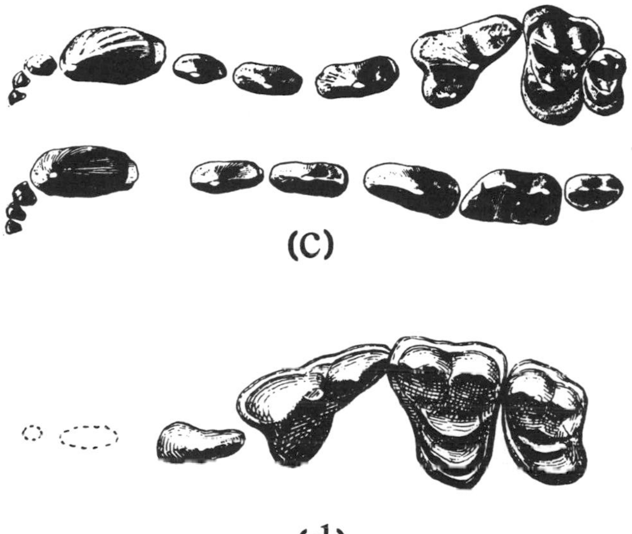

Figure 9.3. Teeth of fossil canids, (a) *Cynodesmus* (upper teeth) (b) *Tomarctos* (upper teeth), (c) *Temnocyon*, (d) *Cephalogale* (upper teeth). [From Matthew, 1930 and Piveteau, 1961]

for the true ancestors of the bears are not amongst the bear-dogs but amongst related, rather similar but smaller forms. It is not easy to decide at exactly what point one should draw the line between bears and dogs and intermediate genera such as *Cephalogale* (figure 9.3) and *Hemicyon* have been allocated sometimes to one, sometimes to the other family: current opinion favours including them in the bears.

The later diversification of the bears appears to have taken place in the Old World and the brown and black bears reached America only in the Pleistocene. The ancestry of the aberrant South American spectacled bear, *Tremarctos*, can be traced back only as far as the North American Pliocene, and this lineage clearly diverged rather early from the line giving rise to the other living species (Kurtén, 1966). A third branch of the original ursid stock gave rise to the subfamily Agriotheriinae, which did not outlast the Pliocene and has left no living descendants.

The Procyonidae may possibly also have originated from the broad-molared canids of the miocene radiation. *Phlaocyon* (figure 9.4), although the structure of its auditory region shows it to be a canid, has exactly the sort of dentition to be expected in a procyonid ancestor. The procyonid lineage developed in North America and extant genera are identifiable in the North American Pliocene. The only fossil procyonids known from the Old World are Pleistocene

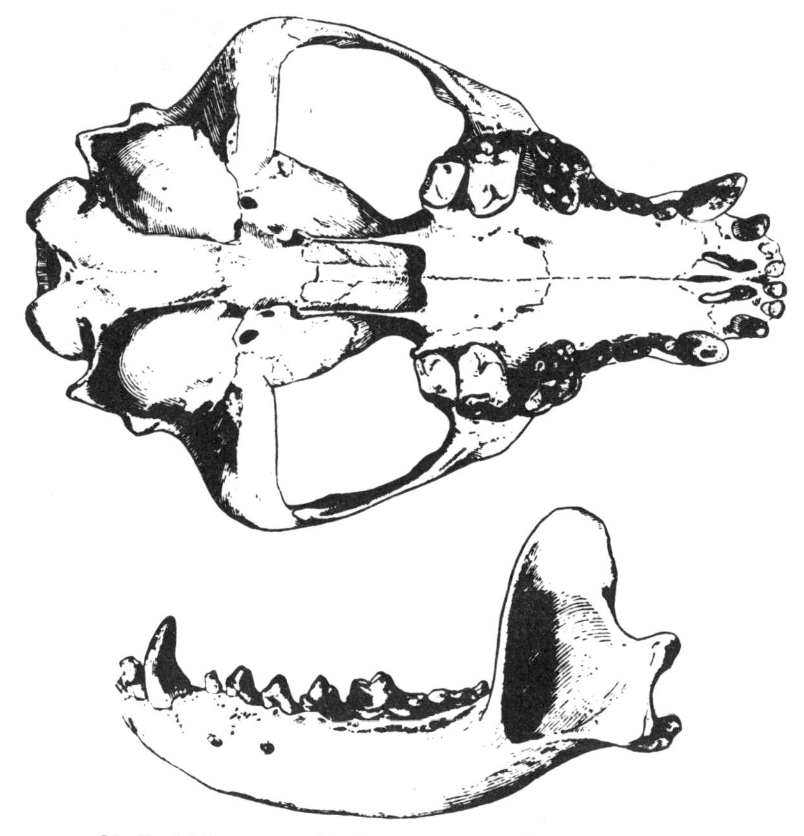

Figure 9.4. Skull of Miocene canid, *Phlaocyon*, with dentition paralleling that of the Procyonidae. [From Piveteau, 1961]

forms, not very different from the modern pandas. Since these are also the only fossils which bear any discernible relationship to the pandas, palaeontology gives little help in trying to decide the affinities of the latter group. The Procyoninae are essentially a tropical group and were apparently not able to cross the Bering straits land bridge. It therefore seems most reasonable to suggest that the pandas originated as an early offshoot from the procyonid line, which was aberrant in becoming adapted to cold conditions and hence able to make the crossing. One may further suppose that, in competition with the canids and mustelids adapted to cold conditions, this group was not an evolutionary success. Failure to discover them as fossils may mean no more than that its members were never numerous and soon became extinct in the New World.

The Mustelidae, the one remaining canoid family, appear to have originated independently of the Canidae and their early stages are difficult to trace, partly because they remained in the ancestral forest habitat. They appear to be a continuation of the same miacid major group as gave rise to the Canidae, showing progressive development of a more carnivorous dentition but without radical change in the post-cranial skeleton. As in the Canidae, not all members of the family remained pure carnivores: a number followed the pathway of reducing the carnassial shear and enlarging the crushing molars in adaptation to a more omnivorous diet. While the Mustelinae are the truly predacious lineage the other subfamilies, the badgers, skunks and otters, represent the diversification of the more omnivorous types. The honey badger, *Mellivora*, is a rather isolated form, with molars much less specialised for crushing than the true badgers and is certainly not closely related to them. More likely it represents an offshoot from the Mustelinae, which has paralleled the badgers in a number of characteristics.

Seals cannot be traced back beyond the Miocene but on the grounds of structural similarities, they are generally believed to have arisen as an offshoot from the early canoid stock and become adapted to an aquatic existence very much as the otters have done, but more completely. Indeed, *Semantor*, from the early Pliocene, now classified as a mustelid, was formerly regarded as a late surviving primitive seal. The decision to classify the Pinnipedia as an independent order is based on the judgement that their aquatic adaptations involve such major structural changes as to warrant this taxonomic recognition and is not a function of how long ago they are estimated to have diverged from their terrestrial ancestors. Sarich's (1969a, b) immunological studies of serum albumens are therefore of no great assistance in deciding the taxonomic issue, although they undoubtedly support the opinion that seals evolved from the Canoidea.

The tropical forest habitat of the Viverridae makes for poor fossilisation and little is therefore known of their history. They do not differ very radically from the ancestral miacids; indeed, Gregory and Hellman (1939) described the latter as 'nothing but primitive forerunners and ancestors of the later civets and their allies' and suggested uniting them all in a single family. This, however, is not a very practical proposal, even if no hard and fast boundary can be drawn and little is known of the earliest viverrids. From the Miocene onwards there are occasional fossils which can be identified as civet-like or genet-like and it has been suggested that the Eocene *Stenoplesictis* may be ancestral to the linsangs. The fossils, however, are not sufficiently abundant to permit any detailed tracing of the lineages of the modern forms. It is nevertheless clear from the characteristics of the living subfamilies that in the Viverridae, very much as in the Mustelidae, diversification has involved the same two processes of elaboration of the carnassial shear in the more predacious species and of the crushing features of the molars in those that became more omnivorous or more vegetarian. In addition to these feeding adaptations, there are skeletal differences related to arboreal or terrestrial life, the latter either in forest or in more open terrain, and it is their adaptations to the various combinations of these different types of habitat and feeding habit which form the basis for the diversity of the extant Viverridae.

The civets and genets are progressive descendants of the basic stock, the genets remaining arboreal and largely carnivorous, the terrestrial civets becoming more omnivorous. The Asiatic palm civets are a group which combine arboreal habits with a diet including fruit as an important item and, in many ways, they parallel the procyonids of the New World tropics. The mongooses, adopting a terrestrial mode of life and thus being free to leave the forest habitat, constitute one of the most distinctive subfamilies: the success of their radiation in Africa may have been determined partly by the virtual absence of small mustelid competitors.

In the isolation of Madagascar a subsidiary radiation took place but here again it is not possible to trace the origins of any of the three Malagasy subfamilies. Amongst these, *Cryptoprocta*, the sole representative of its subfamily, is the most highly adapted of all the viverrids to a purely predacious habit. With its specialised carnassials, post-carnassial teeth reduced to the single small upper molar and its retractile claws, it shows many resemblances to the Felidae. The latter family diverged early from the main viverrid stock and although there is no evidence of any close relationship to *Cryptoprocta*, they must have gone through a rather similar evolutionary stage. Once this point had been reached the Felidae were committed to a purely predacious way of life. With the post-carnassial teeth reduced virtually to zero and in the face of already established competitors amongst other families, any shift to plant food was ruled out. The further evolution of the Felidae is therefore a story of diversification in techniques of hunting and killing different types of prey, living in different habitats. This must have involved modifications in limb skeleton and general body proportions in relation to habitat but in relatively few cases are details of skeletal structure known very fully. More important and more adequately known are the modifications that occurred in the lethal weapons, the canine teeth. Two quite distinct evolutionary pathways were followed. One, relatively conservative, gave rise to the Felinae, the cats as we know them today, with the canines only moderately enlarged and no great disparity in size between upper and lower. The carnassials form effective cutting blades but the upper one, P^4, remains a dual-purpose tooth. Its anterior and internal cusps remain moderately broad and, together with the relatively heavy P^3 and P_4, form a mechanism capable of crushing bones. Since they are with us today we know a considerable amount about the hunting and killing techniques of the Felinae.

The second adaptational syndrome involved concentration on the upper canines as the lethal weapons and gave rise to a great variety of forms known as the sabre-tooths, which are classified as a separate subfamily, the Machairodontinae (plate 16). These became the first successful large felid predators and it is they, not the true cats, that dominate the scene during the Oligocene and Pliocene. Along with the immense development of the upper canines go correlated changes in the rest of the dentition and in skull architecture. The lower canine is reduced until it is only a trifle larger than the third incisor; an obvious necessity if the upper and lower teeth are to be separated widely enough to permit an effective bite. Even so, the jaw has to be opened extremely widely to free the upper canine and the glenoid articulation is specially modified to permit this. As already explained in Chapter 2, the form of the glenoid is related to the arrangement of the jaw muscles and the forces they exert at the articulation. Moreover, the fibres of the temporalis muscles, running from the high occiput to the low coronoid process, are unusually long and can thus provide the necessary amount of stretch. Driving the upper canines home must have required a powerful strike and the reaction when the teeth encountered resistance would have tended to force the skull upwards and so dislocate the animal's own neck. This is counteracted by the ventral neck muscles, the sterno- and cleido-mastoids. These are not only much larger than those of ordinary cats but their insertions on the skull are situated further below the occipital condyles, so that their mechanical advantage is increased. One can hardly imagine a sabre-tooth crushing up bones: the sabres would make it difficult to work large bones into position in the mouth and would be in considerable danger of getting broken in the process. In fact, the premolars are reduced and the carnassials form the most exaggerated shearing blades found in any carnivore, so that the cheek teeth as a whole have a purely cutting function. The dentition is therefore adapted to do two things only: to kill, and to slice meat from the kill. The sabre-tooths must have left rich pickings for any scavenger that could utilise them – a point to which we will return later.

The sabre-tooths do not constitute a single lineage, showing progressive increase in size of the upper canine but include a variety of species amongst which large sabres were obviously evolved independently several times. Their interrelationships are, however, complex and the details are by no means fully elucidated. The same general principles, however, must apply to the use of the large sabres in all cases and it is legitimate to ask the general question – how did the sabre-tooths kill? One can take it that they were specialist killers of large prey: the evolution of such formidable weapons as the sabres is not comprehensible on any other assumption. In many species the sabres are very slender and, in my judgement, quite unsuited to piercing through bone: there would also be grave risk of their being broken if they were used in the feline manner for forcing two vertebrae apart and dislocating the neck of the prey. I therefore believe that they were used for killing by means of a throat bite. The fore limbs of the sabre-tooths were extremely powerful and the prey was probably first pulled down and the sabres then plunged into the throat, where severing of carotid artery, jugular vein or windpipe would have been lethal. The limb structure shows no adaptations for fast running but the herbivores of the Oligocene and Miocene were slower than the swift cursorial modern species and, in any case, the sabre-tooths may have had some skill in ambushing and stalking.

The sabre-tooths include a variety of lineages but by the Pliocene, two major adaptive types had become predominant (Kurtén, 1963) (figure 9.5). In one, typified by the genus *Homotherium*, the sabres are wide, laterally compressed blades, not very greatly elongated and with their edges serrated, much as in a modern bread knife. The upper carnassials are extremely specialised, with no inner lobe and with an extra anterior accessory cusp adding to the length of the cutting blade. The other, typified in the Old World by the genus *Megantereon*, has very elongated but much less flattened canines, either smooth or with only very slight serrations and the upper carnassials are less specialised; a small inner lobe is present and the anterior accessory cusp is absent. The limbs in these latter forms were extremely heavy, with long proximal and short distal segments, well adapted for pulling

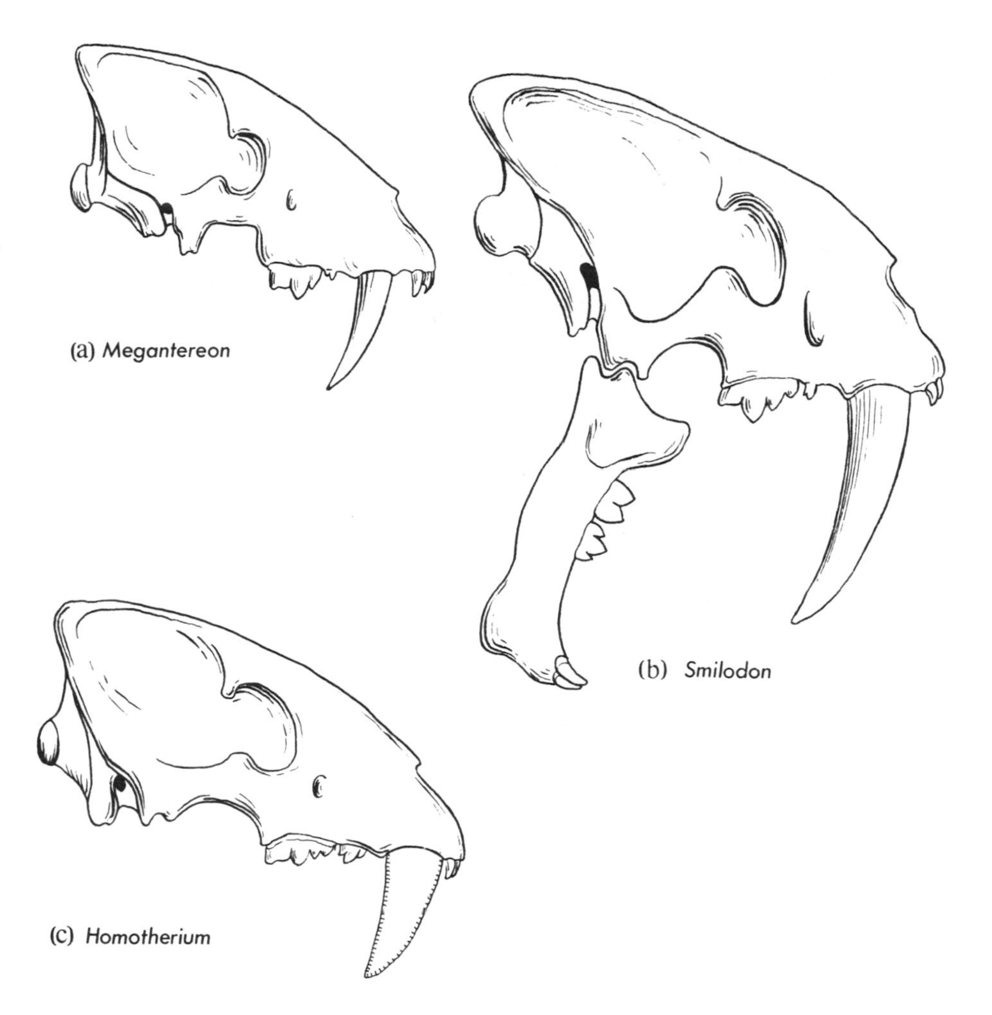

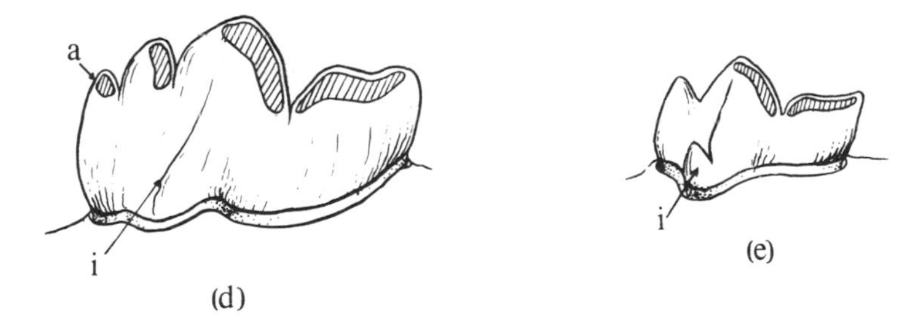

Figure 9.5. Two basic sabre-tooth adaptations: the stabbing *Megantereon* type (a) with *Smilodon* (b) as its extreme form and the slicing *Homotherium* type (c). (d) Advanced sabre-tooth upper carnassial with extra anterior cusp, a: and vestigial inner cusp, i: whole length of tooth shows shearing wear. (e) Felid upper carnassial with inner cusp present: shearing wear on the two posterior cusps only.

down heavy prey but not for swift pursuit. The American Pleistocene sabre-tooth, *Smilodon*, agrees in all essentials with the *Megantereon* type and represents its most extreme form and in the most highly evolved species the upper carnassial is almost as specialised as in *Homotherium*. The American genus *Dinobastis* belongs to the *Homotherium* type and, indeed, Churcher (1966) does not regard the generic distinction as valid. In southern Africa, too, the same two types are present in the early Pleistocene (Ewer, 1955b). It seems obvious that they must represent adaptations to dealing with two different sorts of prey. The most reasonable suggestion is that the *Megantereon–Smilodon* forms specialised in killing large, heavy prey with a tough protective hide. Their canines are essentially stabbing daggers and could have been driven home even through thick hide. Possibly the jaws were then closed firmly and the throat grip held until the prey died of suffocation. The blade-like *Homotherium* canines are slicing rather than stabbing weapons and were more suited for use against thinner-skinned prey. They may have been driven home and then, with a backward pull, used to rip open the victim's throat. Limb structure in the two types of predator accords with such an interpretation but it is not apparent why the carnassials should have been so much more specialised in one than in the other.

Although sabre-tooths occur in Pleistocene deposits in both Old and New Worlds, they did not survive into the Recent period. A number of factors may have contributed to their extinction but the most important was probably the rise of the modern swift ungulate fauna and of predators amongst the Felinae better adapted to hunting them. Early man may have taken a hand in the process both indirectly by competing as a hunter and directly, by deliberate killing of animals that he regarded both as competitors and as a personal danger. The sabre-tooths were smaller-brained as well as slower-moving than the Felinae; factors which may have made them not only less effective as hunters of swift prey but also themselves more easily killed by early human hunters.

The history of the true cats is not well documented in the fossil record. Cats about the size of a large wildcat appear in early Pliocene European deposits and cheetahs in the late Pliocene but the large cats – lions, tigers, leopards, etc. – become common only in the early Pleistocene. It is not exactly clear why they should have appeared on the scene so late and the sabre-tooths so early but it seems most likely that this may have been a question of the evolution of hunting and killing techniques.

The primitive killing bite, found both in viverrids and in mustelids, is aimed at the occiput and, when successful, the back of the skull is crushed in a single bite. If the first bite fails the prey may be shaken, thrown aside and pounced upon anew, or the grip may be shifted by the iterant snapping technique. The early Felidae presumably also used the 'occipital crunch', which is suited to small prey with thin skulls but not to dealing with large prey. The modern Felidae, we know, have evolved a much more precise orientation of the killing bite, with the canines driven in between two vertebrae, thus dislocating the neck. This, although it deals with larger prey than does the occipital crunch, is not suited to very large prey with thick neck muscles, particularly if horns are also present and, as we have seen, the modern feline killers of really large game turn to the throat bite. It seems possible that the sabre-tooth lineage never evolved the dislocation technique but took to the throat bite directly in their first efforts to attack prey too large to be mastered by the occipital crunch and in so doing gave rise to the first successful large felid predators. The Felinae, adapting to medium-sized prey, perfected the complex dislocation technique which, although it permitted them to become adept small to medium-sized predators, did not provide the basis for coping with the large, heavy herbivores on which the sabre-tooths were already preying successfully. It was therefore not until the rise of the swift modern ungulates placed a premium on speed and hunting skill that the Felinae were able to come into their own and displace the sabre-tooths as top predators, despite, rather than because of, their previous evolution of the dislocation technique.

Only the Hyaenidae remain to be considered. They possess highly efficient carnassial teeth, much like those of the Felidae but their distinguishing feature is their heavy, hammer-like premolar teeth. These, combined with very powerful jaw muscles, make it possible for hyaenas to crack bigger bones than any other carnivore can cope with. As Schaller and Lowther (1969) point out, remains of kills are not sufficiently abundant today for any large mammal to exist as a pure scavenger, and modern hyaenas are perfectly capable of making their livings as self-supporting primary predators. Nevertheless, it is a fact that the hyaenid dentition is specifically adapted for bone crushing, which suggests that at a crucial stage in their evolution, scavenging provided a food source sufficiently important to confer decisive selective advantage on the acquisition of such special adaptations.

The evolution of the hyaenas must have been under way by the early Miocene for, although the group did not become numerous until the Pliocene, one rather aberrant advanced species is known from the upper Miocene. By this time sabre-tooths were in existence and, as already pointed out, the remains of their kills must have included abundant bones for any carnivore who could utilise them. This would have provided exactly the conditions necessary to make the typical hyaena bone-crushing adaptations a worthwhile acquisition.

The hyaenids' nearest relatives are the Viverridae and it is very easy to imagine how hyaenas might have evolved from a terrestrial form rather like the modern civet cats, differing from them, however, in having a more strictly carnivorous mode of life. Even in such minor characters as their coarse fur and the mixture of spots and stripes in their colour pattern, the civet cats have a hyaena-like air. The modern civets have no objection to carrion and presumably the ancestral hyaenids were quick to take advantage of the remains of any kills they could discover. The genus *Ictitherium*, as known from Pliocene deposits, may represent a late surviving branch of the actual ancestral stock. It is certainly at an intermediate stage and has exactly the characters to be expected in an early hyaenid. The carnassial teeth, although more advanced than those of civets, are still relatively primitive; the two upper molars, although reduced, are still present, while the premolars show the characteristic enlargement of the central cusps although they are not very massive. From such beginnings a radiation of hyaenids occurred, giving rise to a wealth of species both in Eurasia and in Africa. They did not, however, reach North America and it is interesting to note that in that continent the same period saw the evolution amongst the Canidae of large species with massive skulls, heavy teeth and powerful jaw muscles, all features paralleling the hyaenas. These hyaena-dogs must have played the same role as the hyaenas of the Old World, and Matthew (1930) pointed out their probable ecological association with their sabre-tooth contemporaries. With the decline of the sabre-tooths the hyaena-dogs vanished and the hyaenas, although still with us, became greatly reduced in numbers, a mere three remaining from all the wealth of species that flourished in the early Pleistocene.

In Africa, the majority of the early Pleistocene hyaenids were large, powerful beasts, with very heavy premolars. There were, however, also some smaller species, placed in the genus *Lycyaena*, in which the premolars are secondarily modified as more sectorial teeth, less suited to bone cracking (Ewer, 1955a). These may be species which, not being successful in competition with the already highly evolved scavenging species, were reverting to a more predacious habit. Their appearance is, however, brief. They are found only in the very early Pleistocene and then vanish without trace. The natural conclusion is that, as predators, they met with competition from the large Felidae whose expansion was just starting and so, failing for the second time, they died out. There is, however, another possibility. *Proteles*, the aardwolf, is a hyaenid found only in Africa. We know nothing of its history as a fossil but its ancestor can only have been a small hyaenid that abandoned the mode of life typical of its fellows and became insectivorous. The African *Lycyaenas* are obvious candidates for the ancestral role and there are no other claimants to this position. It seems entirely reasonable that, faced with difficulties both as scavengers and as predators, they should have turned to insects as a food source, finally becoming termite eaters and giving rise to the present day *Proteles*. No fossils have yet been found to controvert or to support this suggestion but, with the extensive expeditions that are now at work investigating the deposits of East Africa in search of our own fossil ancestors, it may not be very long before those of the aardwolf too are brought to light and one more short chapter in the history of the carnivores can be written.

Not very much comparative work has been done on the biochemical characteristics of carnivores and the results do not as yet cast any great light on taxonomic relationships. The early serological investigations of Leone and Wiens (1956) and Pauly and Wolfe (1957) were done when techniques were still very crude and their results are not very illuminating. More recently Seal (1969) has made a comparative study of the electrophoretic mobilities of the haemoglobins of a large number of carnivores, including representatives of every family. The Canoidea are extremely uniform, all possessing a major component with a mobility of 0·85, relative to human haemoglobin A. Of the fifty-four species studied, only seven possessed in addition second major components: these were five species of *Mustela* and two procyonids, *Bassariscus astutus* and *Nasua narica*. The Feloidea, however, are much more diverse and, apart from the fact that the four extant hyaenid species all have a single major component with a relative mobility of 1·17, little taxonomic relationship emerges. Amongst the Viverridae, for instance, five different haemoglobins are found which, in various combinations of one, two or in a few species even three, produce eight different patterns. The Herpestinae, regarded by all taxonomists as a rather uniform and closely knit group, include representatives of no less than six of the eight possible patterns and every one of the five different haemoglobins occurs in this subfamily. The Felidae are only a trifle less confusing, with three different haemoglobins and three different patterns. One pattern is shared by the four species of *Felis* studied, the two lynxes, all four *Leopardus*, the jaguarondi, puma and cheetah: the caracal, serval, two species of *Prionailurus* and the leopard form a second group and the lion, tiger, snow leopard and clouded leopard constitute the remaining one.

Seal and his co-workers (1970) have also made a biochemical study of the relationships of the Ursidae. Since their haemoglobins are identical it was necessary to utilise some other blood protein and serum albumin was selected. Purified serum albumin of the black bear (*Ursus americanus*) was used to immunise rabbits and the reactions of the resulting immune sera tested against the albumins from other species. Those of brown, polar, Himalayan, Malayan and sloth bears proved to be indistinguishable from black bear serum albumin but the most sensitive sera reacted more weakly with *Tremarctos* albumin, indicating that the latter is the most isolated species and diverges most from other bears - a conclusion which is in agreement with the general opinion on the subject. When tested with albumins from other carnivore families, the antisera reacted more strongly with those from Procyonidae than with those of Canidae or Mustelidae and still more weakly with feloid albumins. This shows that bear serum albumin resembles procyonid albumins more closely than those of any other carnivore family, which Seal regards as 'suggesting the derivation of the ursids from an immediate procyonid ancestry'. Palaeontological evidence links the bears clearly with the Canidae but leaves the origin of the Procyonidae in doubt. It would therefore be premature to accept the suggested relationship as definitely established, for it is unsafe to base taxonomic conclusions on the study of a single protein. No one, for instance, would advocate separating *Herpestes* from the other mongooses and uniting it with the civets, on the grounds of haemoglobin similarity. It is therefore desirable to proceed with some caution until considerably more biochemical data have been accumulated.

Another field that might be expected to provide data of taxonomic relevance is the study of chromosome morphology. A considerable body of information now exists and Wurster and Benirschke (1968) provide a convenient summary of all the earlier data for carnivores. Todd and Pressman (1968) deal with the lesser panda; Todd (1970) and Fredga (1970) give details for a few additional members of the Canidae and Herpestinae respectively.

In considering the significance of these data, it would be convenient if one could decide what is most likely to have been the primitive condition and whether low counts have, as a rule, arisen by centric fusion from originally higher numbers or whether the reverse process of karyotypic fissioning has predominated. Either process could have produced the distributions shown in figure 9.6. If a count of 38 or thereabouts is primitive, then the majority of species have retained this condition: on the other hand, if the count of 38 is the result of centric fusion from an originally much higher value, then in the majority of species fusion has virtually run its full course and little further reduction in number is possible. Wurster and Benirschke conclude that centric fusion has been the major factor: Todd is a protagonist of the opposite view. He believes fissioning is a progressive process, enhancing evolutionary potential by increasing the possibili-

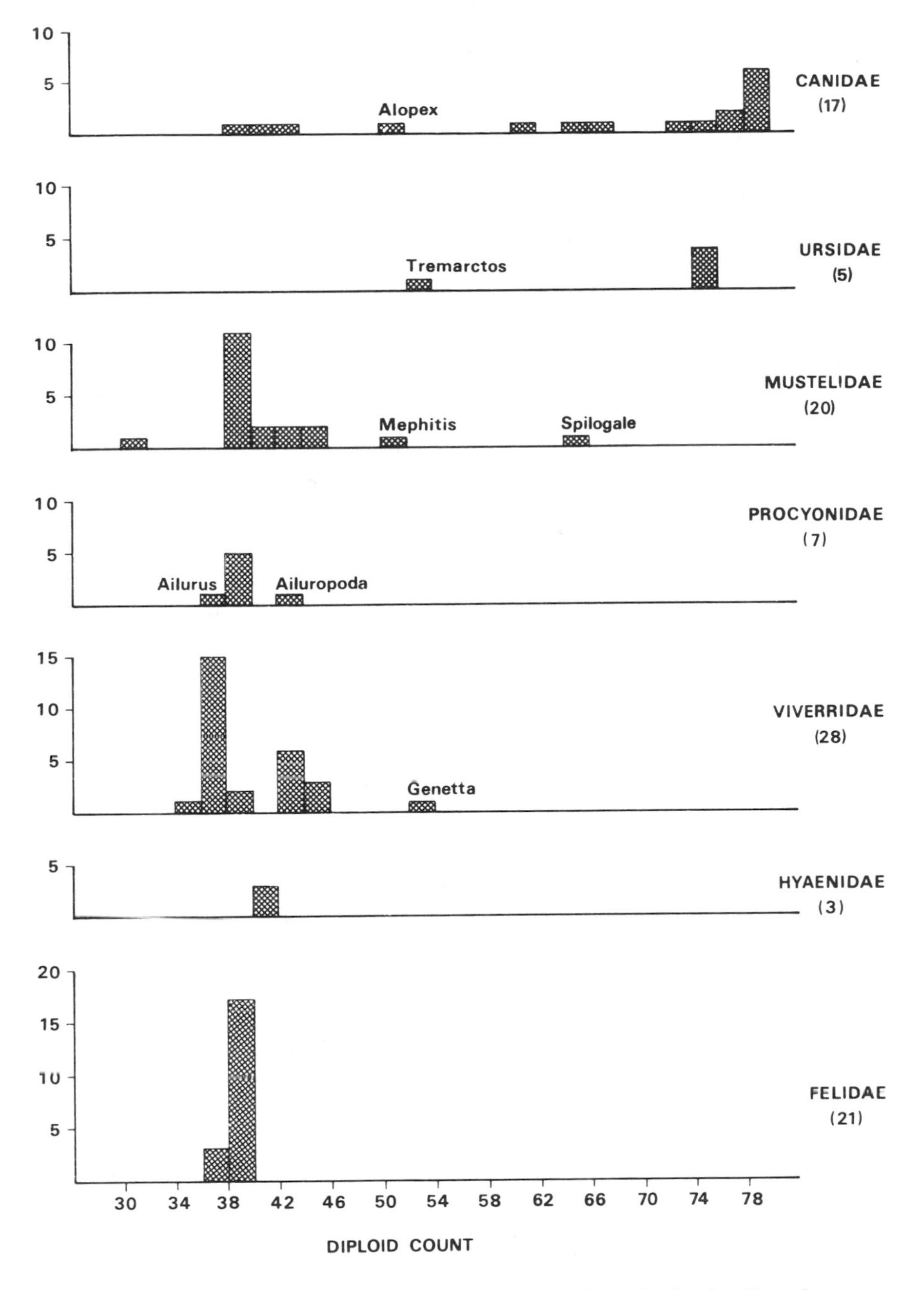

Figure 9.6. Distribution of diploid chromosome numbers in the families of carnivores. Figures in brackets are the number of species investigated in each family.

ties of recombination. On his view, the Canidae and Ursidae are progressive and in the former family it is the genus *Canis* which is characterised by a high count while the 'primitive' low numbers occur in *Vulpes* and *Nyctereutes*. It may be worth noting here that in the Pinnipedia, a group which can hardly be regarded as either primitive or slowly evolving, the chromosome counts are low, ranging from 32 to 36 (Wurster and Benirschke, 1968). On the other hand, if we regard fusion as advanced, then this process must have run virtually to completion in the majority of Mustelidae, Procyonidae, Viverridae and Felidae and only the canids and bears are primitive. Neither view seems particularly seductive: it may be that each process has its own advantages but that there is a tendency to become committed to one or the other course, fusion or fission and to carry it more or less to completion. However that may be, it is clear from figure 9.6 that the distribution of chromosome numbers links the Canidae and Ursidae but divides them sharply from all the other families. In particular the Procyonidae, usually regarded as an offshoot from the canid stock, differ from them just as much as do the Mustelidae. This is exactly the opposite of the conclusion reached by Seal *et al.* (1970) from their comparative studies of serum albumins. As already mentioned, Seal suggested a derivation of the Ursidae directly from the Procyonidae but the former is the only family whose lineage can, in fact, be traced back in the fossil record to an origin from the Canidae. Chromosome morphology thus suggests that the search for procyonid ancestors amongst the early Canidae may be a hopeless quest: both they and the mustelids may have originated independently from the Miacidae. Possibly, therefore, one should draw the fissipede family tree not with the usual three main stems - Canidae (giving rise to Ursidae and Procyonidae), Mustelidae and Viverridae (giving rise to Felidae and Hyaenidae) - but rather with four: Canidae (giving rise only to Ursidae), Procyonidae, Mustelidae and Viverridae (figure 9.7).

Within the family Canidae, *Lycaon*, like *Canis*, has 78 chromosomes. Unfortunately neither *Cuon* nor *Speothos* has yet been studied but it would clearly be of interest to know whether they also have the same number. As may be seen from table 9.1 (p. 382), the genus *Vulpes* is karyotypically very diverse. It would be desirable to have data for more species of this genus; this might suggest relationships both within the genus and with other genera such as *Alopex* and *Fennecus*. As they stand at present, the chromosome numbers do not suggest any useful grouping of the canids into subfamilies.

The lesser panda, with 36 chromosomes, clearly belongs with the procyonids but the count of 42 for the giant panda is equivocal. The difference from the 38 typical of the Procyoninae is not very great and cannot be held to rule out classification as an aberrant procyonid. On

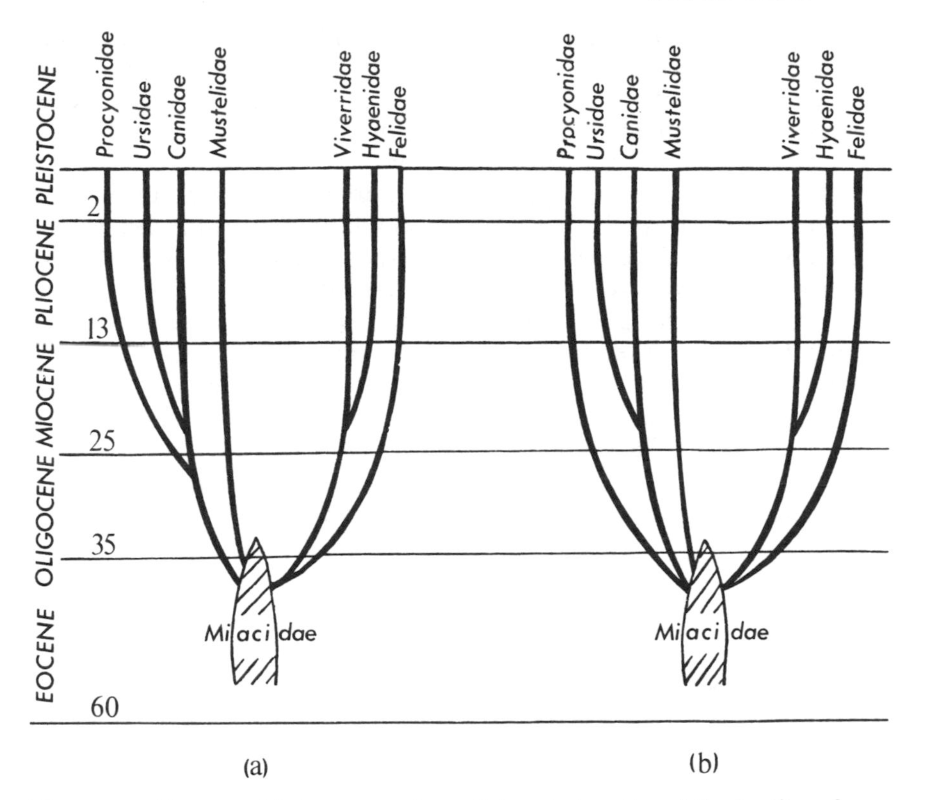

Figure 9.7. Two possible interpretations of the relationships of the families of carnivores. (a) The Procyonidae arising as an offshoot from the Canidae or (b) originating independently from the Miacidae. Figures on the left are millions of years ago.

the other hand, one can also argue that if *Tremarctos*, an aberrant bear, has 52, as against 74 chromosomes for other bears, then the panda may be an even more aberrant bear with an even lower count. The procyonid relationship, however, seems distinctly more probable.

The Herpestinae present an interesting peculiarity. In the seven species of *Herpestes* that have been studied, the male has an odd number of chromosomes, one less than the female. This situation results from the fact that translocation of the Y chromosome has occurred and it has become fused to one of the autosomes. The female therefore has a normal pair of sex chromosomes, the usual two X's, but the male has a single X lacking a partner together with an unequal chromosome pair, consisting of one normal autosome and its fellow with the attached Y. Of the other eight genera which have been investigated, *Atilax* alone has a similar arrangement (Fredga, 1970).

This suggests a close relationship between *Herpestes* and *Atilax*, and it is interesting to note that palaeontology supports this, for a fossil mongoose from a South African Pleistocene deposit appears to represent an intermediate stage in the evolution of the more specialised *Atilax* from an ancestral *Herpestes* (Ewer, 1956b).

The Felidae are chromosomally remarkably uniform with a diploid count of 38 in both small and large cats and also in the cheetah. The only exceptions so far recorded are the ocelots and Geoffroy's cat, with 36. This suggests descent of the latter forms from a common ancestor and indicates that, if we place the ocelots in the genus *Leopardus*, Geoffrey's cat should also be included. Detailed study of chromosome morphology shows similarities between the Bengal cat and the fishing cat, while the jaguarondi shows some unique features. These findings may therefore be held to support the uniting of the Bengal and fishing cats in *Prionailurus* and the placing of the jaguarondi in a genus of its own, *Herpailurus*.

Table 9.1
Diploid chromosome numbers found in the carnivore families

(Where there is polymorphism within a species, the higher number is given.)

Family CANIDAE

78 *Canis lupus*; *C. niger*; *C. latrans*; *C. aureus*; *C. dingo*; *Lycaon pictus*
76 *Chrysocyon brachyurus*; *Atelocynus microtis*
74 *Dusicyon vetulus*
72 *Otocyon megalotis*
66 *Urocyon cinereoargenteus*
64 *Fennecus zerda*
60 *Vulpes bengalensis*
50 *Alopex lagopus*
42 *Nyctereutes procyonoides*
40 *Vulpes ruppelli*
38 *Vulpes vulpes*

Family URSIDAE

74 *Ursus americanus*; *U. arctos*; *Thalarctos maritimus*; *Selenarctos thibetanus*
52 *Tremarctos ornatus*

Family PROCYONIDAE

38 *Procyon lotor*; *Potos flavus*; *Bassariscus astutus*; *Bassaricyon gabbii*; *Nasua nasua*
36 *Ailurus fulgens*
42 *Ailuropoda melanoleuca*

Family MUSTELIDAE

64 *Spilogale putorius*
50 *Mephitis mephitis*
44 *Mustela erminea*; *Meles meles*
42 *Mustela nivalis*; *Gulo gulo*
40 *Mustela putorius*; *Martes flavigula*

38 *Mustela sibirica*; *Martes martes*; *M. foina*; *M. americana*; *M. pennanti*; *Tayra barbara*; *Grison vittatus*; *Melogale moschata*; *Lutra canadensis*; *Lutrogale perspicillata*; *Amblonyx cinerea*
30 *Mustela vison*

Family VIVERRIDAE

52 *Genetta genetta*
44 *Paguma larvata*; *Galidia elegans*; *Herpestes ichneumon*
42 *Paradoxurus hermaphroditus*; *Arctictis binturong*; *Fossa fossa*; *Hemigalus derbyanus*; *Cryptoprocta ferox*; *Herpestes sanguineus*
38 *Civettictis civetta*; *Nandinia binotata*
36 *Viverricula indica*; *Herpestes auropunctatus*; *H. javanicus*, *H. edwardsi*, *H. brachyurus*, *H. fuscus*, *H. urva*; *Crossarchus obscurus*; *Helogale parvula*; *Mungos mungo*; *Ichneumia albicauda*; *Cynictis penicillata*; *Atilax paludinosus*; *Bdeogale* sp.; *Suricata suricatta*
34 *Prionodon linsang*

Family HYAENIDAE

40 *Crocuta crocuta*; *Hyaena hyaena*; *Proteles cristatus*

Family FELIDAE

38 *Felis chaus*; *F. nigripes*; *Prionailurus bengalensis*; *P. viverrinus*; *Profelis auratus*; *P. temmincki*; *Caracal caracal*; *Lynx lynx*; *L. rufus*; *Leptailurus serval*; *Herpailurus yagouaroundi*; *Puma concolor*; *Panthera leo*; *P. tigris*; *P. pardus*; *P. onca*; *Acinonyx jubatus*
36 *Leopardus pardalis*; *L. tigrinus*; *L. wiedi*; *L. geoffroyi*

Chapter 10

Classification and distribution of the living species

THE list of species which follows is based mainly on the following authorities: Ellerman and Morrison-Scott (1951) for Eurasia; Coetzee (1967) and Smithers (1968b) for Africa; Hall and Kelson (1959) for North America and Cabrera and Yepes (1960) for South America. After each family I have added a few notes dealing with points of disagreement or with subsequently appearing papers which suggest the desirability of some modification. Domestic species are not listed but are mentioned in the notes. Species mentioned in the notes are marked with an asterisk.

A. Superfamily Canoidea

Maxillo-turbinals large; ethmo-turbinals short, not reaching front of nasal chamber.

Auditory bulla of the adult formed from a single tympanic bone without a separate entotympanic element.

Baculum large.

(i) *Family* Canidae

Digitigrade, with five toes on fore and four on hind paws; limbs typically long. (*Lycaon* is exceptional in having four digits on the fore paws.)

Jaws long; dentition typically combines cutting carnassials and crushing molars; M_1 with well developed talonid.

*1. *Canis lupus* Linnaeus, 1758; wolf: formerly widespread over the entire Holarctic region and thence south into India. Now extinct in western Europe except for small relict populations in Scandinavia and Spain. Widespread in the USSR and in North America and from southeastern Europe through southwest Asia to India.
*2. *Canis latrans* Say, 1823; coyote: North America from Alaska

south to Costa Rica in Central America and as far east as New York State.

3. *Canis aureus* Linnaeus, 1758; golden jackal: widespread in Asia and southeast Europe; extending into north Africa and east Africa as far south as Tanzania.

4. *Canis adustus* Sundevall, 1846; side-striped jackal: Africa, from Senegal, Ethiopia and Somalia southwards as far as northern South West Africa and northern Zululand.

5. *Canis mesomelas* Schreber, 1775; black-backed jackal: throughout South and east Africa, extending north as far as Ethiopia.

6. *Canis simensis* Rüppell, 1835; Simenian jackal: Ethiopian highlands.

7. *Lycaon pictus* (Temminck, 1820); African hunting dog: Africa south of the Sahara from South West Africa and the Transvaal northwards to Somalia, Ethiopia and Sudan; westwards to Ivory Coast and north through Mali and Niger to southern Algeria.

8. *Cuon alpinus* (Pallas, 1811); red dog or dhole: southern USSR, India, southeast Asia, Sumatra and Java.

9. *Alopex lagopus* (Linnaeus, 1758); arctic fox: circumpolar distribution including the arctic regions of North America, the USSR, Scandinavia, Spitzbergen, Iceland and Greenland.

*10. *Vulpes vulpes* (Linnaeus, 1758); red fox: throughout the Holarctic region and extending into north Africa and southeast Asia.

11. *Vulpes corsac* (Linnaeus, 1768); corsac fox: Afghanistan, southern USSR, northern Manchuria.

12. *Vulpes bengalensis* (Shaw, 1800); Bengal fox: India, extending into Nepal and the Punjab region of Pakistan.

13. *Vulpes ferrilata* Hodgson, 1842; Tibetan sand fox: Tibet and Nepal.

14. *Vulpes cana* Blanford, 1877; Blanford's fox: Afghanistan, extending into northeast Iran and Turkestan and southwards into Pakistan.

15. *Vulpes ruppelli* (Schinz, 1825); sand fox: north Africa, extending eastwards from Morocco as far as Afghanistan.

16. *Vulpes pallida* (Cretzschmar, 1826); sand fox or pale fox: southern fringes of Sahara from Sudan westwards to Senegal.

17. *Vulpes chama* (Smith, 1833); Cape fox or silver jackal: South Africa, South West Africa, Botswana, Angola.

18. *Vulpes velox* (Say, 1823); kit fox or swift fox: Great Plains region of North America, from southwestern Canada southwards to Texas.

19. *Fennecus zerda* (Zimmermann, 1780); fennec: north Africa from Morocco eastwards into Arabia.

20. *Nyctereutes procyonoides* (Gray, 1834); raccoon dog: eastern USSR, China, Japan.

21. *Otocyon megalotis* (Desmarest, 1822); bat-eared fox: South Africa and Angola; east Africa and thence north to Sudan, Ethiopia and Somalia.
22. *Urocyon cinereoargenteus* (Schreber, 1776); grey fox: southern Canada southwards through Central America into northwestern South America.
23. *Urocyon littoralis* (Baird, 1858); island grey fox: islands off the Californian coast.
24. *Dusicyon culpaeolus* (Thomas, 1914): Uruguay.
25. *Dusicyon culpaeus* (Molina, 1782): Andes, from Ecuador southwards.
26. *Dusicyon fulvipes* (Martin, 1837); Chiloe fox: island of Chiloe, off the coast of Chile.
27. *Dusicyon griseus* (Gray, 1837): eastern Argentina and Chile.
28. *Dusicyon gymnocercus* (Thomas, 1914): central part of South America, including Paraguay, southeastern Brazil, Uruguay and northern Argentina from the foothills of the Andes to the Atlantic coast.
29. *Dusicyon inca* (Thomas, 1914): Andean region of Peru, up to an altitude of 4,000 metres.
30. *Dusicyon sechurae* (Thomas, 1900): desert regions of northeastern Peru.
31. *Dusicyon vetulus* (Lund, 1840): central and southern Brazil.
32. *Atelocynus microtus* (Sclater, 1882); small-eared zorro: Amazon and upper Parana and Orinoco basins in Brazil, Peru, Ecuador and Colombia.
33. *Cerdocyon thous* (Linnaeus, 1766); crab-eating fox: South America as far south as northern Argentina and Uruguay.
34. *Chrysocyon brachyurus* (Illiger, 1804); maned wolf: South America east of the Andes, as far south as northern Argentina and Uruguay.
35. *Speothos venaticus* (Lund, 1842); bush dog: Central and South America east of the Andes, from Panama south to Bolivia.

The red wolf of the southern United States used to be regarded as a distinct species and was known as *Canis niger*. Lawrence and Bossert (1967), however, using a discrimination analysis technique on skull measurements chosen so as to reflect the major differences between coyotes and wolves, came to the conclusion that the red wolf should be regarded only as a subspecies of *C. lupus* and I have accepted their opinion.

The picture is further complicated by the fact that, in the southeastern states, the disturbance of the habitat caused by farming has brought together wolves and coyotes which had previously remained ecologically separated, the coyotes in the western plains and the wolves in more woodland areas. As a result, hybridisation appears to be taking place and, as Paradiso (1968) has shown, although the coyotes from western Texas are clearly separable from the red wolves of Arkansas and Louisiana, specimens from eastern Texas, believed to be hybrids, are intermediate between the two and are not clearly separable either from the coyotes to the west or the wolves to the east. This is not altogether surprising, since there is no reason why reproductive isolation should have evolved between closely related animals that were not actually in contact with each other and it cannot be held to invalidate the specific distinction between *Canis latrans* and *C. lupus*.

The eastern coyotes of New England present a rather similar problem. Wolves were eliminated from this area towards the end of the last century and it subsequently became clear that their role as large predators was being taken over by the coyotes, which were bigger and more wolf-like than in other parts of their range. For a time it was considered that the New England canids were in fact a coyote stock which had interbred with domestic dogs. However, as the characteristics of authentic coyote-dog hybrids became better known, their differences from the New England canids became increasingly obvious. The latter have recently been investigated by Silver and Silver (1969), who studied the behaviour and development of a litter of wild-caught pups and their descendants bred in captivity, and by Lawrence and Bossert (1969), who studied skull characteristics, using the discriminant functions which they had previously found would clearly separate the skulls of coyote, wolf and dog. Both sets of workers came to the same conclusion. Silver and Silver regard the New England canids as closest to coyotes but possessing some wolf-like or dog-like characteristics and Lawrence and Bossert are of the opinion that they are a coyote stock into which both dog and wolf genes have been introduced. Examination of the data presented, however, makes it rather difficult to understand how such a conclusion was reached: while the evidence clearly indicates that the stock is basically coyote, there is nothing to suggest an admixture with anything other than wolf. One cannot, of course, completely exclude the possibility of occasional crossing with domestic dog but, as Mengel (1971) points out, the coyote-dog hybrid, breeding as it does much earlier than coyote or wolf and showing no male parental care, is at a considerable reproductive disadvantage and hybrids are more likely to be eliminated than to contribute significantly to the population gene pool. Mengel therefore believes that the New England canids are coyotes with an admixture of wolf, a conclusion with which I am in agreement since it appears to be the one most consonant with the findings of the Silvers and of Lawrence and Bossert. Whatever the genetical situation may ultimately prove to be, there can be few who would not subscribe to the latter authors' taxonomic verdict: in view of the wide range of variability shown and the virtual certainty of hybrid ancestry, there is little point in giving a subspecific name to the New England canids; they should rather simply 'be called *Canis latrans* var. and may be referred to as the eastern coyote'.

Until quite recently, the red fox in America was commonly regarded as distinct from the Eurasian *Vulpes vulpes* and was known as *Vulpes fulva*. Churcher (1959) carried out a comparison based mainly on cranial and dental characters and came to the conclusion that we are here dealing with a case of clinal variation. No discontinuities can be found if we start from western Europe and work eastwards through Asia to Alaska and across America from west to east, and there is no ground for making more than a subspecific distinction between any of the forms encountered. If, however, skulls from western Europe and eastern North America are compared, then the two groups, which form the two ends of the range, are distinguishable. This, however, does not imply that two species are involved and Churcher regards *V. fulva* as a synonym of *V. vulpes*, a verdict which is now generally accepted.

In the eastern United States the fox appears to have had a rather chequered history. Remains from datable archaeological sites in the area suggest that during the last 4,000 years or so its southern limit has fluctuated with changing climatic conditions (Waters, 1964, 1967). The red fox was probably not present south of latitude 40°N when the position was further complicated by the introduction of European red foxes into New England in the eighteenth century by gentlemen interested in fox hunting. How many of them survived and bred is not known but, according to Churcher's analysis, any European genes now included in the American population have not sufficed to make a noticeable difference to the cranial and dental characteristics of the foxes of the eastern region.

The domestic dog, *Canis familiaris* Linnaeus, 1758 is generally believed to have been derived from the wolf, the wild species which it most closely resembles both anatomically (Lawrence and Bossert, 1967) and behaviourally (Scott, 1967). The differences between the two are not great enough to make it necessary to postulate a hypothetical ancestral wild species which has neither survived to the present day nor left any known fossil remains; a possibility given considerable weight by Zeuner (1963). It is generally believed that the ancestral wolf must have been a small subspecies and that domestication occurred somewhere in the Near East, in the southern part of the wolf's Old World range. Higgs and Jarman (1969), however, point out that there is no direct evidence to support this view and that the earliest remains of dogs in the Old World are from Yorkshire, carbon dated at *c.* 7,500 BC. Lawrence (1967), however, has described domestic dog remains from a cave in Idaho with the significantly earlier date of 9,500–8,400 BC. As she points out, if dogs reached the New World from the Old, as is generally assumed, then their domestication must have occurred considerably earlier, around 10,000 to 9,000 BC. This means that the Yorkshire dog remains cannot be regarded as close in time to the origin of domestication and leaves the problem of where this event took place once more a completely open question. Since fertile crossing is possible between most species of the genus *Canis* (Gray, 1954, 1966), wolf ancestry would not preclude some subsequent local admixture of genes from other wild species as a result of occasional accidental crossings.

Wherever and whenever domestication may ultimately prove to have started, it had led to alterations in dentition and skull making it possible to distinguish dog remains from those of wild canids long before any deliberate selection of breeds suited to different purposes was practised. The teeth are relatively smaller and the jaws shorter. The result is that the palate appears relatively wider and more bowed out posteriorly. This characteristic distinguishes the skull of the dog from that of the coyote as well as the wolf (Howard, 1949). This change may represent no more than a decreased selective pressure for effective killing of large prey. Early man may or may not have used the dog as a hunting assistant but it is certainly true to say that the domesticated dog had acquired human help in procuring food. A further change affects the skull contour: the frontal sinuses of the dog are more inflated than those of the wolf and the snout slopes down more, relative to the braincase (Lawrence, 1967; Lawrence and Bossert, 1967). Both these features accentuate the forehead and may have been the result of a human preference for dogs that look intelligent, exercised at first without any thought of deliberately affecting the breed.

The dingo of Australia is regarded as a feral dog, originally brought to the continent by its early human immigrants, but opinions differ as to whether it should be regarded as a subspecies of the ordinary domestic dog or accorded the rank of a separate species, *Canis dingo*.

The Falkland Islands 'dog', exterminated within historic times, did not belong to the genus *Canis* but was actually a species of *Dusicyon, D. australis*. The latter genus requires thorough revision. The species listed are those given by Morris (1965) but how many of them are really valid remains somewhat uncertain.

(*ii*) *Family* Ursidae

Large heavily built animals; plantigrade with long non-retractile claws and very reduced tail.

Anterior premolars reduced, carnassials non-sectorial and both they and the molars are broad, multi-cusped teeth.

*1. *Ursus arctos* Linnaeus, 1758; brown bear: northern and temperate Holarctic region from Sweden eastwards across the USSR and North America. In western Europe only small residual populations remain in Scandinavia, the Pyrenees and northern Italy: farther east the species extends south as far as Syria, Pakistan and northern India.

2. *Ursus americanus* Pallas, 1780; American black bear: formerly widespread in America from Mexico northwards; residual populations survive over much of the range in sparsely populated wooded regions and under protection in National Parks they are numerous and thriving.

*3. *Thalarctos maritimus* (Phipps, 1774); polar bear: circumpolar.

4. *Selenarctos thibetanus* (G.Cuvier, 1823); Asiatic black bear: Japan, eastern Asia from Siberia south to Thailand and westwards through Burma and Nepal to Afghanistan.

5. *Helarctos malayanus* (Raffles, 1822); Malayan sun bear: southeast Asia, Sumatra, Borneo.

6. *Melursus ursinus* (Shaw, 1791); sloth bear: India, Ceylon.

7. *Tremarctos ornatus* (F.Cuvier, 1825); spectacled bear: mountainous regions of northern South America as far south as Peru and Bolivia but now rare except in the most inaccessible parts of the range.

The brown bears of America comprise the kodiak bears of Alaska, Kodiak and other islands off the Alaskan coast and the grizzlies. The latter once extended over most of western North America: they still occur over wide areas in Alaska and western Canada but within the United States they are now restricted to the more mountainous parts of Montana, Idaho and Wyoming, centring in Glacier and Yellowstone National Parks. Although larger than the Old World brown bears, the American bears are now included with them in a single species but were formerly classified separately as *Ursus horribilis*.

The polar bear is closely related to the brown and black bears and fossil evidence suggests that the three diverged only very recently, during the Pleistocene. Kurtén (1966) therefore regards it as unjustifiable to place the polar bear in a separate genus and includes it in *Ursus*. The argument for generic distinction is exactly the same as that for placing the Pinnipedia in an order of their own. Adaptation to a different mode of life has caused the polar bear to diverge very rapidly and very markedly from its relatives and the generic separation reflects this fact. The rate at which two species diverge is determined mainly by the directions and intensities of the selective pressures to which they are exposed and a classification which does not admit this is quite unbiological. A classification based on time relations and not on degree of similarity would, in any case, be quite impractical, since only in a minority of cases is it possible to tell with any certainty how far back in time it is necessary to go in order to find a common ancestor for two extant species. I have therefore retained the genus *Thalarctos*.

(*iii*) *Family* Procyonidae

Plantigrade or semi-plantigrade; generally good climbers; tail typically long and often marked with a pattern of rings.

Cheek teeth generally somewhat broad and flat with carnassials typically non-sectorial.

(*a*) *Subfamily* Procyoninae

1. *Bassariscus astutus* (Lichtenstein, 1827); ringtail or cacomistle: states of Oregon and Colorado and from Texas southwards to southern Mexico.

2. *Jentinkia sumichrasti* (Saussure, 1860); Central American cacomistle: southern Mexico and Central America.

*3. *Bassaricyon gabbii* Allen, 1876; olingo: from Nicaragua south through Central America and into northern and northwestern South America, as far south as Peru and Bolivia.

4. *Potos flavus* Schreber, 1774; kinkajou: southern Mexico, Central America and extending southwards into Brazil as far as the Matto Grosso.

5. *Nasua nasua* (Linnaeus, 1766); ring-tailed coati: wooded areas throughout most of South America.

6. *Nasua narica* (Linnaeus, 1766); coati: southwestern United States and southwards through Central America into western Colombia and western Ecuador.

7. *Nasua nelsoni* Merriam, 1901; island coati: restricted to Cozumel Island, off the east coast of Yucatan.

8. *Nasuella olivacea* (Gray, 1843); little coati or mountain coati: mountainous regions from western Venezuela through Colombia into Ecuador.

9. *Procyon lotor* (Linnaeus, 1758); raccoon: from southern Canada southwards throughout the United States into Central America.

10. *Procyon cancrivorus* (Cuvier, 1798); crab-eating raccoon: Central

America from southern Costa Rica southwards into northern South America.

11. *Procyon insularis* Merriam, 1898; Tres Marias raccoon: Tres Marias Island, off the west coast of Mexico.

12. *Procyon maynardi* Bangs, 1898; Bahama raccoon: Bahama islands.

13. *Procyon minor* Miller, 1911; Guadeloupe raccoon: Guadeloupe Island (West Indies).

14. *Procyon gloveralleni* Nelson and Goldman, 1930; Barbados raccoon: island of Barbados (West Indies).

15. *Procyon pygmaeus* Merriam, 1901; Cozumel Island raccoon: Cozumel Island, off the east coast of Yucatan.

Following Allen's description of *Bassaricyon gabbii* in 1876, six further species were described by various authors. Cabrera (1957) regarded only two of these, in addition to *B. gabbii*, as valid; *B. alleni* Thomas and *B. beddardi* Pocock: most subsequent authors have accepted his opinion. Poglayen-Neuwall and Poglayen-Neuwall (1965), however, from a study of as many specimens as possible (admittedly a far from adequate sample) and from observations on living animals, have come to the conclusion that the supposed differences between the species are not reliable and are of very doubtful specific value. They consider that, in all probability, only the single species, *B. gabbii*, is valid. Clearly a thorough revision of the genus is required and for the meantime I have listed only this one species.

(*b*) *Subfamily* Ailurinae

*16. *Ailurus fulgens* F.Cuvier, 1825; red panda: Sikkim, Nepal and northern Burma and the southern and western Chinese provinces of Yunnan and Szechuan.

*17. *Ailuropoda melanoleuca* (David, 1869); giant panda: central and western China, from Szechuan and Kansu westwards to Tsinghai and Sikiang.

Opinions are divided as to whether the pandas, in particular the giant panda, are more closely related to the Procyonidae or to the Ursidae. There is little doubt that, if the red panda had occurred in South America instead of in Asia, it would be regarded as a somewhat aberrant procyonid, with teeth adapted to an unusually specialised vegetarian diet. This, however, is of little assistance in determining the relationships of the giant panda for, apart from geographical distribution, most of the features which link the two could be attributed to convergent adaptation to extreme vegetarianism. On the other hand, the resemblances between the giant panda and the bears can equally well be regarded as convergent.

The arguments which to me appear to weigh most heavily against relationship to bears are, firstly, the very different structure of the reproductive organs, scent gland and associated marking behaviour; for if the panda is really a bear it is very difficult to see any reason for these differences. Secondly, it is necessary to account for the presence of the enlarged radial sesamoid forming the panda's 'thumb'. A red panda grips a stem in almost exactly the same way as does the giant panda and one has only to watch it doing so to see how advantageous the development of the extra 'thumb' would be: a bear's use of its paws, on the other hand, does not suggest why the sesamoid should ever have become enlarged.

Davis (1964) gives a very detailed description of the anatomy of the giant panda but he does not assess the evidence he assembles in a critical manner: he appears simply to brush aside the features which do not support his opinion that the giant panda is an aberrant bear. Morris and Morris (1966) give a more balanced treatment and come to what I consider the most rational conclusion; to wit, that the characteristics of the giant panda accord best with the view that it first became adapted to a diet mainly of bamboo and subsequently attained large size (and therewith its bear-like characteristics), rather than that it was originally a large bear-like animal which latterly specialised as a bamboo eater. I have therefore retained both the red and giant pandas in their traditional place as a separate subfamily of procyonids.

(*iv*) *Family* Mustelidae

Typically long-bodied animals with short legs, semi-plantigrade or digitigrade.

Carnassials typically highly sectorial but secondarily modified for crushing in a number of subfamilies; upper molars with a broad inner lobe, giving them a rectangular rather than a triangular outline.

Anal sac secretion generally strong smelling and used defensively in many species.

(*a*) *Subfamily* Mustelinae

1. *Mustela erminea* Linnaeus, 1758; stoat or ermine: Holarctic region from Ireland across Europe and the USSR to Japan and North America; Algeria and the Middle East as far east as Afghanistan and Kashmir.

2. *Mustela nivalis* Linnaeus, 1766; weasel: Palaearctic region from Europe across the whole of Asia; Sardinia, Malta, Crete and across north Africa from Morocco to Egypt. Absent from Ireland.

*3. *Mustela rixosa* (Bangs, 1896); least weasel or pygmy weasel: North America.

4. *Mustela frenata* Lichtenstein, 1831; long-tailed weasel: North America and southwards through Central America into northern South America.

5. *Mustela altaica* Pallas, 1811; alpine weasel: mountainous regions of southern and eastern USSR and southwards to Kashmir, Tibet, outer Mongolia and northern China.

6. *Mustela sibirica* Pallas, 1773; Siberian weasel: USSR from west of the Urals eastwards, extending north to the limits of the forest zone and south to the Himalayas; throughout China, Manchuria, Tibet and northern Burma; Java, Taiwan, Japan.

7. *Mustela kathiath* Hodgson, 1835; yellow-bellied weasel: Himalayas from northern Pakistan through Nepal to Burma; western and southern China and Indo-Chinese peninsula.

8. *Mustela strigidorsa* Gray, 1853; back-striped weasel: from Nepal eastwards through northern Burma into Indo-Chinese peninsula.

9. *Mustela lutreola* Linnaeus, 1761; European mink: Europe, from France eastwards to northern Caucasus.

10. *Mustela vison* Linnaeus, 1777; American mink: North America from Alaska and Canada south through most of the United States.

11. *Mustela putorius* Linnaeus, 1758; polecat: Holarctic region from Britain eastwards across Europe and the USSR, southwards to Mongolia, Tibet and northern China; Morocco.

12. *Mustela nigripes* (Audubon and Bachman, 1851); black-footed ferret: North America.

13. *Mustela lutreolina* Robinson and Thomas, 1917; Java weasel: Java.

14. *Mustela nudipes* Cuvier, 1821; bare-foot weasel: southeast Asia; Sumatra, Borneo.

*15. *Grammogale africana* Desmarest, 1818; belly-striped or South American weasel: northeastern Peru and also recorded from the Para region, near the mouth of the Amazon.

16. *Vormela peregusna* (Güldenstaedt, 1770); marbled polecat: Rumania eastwards through Asia minor to Afghanistan and through the Himalayan region to Mongolia.

17. *Martes martes* (Linnaeus, 1758); pine marten: throughout Europe (including Ireland) and eastwards through the USSR into western Siberia; Sardinia, Balearic Islands.

18. *Martes foina* (Erxleben, 1777); stone marten: Europe (excluding

Britain) and eastwards through southern USSR; Asia minor eastwards to Afghanistan; Tibet, Mongolia, Manchuria.

19. *Martes americana* (Turton, 1806); American marten: Alaska, Canada and United States.

20. *Martes pennanti* (Erxleben, 1777); fisher: North America. Exterminated over most of the continent but in recent years protection has resulted in an increase in various parts of the eastern United States.

21. *Martes melampus* (Wagner, 1840); Japanese marten: Japan and Korea.

22. *Martes zibellina* (Linnaeus, 1758); sable: formerly widespread over most of Siberia, Mongolia, Manchuria and Japan but now surviving only as small scattered isolated populations.

23. *Martes flavigula* (Boddaert, 1758); yellow-throated marten: eastern Siberia, Korea, Manchuria; southwards through China to Indo-Chinese and Malayan peninsulas and westwards through Tibet to Kashmir; Sumatra, Java, Borneo; Taiwan.

24. *Martes gwatkinsi* Horsfield, 1851; south Indian yellow-throated marten: southern India.

25. *Gulo gulo* (Linnaeus, 1758); wolverine; Holarctic taiga and forest tundra from Scandinavia eastwards through Siberia and North America, extending south into Mongolia and northern Manchuria.

*26. *Tayra* (= *Eira*) *barbara* (Linnaeus, 1758); tayra: southern Mexico southwards through Central America into South America as far south as Paraguay and northern Argentina; Trinidad.

*27. *Grison* (= *Galictis*) *vittatus* (Schreber, 1777); grison: southern Mexico and southwards through Central America to Peru and Brazil.

28. *Grison cuja* (Molina, 1782); little grison: central and southern America, usually at higher altitudes than *G. vittatus*.

29. *Lyncodon patagonicus* (D'Orbigny and Gervais, 1847); Patagonian weasel: Argentina and Chile.

30. *Ictonyx striatus* (Perry, 1810); striped polecat or stink muishond: semi-arid to arid regions throughout Africa south of the Sahara from Senegal and Ethiopia southwards.

31. *Poecilogale albinucha* (Gray, 1864); white-naped weasel or slang muishond: Africa from Congo, Uganda and Tanzania southwards.

32. *Poecilictis libyca* (Hempricht and Ehrenberg, 1832); Libyan striped weasel: fringes of the Sahara from Morocco to Egypt on the north and from northern Nigeria to Sudan on the south.

(*b*) *Subfamily* Mellivorinae

33. *Mellivora capensis* (Schreber, 1776); honey-badger or ratel: borders of the Sahara from Morocco, Senegal and Niger eastwards to Sudan and Somalia; thence south to Cape Province and east through Arabia and the Near East to India as far south as Madras.

(*c*) *Subfamily* Melinae

34. *Meles meles* (Linnaeus, 1758); European badger: throughout Palaearctic region, including Great Britain and Ireland; as far south as Israel, Iran and southern China; Japan.

35. *Arctonyx collaris* F.Cuvier, 1825; hog badger: China, northeast India, Burma and Indo-Chinese peninsula; Sumatra.

36. *Mydaus javanensis* F.Cuvier, 1821; Malayan stink-badger: Java, Sumatra, Borneo and North Natuna Island.

37. *Suillotaxus marchei* (Huet, 1887); Philippines badger: islands of Palawan and Calamianes in the Philippines.

38. *Taxidea taxus* (Schreber, 1777); American badger: North America from southwestern Canada to central Mexico.

39. *Melogale personata* Geoffroy, 1831; Burmese ferret badger: Nepal, eastwards through Burma and the Indo-Chinese peninsula.

*40. *Melogale moschata* (Gray, 1831); Chinese ferret badger: China, from Szechuan southwards to Burma and the Indo-Chinese peninsula; Taiwan.

41. *Melogale orientalis* (Horsfield, 1821); Javanese ferret badger: Java and Borneo.

(*d*) *Subfamily* Mephitinae

42. *Mephitis mephitis* (Linnaeus, 1758); striped skunk: North America, from southern Canada south to northern Mexico.

43. *Mephitis macroura* Lichtenstein, 1832; hooded skunk: southwest United States and Central America.

*44. *Spilogale putorius* (Linnaeus, 1758); spotted skunk; North America, from southwest Canada southwards through Mexico into Central America as far south as northern Costa Rica.

*45. *Conepatus leuconotus* (Lichtenstein, 1832); hog-nosed skunk: southwest United States and Central America.

*46. *Conepatus semistriatus* Boddaert, 1784; Amazonian skunk: central and northern South America.

*47. *Conepatus chinga* Molina, 1782: Chile, Peru, northwest Argentina and Bolivia.

*48. *Conepatus humboldti* Gray, 1837; Patagonian skunk: from northeast Argentina and Paraguay south to the Straits of Magellan.

(*e*) *Subfamily* Lutrinae

49. *Lutra lutra* (Linnaeus, 1758); European otter: widely distributed over most of Europe and Asia; Ceylon, Sumatra, Java, Taiwan, Japan; Morocco and Algeria.

50. *Lutra sumatrana* (Gray, 1865); hairy-nosed otter: Sumatra, northern Borneo, Malay peninsula and South Vietnam.

51. *Lutra canadensis* (Schreber, 1776); Canadian river otter: North America from Canada south throughout the United States.
52. *Lutra annectens* Major, 1897; Central American otter: Central America and southwards to Ecuador.
53. *Lutra enudris* Cuvier, 1823: Trinidad and northern South America from Venezuela and Guiana south into the Amazon basin in Brazil.
54. *Lutra felina* (Molina, 1782); sea cat: west coast of South America from Ecuador south to the Straits of Magellan.
55. *Lutra incarum* Thomas, 1908: western Peru and possibly Bolivia.
56. *Lutra mesopetes* (Cabrera, 1924): Costa Rica.
57. *Lutra platensis* Waterhouse, 1838: Plate River system in Argentina, Uruguay, Paraguay and Brazil.
58. *Lutra provocax* Thomas, 1908: southern Chile and western part of southern Argentina.
59. *Lutra maculicollis* Lichtenstein, 1835; spotted-necked otter: Africa, from Liberia in the west to Ethiopia and Somalia in the east and southwards to eastern Cape Province of South Africa.
60. *Lutrogale perspicillata* (Geoffroy, 1826); smooth-coated otter: throughout southeast Asia from India to southwest China, including Malayan and Indo-Chinese peninsulas; Java, Borneo, Sumatra; Tigris marshes of Iraq.
61. *Amblonyx cinerea* (Illiger, 1815); oriental small-clawed otter: India, Burma, southern China and southeast Asia; Sumatra, Java, Borneo, Palawan Island.
62. *Aonyx capensis* (Schinz, 1821); Cape clawless otter: Africa from the Cape north to Ethiopia and westwards to Senegal.
63. *Aonyx congica* (Lönnberg, 1910); Congo clawless otter: Congo basin.
*64. *Aonyx microdon* Pohle, 1920; Cameroon otter: Cameroon and eastern Nigeria.
*65. *Aonyx philippsi* (Hinton, 1921): from Lake Kivu through Ruanda into southwest Uganda and Burundi.
66. *Pteronura brasiliensis* (Gmelin, 1788); giant otter: major river systems of South America east of the Andes, as far south as northern Argentina.
67. *Enhydra lutris* (Linnaeus, 1758); sea otter: formerly widespread from Kurile Islands across the Aleutian chain and southwards down the west coast of North America to California. Residual populations remain in Kuriles, Aleutians, south coast of Alaska and west coast of California.

Mustela rixosa and *M. nivalis* are very closely related. Some authorities refer specimens from eastern Siberia to *M. rixosa* rather than to

M. nivalis and Hall (1951) was of the opinion that *M. rixosa* extends westwards through the USSR to central Europe, where it gives place to *M. nivalis.* Reichstein (1957), however, on the basis of skull characters, found it impossible to divide the European material into two species and suggested that *M. rixosa* is probably not a distinct species, an opinion accepted by Jones (1964). Possibly, like the red fox, the weasels form a circumpolar cline but I have for the meantime retained *M. rixosa* until it is clear that there is no discontinuity at any point in the range to warrant the recognition of more than one species.

The ferret, *Mustela furo* Linnaeus, 1758, is a domesticated, generally albino, form of the polecat. According to Ashton and Thomson (1955) it resembles the European subspecies of *M. putorius* more closely than the Asiatic and was therefore probably derived from the former. Aristotle mentions that polecats are easily tamed but there is no suggestion of true domestication. Thomson (1951) regarded the earliest certain mention of ferrets as being in Strabo's *Geographica,* dating back to the beginning of the Christian era. Strabo describes how rabbits can easily be killed by using an animal believed to be a ferret to flush them from their burrows. His account agrees very well with the practice of ferreting but he must surely have been mistaken in attributing an African origin to the ferrets. It is not correct to say that Strabo describes their being used to control a rabbit plague on the Balearic islands: he merely mentions the plague as something quite out of the ordinary but does not say whether control by ferrets was ever attempted.

Grammogale was originally described from a specimen taken from the Lisbon Museum in 1808 by Napoleon's troops and brought to Paris. There was no indication as to where the animal had been caught and, in view of Portugal's considerable possessions in Africa, Desmarest concluded that it must have originated in that continent. He therefore gave it the inappropriate specific designation, *africana* and its true origin was not established until very much later.

The generic names *Tayra* and *Grison* were given by Oken, 1816 and are technically invalid. Strictly speaking, therefore, *Grison* must become *Galictis* Bell, 1826, while *Tayra* gives place to *Eira.* There is, however, no doubt about what animals Oken meant and since 'tayra' and 'grison' have become accepted as common names the change has little to recommend it; furthermore, *Galictis* has a confusing similarity to *Galidictis* and *Galidia,* members of the Viverridae. I have therefore followed Walker (1964) in retaining Oken's names.

Everts (1968) has made a study of the southeast Asian ferret badgers. He considers that *Melogale moschata* should be placed in a separate genus from *M. personata,* largely on the basis of its less molariform upper carnassial. He therefore revives Gray's genus

Helictis for *H. moschata* but the differences he describes do not seem to me to warrant this. He also regards the badgers from Borneo as specifically distinct, on the basis of a discrimination analysis involving skull measurements. The number of skulls available, however, was very small and the value of the distinction is therefore doubtful. Morris (1965) unites the Java and Borneo badgers in *M. orientalis* and I have followed him, although it seems quite possible that more adequate material might show that the Borneo form is closest to the mainland *M. moschata*.

Van Gelder (1959), in his revision of the genus *Spilogale*, concludes that *S. pygmaea* is specifically distinct from *S. putorius*. *S. pygmaea* occurs along the Pacific coast of Mexico from Rosario to the isthmus of Tehuantepec; its distinguishing features are its small size and certain details of its colour pattern. In both, however, it closely approaches its neighbour, the subspecies *tropicalis* of *S. putorius*. *S. pygmaea* was based originally on only four specimens, to which a fifth has subsequently been added (Greer and Greer, 1970) and its validity seems very questionable: I have therefore not listed it as a separate species.

Cabrera (1957) lists five South American species of *Conepatus*, to which Hall and Kelson (1959) added a further two from Central America and southern North America. Kipp (1965), working on material from the southern half of South America, reached the conclusion that in this area only two groups were separable: an eastern lowland and a western Andean form. Although not fully convinced that even these were specifically distinct, she provisionally placed them in *C. humboldti* and *C. chinga* respectively, thus sinking two of the species recognised by Cabrera – *C. castaneus* D'Orbigny and Gervais, 1847 and *C. rex* Thomas, 1898. Material covering the more northerly part of the range was not available and for the meantime therefore *C. semistriatus* must be retained as the designation of the animals from northern South America but I have followed Morris (1965) in uniting the North to Central American forms in the single species, *C. leuconotus*.

The status of *Aonyx philippsi* and *A. microdon* is not fully clear. Coetzee (1967) regards them both as subspecies of *A. congica* but Harris (1968) classifies both as separate species.

B. Superfamily Feloidea

Ethmo-turbinals long, extending to the anterior part of the nasal chamber and overlying the relatively small maxillo-turbinals.

Auditory bulla formed from both ecto- and entotympanic bones and divided by a partition where the two meet.

(*v*) *Family* Viverridae

Limbs relatively short, semi-plantigrade or digitigrade; claws typically non-retractile but there are some exceptions to this.

Head rather long and low, with jaws not shortened; carnassials not highly specialised and upper molars triangular in outline.

Scent glands highly developed but the secretion is rarely nauseous and only in a few species is it used defensively.

Spotted, blotched or striped colour patterning is common.

(*a*) *Subfamily* Viverrinae

1. *Viverra zibetha* Linnaeus, 1758; large Indian civet: China from Szechuan southwards; Burma westwards to Nepal; Malayan and Indo-Chinese peninsulas.

2. *Viverra megaspila* Blyth, 1862; large-spotted civet: Indo-Chinese and Malayan peninsulas; Burma and eastern India as far south as Madras.

3. *Viverra tangalunga* Gray, 1832; oriental civet: Sumatra, Borneo, Celebes, Philippines.

*4. *Civettictis civetta* (Schreber, 1778); African civet: Africa, from Senegal eastwards to Somalia and thence southwards to northern Zululand, northern Botswana and northern South West Africa.

5. *Viverricula indica* (Desmarest, 1817); small Indian civet: southern China, Burma, India, Malayan and Indo-Chinese peninsulas; Sumatra, Java, Bali, Hainan, Taiwan. Introduced into Zanzibar and Pemba Islands, off the coast of Tanzania.

*6. *Genetta genetta* (Linnaeus, 1758); European or small-spotted genet: Africa, fringing the Sahara from Libya through Algeria and Morocco to Senegal, eastwards to Ethiopia, thence southwards to Cape Province and east to Arabia and Palestine; Balearic Islands; Spain, France; has been recorded from Switzerland, Germany and Belgium.

7. *Genetta tigrina* (Schreber, 1778); large-spotted genet: widespread over most of Africa south of the Sahara.

8. *Genetta servalina* Pucheran, 1855; servaline genet: Congo basin, east into Kenya and north through Gabon, Cameroon and into southern Nigeria.

9. *Genetta victoriae* Thomas, 1901; giant forest genet: dense forests of the Congo basin, north of the river.

10. *Genetta pardina* I.Geoffroy, 1832; pardine or forest genet: west Africa from Gambia to Cameroon.

11. *Genetta angolense* Bocage, 1882; Angolan genet: northern and northeastern Angola, southern Congo, northwest Zambia.

12. *Genetta abyssinica* (Rüppell, 1836); Abyssinian genet: highlands of Ethiopia.

13. *Genetta villiersi* (Dekeyser, 1949); false genet: Guinea savanna, from Senegal eastwards to the area south of Lake Chad.

14. *Genetta johnstoni* Pocock, 1907; Johnston's genet: known only from one locality in Liberia.

15. *Osbornictis piscivora* Allen, 1919; fishing genet: known only from a small area in eastern parts of Kisangani and Kibale-Ituri districts of Congo.

*16. *Poiana richardsoni* (Thomas, 1842); oyan: island of Fernando Po; dense forest regions of Sierra Leone and Ivory coast and of Gabon, Cameroon and northern Congo.

17. *Prionodon linsang* (Hardwicke, 1821); banded linsang: Malayan peninsula; Sumatra, Java, Borneo.

18. *Prionodon pardicolor* Hodgson, 1842; spotted linsang: Nepal, Assam, northern Burma, Indo-Chinese peninsula.

(*b*) *Subfamily* Paradoxurinae

19. *Nandinia binotata* (Gray, 1830); two-spotted palm civet: widespread in the forests of west and central Africa, from Guinea to Sudan in the north and from Angola to Mozambique in the south; Fernando Po.

20. *Arctogalidia trivirgata* (Gray, 1832); small-toothed palm civet: Assam, Burma, Malayan and Indo-Chinese peninsulas; Sumatra, Java, Borneo.

21. *Paradoxurus hermaphroditus* (Pallas, 1777); common palm civet or toddy cat: greater parts of southeast Asia, Nepal, Assam, Burma, southern China, Malayan and Indo-Chinese peninsulas; Sumatra, Java and Sunda Islands eastwards to Timor; Borneo, Celebes, Ceram and Kei Islands, Philippines.

22. *Paradoxurus zeylonensis* (Pallas, 1777); golden palm civet: Ceylon.

23. *Paradoxurus jerdoni* Blanford, 1885; Jerdon's palm civet: southern India.

24. *Paguma larvata* (Hamilton-Smith, 1827); masked palm civet: China, Assam, Kashmir, Burma; Andaman Islands; Malayan and Indo-Chinese peninsulas; Sumatra, Borneo; Hainan, Taiwan. Introduced into Japan.

25. *Macrogalidia musschenbroeki* (Schlegel, 1879); brown palm civet: Celebes.

26. *Arctictis binturong* (Raffles, 1821); binturong: Burma, Malayan and Indo-Chinese peninsulas; Sumatra, Java, Borneo, Palawan Island.

(*c*) *Subfamily* Hemigalinae

27. *Hemigalus derbyanus* (Gray, 1837); banded palm civet: Malayan peninsula; Sumatra, Borneo.
28. *Hemigalus hosei* Thomas, 1912; Hose's palm civet: Borneo.
29. *Chrotogale owstoni* Thomas, 1912; Owston's banded civet: Indo-Chinese peninsula.
30. *Cynogale bennetti* Gray, 1837; otter civet: Malayan and Indo-Chinese peninsulas; Sumatra, Borneo.
31. *Fossa fossa* Gray, 1864; fanaloka: Madagascar.
32. *Eupleres goudoti* Doyère, 1835: Madagascar.
33. *Eupleres major* Lavauden, 1929: Madagascar.

(*d*) *Subfamily* Galidiinae

34. *Galidia elegans* I.Geoffroy, 1839; ring-tailed mongoose: Madagascar.
35. *Galidictis striata* I.Geoffroy, 1839; broad-striped mongoose: Madagascar.
36. *Galidictis fasciata* (Gmelin, 1790): Madagascar.
37. *Galidictis ornata* Pocock, 1915: Madagascar.
38. *Mungoictis lineata* Pocock, 1915; narrow-striped mongoose: Madagascar.
39. *Mungoictis substriata* Pocock 1915: Madagascar.
40. *Salanoia olivacea* (I.Geoffroy, 1839): Madagascar.
41. *Salanoia unicolor* (I.Geoffroy, 1839); brown-tailed mongoose: Madagascar.

(*e*) *Subfamily* Herpestinae

42. *Herpestes ichneumon* (Linnaeus, 1758); Egyptian mongoose: throughout most of Africa except for the Congo and west African forest regions and the Sahara; Spain; Palestine. Introduced into Madagascar.
43. *Herpestes naso* de Winton, 1901; long-nosed mongoose: from southeast Nigeria southwards to the Congo river and east to Semliki valley.
44. *Herpestes pulverulentus* Wagner, 1839; Cape grey mongoose: southern Angola, South West Africa, South Africa south of the Vaal River; Lesotho.
45. *Herpestes sanguineus* Rüppell, 1835; slender or red mongoose semi-arid and open savanna regions of Africa south of the Sahara.
46. *Herpestes auropunctatus* (Hodgson, 1836); small Indian mongoose: northern Arabia eastwards through Near East to Afghanistan, Nepal, Assam, Burma, Malayan peninsula and Thailand; southwards

through Pakistan into northern India; Hainan. Introduced into West Indies, Hawaii and Mafia Island off the coast of Tanzania.

*47. *Herpestes javanicus* (Geoffroy, 1818); Java mongoose: Malayan and Indo-Chinese peninsulas; Java.

48. *Herpestes edwardsi* (Geoffroy, 1818); Indian grey mongoose: from Arabia eastwards to Nepal and Assam and southwards through Pakistan and India; Ceylon. Introduced to Malaya.

49. *Herpestes smithi* Gray, 1837; ruddy mongoose: southern India; Ceylon.

50. *Herpestes fuscus* Waterhouse, 1838; Indian brown mongoose: southern India; Ceylon.

51. *Herpestes vitticollis* Bennett, 1835; striped-necked mongoose: southern India; Ceylon.

*52. *Herpestes urva* (Hodgson, 1836); crab-eating mongoose: southern China, Nepal, Assam, Burma, Indo-Chinese peninsula; Taiwan; Hainan; Sumatra, Borneo; Philippines.

53. *Herpestes brachyurus* Gray, 1836; short-tailed mongoose: Malayan peninsula; Sumatra, Java, Borneo, Palawan Island.

54. *Mungos mungo* (Gmelin, 1788); banded mongoose: widespread in semi-arid and savanna regions of Africa south of the Sahara, apart from southern South West Africa and the western Cape Province of South Africa.

55. *Mungos gambianus* (Ogilby, 1835); Gambian mongoose: savanna regions of west Africa from Gambia to Nigeria.

56. *Crossarchus obscurus* F.Cuvier, 1825; kusimanse or long-nosed mongoose: forest regions of west Africa from Sierra Leone to Cameroon.

57. *Crossarchus ansorgei* Thomas, 1910; Angolan mongoose: Angola northwards, possibly as far as Congo river.

58. *Crossarchus alexandri* Thomas and Wroughton, 1907; Congo kusimanse: forest regions of the Congo.

59. *Liberiictis kuhni* Hayman, 1958: known only from one locality in Liberia.

60. *Helogale parvula* Sundevall, 1846; dwarf mongoose: Ethiopia southwards through east Africa as far as northern Zululand.

61. *Helogale hirtula* Thomas, 1904: southern Ethiopia southwards into northern Kenya.

62. *Dologale dybowskii* (Pousargues, 1894); Pousargues's mongoose: open savanna regions from northeast Congo through western Uganda to southern Sudan.

63. *Bdeogale crassicaudata* Peters, 1852; bushy-tailed mongoose: from central Mozambique and eastern Zambia northwards into eastern Kenya.

64. *Bdeogale nigripes* Pucheran, 1855; black-legged mongoose: eastern Nigeria and southwards to northern Angola.
65. *Bdeogale jacksoni* (Thomas, 1894); Jackson's mongoose: central Kenya and southeastern Uganda.
66. *Rhyncogale melleri* (Gray, 1865); Meller's mongoose: central East Africa, Zambia, Malawi, Mozambique.
67. *Ichneumia albicauda* (Cuvier, 1829); white-tailed mongoose: widespread throughout Africa south of the Sahara from Senegal to Sudan and southwards to northern South West Africa and the eastern Cape Province of South Africa.
68. *Atilax paludinosus* (Cuvier, 1831); marsh mongoose: widespread in the less arid regions of Africa south of the Sahara from Sierra Leone to Ethiopia and southwards to Cape Province in South Africa.
69. *Cynictis penicillata* (Cuvier, 1829); yellow mongoose: South Africa, Botswana, South West Africa and southern Angola.
70. *Paracynictis selousi* (de Winton, 1896); Selous's mongoose: central Africa, in a belt from southern Angola and northern South West Africa in the west through northern Botswana, southern Zambia, Rhodesia and northern Transvaal to Mozambique.
71. *Suricata suricatta* (Schreber, 1777); suricate or stokstert meerkat: South Africa, South West Africa and Botswana.

(*f*) *Subfamily* Cryptoproctinae
72. *Cryptoprocta ferox* Bennet, 1833; fossa: Madagascar.

I have retained Pocock's genus *Civettictis*, although many modern authors include the African civet in *Viverra*. Pocock (1915e) based the genus *Civettictis* on differences in the feet, which are distinctly more primitive in the African species and on the detailed structure of the perfume gland, which is somewhat simpler. He also notes that there are differences in the shape of the rhinarium. Recently Petter (1969) has pointed out that the dentition of the African civet differs very markedly from that of *Viverra zibetha*, being much more modified for crushing. She expresses the opinion that the dental differences alone would justify placing the African species in a separate genus but unfortunately she does not include a comparison with *V. megaspila* and *V. tangalunga*.

The taxonomy of the genets is at present very confused. The species listed are those given by Coetzee (1967) but the genus is in need of thorough revision and it is likely that when this is done a number of changes will have to be made.

Poiana is normally referred to as the 'African linsang', but since it is nowadays considered to be more closely related to genets than to the

Asiatic linsangs this name is no longer appropriate. I consider that its use should be discontinued and the native name 'oyan', quoted by Bates (1905), used instead.

In 1919 Allen described a mongoose from the Congo which he named *Xenogale microdon*. Hayman (in Sanderson, 1940) showed that this was the same species as de Winton's *Herpestes naso* from south-eastern Nigeria. While Hayman's synonym has not been disputed, the correctness of the original classification as a species of *Herpestes* is more open to question. Orts (1970) points out that the differences from *Herpestes ichneumon* are considerable and revives the genus *Xenogale* to include both the Nigerian and the Congo animals. He does not, however, make any comparison with other species of *Herpestes* and, as Hayman himself pointed out, when these are taken into account generic distinction becomes difficult. Clearly a study of chromosome morphology might go far to settle the issue, since all species of *Herpestes* so far examined have the curious arrangement of a translocated Y chromosome in the male (see Chapter 9). In the meantime, I have accepted Hayman's opinion.

Herpestes semitorquatus, from Sumatra and Borneo, is listed as a separate species by Morris (1965), but Ellerman and Morrison-Scott (1951) regard it as very possibly only subspecifically distinct from *H. urva*. The same is possibly true of *H. palustris*, a swamp-dwelling, fish-eating form, described by Ghose (1965) from the neighbourhood of Calcutta. Ghose distinguishes *H. palustris* from *H. auropunctatus* but makes no comparison either with *H. urva* or with *H. vitticollis*, the two species which it resembles in its feeding habits and also in its ability to emit an obnoxious odour from its anal glands. Pending further clarification of their status I have not listed either of these forms as distinct species.

The ease with which mongooses become tame and accept human companionship, together with their value in the control of rodents and snakes, led to their being kept as house pets both in ancient Egypt and in India. In Egypt, *Herpestes ichneumon* ultimately attained the status of a sacred animal but no true domestication appears to have occurred. Hinton and Dunn (1967) give an account of what is known of the mongoose in ancient Egypt and also recount some of the Indian folk tales in which a mongoose figures – including the original version of the Welsh tale of the dog Gelert, in which the protagonists are mongoose and cobra instead of dog and wolf.

(*vi*) *Family* Hyaenidae

Medium-sized to large animals, with relatively large heads and with forequarters heavier than hindquarters; legs long; digitigrade, with four digits on fore and hind limbs.

Carnassials highly sectorial and post-carnassial teeth absent or vestigial; premolars large with heavy conical main cusps.

(*Proteles* is aberrant: there are five toes on the fore foot; the dentition is reduced and the head and shoulders are not large relative to the hindquarters.)

(*a*) *Subfamily* Hyaeninae

1. *Hyaena hyaena* (Linnaeus, 1758); striped hyaena: India, southwestern USSR and throughout the Near East into Asia Minor; Arabia and north Africa as far west as Senegal and extending southwards through Egypt as far as Tanzania.
2. *Hyaena brunnea* Thunberg, 1820; brown hyaena: Africa from southern Mozambique and Rhodesia southwards through South West Africa, Botswana and South Africa.
3. *Crocuta crocuta* (Erxleben, 1777); spotted hyaena: formerly widespread throughout Africa south of the Sahara but now killed out in the southern parts of South Africa.

(*b*) *Subfamily* Protelinae

4. *Proteles cristatus* (Sparrman, 1783); aardwolf: arid or semi-arid regions throughout southern Africa, from southern Angola southwards and in the east extending northwards through southern Zambia, Tanzania and Kenya into Somalia, Ethiopia and Sudan.

(*vii*) *Family* Felidae

Digitigrade, typically with sharp retractile claws generally protected by lobes of skin, forming sheaths.

Braincase large, jaws short; only a single small upper molar; carnassials highly sectorial; canines with a fine longitudinal groove.

Penis short and baculum small or absent; apart from the anal sacs, no specialised cutaneous glands in the ano-genital area.

Ears with bursa well developed; inter-ramal vibrissae absent; colour patterns of spots, blotches or stripes very common.

1. *Felis silvestris* Schreber, 1777; European wild cat: Europe, including Britain; southwest USSR; Asia Minor.

*2. *Felis libyca* Forster, 1780; African wild cat: Arabia and the Near East, southern USSR and eastwards into western China, Afghanistan, Pakistan, northern and central India; Sardinia, Corsica and Majorca; north Africa from Morocco to Egypt and throughout the savanna regions of Africa south of the Sahara.

3. *Felis chaus* Güldenstaedt, 1776; jungle cat: Egypt, Asia Minor and the Near East and eastwards throughout Afghanistan, Nepal, Kashmir, Assam and Burma to southern China and the Indo-Chinese peninsula; southwards through Pakistan and India; Ceylon.

4. *Felis bieti* Milne-Edwards, 1892; Chinese desert cat: from southern Mongolia through Kansu to Szechuan in central China.

*5. *Felis margarita* Loche, 1858; sand cat: Algeria, western Libya and northern Niger; southwest Sudan; Sinai and Arabia and desert regions of Turkmenia and Uzbekistan.

6. *Felis nigripes* Burchell, 1824; black-footed cat: South Africa, central and southern Botswana and westwards into eastern South West Africa.

7. *Leptailurus serval* (Schreber, 1776); serval cat: Algeria and widespread through Africa south of the Sahara.

8. *Prionailurus bengalensis* (Kerr, 1792); Bengal cat or leopard cat: throughout southern and eastern Asia: from India northwards through Nepal and Tibet to China, Manchuria, eastern Siberia and Korea and eastwards through Assam to Burma and the Malayan and Indo-Chinese peninsulas; Sumatra, Java, Borneo; Philippines; Hainan, Taiwan and Japan.

9. *Prionailurus rubiginosa* (Geoffroy, 1831); rusty-spotted cat: southern India and Ceylon.

10. *Prionailurus viverrinus* (Bennett, 1833); fishing cat: Ceylon, India, Indo-Chinese peninsula; Sumatra and Java.

*11. *Mayailurus iriomotensis* Imaizumi, 1967: known only from Iriomote Island, off the east coast of Taiwan.

12. *Ictailurus planiceps* (Vigors and Horsfield, 1827); flat-headed cat: Thailand, Malayan peninsula; Sumatra, Borneo.

13. *Otocolobus manul* (Pallas, 1776); Pallas's cat: Iran eastwards throughout southern USSR, Afghanistan, Kashmir and Tibet to western China.

14. *Pardofelis marmorata* (Martin, 1837); marbled cat: Nepal, Assam, northern Burma, Malayan and Indo-Chinese peninsulas; Sumatra, Borneo.

15. *Pardofelis badia* Gray, 1874; bay cat: Borneo.

16. *Profelis temmincki* (Vigors and Horsfield, 1826); Asiatic golden cat: central and southern China, Tibet, Nepal, Assam, Burma, Malayan and Indo-Chinese peninsulas; Sumatra.

17. *Profelis aurata* (Temminck, 1827); African golden cat: West Africa, from Sierra Leone eastwards to Cameroon and forests of the Congo, Uganda and western Kenya.

*18. *Caracal caracal* (Schreber, 1776); caracal: north Africa from Morocco eastwards to Egypt, thence southwards throughout Africa, excluding the forest areas of west and central Africa and eastwards through Arabia and the Near East to Afghanistan, Pakistan, northern and central India.

19. *Puma concolor* (Linnaeus, 1771); puma or mountain lion: mountainous regions of western America from British Columbia to southern Argentina.

20. *Leopardus pardalis* (Linnaeus, 1758); ocelot: Central and South America.

*21. *Leopardus tigrinus* (Schreber, 1777); tiger ocelot: Central and South America east of the Andes to the southern limits of Brazil.

22. *Leopardus wiedi* (Schinz, 1821); tree ocelot: Central and South America.

*23. *Leopardus geoffroyi* (D'Orbigny and Gervais, 1843); Geoffroy's ocelot: South America from southeastern Bolivia south through Paraguay, Uruguay and most of Argentina.

24. *Oncifelis guigna* (Molina, 1782); kodkod or hũina: southern parts of South America.

25. *Lynchailurus colocolo* (Molina, 1782); pampas cat: southern parts of South America.

26. *Oreailurus jacobita* Cornalia, 1865; mountain cat: high Andean regions of northeastern Argentina and Chile.

27. *Herpailurus yagouaroundi* (Desmarest, 1820); jaguarondi: from Texas through Central and South America to northern Argentina.

28. *Lynx lynx* (Linnaeus, 1758); lynx: Holarctic region from Scandinavia eastwards through USSR, Tibet, Mongolia, Manchuria, North America. Formerly widespread in Europe but now restricted to residual populations in the less inhabited areas.

*29. *Lynx pardina* (Temminck, 1824); Spanish lynx: restricted to the wilder parts of the Iberian peninsula.

30. *Lynx rufus* (Güldenstaedt, 1776); bobcat: southern Canada and throughout the United States southwards to Mexico.

31. *Panthera leo* (Linnaeus, 1758); lion: southern fringe of Sahara from west Africa to Sudan and Somalia, thence southwards through most of Africa, excluding the Congo forest regions, as far south as northern Natal. Formerly widespread in Africa to the southern limits of the continent as well as in eastern Europe, the Near East and probably India; present distribution in India restricted to the Gir peninsula.

32. *Panthera tigris* (Linnaeus, 1758); tiger: southern USSR, Manchuria, China, Nepal, Assam, Burma and throughout most of India, Malayan and Indo-Chinese peninsulas; Sumatra, Java.

33. *Panthera pardus* (Linnaeus, 1758); leopard: Senegal eastwards to Sudan, Ethiopia and Somalia and thence southwards through most of Africa; north Africa from Morocco to Egypt; Arabia, Asia minor and the Near East and thence eastwards through southern USSR,

Manchuria, China, Tibet and southwards through India; Ceylon; Burma, Malayan and Indo-Chinese peninsulas; Java.

34. *Panthera onca* (Linnaeus, 1758); jaguar: southwest United States, southwards through Central America into South America, the southern limit being approximately the Rio Negro in Argentina.

35. *Panthera uncia* (Schreber, 1776); snow leopard or ounce: Kashmir, Tibet and mountainous regions of Turkestan northwards to the Altai mountains.

36. *Neofelis nebulosa* (Griffith, 1821); clouded leopard: southern China, Nepal, Burma, Malayan and Indo-Chinese peninsulas; Sumatra, Borneo; Taiwan.

37. *Acinonyx jubatus* (Schreber, 1776); cheetah: grassland, savanna and open woodland areas over most of Africa south of the Sahara. Formerly common but now rare except under protection in National Parks. In India, where the cheetah was formerly widespread, the last was killed in 1952: there are a few survivors in southwest Iran and possibly in Turkmenia and there are isolated records from localities in the Near East and in north Africa.

The nomenclature of the Felidae presents a major problem and almost the only point on which there is universal agreement is that the cheetah requires a genus of its own: it is also commonly felt that the clouded leopard is sufficiently isolated to be accorded the same honour. Many workers merely divide all other cats into two genera: *Panthera*, for the large species in which the median part of the hyoid suspensorium is not ossified, and *Felis*, for the smaller ones in which the hyoid is ossified in the normal manner (Pocock, 1916c). This, however, is unsatisfactory, for the former possibly and the latter certainly is a heterogeneous assemblage of species by no means equally closely related to each other. The attempt to circumvent the difficulty by a welter of subgenera does not appeal to me as a satisfactory solution either.

As Leyhausen (1969) points out, a satisfactory classification will be arrived at only when each and every cat species has been thoroughly studied. He estimates that his own work in this field may take another ten years to complete and it is therefore necessary to adopt some system of nomenclature in the meantime, even if it is admittedly provisional. Leyhausen (1969) suggests that amongst the large cats, the tiger and the snow leopard may not be very closely related to the lion, leopard and jaguar but for the present there is no alternative to retaining them all in *Panthera*. Amongst the smaller cats, the genus *Felis* as defined by Pocock (1951) constitutes a group of closely related species and I have adopted his usage of *Felis*. Amongst the remaining small cats a few groups of closely related species may be united in genera of their own. Of these, the clearest are *Prionailurus* for the Bengal, rusty-spotted and fishing cats; *Profelis* for the two golden cats and *Leopardus* for the ocelots and Geoffroy's 'cat'. Quite possibly other species should be included in each of these groups. *Mayailurus iriomotensis*,[1] for instance, is described by Imaizumi (1967) as very similar to *Prionailurus bengalensis* and probably should be included in *Prionailurus*. However, pending a clearer understanding of their status I have preferred to leave the species other than those just mentioned in separate genera.

The spelling adopted for *Felis libyca* is that now generally accepted. Forster originally wrote '*lybica*', which was retained by Pocock and by a number of other workers following him. As Ellerman and Morrison-Scott (1951) point out, Forster's spelling was surely not deliberate but simply a mistake for the correct *libyca*, which is now generally used. Some workers regard *F. libyca* as only subspecifically distinct from *F. silvestris*.

The sand cats are highly adapted to desert conditions and their discontinuous distribution reflects this fact. Their range runs across the north African Saharan region and into Arabia and they also occur further to the east in the deserts of Turkmenia and Uzbekistan. Pocock (1938) originally regarded the animals from the latter region as belonging to a separate species: this he named *Felis thinobius* and interpreted its similarities to *F. margarita* as convergent adaptations to desert life. In his later (1951) revision of the genus *Felis*, however, he accepted the opinion of Heptner and Dementiev (1937) that the eastern sand cats are only subspecifically distinct. Clearly a detailed study, including observations on living animals from both parts of the range, is necessary to decide the issue and, although Pocock's original view may well prove to be correct, I have not listed *F. thinobius* as a separate species.

The Spanish lynx is often regarded as only subspecifically distinct from *Lynx lynx* despite its small size and very well marked spotting. Kurtén (1968), however, points out that the palaeontological evidence strongly supports the view that the two are distinct species. In the Pleistocene the range of the Spanish lynx extended farther north than at present and overlapped with that of *L. lynx*. The two nevertheless remained distinct; intermediates are not found and interbreeding apparently did not occur.

The caracal is often included in the genus *Lynx* but I have never been able to see any striking resemblance between the two and have left the caracal in a genus of its own.

Leyhausen (1963) points out that the names ocelot, margay and

[1] An account by an anonymous author of this recently discovered species is given in *Animals* 1968, **10**, 501–3.

tiger-cat have been used rather indiscriminately for members of the genus *Leopardus* and proposes the adoption of the following common names: ocelot for *L. pardalis*, tree ocelot for *L. wiedi* and ocelot cat for *L. tigrinus*. While agreeing with the first two, I find 'ocelot cat' confusing, especially in view of the erratic way in which the addition of 'cat' to a common name is made in other cases, for example serval or serval cat and civet or civet cat. I therefore prefer to call *L. tigrinus* the tiger ocelot and, if Geoffroy's cat is to be included in *Leopardus*, it should become Geoffroy's ocelot.

The domestic cat, *Felis catus* Linnaeus, 1758, is most closely related to *F. libyca* and *F. silvestris*. The earliest known domestic cats are those of ancient Egypt, where they appear as sacred animals from about 1600 BC onwards. Zeuner (1963) traces what is known of the later spread of the domestic cat to other regions, including Europe. In Egypt their religious significance resulted in cats being mummified and buried in vast numbers. The possible scientific value of mummified cats was not at first realised, and it is a little horrifying to learn that a single skull in the British Museum collection represents the sole survivor of a consignment of nineteen tons, shipped to England around the beginning of the century, to be used as fertiliser. However, 190 skulls from a subsequent excavation at Gizeh, dating from 600–200 BC, have been studied by Morrison-Scott (1952). He found that of these, all except three agree with *Felis libyca*. The three exceptions he identified as *F. chaus*: their inclusion would seem to be incidental and can hardly be taken to suggest that this species was also domesticated.

Egyptian paintings usually portray a predominantly ginger coloured animal with dark markings on ears and fore limbs and a long dark-ringed tail. The typical colouring of the modern European domestic cat suggests that there may have been some subsequent crossing with the European wild cat.

Envoi

During the course of human development, man's relations with the carnivores have passed through several stages. At first they were simple; early man and the larger carnivores were both predators and their relations must have been much the same as those amongst the larger carnivores of today: they competed for food and they probably stole each other's kills and killed each other or each other's young as and when opportunity offered. They coexisted, nevertheless, because their interrelations were not basically different from those between other competing species.

The development of increasingly effective weapons and hunting methods changed the picture and gradually made man the dominant form: the growth of agriculture and the raising of domestic stock gave him an added reason for attacking and attempting to eliminate the competitors that now appeared in a new guise, as actual or suspected stock thieves. Gradually the period of ever-accelerating destruction got under way: a senseless, unplanned, uncomprehending attempt to eliminate some species and to exploit others, without preliminary cost-benefit analysis or evaluation of the rate of cropping the stocks could be expected to stand. Present apparent gain, without thought for the future or full analysis of the situation, was the order of the day and there is no need to rehearse the sorry story of the destruction of wildlife in general, or Carnivora in particular, on one continent after another.

With the present century a new attitude came gradually into existence: the concept of conservation began to take shape. The idea of conservation is based on two simple fundamental principles. Firstly, that there are limits to the energy and nutrients available for the maintenance of living organisms and secondly that during the long course of evolution, interrelated systems of extraordinary complexity have been slowly built up, whereby these resources are continually cycled through the bodies of the living organisms. We have come to realise that, if you remove one member from a balanced working system of such complexity, there may be far-reaching effects which we failed to foresee and which may be deleterious to the system as a whole: we have come to realise too that predators are an essential part of such a system and that if we remove them, we may create new problems. A coyote, for instance, may now and then kill a calf or a lamb but – quite apart from the question of how many of these were weaklings doomed to die in any case – he also acts as a rodent control: one that costs nothing, requires no maintenance and ultimately returns to the environment everything that he borrowed from it. Remove him and the rodents multiply; the grazing deteriorates and the range can carry not more but fewer stock than it did when the coyote kept the balance. The story of pasture deterioration in the Valle de la Trinidad, following coyote destruction, is a classical example (Huey, 1937). Another manifestation of the importance of predators in rodent control has recently come to light in South America, where uncontrolled destruction of the smaller felids for the fur trade has been followed by epidemics of rodent-transmitted diseases in the human population.

Conservation, of course, has its aesthetic as well as its practical side. Man is not only a part of nature; he also has the ability to stand back and contemplate it, to be at once apart and involved. There is something inherently beautiful to the human eye in the precision and ease of a carnivore's movements, and the cry of a jackal or the howl of a wolf wakes an answering chord in the human imagination. To do away with the wilds is to deprive ourselves of a source of enjoyment which, once gone, can never be replaced.

I would have liked to have closed this book by expanding these themes and by noting hopefully the various moves that have been made and are being made to implement conservation and to limit destruction. Only a few years ago this would have seemed right and natural but today it seems futile, if not cynical; for it is too late for expositions of the advantages of conservation to have any relevance. In the face of exponentially expanding human population, coupled with ever-increasing ability for the individual to destroy his habitat, what point is there in pleading for conservation? To take just one example: legislation to protect the sea otter has resulted in re-establishing this species after its near extinction by man. Protection started in 1911 and since then there have been continuing reports of increase in residual populations and reappearance of otters in areas where they had been exterminated (Bolin, 1938; Lensink, 1960; Pedersen and Stout, 1963; Jones, 1965). The world population of sea otters in 1965 was estimated as about 35,000–36,000 (Barabash-Nikiforov, 1969; Kenyon, 1969) – but what chance has the sea otter against oil pollution? Once the fur is soiled by oil, it loses its insulating air layer; the animal chills and dies and no conservation legislation can save it.

If present trends continue unchecked the tale will be the same everywhere - protection powerless against pollution and the habitat available for other species continually eroded by ever-growing human utilisation.

Whether we like it or not, our planet is no more than a manned space vehicle and all its resources are finite. There are therefore only two alternatives before us:
(i) We may first limit and then reduce our own population by deliberate voluntary control. We will then be able to create a world in which a restricted number of people can live in very considerable freedom, in a rich and varied environment, along with a vast number of other species of organisms, both plant and animal: a world in which there will be room for the wonderful and beautiful creatures which I have tried to describe in this book.
(ii) We may continue to expand, utilising our technological skill to organise our environment, until we reach the ultimate thermodynamic limit set by the rate at which the metabolic heat we generate can be dissipated (Fremlin, 1964). We will then live in small groups, imprisoned, without freedom to move from place to place, living on the nutrient broth produced by synthetic organisms from recycling our own wastes; permitted to reproduce one baby each time someone dies and is fed back into the machine to be recycled to us in our broth. We and our synthetic yeasts or bacteria will then be the only living organisms left on our space ship. Fremlin has worked out in some detail the way in which this nightmare world could be operated.

On the other hand, while continuing to expand, we may leave it to the natural disasters of disease, famine and mutual slaughter to decide our problems for us. Most likely, the two processes of technological control and *laissez-faire* will operate in parallel in different parts of the world, with one ultimately taking the lead and swamping the other. Even if this does not lead to the Fremlin world, our technological skill is such that true major disasters will probably not occur until after we have destroyed much of our habitat and greatly reduced the carrying capacity of our space vehicle, so that ultimate stabilisation may well be at a lower population density than would be required under possibility (i). The earth we have by then left ourselves to inherit will have neither the beauty nor the potentialities it has today, and who can predict how many other species will survive to share it with us?

The choice is before us and must be made. Willy-nilly, there is one simple way in which we not only may, but are obliged to cast our vote. To refrain from having more than two children is to vote for alternative (i): to have more offspring is to cast a vote in favour of the second alternative, with its two possible ultimate outcomes, as the future we are preparing for them.

I count myself among the fortunate: I have lived to see the cheetah and the leopard, to camp where lions roared, to share a house with mongooses and a laboratory with civet cats - but it would be good to feel confident that my grandchildren could do the same, and not to fear that they will say: 'Just think - all those lovely animals were actually alive in Granny's day: it must have been wonderful.' It was.

References

ANON. (1964) Ratel kills python. *Afr. wild Life* **18**, 37

ANON. (1966) Small-eared cat. *Korean Nature* **4**, 8

ABLES, E.D. (1965) An exceptional fox movement. *J. Mammal.* **46**, 102

—— (1969) Home-range studies of red foxes (*Vulpes vulpes*). *Ibid.* **50**, 108–20

ADAMSON, G.A.G. (1964) Observations on lions in Serengeti National Park, Tanganyika. *East Afr. wildl. J.* **2**, 160–1

—— (1968) *Bwana Game* (Collins & Harvill Press: London)

ADAMSON, J. (1960) *Born free* (Collins & Harvill Press: London)

—— (1969) *The spotted sphinx* (Collins & Harvill Press: London)

ADLER, N.T. & ZOLOTH, S.R. (1970) Copulatory behavior can inhibit pregnancy in female rats. *Science* **168**, 1480–2

ADRIAN, E.D. (1955) Synchronised activity in the vomero-nasal nerves with a note on the function of the organ of Jacobsen. *Pflüg. Arch.* **260**, 188–92

ALBIGNAC, R. (1969a) Naissance et élevage en captivité de jeunes *Cryptoprocta ferox*, viverridés malgaches. *Mammalia* **33**, 93–7

—— (1969b) Notes éthologiques sur quelques carnivores malgaches: le *Galidia elegans* I. Geoffroy. *Terre et Vie* **23**, 202–15

—— (1970) Notes éthologiques sur quelques carnivores malgaches: le *Fossa fossa* (Schreber). Ibid. **24**, 383–93

ALBONE, E.S. & FOX, M.W. (1971) Anal gland secretion of the red fox. *Nature Lond.* **233**, 569–70

ALDOUS, S.E. & MANWEILER, J. (1942) The winter food habits of the short-tailed weasel in northern Minnesota. *J. Mammal.* **23**, 250–5

ALEXANDER, A.J. & EWER, R.F. (1959) Observations on the biology and behaviour of the smaller African polecat. *Afr. wild Life* **13**, 313–20

ALIEV, F.F. (1966) Enemies and competitors of the nutria in USSR. *J. Mammal.* **47**, 353–5

ALIKHAN, I. (1938) Method adopted by leopards in hunting monkeys. *J. Bombay nat. Hist. Soc.* **40**, 555–7

ALLEN, D.L. (1938) Notes on the killing technique of the New York weasel. *J. Mammal.* **19**, 225–9

—— (1939) Winter habits of Michigan skunks. *J. Wildl. Mgmt.* **3**, 212–28

—— & SHAPTON, W.W. (1942) An ecological study of winter dens, with special reference to the eastern skunk. *Ecology* **23**, 59–68

ALLEN, E.R. & NEILL, W.T. (1956) Albinistic sibling raccoons from Florida. *J. Mammal.* **37**, 120

ALLEN, J. A. (1919) Preliminary notes on African Carnivora. *Ibid.* **1**, 23–32
—— (1924) Carnivora collected by the American Museum Congo expedition. *Bull. Amer. Mus. nat. Hist.* **47**, 73–281
ANSELL, W.F.H. (1960a) The African striped weasel, *Poecilogale albinucha* (Gray). *Proc. zool. Soc. Lond.* **134**, 59–64
—— (1960b) The breeding of some larger mammals in Northern Rhodesia. *Ibid.* **134**, 251–74
—— (1960c) *Mammals of Northern Rhodesia* (Govt. Printer: Lusaka)
—— (1964) Addenda and corrigenda to *Mammals of Northern Rhodesia. Puku* **2**, 14–52
—— (1965) Addenda and corrigenda to *Mammals of Northern Rhodesia*, No. 2. *Puku* **3**, 1–14
ANTHONY, R.L. & ILIESCO, G.M. (1926) Etude sur les cavités nasales chez des carnassiers. *Proc. zool. Soc. Lond.* **1926** (2), 989–1015
ANTONIUS, O. (1943) Nachtrag zu 'Symbolhandlungen und Verwandtes bei Säugetieren.' *Z. Tierpsychol.* **5**, 38–42
APPELBERG, B. (1958) Species differences in the taste qualities mediated through the glossopharyngeal nerve. *Acta physiol. Scand.* **44**, 129–37
ARATA, A.A. (1965) The os clitoridis of *Bassariscus. J. Mammal.* **46**, 523
ARONSON, L.R. & COOPER, M.L. (1967) Penile spines of the domestic cat: their endocrine-behavior relations. *Anat. Rec.* **157**, 71–8
ASDELL, S.A. (1946) *Patterns of mammalian reproduction* (Constable & Co.: London)
—— (1966) Evolutionary trends in physiology of reproduction. *Symp. zool. Soc. Lond.* **15**, 1–13
ASHBROOK, F.G. (1925) Blue fox farming in Alaska. *Bull. U.S. Dept. Agric.* **No. 1350**, 1–35
ASHTON, E.H. & THOMSON, A.P.D. (1955) Some characters of the skulls and skins of the European polecat, the Asiatic polecat and the domestic ferret. *Proc. zool. Soc. Lond.* **125**, 317–33
ASTLEY-MABERLEY, C.T. (1955) The African civet. *Afr. wild Life* **9**, 55–8
AUERLICH, R.J. & SWINDLER, D.R. (1968) The dentition of the mink (*Mustela vison*). *J. Mammal.* **49**, 488–94
AVARGUES, M. & GOUDEAU, H. (1962) Variations nycthémérales de la température chez le fennec (*Fennecus zerda* (Zimm.)). C. R. Soc. Biol. Paris **156**, 290–2
AZZAROLI, L. & SIMONETTA, A.M. (1966) Carnivori della Somalia ex-italiana. *Monitore zool. Ital.* **74**, suppl., 102–95
BAEVSKY, U.B. (1963) The effect of embryonic diapause on the nuclei and mitotic activity of mink and rat blastocysts. In *Delayed Implantation,* ed. A. C. Enders (Chicago University Press), 141–53
BAKER, B.E., HARINGTON, C.R. & SYMES, A.L. (1963) Polar bear milk. I Gross composition and fat constitution. *Canad. J. Zool.* **41**, 1035–9
BALDWIN, P.H., SCHWARTZ, C.W. & SCHWARTZ, E.R. (1952) Life history and economic status of the mongoose in Hawaii. *J. Mammal.* **33**, 335–56
BALESTRA, F.A. (1962) The man-eating hyenas of Mlanje. *Afr. wild Life* **16**, 25–7
BALLIET, R.F. & SCHUSTERMAN, R.J. (1971) Underwater and aerial vision in the Asian 'Clawless' otter (*Amblonyx cineria cineria*). *Nature Lond.* **234**, 305–6

BALPH, D.F. (1961) Underground concealment as a method of predation. *J. Mammal.* **42**, 423–4
BANFIELD, A.W.F. (1953) Range of individual wolves (*Canis lupus*). *Ibid.* **34**, 389–90
BANNIKOV, A.G. (1964) Biologie du chien viverrin en URSS. *Mammalia* **28**, 1–39
—— (1967) Bears and burunduks of Barguzinski. *Animals* **9**, 656–60
BARABASH-NIKIFOROV, I. (1935) The sea otters of the Commander Islands. *J. Mammal.* **16**, 255–61
—— (1938) Mammals of the Commander Islands and the surrounding sea. *Ibid.* **19**, 423–9
—— (1962) Der Seeotter oder Kalan (*Enhydra lutris* L.) *Neue Brehm-Bucherei* (A. Ziemsen-Verlag. Wittenberg)
—— (1969) The Russian sea otter. *Animals* **12**, 156–8
BARASH, D.P. (1971) Cooperative hunting in the lynx. *J. Mammal.* **52**, 480
BARCLAY, O.R. (1953) Some aspects of the mechanics of mammalian locomotion. *J. exp. Biol.* **30**, 116–20
BARNETT, C.H. & NAPIER, J.R. (1953) The rotatory mobility of the fibula in eutherian mammals. *J. Anat.* **87**, 11–21
BARNUM, C.C. (1930) Rat control in Hawaii. *Hawaii Plant. Rec.* **34**, 421–43
BARONE, R. (1967) La myologie du lion (*Panthera leo*). *Mammalia* **31**, 459–514
—— & LOMBARD, M. (1966) Organe de Jacobson, nerf voméro-nasal et nerf terminal du chien. *Bull. Soc. Sci. vet. Med. comp. Lyon* **1966**, 257–70
BARTON, C.J. (1968) Expedition sea otter. *Animal Kingd.* **63**, 24–7
BARTOSHUK, L.M., HARNED, M.A. & PARKER, L.H. (1971) Taste of water in the cat: effects on sucrose preference. *Science* **171**, 699–701
BASSETT, C.F., PEARSON, O.P. & WILKE, F. (1944) The effect of artificially-increased length of day on molt, growth and priming of silver fox pelts. *J. exp. Zool.* **96**, 77–83
—— & LLEWELLYN, L.M. (1948) The molting and fur growth pattern in the adult silver fox. *Amer. midl. Nat.* **39**, 597–601
——, TRAVIS, H.F., ABERNATHY, R.P. & WARNER, R.G. (1959) Relationship between age at separation and subsequent breeding performance of growing male mink. *J. Mammal.* **40**, 247–8
BATES, G.L. (1905) Notes on the mammals of southern Cameroons and the Benito. *Proc. zool. Soc. Lond.* **1905**, 65–85
BAUDY, R.E. (1971) Notes on breeding felids at the Rare Feline Breeding Center. *Int. Zoo Yb.* **11**, 121–3
BEACH, F.A. (1968) Coital behavior in dogs. III Effects of early isolation on mating in males. *Behaviour* **30**, 218–38
BEIDLER, L.M., FISHMAN, I.Y. & HARDIMAN, C.W. (1955) Species differences in taste receptors. *Amer. J. Physiol.* **181**, 235–9
VAN BEMMEL, A.C.V. (1968) Breeding tigers, *Panthera tigris*, at Rotterdam Zoo. *Int. Zoo Yb.* **8**, 60–3
BENNETT, E.T. (1835) Notice of a mammiferous animal from Madagascar, constituting a new form among the viverridous Carnivora. *Trans. zool. Soc. Lond.* **1**, 137–40
BENNETT, L.J., ENGLISH, P.F. & WATTS, R.L. (1943) The food habits of the black bear in Pennsylvania. *J. Mammal.* **24**, 25–31

BERGMAN, S. (1936) Observations on the Kamchatkan bear. *Ibid.* **17**, 115–20

BERKLEY, M.A. & WATKINS, D.W. (1971) Visual activity of the cat estimated from evoked cerebral potentials. *Nature Lond. New Biol.* **234**, 91–2

BERKOVITZ, B.K.B. (1968) Supernumerary deciduous incisors and the order of eruption of the incisor teeth in the albino ferret. *J. Zool.* **155**, 445–9

BERNER, A. & GYSEL, L.W. (1967) Raccoon use of large tree cavities and ground burrows. *J. Wildl. Mgmt.* **31**, 706–14

BERRIE, P.M. (in press) Ecology and status of the Canada lynx in Interior Alaska. In *Ecology and Conservation of the world's cats*, ed. R.L. Eaton

BIDER, J.R., THIBAULT, P. & SARRAZIN, R. (1968) Schèmes dynamiques spatio-temporels de l'activité de *Procyon lotor* en relation avec le comportement. *Mammalia* **32**, 137–63

BIGALKIE, R. (1961) The size of the litter of the wild dog, *Lycaon pictus* (Temminck). *Flora & Fauna* **1961**, 9–15

BIGGERS, J.D. & CREED, R.F.S. (1962) Two morphological types of placenta in the raccoon. *Nature Lond.* **194**, 103–4

BIRDSEYE, C. (1956) Observations on a domesticated Peruvian desert fox, *Dusicyon*. *J. Mammal.* **37**, 284–7

BIRKENMEIER, E. & BIRKENMEIER, E. (1971) Hand-rearing the leopard cat, *Felis bengalensis borneoensis*. *Int. Zoo Yb.* **11**, 118–21

BISHOP, S.C. (1923) Note on the nest and young of the small brown weasel. *J. Mammal.* **4**, 26–7

BISSONNETTE, T.H. (1935) Relations of hair cycles in ferrets to changes in the anterior hypophysis and to light cycles. *Anat. Rec.* **63**, 159–68

—— (1943) Some recent studies on photoperiodicity in animals. *Trans. N. Y. Acad. Sci.* **5**, 43–51

BLACKMAN, M.W. (1911) The anal glands of *Mephitis mephitica*. *Anat. Rec.* **5**, 491–515

BLONK, H.L. (1965) Einige Bemerkungen über das Fellmuster bei einen Surinam-Puma, *Puma concolor discolor* (Schreber 1775). *Säugetierk. Mitt.* **13**, 39–40

BOJER, M.W. (1834) Letter. *Proc. zool. Soc. Lond.* **1834**, 13

BOLIN, R.L. (1938) Reappearance of the southern sea otter along the California coast. *J. Mammal.* **19**, 301–3

BONNIN-LAFFARGUE, M. & CANIVENC, R. (1961) Etude de l'activité du blaireau européen (*Meles meles* L.). *Mammalia* **25**, 476–84

BOOLOOTIAN, R.A. (1965) Report to Senate Permanent Factfinding Committee on Natural Resources

DE BOOM, H.P.A. (1957) Supposed hermaphroditism of the hyena. *Afr. wild Life* **11**, 284–6

BOOTH, A.H. (1959) On the mammalian fauna of the Accra plain. *J. West Afr. Sci. Ass.* **5**, 26–36

—— (1960) *Small mammals of West Africa* (Longmans, Green & Co: London)

BOTHMA, J. DU P. (1959) Notes on the stomach contents of certain Carnivora (Mammalia) from the Kalahari Gemsbok Park. *Koedoe* **9**, 37–9

—— (1965) Random observations on the food habits of certain Carnivora (Mammalia) in southern Africa. *Fauna & Flora* **16**, 16–22

—— (1966) Food of the silver fox, *Vulpes chama*. *Zoologica Africana* **2**, 205–9

BRAIN, C.K. (1970) New finds at the Swartkrans Australopithecine site. *Nature Lond.* **225**, 1112–19

BRAND, D.J. (1963) Records of mammals bred in the National Zoological Gardens of South Africa during the period 1908–1960. *Proc. zool. Soc. Lond.* **140**, 617–59

—— & CULLEN, L. (1967) Breeding of the Cape hunting dog, *Lycaon pictus*, at Pretoria Zoo. *Int. Zoo Yb.* **7**, 124–6

BRANNON, P.A. (1923) Cacomixtl in Alabama. *J. Mammal.* **4**, 54

BROOM, R. (1949) Notes on the milk dentition of the lion, leopard and cheetah. *Ann. Transv. Mus.* **21**, 183–5

BROSSET, A. (1968) Observations sur l'éthologie du Tayra, *Eira barbara* (Carnivore). *Terre et Vie* **1968**, 29–50

BROWN, C.E. (1936) Rearing wild animals in captivity, and gestation periods. *J. Mammal.* **17**, 10–13

BROWNLEE, R.G., SILVERSTEIN, R.M., MÜLLER-SCHWARZE, D. & SINGER, A.G. (1969) Isolation, identification and function of the chief component of the male tarsal scent in black-tailed deer. *Nature Lond.* **221**, 284–5

BROWNLOW, A. L'E. (1940) Crab-eating mongoose (*Herpestes urva* (Hodgs.)) in captivity. *J. Bombay nat. Hist. Soc.* **41**, 893–4

BUDGETT, H.M. (1933) *Hunting by scent* (Eyre & Spottiswoode: London)

BURKHOLDER, B. (1959) Movements and behavior of a wolf pack in Alaska. *J. Wildl. Mgmt.* **23**, 1–11

BURNESS, G.(1970) Seven-year watch on a white badger. *Animals* **12**, 404–11

BURNS, M. (1969) The mutual behaviour of sheep and sheep dogs in Ghana. *Trop. Agric. Trinidad* **46**, 91–102

BURNS, R.D. (1960) Stomach contents of a kit fox. *Ecology* **41**, 365

BURROWS, R. (1968) *Wild foxes* (David & Charles: Newton Abbot, Devon)

BURT, W.H. (1960) Bacula of north American mammals. *Misc. Publ. Mus. Zool. Univ. Michigan* **113**, 1–76

BUTLER, P.M. (1939) Studies of the mammalian dentition. Differentiation of the post-canine dentition. *Proc. zool. Soc. Lond. B* **109**, 1–36

—— *et al.* (1967) 'Mammalia' in *The Fossil Record*, ed. Harland, W.B. *et al.* (Geological Society: London)

CABRERA, A. (1957) Catalog do los mamiferos de America del Sur. I. *Rev. Mus. Argent. Cienc. nat. Zool.* **4**, 1–307

—— & YEPES, J. (1960) *Mamiferos Sud-Americanos* (2nd edn). (Historia Natural Eidar, Compania Argentina de Editores: Buenos Aires)

CADE, C.E. (1967) Notes on breeding the Cape hunting dog, *Lycaon pictus*, at Nairobi Zoo. *Int. Zoo Yb.* **7**, 122–3

—— (1968) A note on breeding the caracal lynx, *Felis caracal*, at Nairobi Zoo. *Ibid.* **8**, 45

CAGLE, F.R. (1949) Notes on the raccoon, *Procyon lotor megalodous* Lowery. *J. Mammal.* **30**, 45–7

CALVIN, L.O. (1969) A brief note on the birth of snow leopards, *Panthera uncia*, at Dallas Zoo. *Int. Zoo Yb.* **9**, 96

CANIVENC, R. (1966) A study of progestation in the European badger (*Meles meles* L.) *Symp. zool. Soc. Lond.* **15**, 15–26

—— & BONNIN-LAFFARGUE, M. (1963) Inventory of problems raised by the

delayed ova implantation in the European badger (*Meles meles* L.). In *Delayed Implantation*, ed. A.C. Enders (University of Chicago Press), 115–24
CANSDALE, G.S. (1960) *Animals of West Africa* (Longmans, Green & Co.: London)
CARR, N. (1962) *Return to the wild*, 1965 edition (Fontana Books: London)
CARTER, Q.E. (1956) Ntini, the clawless otter (*Aonyx capensis*). *Afr. wild Life* **10**, 300
DE CARVALHO, C.T. (1968) Comparative growth rates of hand-reared big cats. *Int. Zoo Yb.* **8**, 56–9
CHAMPION, F.W. (1936) Ratels and corpses. *J. Bombay nat. Hist. Soc.* **39**, 159–61
CHAPMAN, J.A., ROMER, J.I. & STARK, J. (1955) Ladybird beetles and army cutworm adults as food for grizzly bears in Montana. *Ecology* **36**, 156–8
CHAPUIS, G. (1966) Contribution à l'étude de l'artère carotide interne des carnivores. *Mammalia* **30**, 82–96
CHARLESWORTH, — (1841) Report of meeting of Zoological Society on July 13th. *Proc. zool. Soc. Lond.* **1841**, 60
CHESEMORE, D.L. (1968) Distribution and movements of white foxes in northern and western Alaska. *Canad. J. Zool.* **46**, 849–54
CHESLER, P. (1969) Maternal influence in learning by observation in kittens. *Science* **166**, 901–3
CHESTERS, K.I.M., GNAUCK, F.R. & HUGHES, N.F. (1967) 'Angiospermae', in *The Fossil Record*, ed. W.B. Harland *et al.* (Geological Society: London)
CHURCHER, C.S. (1959) The specific status of the New World red fox. *J. Mammal.* **40**, 513–20
—— (1966) The affinities of *Dinobastis serus* Cope, 1893. *Quaternaria*, **8**, 263–75
CLIFT, C.E. (1967) Notes on breeding and rearing a kinkajou, *Potos flavus*, at Syracuse Zoo. *Int. Zoo Yb.* **7**, 126–7
COBNUT, G. (1955) Foxes and myxomatosis in Kent. *Oryx* **3**, 156–7
COETZEE, C.G. (1967) Preliminary identification manual for African mammals 7: Carnivora (excluding the family Felidae) (Smithsonian Institution: Washington)
COLE, L.W. (1907) Concerning the intelligence of racoons. *J. comp. Neurol. Psychol.* **17**, 211–62
—— & LONG, F.M. (1909) Visual discrimination in raccoons. *J. comp. Neurol.* **19**, 657–83
CONDE, B. & SCHAUENBERG, P. (1969) Reproduction du chat forestier d'Europe Felis silvestris Schreber) en captivité. *Rev. Suisse Zool.* **76**, 183–210
COTT, H.B. (1940) *Adaptive colouration in animals* (Methuen & Co: London)
—— (1953) The palatability of the eggs of birds: illustrated by experiments on the food preferences of the ferret (*Putorius furo*) and cat (*Felis catus*); with notes on other egg-eating Carnivora. *Proc. zool. Soc. Lond.* **123**, 123–41
COTTRAM, C., NELSON, A.L. & CLARKE, T.E. (1939) Notes on early winter food habits of the black bear in George Washington National Forest. *J. Mammal.* **20**, 310–14
COWAN, I.MCT. & MACKAY, R.H. (1950) Food habits of the marten (*Martes americana*) in the Rocky Mountain region of Canada. *Canada Field Nat.* **64**, 100–4
CRABB, W.D. (1941) Food habits of the prairie spotted skunk in southeastern Iowa. *J. Mammal.* **22**, 349–64
—— (1948) The ecology and management of the prairie spotted skunk in Iowa. *Ecol. Monogr.* **18**, 202–32

CRANDALL, L.S. (1964) *The management of wild mammals in captivity* (University of Chicago Press)

CRAWFORD, E.C. (1962) Mechanical aspects of panting in dogs. *J. appl. Physiol.* **17**, 249–51

CRISLER, L. (1956) Observations of wolves hunting caribou. *J. Mammal.* **37**, 337–46

—— (1959) *Arctic wild* (Secker & Warburg: London)

CROMPTON, A.S. & HIIEMÄE, K. (1970) Molar occlusion and mandibular movements during occlusion in the American opossum, *Didelphis marsupialis* L. *Zool. J. Linn. Soc.* **49**, 21–47

CROSS, E.C. (1941) Color phases of the red fox (*Vulpes fulva*) in Ontario. *J. Mammal.* **22**, 25–39

CUNNINGHAM, D.J. (1905) Cape hunting dogs (*Lycaon pictus*) in the gardens of the Royal Zoological Society of Ireland. *Proc. Roy. Soc. Edinburgh* **25**, 843–8

CUTTER, W.L. (1958a) Denning of the swift fox in northern Texas. *J. Mammal.* **39**, 70–4

—— (1958b) Food habits of the swift fox in northern Texas. *Ibid.* **39**, 527–32

CUYLER, W.K. (1924) Observations on the habits of the striped skunk (*Mephitis mesomelas varians*). *Ibid.* **5**, 180–9

DALQUEST, W.W. & ROBERTS, J.H. (1951) Behavior of young grisons in captivity. *Amer. midl. Nat.* **46**, 359–66

DATHE, H. (1963) Beitrag zur Fortpflanzungsbiologie des Malaienbären *Helarctos m. malayanus* (Raffl.). *Z. Säugetierk.* **28**, 155–62

—— (1966) Einige Bemerkungen zur Zucht des Malaienbären *Helarctos malayanus* (Raffl.). *Zool. Gart. Lpz.* **32**, 193–8

—— (1968) Breeding the Indian leopard cat, *Felis bengalensis*, at East Berlin Zoo. *Int. Zoo Yb.* **8**, 42–4

—— (1970) A second generation birth of captive sun bears, *Helarctos malayanus*, at East Berlin Zoo. *Ibid.* **10**, 79

DAVID, R. (1967) A note on the breeding of the large Indian civet, *Viverra zibetha*, at Ahmeddabad Zoo. *Ibid.* **7**, 131

DAVIS, D.D. (1949a) The female external genitalia of the spotted hyena. *Fieldiana Zool.* **31**, 277–83

—— (1949b) The shoulder architecture of bears and other carnivores. *Ibid.* **31**, 285–305

—— (1964) The giant panda. A morphological study of evolutionary mechanisms. *Fieldiana Zool. Mem.* **3**, 1–339

DAVIS, H.B. (1907) The raccoon: a study in animal intelligence. *Amer. J. Psychol.* **18**, 447–89

DAW, N.W. & PEARLMAN, A.L. (1969) Cat colour vision: one cone process or several? *J. Physiol.* **201**, 745–64

—— & —— (1970) Cat colour vision: evidence for more than one process. *J. Physiol.* **211**, 125–37

DAY, M.G. (1968) Food habits of British stoats (*Mustela erminea*) and weasels (*Mustela nivalis*). *J. Zool.* **155**, 485–97

DEANE, N.N. (1962) The spotted hyaena, *Crocuta crocuta crocuta*. *Lammergeyer* **2**, 26–43

DEANSLEY, R. (1935) The reproductive processes of certain mammals IX Growth and reproduction in the stoat (*Mustela erminea*). *Philos. Trans. B.* **225**, 459–92

—— (1943) Delayed implantation in the stoat (*Mustela mustela*). *Nature Lond.* **151**, 365–66
—— (1944) The reproductive cycle of the female weasel (*Mustela nivalis*). *Proc. zool. Soc. Lond.* **114**, 339–49
DEARBORN, N. (1932) Foods of some predatory fur-bearing animals in Michigan. *Bull. Sch. For. Conserv. Univ. Michigan* **1**, 1–52
DEMENTIEV, G.P. (1956) Nouvelles données sur le chat désertique *Felis margarita* Loche. *Mammalia* **20**, 217–22
DEXTER, R.W. (1951) Food of a crippled red fox. *J. Mammal.* **32**, 464
DIAMOND, M. (1970) Intromission pattern and species vaginal code in relation to induction of pseudopregnancy. *Science* **169**, 995–7
DIDIER, R. (1946) Etude systématique de l'os pénien des mammifères: Carnivores, Canidés. *Mammalia* **10**, 78–91
—— (1947, 1948) Etude systématique de l'os pénien des mammifères: Famille des Mustélidés. *Ibid.* **11**, 30–43, 139–52 and **12**, 67–93
—— (1949) Etude systématique de l'os pénien des mammifères: Famille des Félidés. *Ibid.* **13**, 17–37
—— (1950) Etude systématique de l'os pénien des mammifères. Procyonidés. Ursidés. *Ibid.* **14**, 78–94
DITTRICH, L. & KRONBERGER, H. (1963) Biologisch-anatomisch Untersuchungen über die Fortpflanzungsbiologie des Braunbären (*Ursus arctos* L.) und anderer Ursiden in Gefangenschaft. *Z. Säugetierk* **28**, 129–55
DIXON, J. (1925) Food predilections of predatory and fur-bearing mammals. *J. Mammal.* **6**, 34–46
DOBRORUKA, L.J. (1968) A note on the gestation period and rearing of the young in the leopard, *Panthera pardus*, at Prague Zoo. *Int. Zoo Yb.* **8**, 65
DODDS, D.G. (1955) Food habits of the Newfoundland red fox. *J. Mammal.* **36**, 291
DOMINIS, J. & EDEY, M. (1968) *The cats of Africa.* (Time-Life Books Inc.: New York)
DONOVAN, C.A. (1967) Some clinical observations on sexual attraction and deterrence in dogs and cattle. *Vet. Med.* **62**, 1047–51
—— (1969) Canine anal glands and chemical signals (pheromones). *J. Amer. vet. med. Ass.* **155**, 1995–6
DOUGLAS, M.J.W. (1965) Notes on the red fox (*Vulpes vulpes*) near Braemar, Scotland. *J. Zool.* **147**, 228–33
DOUTT, J.K. (1967) Polar bear dens on the Twin Islands, James Bay, Canada. *J. Mammal.* **48**, 468–71
DÜCKER, G. (1957) Farb- und Helligkeitssehen und Instinkte bei Viverriden und Feliden. *Zool. Beitr. Berlin* **3**, 25–99
—— (1959) Untersuchungen an der Retina einiger Viverriden. *Z. Zellforsch.* **51**, 43–9
—— (1960) Beobachtungen über das Paarungsverhalten des Ichneumons (*Herpestes ichneumon* L.). *Z. Säugetierk.* **25**, 47–51
—— (1962) Brutpflegverhaltung und Ontogenese des Verhaltens bei Surikaten (*Suricata suricatta* Schreb., Viverridae). *Behaviour* **14**, 305–40
—— (1965a) Das Verhalten der Viverriden. *Handb. Zool. Berl.* **8** (38), 10, 1–48
—— (1965b) Colour vision in mammals. *J. Bombay nat. Hist. Soc.* **61**, 572–86
—— (1968) Beobachtungen am kleinen Grison, *Galictis* (*Grisonella*) *cuja* (Molina). *Z. Säugetierk.* **33**, 288–97

DUNTON, S. (1960) Zoo news. *Animal Kingd.* **63**, 17

EAST, K. & LOCKIE, J.D. (1964) Observations on a family of weasels (*Mustela nivalis*) bred in captivity. *Proc. zool. Soc. Lond.* **143**, 359–63

—— (1965) Further observations on weasels (*Mustela nivalis*) and stoats (*Mustela erminea*) born in captivity. *J. Zool.* **147**, 234–8

EATON, R.L. (1969a) The cheetah. *Africana* **3** (10), 19–23

—— (1969b) The social life of the cheetah. *Animals* **12**, 172–5

—— (1969c) Cooperative hunting by cheetahs and jackals and a theory of domestication of the dog. *Mammalia* **33**, 87–92

—— (1970a) Notes on the reproductive biology of the cheetah, *Acinonyx jubatus. Int. Zoo Yb.* **10**, 86–9

—— (1970b) Hunting behavior of the cheetah. *J. Wildl. Mgmt.* **34**, 56–67

—— (1970c) Group interactions, spacing and territoriality in cheetahs. *Z. Tierpsychol.* **27**, 481–91

—— (1970d) The predatory sequence, with emphasis on killing behavior and its ontogeny, in the cheetah (*Acinonyx jubatus* Schreber). *Ibid.* **27**, 492–504

EBERHARD, T. (1954) Food habits of Pennsylvania house cats. *J. Wildl. Mgmt.* **18**, 284–6

EBERT, E.E. (1968) A food habits study of the southern sea otter, *Enhydra lutris nereis. Calif. Fish Game* **54**, 33–42

ECCLES, R.M., PHILLIPS, C.G. & WU, C-P. (1968) Motor innervation, motor unit organisation and afferent innervation of M. extensor digitorum communis of the baboon's forearm. *J. Physiol.* **198**, 179–92

EDMOND-BLANC, F. (1957) Observations sur le comportement de la panthère et du lion. *Mammalia* **21**, 452–3

EDWARDS, R.L. (1955) Observations on the ring-tailed cat. *J. Mammal.* **36**, 292–3

EGOSCUE, H.J. (1956) Preliminary studies of the kit fox in Utah. *Ibid.* **37**, 351–7

EIBL-EIBESFELDT, I. (1950) Über die Jugendentwicklung des Verhaltens eines männlichen Dachses (*Meles meles* L.) unter besonderer Berücksichtigung des Spieles. *Z. Tierpsychol.* **7**, 327–55

EISENBERG, J.F. & GOULD, E. (1970) The tenrecs: a study in mammalian behavior and evolution. *Smithsonian Contributions to Zoology* **27**, 1–138

—— & LOCKHART, M. (1972) An ecological reconnaissance of Wilpattu National Park, Ceylon. *Smithsonian Contributions to Zoology*, **101**

EISNER, T. (1968) Mongoose and millipede. *Science* **160**, 1367

—— & DAVIS, J.A. (1967) Mongoose throwing and smashing millipedes. *Ibid.* **155**, 577–9

ELDER, W.H. (1951) The baculum as an age criterion in mink. *J. Mammal.* **32**, 43–50

ELLENBERGER, W. & BAUM, H. (1891) *Systematische und topographische Anatomie des Hundes* (Parey: Berlin)

ELLERMAN, J.R. & MORRISON-SCOTT, T.C.S. (1951) *Checklist of palaearctic and Indian mammals* (British Museum of Natural History: London)

ELMHIRST, R. (1938) Food of the otter in the marine littoral zone. *Scott. Nat.* **1938**, 99–102

ELOFF, F.C. (1964) On the predatory habits of lions and hyaenas. *Koedoe* **7**, 105–12

—— (in press) Ecology and behavior of the Kalahari lion. In *Ecology and conservation of the world's cats*, ed. R.L. Eaton

ELSEY, C.A. (1954) A case of cannibalism in Canada lynx (*Lynx canadensis*). *J. Mammal.* **35**, 129

ELTON, C. & NICHOLSON, M. (1942) The ten-year cycle in numbers of the lynx in Canada. *J. Anim. Ecol.* **11**, 215–44

ENDERS, R.K. (1952) Reproduction in the mink (*Mustela vison*). *Proc. Amer. philos. Soc.* **96**, 691–755

—— & ENDERS, A.C. (1963) Morphology of the female reproductive tract during delayed implantation in the mink. In *Delayed Implantation*, ed. A. C. Enders (Chicago University Press), 129–39

ENGLUND, J. (1965a) Studies on food ecology of the red fox (*Vulpes vulpes*) in Sweden. *Viltrevy* **3**, 377–485

—— (1965b) Myxomatosis in Gotland. *Ibid.* **3**, 487–506

—— (1965c) The diet of foxes (*Vulpes vulpes*) on the island of Gotland since myxomatosis. *Ibid.* **3**, 507–30

ERICKSON, A.W. & MILLER, L.H. (1963) Cub adoption in the brown bear. *J. Mammal.* **44**, 584–5

ERLINGE, S. (1967a) Home range of the otter, *Lutra lutra* L., in southern Sweden. *Oikos* **18**, 186–209

—— (1967b) Food habits of the fish-otter, *Lutra lutra* L., in south Swedish habitats. *Viltrevy* **4**, 372–443

—— (1968a) Territoriality of the otter, *Lutra lutra* L. *Oikos* **19**, 81–98

—— (1968b) Food studies on captive otters, *Lutra lutra* L. *Ibid.* **19**, 259–70

—— (1969) Food habits of the otter, *Lutra lutra* L., and the mink, *Mustela vison* Schreber, in a trout water in southern Sweden. *Ibid.* **20**, 1–7

ERRINGTON, P.L. (1935) Food habits of mid-west foxes. *J. Mammal.* **16**, 192–200

—— (1936) Notes on food habits of southern Wisconsin house cats. *Ibid.* **17**, 64–5

—— (1937a) Food habits of Iowa red foxes during a drought summer. *Ecology* **18**, 53–61

—— (1937b) Summer food habits of the badger in northwestern Iowa. *J. Mammal.* **18**, 213–16

—— (1967) *Of predation and life* (Iowa State University Press: Ames, Iowa)

—— & BERRY, R.M. (1937) Tagging studies of red foxes. *J. Mammal* **18**, 203–5

ESTES, R.D. & GODDARD, J. (1967) Prey selection and hunting behavior of the African wild dog. *J. Wildl. Mgmt.* **31**, 52–70

EVANS, W.E. (1969) 'Marine mammal communication: social and ecological factors': chapter 11 in *The biology of marine mammals*, ed. H.T.A. Andersen (Academic Press: New York)

EVERTS, W. (1968) Beitrag zur Systematik des Sonnendachse. *Z. Säugetierk.* **33**, 1–19

EWER, R.F. (1954) Some adaptive features in the dentition of hyaenas. *Ann. Mag. nat. Hist.* **7**, 188–94

—— (1955a) The fossil carnivores of the Transvaal caves. The Lycyaenas of Sterkfontein and Swartkrans, together with some general considerations of the Transvaal fossil hyaenids. *Proc. zool. Soc. Lond.* **124**, 839–57

—— (1955b) The fossil carnivores of the Transvaal caves: Machairodontinae. *Ibid.* **125**, 587–615

—— (1956a) The fossil carnivores of the Transvaal caves: Canidae. *Ibid.* **126**, 97–119

—— (1956b) The fossil carnivores of the Transvaal caves: two new viverrids, together with some general considerations. *Ibid.* **126**, 259–74

—— (1959) Suckling behaviour in kittens. *Behaviour* **15**, 146–62

—— (1961) Further observations on suckling behaviour in kittens, together with some general considerations of the interrelations of innate and acquired responses. *Ibid.* **18**, 247–60

—— (1963a) A note on the suckling behaviour of the viverrid, *Suricata suricatta* (Schreber). *Anim. Behav.* **11**, 599–601

—— (1963b) The behaviour of the meerkat, *Suricata suricatta* (Schreber). *Z. Tierpsychol.* **20**, 570–607

—— (1969) The 'instinct to teach'. *Nature Lond.* **223**, 698

EYRE, M. (1963) A tame otter. *Afr. wild Life* **17**, 49–53

FAIRALL, N. (1968) The reproductive seasons of some mammals in the Kruger National Park. *Zoologica Africana* **3**, 189–210

FAIRLEY, J.S. (1969a) Some field observations on the fox in Northern Ireland. *Irish Nat. J.* **16**, 189–92

—— (1969b) Tagging studies of the red fox, *Vulpes vulpes*, in northeast Ireland. *J. Zool.* **159**, 527–32

—— (1971) Notes on the breeding of the fox (*Vulpes vulpes*) in county Galway, Ireland. *Ibid.* **164**, 262–3

FAUST, R. & SCHERPNER, C. (1967) A note on the breeding of the maned wolf, *Chrysocyon brachyurus*, at Frankfurt Zoo. *Int. Zoo Yb.* **7**, 119

FELLNER, K. (1968) Erst natürliche Aufzucht von Nebelpardern (*Neofelis nebulosa*) in einem Zoo. *Zool. Gart. Lpz.* **35**, 105–37

FEY, V. (1964) The diet of leopards. *Afr. wild Life* **18**, 105–8

FICHTER, E., SCHILDMAN, G. & SATHER, J.H. (1955) Some feeding patterns of coyotes in Nebraska. *Ecol. Monogr.* **25**, 1–37

FIEDLER, W. (1957) Beobachtungen zum Markierungsverhalten einiger Säugetiere. *Z. Säugetierk.* **22**, 57–76

FISCHER, C.E.C. (1921) The habits of the grey mongoose. *J. Bombay nat. Hist. Soc.* **28**, 274

FISHER, E.M. (1939) Habits of the southern sea otter. *J. Mammal.* **20**, 21–36

—— (1940) Early life of a sea otter pup. *Ibid.* **21**, 132–7

—— (1941) Notes on the teeth of the sea otter. *Ibid.* **22**, 428–33

—— (1942) *The osteology and myology of the California sea otter* (Stanford University Press: California)

FISHER, H.I. (1951) Notes on the red fox (*Vulpes vulpes*) in Missouri. *J. Mammal.* **32**, 296–9

FLORIO, P.L. & SPINELLI, L. (1967) Successful breeding of a cheetah, *Acinonyx jubatus*, in a private zoo. *Int. Zoo Yb.* **7**, 150–2

—— & —— (1968) Second successful breeding of cheetahs, *Acinonyx jubatus*, in a private zoo. *Ibid.* **8**, 76–8

FLOWER, W.H. (1869a) On the value of the characters of the base of the cranium in the classification of the Carnivora. *Proc. zool. Soc. Lond.* **1869**, 4–37

—— (1869b) On the anatomy of the proteles, *Proteles cristatus* (Sparrman). *Ibid.* **1869**, 474–96

—— (1872) Note on the anatomy of the two-spotted paradoxure (*Nandinia binotata*). *Ibid.* **1872**, 683–4

—— (1879) On the caecum of the red wolf (*Canis jubatus* Desm.). *Ibid.* **1879**, 766–7
—— (1880) On the bush dog (*Icticyon venaticus* Lund). *Ibid.* **1880**, 70–6
FLYGER, V. & TOWNSEND, M.R. (1968) The migration of polar bears. *Sci. Amer.* **218**, 108–116
FOLK, G.E. (1967) Physiological observations of subarctic bears under winter den conditions: in *Mammalian hibernation III*, ed. Fisher *et al.* (Oliver & Boyd; London), 75–85
FOOTT, J.O. (1970) Nose scars in female sea otters. *J. Mammal.* **51**, 621–2
FOSTER, J.B. & KEARNEY, D. (1967) Nairobi National Park game census, 1966. *E. Afr. Wildl. J.* **5**, 112–20
FOX, M.W. (1969a) The anatomy of aggression and its ritualisation in Canidae: a developmental and comparative study. *Behaviour* **35**, 242–58
—— (1969b) Ontogeny of prey-killing behavior in Canidae. *Ibid.* **35**, 259–72
—— (1971a) Socio-infantile and socio-sexual signals in canids: a comparative and ontogenetic study. *Z. Tierpsychol.* **28**, 185–210
—— (1971b) The development and temporal sequencing of agonistic behavior in the coyote (*Canis latrans*). *Ibid.* **28**, 262–78
FREDGA, K. (1970) Unusual sex chromosome inheritance in mammals. *Philos. Trans. B.* **259**, 15–36
FREI, P. (1968) Über das Verhalten von Zoo-Bären bei Witterungsnahme freilebender Artgenossen. *Säugetierk. Mitt.* **16**, 56–7
FREMLIN, J.H. (1964) How many people can the world support? *New Scientist* **24**, 285–7
FRERE, A.G. (1928) Breeding habits of the common mongoose (*Herpestes edwardsi*). *J. Bombay nat. Hist. Soc.* **33**, 426–8
FRIEDMANN, H. (1955) The honey-guides. *Bull. U.S. nat. Mus.* **208**, 1–292
FRILEY, C.E. (1949) Age determination, by use of the baculum, in the river otter, *Lutra c. canadensis* Schreber. *J. Mammal.* **30**, 102–110
FRUEH, R.J. (1968) A note on breeding snow leopards, *Panthera uncia*, at St Louis Zoo. *Int. Zoo Yb* **8**, 74–6
FULLER, W.A. (1966) The biology and management of the bison of Wood Buffalo National Park. *Wildl. Mgmt. Bull. Can.* (1), No. **16**, 1–52
FUNDERBURG, J.B. (1961) Erythristic raccoons from North Carolina. *J. Mammal.* **42**, 270–1
GAILLARD, — (1969) Sur la présence du chat doré (*Felis aurata* Temminck) et du caracal (*Felis caracal* Schreber) dans le sud du Sénégal. *Mammalia* **33**, 350–1
GANDER, F.F. (1928) Period of gestation in some American mammals. *J. Mammal.* **9**, 75
—— (1965) Spotted skunks make interesting neighbours. *Animal Kingd.* **68**, 104–8
—— (1966) Raccoons. *Ibid.* **69**, 84–9
GARROD, A.H. (1873) Notes on the anatomy of the binturong (*Arctictis binturong*). *Proc. zool. Soc. Lond.* **1873**, 196–202
—— (1878) Note on the anatomy of the binturong (*Arctictis binturong*). *Ibid.* **1878**, 142
GASHWILER, J.S., ROBINETTE, W.L. & MORRIS, O.W. (1961) Breeding habits of bobcats in Utah. *J. Mammal.* **42**, 76–84
GASPARD, M. (1964) La région de l'angle mandibulaire chez les Canidae. *Mammalia* **28**, 249–329

Gauthier-Pilters, H. (1962) Beobachtungen an Feneks (*Fennecus zerda* Zimm.). *Z. Tierpsychol.* **19**, 440–64

—— (1966) Einige Beobachtungen über das Spielverhalten beim Fenek (*Fennecus zerda* Zimm.). *Z. Säugetierk.* **31**, 337–50

—— (1967) The fennec. *Afr. wild Life* **21**, 117–25

Van Gelder, R.G. (1953) The egg-opening technique of a spotted skunk. *J. Mammal.* **34**, 255–6

—— (1959) A taxonomic revision of the spotted skunks (genus *Spilogale*). *Bull. Amer. Mus. nat. Hist.* **117**, 229–392

Gensch, W. (1962) Successful rearing of the binturong, *Arctictis binturong* Raffl. *Int. Zoo Yb.* **4**, 79–80

—— (1966) Nochmals zur Nachzucht des Binturong (*Arctictis binturong* Raffl.) im zoologischen Garten Dresden. *Zool. Gart. Lpz.* **33**, 126–8

—— (1968) Notes on breeding timber wolves, *Canis lupus occidentalis*, at Dresden Zoo. *Int. Zoo Yb.* **8**, 15–16

Gerell, R. (1967) Food selection in relation to habitat in mink (*Mustela vison* Schreber) in Sweden. *Oikos* **18**, 233–46

—— (1970) Home ranges and movements of the mink, *Mustela vison*, in southern Sweden. *Ibid.* **21**, 160–73

Gewalt, W. (1959) Beiträge zur Kenntnis des optischen Differenzierungsvermogens einiger Musteliden mit besonderer Berücksichtigung des Farbensehens. *Zool. Beitr.* **5**, 117–75

Ghose, R.K. (1965) A new species of mongoose (Mammalia: Carnivora: Viverridae) from west Bengal, India. *Proc. zool. Soc. Calcutta* **18**, 173–8

Ginsburg, L. (1961) Plantigrade et digitigrade chez les carnivores fissipèdes. *Mammalia* **25**, 1–21

Goethe, F. (1940) Beiträge zur Biologie des Iltis. *Z. Säugetierk.* **15**, 180–223

—— (1950) Vom Leben des Mauswiesels (*Mustela n. nivalis* L.). *Zool. Gart. Lpz.* **17**, 193–204

Gossow, H. (1970) Vergleichende Verhaltensstudien an Marderartigen. I Über Lautäusserungen und zum Beuteverhalten. *Z. Tierpsychol.* **27**, 405–80

Gowda, C.D.K. (1967) A note on the birth of caracal lynx, *Felis caracal*, at Mysore Zoo. *Int. Zoo Yb.* **7**, 133

—— (1968) A note on the mating behaviour of tigers, *Panthera t. tigris*, at Mysore Zoo. *Ibid.* **8**, 63–4

Grafton, R.N. (1965) Food of the black-backed jackal: a preliminary report. *Zoologica Africana* **1**, 41–54

Graham, A. (1966) East African Wild Life Society cheetah survey: extracts from the report by Wildlife Services. *E. Afr. Wildl. J.* **4**, 50–5

Granit, R. (1947) *Sensory mechanisms of the retina* (Oxford University Press)

Grassé, P.P. (1955) *Traité de Zoologie,* **17** (1) (Masson et Cie: Paris)

—— (1967) *Traité de Zoologie,* **16** (1) (Masson et Cie: Paris)

Gray, A.P. (1954) *Mammalian hybrids* (Commonwealth agric. Bureaux: Farnham Royal)

—— (1966) *Supplementary bibliography on mammalian hybrids* (Commonwealth Bureau of Animal Breeding & Genetics)

Gray, J.E. (1848) Description of a new species of *Galidictis* from Madagascar. *Proc. zool. Soc. Lond.* **1848**, 21–3

—— (1865) Note on the habits of the kinkajou (*Cercoleptes caudivolvulus*). *Ibid.* **1865**, 680

GREEN, D.D. (1947) Albino coyotes are rare. *J. Mammal.* **28**, 63

GREER, J.K. & GREER, M. (1970) Record of the pygmy spotted skunk (*Spilogale pygmaea*) from Colima, Mexico. *Ibid.* **51**, 629–30

GREER, K.R. (1955) Yearly food habits of the river otter in the Thompson Lakes region, northwestern Montana, as indicated by scat analysis. *Amer. midl. Nat.* **54**, 299–313

GREGORY, W.K. (1951) *Evolution emerging* (Macmillan: New York)

—— & HELLMAN, M. (1939) On the evolution and major classification of the civets (Viverridae) and allied fossil and recent Carnivora: a phylogenetic study of the skull and dentition. *Proc. Amer. Philos. Soc.* **81**, 309–92

GRIMPE, G. (1923) Neues über die Geschlechtsverhaltnisse der gefleckten Hyäne (*Crocotta crocuta* Erxl.). *Verhl. dtsch. zool. Ges. Berlin* **28**, 77–8

GUDGER, E.W. (1946) Does the jaguar use his tail as a lure in fishing? *J. Mammal.* **27**, 37–49

GUGGISBERG, C.A.W. (1955) *Das Tierleben der Alpen* (Hallwag: Berlin)

—— (1960) *Simba* (Hallwag; Berlin) English translation 1963 (Chilton Books)

GUILDAY, J.E. (1949) Winter foods of Pennsylvania mink. *Penn. Game News* **20**, 32

—— (1962) Supernumerary molars of *Otocyon*. *J. Mammal.* **43**, 455–62

GUNTER, R. (1951) The absolute threshold for vision in the cat. *J. Physiol.* **114**, 8–15

HAGLUND, B. (1966) Winter habits of lynx (*Lynx lynx* L.) and wolverine (*Gulo gulo* L.) as revealed by tracking in the snow. *Viltrevy* **4**, 81–310

HALL, E.R. (1926) The abdominal skin gland of *Martes*. *J. Mammal.* **7**, 227–9

—— (1940) Supernumerary and missing teeth in wild mammals of the orders Insectivora and Carnivora, with some notes on diseases. *J. dental Res.* **19**, 103–42

—— (1951) American weasels. *Publ. Mus. nat. Hist. Univ. Kansas* **4**, 1–466

—— & KELSON, K.R. (1959) *Mammals of North America* (Ronald Press: New York)

HALL, K.R.L. & SCHALLER, G.B. (1964) Tool-using behaviour of the California sea otter. *J. Mammal.* **45**, 287–98

HAMILTON, W.J. (1933) The weasels of New York. *Amer. midl. Nat.* **14**, 289–344

—— (1936a) Food habits of mink in New York. *J. Mammal.* **17**, 169

—— (1936b) Seasonal food of skunks in New York. *Ibid.* **17**, 240–6

—— (1937) Winter activity of the skunk. *Ecology* **18**, 326–7

—— (1940) The summer food of minks and raccoons of the Montezuma Marsh, New York. *J. Wildl. Mgmt.* **4**, 80–4

—— (1951) Warm-weather foods of the raccoon in New York State. *J. Mammal.* **32**, 341–4

—— (1959) Foods of mink in New York. *N.Y. Fish Game J.* **1959**, 77–85

—— (1961) Late fall, winter and early spring foods of 141 otters from New York. *Ibid.* **8**, 106–9

—— & EADIE, W.R. (1964) Reproduction in the otter, *Lutra canadensis*. *J. Mammal.* **45**, 242–52

HAMMOND, J. & MARSHALL, F.H.A. (1930) Oestrus and pseudopregnancy in the ferret. *Proc. roy. Soc. B.* **105**, 607–30

HANGEN, O. (1954) Longevity of the raccoon in the wild. *J. Mammal.* **35**, 439

HARINGTON, C.R. (1968) Denning habits of the polar bear (*Ursus maritimus* Phipps). *Canad. Wildl. Service Rept. Ser.* **5**, 1–30

HARRIS, C.J. (1968) *Otters* (Weidenfeld & Nicolson: London)
HARRISON, R.J. (1963) 'A comparison of factors involved in delayed implantation in badgers and seals in Great Britain': In *Delayed Implantation*, ed. A.C. Enders (Chicago University Press). 99–114
HART, J.S. (1956) Seasonal changes in insulation of the fur. *Canad. J. Zool.* **34**, 53–7
HARTMAN, L. (1964) The behaviour and breeding of captive weasels (*Mustela nivalis* L.). *New Zealand J. Sci.* **7**, 147–56
HARVEY, N.E. & MACFARLANE, W.V. (1958) The effects of day length on the coat-shedding cycles, body weight and reproduction of the ferret. *Aust. J. biol. Sci.* **11**, 187–99
HAWBECKER, A.C. (1943) Food of the San Joaquim kit fox. *J. Mammal.* **24**, 499
HAWLEY, V.D. & NEWBY, F.E. (1957) Marten home ranges and population fluctuations. *Ibid.* **38**, 174–84
HEIDEMANN, G. & VAUK, G. (1970) Zur Nahrungsökologie 'wildernder' Hauskatzen *Felis silvestris* f. *catus* Linné, 1758). *Z. Säugetierk.* **35**, 185–90
HEIDT, G.A., PETERSEN, M.K. & KIRKLAND, G.L. (1968) Mating behavior and development of least weasels (*Mustela nivalis*) in captivity. *J. Mammal.* **49**, 413–19
HEIMBURGER, N. (1959) Das Markierungsverhalten einiger Caniden. *Z. Tierpsychol.* **16**, 104–13
—— (1961) Beobachtungen an handaufgezogenen Wildcaniden (Wölfin und Schakalin) und Versuche über ihre Gedächtnisleistungen. *Ibid.* **18**, 265–84
HEIT, W.S. (1944) Food habits of red foxes of the Maryland marshes. *J. Mammal.* **25**, 55–8
HENSEL, R.J., TROYER, W.A. & ERICKSON, A.W. (1969) Reproduction in the female brown bear. *J. Wildl. Mgmt.* **33**, 357–65
HEPTNER, W. & DEMENTIEV, G. (1937) Sur les relations mutuelles et la position systématique des chats désertiques, *Eremaelurus thinobius* Ognev et *Felis margarita* Loche. *Mammalia* **1**, 227–42
HERRERO, S. (1970) A black bear and her cub. *Animals* **12**, 444–7
HERSHKOVITZ, P. (1969) The evolution of mammals on southern continents. VI The recent mammals of the neotropical region: a zoogeographic and ecological review. *Q. Rev. Biol.* **44**, 1–70
HERTER, K. & OHM-KETTNER, I.D. (1954) Über die Aufzucht und das Verhalten zweier Baummarder (*Martes martes* L.). *Z. Tierpsychol.* **11**, 113–37
HESS, J.K. (1971) Hand-rearing polar bear cubs, *Thalarctos maritimus*, at St Paul Zoo. *Int. Zoo Yb.* **11**, 102–7
HEWSON, R. (1969) Couch building by otters, *Lutra lutra*. *J. Zool.* **159**, 524–7
HIBBEN, F.C. (1939) The mountain lion and ecology. *Ecology* **20**, 584–6
HIGGS, E.S. & JARMAN, M.R. (1969) The origins of agriculture: a reconsideration. *Antiquity* **43**, 31–41
HILDEBRAND, M. (1952) The integument in Canidae. *J. Mammal.* **33**, 419–28
—— (1954a) Incisor tooth wear in sea otter. *Ibid.* **35**, 595
—— (1954b) Comparative morphology of the body skeleton in recent Canidae. *Univ. California Publ. Zool.* **52** (5), 399–470
—— (1959) Motions of the running cheetah and horse. *J. Mammal.* **40**, 481–95
—— (1961) Further studies on locomotion of the cheetah. *Ibid.* **42**, 84–91
—— (1965) Symmetrical gaits of horses. *Science* **150**, 701–8
—— (1966) Analysis of the symmetrical gaits of tetrapods. *Folia Biotheoretica* **6**, 1–22

—— (1968) Symmetrical gaits of dogs in relation to body build. *J. Morph.* **124**, 353–60

HILL, W.C.O. (1948) Rhinoglyphics: epithelial sculpturing of the mammalian rhinarium. *Proc. zool. Soc. Lond.* **118**, 1–35

HINTON, H.E. & DUNN, A.M.S. (1967) *Mongooses, their natural history and behaviour* (Oliver & Boyd: Edinburgh & London)

HIRST, S.M. (1969) Predation as a regulating factor of wild ungulate populations in a Transvaal lowveld nature reserve. *Zoologica Africana* **4**, 199–230

HOCK, R.J. (1960) Seasonal variations in physiologic function of black bears. *Bull. Mus. comp. Zool. Harvard* **124**, 155–73

—— & Larson, A.M. (1966) Composition of black bear milk. *J. Mammal.* **47**, 539–40

HOFFMEISTER, D.F. & WINKLEMANN, R. (1958) The os clitoridis in the badger, *Taxidea taxus. Trans. Illinois Acad. Sci.* **50**, 233–4

HOPWOOD, A.T. (1947) Contributions to the study of some African mammals. III Adaptations in the bones of the fore limb of the lion, leopard, and cheetah. *J. Linn. Soc. Zool.* **41**, 259–71

HORNOCKER, M.G. (1969) Winter territoriality in mountain lion. *J. Wildl. Mgmt.* **33**, 457–64

HOUGH, J.R. (1948) The auditory region in some members of the Procyonidae, Canidae and Ursidae. Its significance in the phylogeny of the Carnivora. *Bull. Amer. Mus. nat. Hist.* **92**, 71–118

HOWARD, W.E. (1949) A means to distinguish skulls of coyotes and domestic dogs. *J. Mammal.* **30**, 169–71

HOWELL, A.H. (1920) The Florida spotted skunk as an acrobat. *Ibid.* **1**, 88

HUEY, L.M. (1937) El Valle de la Trinidad, the coyote poisoner's proving ground. *Ibid.* **18**, 74–6

HUFF, J.N. & PRICE, E.O. (1968) Vocalisations of the least weasel, *Mustela nivalis. Ibid.* **49**, 548–50

HUNT, H. (1967) Growth rate of a new-born, hand-reared jaguar, *Panthera onca*, at Topeka Zoo. *Int. Zoo Yb.* **7**, 147–50

HURRELL, E. (1963) *Watch for the otter.* (Country Life: London)

HURRELL, H.G. (1955) Ranch bred pine martens. *Proc. zool. Soc. Lond.* **125**, 466–7

HUXLEY, T.H. (1880) On the cranial and dental characters of the Canidae. *Ibid.* **1880**, 238–88

IMAIZUMI, Y. (1967) A new genus and species of cat from Iriomote, Ryuku Islands. *J. Mammal. Soc. Japan* **3**, 75–105

INGLES, J.M. (1965) Zambian mammals collected for the British Museum (Natural History). *Puku* **3**, 75–86

IRVING, L. (1964) Terrestrial animals in cold: birds and mammals: in *Handbook of Physiology*, **4**, 361–77 (Amer. Physiol. Soc.: Washington)

ISHUNIN, G.I. (1965) On the biology of *Felis chaus chaus* Güldenstaedt in south Uzbekistan. *Zool. Zh.* **44**, 630–2

JACKLEY, A.M. (1938) Badgers feed on rattlesnakes. *J. Mammal.* **19**, 374–5

JACOBI, A. (1938) Der Seeotter. *Monogr. Wildsäuget.* **6**, 1–93

JAYAKAR, S.D. & SPURWAY, H. (1968) Notes on the common palm civet or toddy cat, *Paradoxurus hermaphroditus* (Pallas), with special reference to the age at shedding of the milk teeth. *J. Bombay nat. Hist. Soc.* **65**, 211–13

JEFFERIES, D.J. & PENDLEBURY, J.B. (1968) Population fluctuations of stoats, weasels and hedgehogs in recent years. *J. Zool.* **156**, 513–17

JERGE, C.R. (1963) Organisation and function of the trigeminal mesencephalic nucleus. *J. Neurophysiol.* **26**, 379–92

JEWELL, P.A. (1952) The anastomoses between internal and external carotid circulations in the dog. *J. Anat.* **86**, 83–94

JOHANSSON, I. (1960) Inheritance of color phases in ranch-bred blue foxes. *Hereditas* **46**, 753–66

JOHNSON, C.E. (1921) The 'hand-stand' habit of the spotted skunk. *J. Mammal.* **2**, 87–9

JOHNSON, D.R. (1968) Coat-colour polymorphism in North Plains red fox populations. *Canad. J. Zool.* **46**, 608–10

JOHNSON, M.L., KENYON, K.W. & BROSSEAU, C. (1967) Notes on a captive sea otter, *Enhydra lutris. Int. Zoo Yb.* **7**, 208–9

JOHNSON, W.J. (1970) Food habits of the red fox in Isle Royale National Park, Lake Superior. *Amer. midl. Nat.* **84**, 568–72

JONES, J.K. (1964) Distribution and taxonomy of mammals of Nebraska. *Univ. Kansas Publ. Mus. nat. Hist.* **16**, 1–356

JONES, M.L. (in press) The snow leopard in captivity: in *Ecology and conservation of the world's cats*, ed. R.L. Eaton

JONES, O.G. (1952) Zoo babies. *Zoo Life* **7**, 112–18

JONES, R.D. (1951) Present status of the sea otter in Alaska. *Trans. N. Amer. Wildl. Conf.* **16**, 376–83

—— (1965) Sea otters in the Near Islands, Alaska. *J. Mammal.* **46**, 702

JORDAN, P.A., SHELTON, P.C. & ALLEN, D.L. (1967) Numbers, turnover and social structure of the Isle Royale wolf population. *Amer. Zool.* **7**, 233–52

JOSLIN, P.W.B. (1967) Movements and home sites of timber wolves in Algonquin Park. *Ibid.* **7**, 279–88

JURGENSON, P.B. (1954) On the influence of marten (*Martes martes* L.) on the numbers of squirrels (*Sciurus vulgaris* L.) in the northern taiga. *Zool. Zh.* **33**, 166–73

KAINER, R.A. (1954) The gross anatomy of the digestive system of the mink. I The headgut and the foregut. *Amer. J. vet. Res.* **15**, 82–90

KALMUS, H. (1955) The discrimination by the nose of the dog of individual human odours and in particular of the odours of twins. *Brit. J. anim. Behav.* **3**, 25–31

KAUFMANN, J.H. (1962) Ecology and social behavior of the coati, *Nasua narica*, on Barro Colorado Island, Panama. *Univ. Calif. Publ. Zool.* **60**, 95–222

—— & KAUFMANN, A. (1963) Some comments on the relationship between field and laboratory studies of behaviour, with special reference to coatis. *Anim. Behav.* **11**, 464–9

—— & —— (1965) Observations on the behavior of tayras and grisons. *Z. Säugetierk.* **30**, 146–55

KELKER, G.H. (1937) Insect food of skunks. *J. Mammal.* **18**, 164–70

KELSALL, J.P. (1957) Continued barren-ground caribou studies. *Canad. wildl. Serv. Wildl. Mgmt. Bull.* **12**

—— (1968) The caribou. *Canad. wildl. Serv. Monogr.* **3**

KENYON, K.W. (1959) The sea otter. *Ann. Repts. Smithsonian Inst.* **1958**, 399–407

—— (1963) Recovery of a fur bearer. *Nat. Hist.* **72** (9), 12–21

—— (1969) *The sea otter in the eastern Pacific Ocean* (U.S. Dept. Interior, Bureau of sport, fisheries and wild life: North American Fauna No. **68**, 1–352)

VON KETELHODT, H.F. (1965) The Aardwolf. *Afr. wild Life* **19**, 23–4

—— (1966) Der Erdwolf, *Proteles cristatus* (Sparrman, 1783). *Z. Säugetierk.* **31**, 300–8

KILGORE, D.L. (1969) An ecological study of the swift fox (*Vulpes velox*) in the Oklahoma panhandle. *Amer. midl. Nat.* **81**, 512–34

KING, J.E. (1964) *Seals of the world* (British Museum of Natural History: London)

——, BECKER, R.F. & MARKEE, J.E. (1964) Studies on olfactory discrimination in dogs: (3) Ability to detect human odour trace. *Anim. Behav.* **12**, 311–15

KIPP, H. (1965) Beitrag zur Kenntnis der Gattung *Conepatus* Molina, 1782. *Z. Säugetierk.* **30**, 193–232

KLEIMAN, D.G. (1966a) The comparative social behavior of the Canidae. *Amer. Zool.* **6**, 335

—— (1966b) Scent marking in the Canidae. *Symp. zool. Soc. Lond.* **18**, 167–77

—— (1967) Some aspects of social behavior in the Canidae. *Amer. Zool.* **7**, 365–72

—— (1968) Reproduction in the Canidae. *Int. Zoo Yb.* **8**, 3–8

—— & EISENBERG, J.F. (in press) Comparison of canid and felid social systems from an evolutionary point of view.

KNAPPE, H. (1964) Zur Funktion des Jacobsonschen Organs (Organon vomeronasale Jacobsoni). *Zool. Gart. Lpz.* **28**, 188–94

KNOPF, F.L. & BALPH, D.L. (1969) Badgers plug burrows to confine prey. *J. Mammal.* **50**, 635–6

KNUDSEN, G.J. & HALE, J.B. (1968) Food habits of otters in the Great Lakes region. *J. Wildl. Mgmt.* **32**, 89–93

KOENIG, L. (1970) Zur Fortpflanzung und Jugendentwicklung des Wüstenfuchses (*Fennecus zerda* Zimm. 1780). *Z. Tierpsychol.* **27**, 205–46

KOLENOSKY, G.B. & JOHNSTON, D.H. (1967) Radio-tracking timber wolves in Ontario. *Amer. Zool.* **7**, 289–303

KORSCHGEN, L.J. (1958) December food habits of mink in Missouri. *J. Mammal.* **39**, 521–7

KRALIK, S. (1967) Breeding the caracal lynx, *Felis caracal*, at Brno Zoo. *Int. Zoo Yb.* **7**, 132

KROTT, P. (1959) Der Vielfrass (*Gulo gulo* L. 1758). *Monogr. Wildsäuget.* **13**, 1–159

—— (1961) Der gefährliche Braunbär (*Ursus arctos* L. 1758). *Z. Tierpsychol.* **18**, 245–56

—— & KROTT, G. (1963) Zum Verhalten des Braunbären (*Ursus arctos* L. 1758) in den Alpen. *Ibid.* **20**, 160–206

KRUUK, H. (1966) Clan-system and feeding habits of spotted hyaenas (*Crocuta crocuta* Erxl.). *Nature Lond.* **209**, 1257–8

—— (1968) Hyaenas, the hunters nobody knows. *Nat. Geogr.* **134**, 44–57

—— & TURNER, M. (1967) Comparative notes on predation by lion, leopard, cheetah and wild dog in the Serengeti area, East Africa. *Mammalia* **31**, 1–27

KÜHME, W. (1964) Die Ernährungsgemeinschaft der Hyänenhundes (*Lycaon pictus lupinus* Thomas 1902). *Naturwissenschaften* **20**, 495

—— (1965a) Communal food distribution and division of labour in African hunting dogs. *Nature Lond.* **205**, 443–4

—— (1965b) Über die soziale Bindung innerhalt eines Hyänenhund-Rudels. *Naturwissenschaften* **23**, 567–8

—— (1965c) Freilandstudien zur Soziologie des Hyänenhundes (*Lycaon pictus lupinus* Thomas 1902). *Z. Tierpsychol.* **22**, 495–541

KUNC, L. (1970) Breeding and rearing the northern lynx, *Felis l. lynx*, at Ostrava Zoo. *Int. Zoo Yb.* **10**, 83–4

KURTÉN, B. (1963) Notes on some Pleistocene mammal migrations from the Palaearctic to the Nearctic. *Eiszeitalter Gegenw.* **14**, 96–103

—— (1966) Pleistocene bears of North America. I Genus *Tremarctos*, spectacled bears. *Acta zool. Fenn.* **115**, 100–20

—— (1968) *Pleistocene mammals of Europe* (Weidenfeld & Nicolson: London)

LAGLER, K.F. & OSTENSON, B.T. (1942) Early spring food of the otter in Michigan. *J. Wildl. Mgmt.* **6**, 244–54

LAMPREY, H.F. (1963) Ecological separation of the large mammal species in the Tarangire Game Reserve, Tanganyika. *E. Afr. Wildl. J.* **1**, 63–92

LANG, E.M. (1958) Zur Haltung des Strandwolfes (*Hyaena brunnea*). *Zool. Gart. Lpz.* **24**, 81–91

LANGGUTH, A. (1969) Die südamerikanischen Canidae unter besonderer Berücksichtigung des Mähnenwolfes, *Chrysocyon brachyurus* Illiger. *Z. wiss. Zool.* **179**, 1–188

LAUER, B.H., KUYT, E. & BAKER, B.E. (1969) Wolf milk. I Arctic wolf (*Canis lupus arctos*) and husky milk: gross composition and fatty acid composition. *Canad. J. Zool.* **47**, 99–102

VAN LAWICK-GOODALL, J. & VAN LAWICK, H. (1966) Use of tools by the Egyptian vulture (*Neophron percnopterus*). *Nature Lond.* **212**, 1468–9

—— & —— (1970) *Innocent killers* (Collins: London)

LAWRENCE, B. (1967) Early domestic dogs. *Z. Säugetierk.* **32**, 44–59

—— & BOSSERT, W.H. (1967) Multiple character analysis of *Canis lupus*, *latrans* and *familiaris*, with a discussion of the relationships of *Canis niger*. *Amer. Zool.* **7**, 223–32

—— & —— (1969) Cranial evidence for hybridization in New England. *Canis*. *Breviora* **330**, 1–13

LAY, D.M., ANDERSON, J.A.W. & HASSINGER, J.D. (1970) New records of small mammals from west Pakistan. *Mammalia* **34**, 98–106

LAYNE, J.N. (1958) Reproductive characteristics of the grey fox in southern Illinois. *J. Wildl. Mgmt.* **22**, 157–62

LEAKEY, L.S.B. (1969) *Animals of East Africa* (National Geographic Society: Washington)

LEITCH, I., HYTTEN, F.E. & BILLEWICZ, W.Z. (1959) The maternal and neonatal weights of some mammalia. *Proc. zool. Soc. Lond.* **133**, 11–28

LENSINK, C.J. (1960) Status and distribution of sea otters in Alaska. *J. Mammal.* **41**, 172–82

——, SKOOG, R.O. & BUCKLEY, J.L. (1955) Food habits of marten in interior Alaska and their significance. *J. Wildl. Mgmt.* **19**, 364–8

LEONE, C.A. & WIENS, A.L. (1956) Comparative serology of carnivores. *J. Mammal.* **37**, 11–23

LESLIE, G. (1971) Further observations on the oriental short-clawed otter, *Amblonyx cinerea*, at Aberdeen Zoo. *Int. Zoo. Yb.* **11**, 112–13

LESOWSKI, J. (1963) Two observations of cougar cannibalism. *J. Mammal.* **44**, 586
LEVER, R.J.A. (1959) The diet of the fox since myxomatosis. *J. anim. Ecol.* **28**, 359–75
LEWIS, R.E., LEWIS, J.H. & ATALLAH, S.I. (1968) A review of Lebanese mammals: Carnivora, Pinnipedia, Hyracoidea and Artiodactyla. *J. Zool.* **154**, 517–31
LEYHAUSEN, P. (1956) Verhaltensstudien an Katzen. *Z. Tierpsychol. Beiheft* **2**, 1–120
—— (1963) Über südamerikanische Pardelkatzen. *Ibid.* **20**, 627–40
—— (1965a) The communal organisation of solitary mammals. *Symp. zool. Soc. Lond.* **14**, 249–63
—— (1965b) Über die Funktion der relativen Stimmungshierarchie (dargestellt am Beispiel der phylogenetischen und ontogenetischen Entwicklung des Beutefangs von Raubtieren). *Z. Tierpsychol.* **22**, 412–94
—— (1969) Further comment on the proposed conservation of *Panthera* Oken, 1816 (Mammalia, Carnivora). *Bull. zool. Nomencl.* **25**, 130
—— & WOLFF, R. (1959) Das Revier einer Hauskatze. *Z. Tierpsychol.* **16**, 666–70
—— & FALKENA, M. (1966) Breeding the Brazilian ocelot-cat, *Leopardus tigrinus* in captivity. *Int. Zoo Yb.* **6**, 176–8
—— & TONKIN, B. (1966) Breeding the black-footed cat, *Felis nigripes*, in captivity. *Ibid.* **6**, 178–82
LIERS, E.E. (1951) Notes on the river otter (*Lutra canadensis*). *J. Mammal.* **32**, 1–9
LIMBAUGH, C. (1961) Observations on the California sea otter. *Ibid.* **42**, 271–3
LINDEMANN, W. (1953) Einiges über die Wildkatzen der Ostkarpathen (*Felis s. silvestris* Schreber, 1777). *Säugetierk. Mitt.* **1**, 73–4
—— (1955) Über die Jugendentwicklung beim Luchs (*Lynx l. lynx* Kerr) und bei der Wildkatze (*Felis s. silvestris* Schreb.). *Behaviour* **8**, 1–45
LINHART, S.B. (1968) Dentition and pelage in the juvenile red fox (*Vulpes vulpes*). *J. Mammal.* **49**, 526–8
LINN, I. & DAY, M.G. (1966) Identification of individual weasels, *Mustela nivalis*, using the ventral pelage pattern. *J. Zool.* **148**, 583–5
LLEWELLYN, L.M. & UHLER, F.M. (1952) The foods of fur animals of the Patuxent Research Refuge, Maryland. *Amer. midl. Nat.* **48**, 193–203
LOCKIE, J.D. (1959) The estimation of the food of foxes. *J. Wildl. Mgmt.* **23**, 224–7
—— (1961) The food of the pine marten, *Martes martes*, in west Ross-shire, Scotland. *Proc. zool. Soc. Lond.* **136**, 187–95
—— (1966) Territory in small carnivores. *Symp. zool. Soc. Lond.* **18**, 143–65
LONG, C.A. (1969) Gross morphology of the penis in seven species of the Mustelidae. *Mammalia* **33**, 145–60
—— & FRANK, T. (1968) Morphometric variation and function in the baculum, with comments on correlation of parts. *J. Mammal.* **49**, 32–43
—— & SHIREK, L.R. (1970) Variation and correlation in the genital bones of ranch mink. *Z. Säugetierk.* **35**, 252–5
LONGLEY, W.H. (1962) Movements of red fox. *J. Mammal.* **43**, 107
LÖNNBERG, E. (1902) On the female genital organs of *Cryptoprocta*. *Bih. Svenska. Akad.* **28** (4), 1–11
LOUWMAN, J.W.W. (1970) Breeding the banded palm civet and the banded linsang, *Hemigalus derbyanus* and *Prionodon linsang* at Wassenaar Zoo. *Int. Zoo Yb.* **10**, 81–2
—— & VAN OYEN, W.G. (1968) A note on breeding Temminck's golden cat, *Felis temminckii*, at Wassenaar Zoo. *Ibid.* **8**, 47–9

LUDWIG, J. (1965) Beobachtungen über das Spiel bei Boxern. *Z. Tierpsychol.* **22**, 813–38

LUICK, J.R., PARKER, H.R. & ANDERSEN, A.C. (1960) Composition of beagle dog milk. *Amer. J. Physiol.* **199**, 731–2

LULL, R.S. (1929) *Organic evolution* (Macmillan: New York)

LUND, H.M. (1962) The red fox in Norway. II The feeding habits of the red fox in Norway. *Medd. St. Viltundersok.* **12**, 1–75

LYALL-WATSON, M. (1963) A critical re-examination of food 'washing' behaviour in the raccoon. *Proc. zool. Soc. Lond.* **141**, 371–93

LYDDEKER, R. (1893–4) *Royal Natural History*, 1 & 2 (Warne & Co: London)

—— (1896) *A hand-book to the Carnivora* (Lloyd Ltd: London)

—— (1912) On the milk dentition of the ratel. *Proc. zool. Soc. Lond.* **1912**, 221–4

LYNCH, G.M. (1967) Long range movement of a raccoon in Manitoba. *J. Mammal.* **48**, 656–60

MCINTOSH, D.L. (1963a) Food of the fox in the Canberra district. *C.S.I.R.O. Wildl. Res.* **8**, 1–20

—— (1963b) Reproduction and growth of the fox in the Canberra district. *Ibid.* **8**, 132–41

MACINTYRE, G.T. (1966) The Miacidae (Mammalia, Carnivora). I The systematics of *Ictidopappus* and *Protictis*. *Bull. Amer. Mus. nat. Hist.* **131**, 119–209

MCMURRY, F.B. & SPERRY, C.C. (1941) Food of feral house cats in Oklahoma, a progress report. *J. Mammal.* **22**, 185–90

MCTOLDRIDGE, E.R. (1969) Notes on breeding ring-tailed coatis, *Nasua nasua*, at Santa Barbara Zoo. *Int. Zoo Yb.* **9**, 89–90

MAKACHA, S. & SCHALLER, G.B. (1969) Observations on lions in the Lake Manyara National Park, Tanzania. *E. Afr. Wildl. J.* **7**, 99–103

MALLINSON, J.J.C. (1969) Notes on breeding the African civet, *Viverra civetta*, at Jersey Zoo. *Int. Zoo Yb.* **9**, 92–3

MANTER, J.T. (1938) The dynamics of quadrupedal walking. *J. exp., Biol.* **15**, 522–40

MANTON, V.J.A. (1970) Breeding cheetahs, *Acinonyx jubatus*, at Whipsnade Park. *Int. Zoo Yb.* **10**, 85–6

—— (1971) A further report on breeding cheetahs, *Acinonyx jubatus*, at Whipsnade Park. *Ibid.* **11**, 125–6

MARKLEY, M.H. & BASSETT, C.F. (1942) Habits of captive marten. *Amer. midl. Nat.* **28**, 604–16

MARLOW, B.J.G. (1958) A survey of the marsupials of New South Wales. *C.S.I.R.O. Wildl. Res.* **3**, 71–114

MARMA, B.B. & YUNCHIS, V.V. (1968) Observations on the breeding, management and physiology of snow leopards, *Panthera u. uncia*, at Kaunas Zoo from 1962 to 1967. *Int. Zoo Yb.* **8**, 66–74

MARSHALL, W.H. (1946) Winter food habits of the pine marten in Montana. *J. Mammal.* **27**, 83–4

—— (1963) The ecology of mustelids in New Zealand. *New Zealand D.S.I.R. Information Series*, **38**, 1–32

MARSTON, M.A. (1942) Winter relations of bobcats to white-tailed deer in Maine. *J. Wildl. Mgmt.* **6**, 328–37

MARTIN, S.J (1929) On the Himalayan palm-civet. *J. Bombay nat. Hist. Soc.* **33**, 703

MARTINS, T. (1949) Disgorging of food to the puppies by the lactating dog. *Physiol. Zool.* **22**, 169–72

MATHESON, C. (1963) The distribution of the red polecat in Wales. *Proc. zool. Soc. Lond.* **140**, 115–20

MATSON, J.R. (1948) Cats kill deer. *J. Mammal.* **29**, 69–70

—— (1954) Observations on the dormant phase of a female black bear. *Ibid.* **35**, 28–35

MATTHEW, W.D. (1930) The phylogeny of dogs. *Ibid.* **11**, 117–38

MATTHEWS, L.H. (1939) Reproduction in the spotted hyaena, *Crocuta crocuta* (Erxl.). *Philos. Trans. B.* **230**, 1–78

MAXWELL, G. (1960) *Ring of bright water* (Longmans: London)

MAYNE, R. (1963) Just a skunk. *Afr. wild Life* **17**, 159–64

MAZAK, V. (1968) Note sur les caractères crâniens de la sousfamille des Pantherinae (Carnivora: Felidae). *Mammalia* **32**, 670–76

MEAD, R.A. (1968a) Reproduction in western forms of the spotted skunk (genus *Spilogale*). *J. Mammal.* **49**, 373–90

—— (1968b) Reproduction in eastern forms of the spotted skunk (genus *Spilogale*). *J. Zool.* **156**, 119–36

MECH, D. (1965) Mink: the weasel of the waterways. *Animal Kingd.* **68**, 87–90

MECH, L.D. (1966) The wolves of Isle Royale. *Fauna nat. Parks U.S.* **7**

—— (1970) *The wolf* (American Museum of Natural History: New York)

——, TESTER, J.R. & WARNER, D.W. (1966) Fall daytime resting habits of raccoons as determined by telemetry. *J. Mammal.* **47**, 450–66

—— & TURKOWSKI, F.J. (1966) Twenty-three raccoons in one winter den. *Ibid.* **47**, 529–30

MELLO, N.K. & PETERSON, N.J. (1964) Behavioral evidence for color discrimination in cat. *J. Neurophysiol.* **27**, 323–33

MENGEL, R.M. (1971) A study of dog-coyote hybrids and implications concerning hybridization in *Canis*. *J. Mammal.* **52**, 316–36

VAN MENSCH, P.J.A. & VAN BREE, P.J.H. (1969) On the African golden cat, *Profelis aurata* (Temminck, 1827). *Biologia Gabonica* **5**, 235–69

MEYER, D.R. & ANDERSON, R.A. (1965) Colour discrimination in cats: in *Colour vision*, ed. A.U.S. de Reuck & J. Knight (Churchill Ltd.: London)

MEYER-HOLZAPFEL, M. (1957) Das Verhalten der Bären (Ursidae). *Handb. Zool. Berl.* **8** (10), 17, 1–28

—— (1968) Breeding the European wild cat, *Felis s. silvestris*, at Berne Zoo. *Int. Zoo Yb.* **8**, 31–8

MICHELS, K.M., FISCHER, B.E. & JOHNSON, J.I. (1960) Raccoon performance on color discrimination problems. *J. comp. physiol. Psychol.* **53**, 379–80

MIDDLETON, A.D. (1935) *Meles meles*: food from stomach examination. *J. anim. Ecol.* **4**, 274

—— (1954) Rural rat control: in *Control of rats and mice*, ed. D. Chitty **1**, 414–48

MILLER, M.E., CHRISTENSEN, G.C. & EVANS, H.E. (1964) *Anatomy of the dog* (Saunders Co.: Philadelphia)

MILNER, C. (1967) Badger damage to upland pasture. *J. Zool.* **153**, 544–6

MINCKLEY, W.L. (1966) Coyote predation on aquatic turtles. *J. Mammal.* **47**, 137

MITCHELL, B.L., SHENTON, J.B. & UYS, J.C.M. (1965) Predation on large mammals in the Kafue National Park, Zambia. *Zoologica Africana* **1**, 297–318

MITCHELL, P.C. (1905) On the intestinal tract of mammals. *Trans. zool. Soc. Lond.* **17**, 437–536

MIVART, ST.G. (1881) *The cat* (Murray: London)

—— (1882a) On the classification and distribution of the Aeluroidea. *Proc. zool. Soc. Lond.* **1882**, 135–208

—— (1882b) Notes on some points in the anatomy of the Aeluroidea. *Ibid.* **1882**, 459–520

MONTGOMERY, G.G. (1969) Weaning of captive raccoons. *J. Wildl. Mgmt.* **33**, 154–9

——, LANG, J.W. & SUNQUIST, M.E. (1970) A raccoon moves her young. *J. Mammal.* **51**, 202–3

MORRIS, D. (1965) *The mammals* (Hodder & Stoughton: London)

MORRIS, R. & MORRIS, D. (1966) *Men and pandas* (Hutchinson: London)

MORRISON-SCOTT, T.C.S. (1952) The mummified cats of ancient Egypt. *Proc. zool. Soc. Lond.* **121**, 861–7

MORTIMER, M.A.E. (1963) Notes on the biology and behaviour of the spotted-necked otter (*Lutra maculicollis*). *Puku* **1**, 192–206

MOULTON, D.G., ASHTON, E.H. & EAYRS, J.T. (1960) Studies in olfactory acuity. 4) Relative detectability of n-aliphatic acids by the dog. *Anim. Behav.* **8**, 117–28

MUNDY, K.R.D. & FLOOK, D.R. (1964) Notes on the mating activity of grizzly and black bears. *J. Mammal.* **45**, 637–8

MURIE, A. (1935) A weasel goes hungry. *Ibid.* **16**, 321–2

—— (1936) Following fox trails. *Misc. Publ. Zool. Univ. Michigan* **32**, 1–45

—— (1937) Some food habits of the black bear. *J. Mammal.* **18**, 238–40

—— (1940) Ecology of the coyote in the Yellowstone. *Bull. Fauna nat. Parks U.S.* **4**, 1–206

—— (1944) The wolves of Mount McKinley. *Fauna Nat. Parks U.S. Fauna Series* **5**, 1–238

—— (1951) Coyote food habits on a southwestern cattle range. *J. Mammal.* **32**, 291–5

—— (1961) Some food habits of the marten. *Ibid.* **42**, 516–21

MURIE, J. (1871) On the female generative organs, viscera and fleshy parts of *Hyaena brunnea* Thunberg. *Trans. zool. Soc. Lond.* **7**, 503–12

MURIE, O.J. (1940) Notes on the sea-otter. *J. Mammal.* **21**, 119–131

—— (1945) Notes on coyote food habits in Montana and British Columbia. *Ibid.* **26**, 33–40

MUUL, I. & LIM, BOO-LIAT (1970) Ecological and morphological observations on *Felis planiceps*. *J. Mammal.* **51**, 806–8

NAAKTGEBOREN, C. (1968) Some aspects of parturition in wild and domestic Canidae. *Int. Zoo Yb.* **8**, 8–13

NADER, I.A. & MARTIN, R.L. (1962) The shrew as prey of the domestic cat. *J. Mammal.* **43**, 417

NEAL, E. (1948) *The badger* (Collins: London)

—— (1970a) The banded mongoose, a little known carnivore. *Animals* **13**, 29–31

—— (1970b) The banded mongoose, *Mungos mungo* Gmelin. *E. Afr. Wildl. J.* **8**, 63–71

NEFF, W.D. & HIND, J.E. (1955) Auditory thresholds of the cat. *J. acoust. Soc. Amer.* **27**, 480–3

—— & DIAMOND, I.T. (1958) The neural basis of auditory discrimination. In

Biological and biochemical bases of behaviour, ed. H.F. Harlow & C.N. Woolsey (University of Wisconsin Press), 101–26

Negus, V.E. (1956) The organ of Jacobson. *J. Anat.* **90**, 515–19

Neill, W.T. (1953) Two erythristic raccoons from Florida. *J. Mammal.* **34**, 500

Nellis, C.H. & Keith, L.B. (1968) Hunting activity and success of lynxes in Alberta. *J. Wildl. Mgmt.* **32**, 718–22

—— & Wetmore, S.P. (1969) Long-range movement of lynx in Alberta. *J. Mammal.* **50**, 640

Neumann, F. & Schmidt, H.D. (1959) Optische Differenzierungsleistungen von Musteliden. Versuche an Frettchen und Iltisfrettchen. *Z. vergl. Physiol.* **42**, 199–205

Norris, T. (1969) Ceylon sloth bear. *Animals* **12**, 300–3

North, W.J. & Pearse, J.S. (1970) Sea urchin population explosion in southern Californian waters. *Science* **167**, 209

Novikov, G.A. (1956) *Carnivorous mammals of the fauna of the USSR* (English Translation, Israel Program for Scientific Translations: Jerusalem, 1962)

Oeming, A.F. (1962) The friendly lynx. *Zoonooz* **35** (4), 3–7

Orians, G.H. & Pfeiffer, E.W. (1970) Ecological effects of the war in Vietnam. *Science* **168**, 544–54

Orts, S.G. (1970) Le *Xenogale* de J.A. Allen (Carnivora, Viverridae) au sujet d'une capture effectuée au Kivu. *Rev. Zool. Bot. Afr.* **82**, 174–86

Osgood, F.L. (1936) Earthworms as supplementary food of weasels. *J. Mammal.* **17**, 64

Oxnard, C.E. (1968) The architecture of the shoulder in some mammals. *J. Morph.* **126**, 249–90

Ozoga, J.J. & Harger, E.M. (1966) Winter activities and feeding habits of northern Michigan coyotes. *J. Wildl. Mgmt.* **30**, 809–18

Palen, G.F. & Goddard, G.V. (1966) Catnip and oestrous behaviour in the cat. *Anim. Behav.* **14**, 372–7

Paradiso, J.L. (1966) Notes on supernumerary and missing teeth in the coyote. *Mammalia* **30**, 120–8

—— (1968) Canids recently collected in east Texas, with comments on the taxonomy of the red wolf. *Amer. midl. Nat.* **80**, 529–34

Pauly, L.K. & Wolfe, H.R. (1957) Serological relationships among members of the order Carnivora. *Zoologica* **42**, 159–66

Pavlov, I.P. (1928) *Lectures on conditioned reflexes* (International Publishers: New York)

Pearlman, A.L. & Daw, N.W. (1970) Opponent color cells in the cat lateral geniculate nucleus. *Science* **167**, 84–6

Pearson, O.P. & Enders, R.K. (1944) Duration of pregnancy in certain mustelids. *J. exp. Zool.* **95**, 21–35

—— & Baldwin, P.H. (1953) Reproduction and age structure of a mongoose population in Hawaii. *J. Mammal.* **34**, 436–47

Pedersen, A. (1962) *Polar animals* (Harrap: London)

Pedersen, R.J. & Stout, J. (1963) Oregon sea otter sighting. *J. Mammal.* **44**, 415

Pemberton, C.E. (1925) The field rat in Hawaii and its control. *Bull. Hawaii Sug. Ass. ent. Ser.* **17**, 1–46

PETERSON, F.A., HEATON, W.C. & WRUBLE, S.D. (1969) Levels of auditory response in fissiped carnivores. *J. Mammal.* **50**, 566–78

PETRABORG, W.H. & GUNVALSON, V.E. (1962) Observations on bobcat mortality and bobcat predation on deer. *Ibid.* **43**, 430–1

PETTER, F. (1957) La reproduction du fennec. *Mammalia* **21**, 307–9

—— (1959) Reproduction en captivité du zorille du sahara, *Poecilictis libyca*. *Ibid.* **23**, 378–80

PETTER, G. (1969) Interprétation évolutive des caractères de la denture des viverridés africains. *Ibid.* **33**, 607–25

PFAFFMAN, C. (1955) Gustatory nerve impulses in rat, cat and rabbit. *J. Neurophysiol.* **18**, 429–40

PHILLIPS, R.L. & MECH, L.D. (1970) Homing behavior of a red fox. *J. Mammal.* **51**, 621

PHILLIPSON, A.T. (1947) The production of fatty acids in the alimentary tract of the dog. *J. exp. Biol.* **23**, 346–9

PIENAAR, U. de V. (1969) Predator-prey relationships amongst the larger mammals of the Kruger National Park. *Koedoe* **12**, 108–76

VAN DER PIJL, L. (1969) *Principles of dispersal in higher plants* (Springer: Berlin)

PILLERI, G. (1967) Retinalfalten im Auge von Wassersäugetieren. *Experientia* **23**, 54–5

PIMENTAL, D. (1955) Biology of the Indian mongoose in Puerto Rico. *J. Mammal.* **36**, 62–8

PIMLOTT, D.H. (1967) Wolf predation and ungulate populations. *Amer. Zool.* **7**, 267–78

PIVETEAU, J. (1961) Traité de Paléontologie, **6** (1) (Masson et Cie: Paris)

PLATT, A.P. (1968) Selective predation by a mink on woodland jumping mice confined in live traps. *Amer. midl. Nat.* **79**, 539–40

POCOCK, R.I. (1908) Warning colouration in the musteline Carnivora. *Proc. zool. Soc. Lond.* **1908**, 944–59

—— (1911) Some probable and possible instances of warning characteristics amongst insectivorous and carnivorous mammals. *Ann. Mag. nat. Hist.* **8**, 750–7

—— (1914a) On the feet of domestic dogs. *Proc. zool. Soc. Lond.* **1914** (1), 478–84

—— (1914b) On the facial vibrissae of mammalia. *Ibid.* **1914**, 889–912

—— (1914c) On the feet and other external features of the Canidae and Ursidae. *Ibid.* **1914**, 913–41

—— (1915a) On some of the external characters of *Cynogale bennettii*, Gray. *Ann. Mag. nat. Hist.* **15**, 351–60

—— (1915b) On some of the external characters of the palm-civet (*Hemigalus derbyanus*, Gray) and its allies. *Ibid.* **16**, 153–62

—— (1915c) On some of the external characters of the genus *Linsang*, with notes upon the genera *Poiana* and *Eupleres*. *Ibid.* **16**, 341–51

—— (1915d) On some external characters of *Galidia*, *Galidictis* and related genera. *Ibid.* **16**, 351–6

—— (1915e) On the feet and glands and other external characters of the Viverrinae, with the description of a new genus. *Proc. zool. Soc. Lond.* **1915**, 131–49

—— (1915f) On the feet and glands and other external characters of the paradoxurine genera *Paradoxurus, Arctictis, Arctogalidia* and *Nandinia*. *Ibid.* **1915**, 387–412

—— (1916a) On some of the external structural characters of the striped hyaena (*Hyaena hyaena*) and related genera and species. *Ann. Mag. nat. Hist.* **17**, 330–43

—— (1916b) On some of the external characters of *Cryptoprocta*. *Ibid.* **17**, 413–25

—— (1916c) On the hyoidean apparatus of the lion (*F. leo*) and related species of the Felidae. *Ibid.* **18**, 222–9

—— (1916d) On the tooth-change, cranial characters and classification of the snow-leopard or ounce (*Felis uncia*). *Ibid.* **18**, 306–13

—— (1916e) On some of the cranial and external characters of the hunting leopard or cheetah (*Acinonyx jubatus*). *Ibid.* **18**, 419–29

—— (1916f) The tympanic bulla in hyaenas. *Proc. zool. Soc. Lond.* **1916**, 303–7

—— (1916g) On the external characters of the mongooses (Mungotidae). *Ibid.* **1916**, 349–74

—— (1916h) The alisphenoid canal in civets and hyaenas. *Ibid.* **1916**, 442–5

—— (1917) On the external characters of the Felidae. *Ann. Mag. nat. Hist.* **19**, 113–36

—— (1918a) The baculum or os penis of some genera of Mustelidae. *Ibid.* **1**, 307–12

—— (1918b) Further notes on some external characters of the bears (Ursidae). *Ibid.* **1**, 375–84

—— (1920a) On the external characters of the ratel (*Mellivora*) and the wolverine (*Gulo*). *Proc. zool. Soc. Lond.* **1920**, 179–87

—— (1920b) On the external and cranial characters of the European badger (*Meles*) and of the American badger (*Taxidea*). *Ibid.* **1920**, 423–36

—— (1921a) The external characters and classification of the *Procyonidae*. *Ibid.* **1921**, 389–422

—— (1921b) On the external characters of some species of Lutrinae (otters). *Ibid.* **1921**, 535–46

—— (1921c) On the external characters and classification of the Mustelidae. *Ibid.* **1921**, 803–37

—— (1923) On the feet and rhinarium of the polar bear (*Thalarctos maritimus*). *Ibid.* **1923**, 159–60

—— (1925) The external characters of an American badger (*Taxidea taxus*) and an American mink (*Mustela vison*), recently exhibited in the Society's Gardens. *Ibid.* **1925**, 17–27

—— (1926) The external characters of the Patagonian weasel (*Lyncodon patagonicus*). *Ibid.* **1926**, 1085–94

—— (1927a) The external characters of the South African striped weasel (*Poecilogale albinucha*). *Ibid.* **1927**, 125–33

—— (1927b) Description of a new species of cheetah (*Acinonyx*). *Ibid.* **1927**, 245–52

—— (1927c) The external characters of a bush-dog (*Speothos venaticus*) and of a maned wolf (*Chrysocyon brachyurus*), exhibited in the Society's Gardens. *Ibid.* **1927** (1), 307–21

—— (1928a) The structure of the auditory bulla in the Procyonidae and the Ursidae, with a note on the bulla of *Hyaena*. *Ibid.* **1928**, 963–74

—— (1928b) Some external characters of the giant panda (*Ailuropoda melanoleuca*). *Ibid.* **1928**, 975–81
—— (1928c) Some external characters of the sea-otter (*Enhydra lutris*). *Ibid.* **1928**, 983–91
—— (1933) The rarer genera of oriental Viverridae. *Ibid.* **1933**, 969–1035
—— (1934) The palm civets or 'toddy cats' of the genera *Paradoxurus* and *Paguma* inhabiting British India. *J. Bombay nat. Hist. Soc.* **36**, 855–77 and **37**, 172–92, 314–46
—— (1938) The Algerian sand cat (*Felis margarita* Loche) *Proc. zool. Soc. Lond.* 108B, 41–6
—— (1939) *The fauna of British India, I* (Taylor & Francis: London)
—— (1951) *Catalogue of the genus Felis* (British Museum of Natural History: London)
POGLAYEN-NEUWALL, I. (1962) Beiträge zu einem Ethogram des Wickelbären (*Potos flavus* Schreber). *Z. Säugetierk.* **27**, 1–44
—— & POGLAYEN-NEUWALL, I. (1965) Gefangenschaftsbeobachtungen an Makibären (*Bassaricyon* Allen 1876). *Ibid.* **30**, 321–66
POHLE, A. (1967) Beiträge zur Ethologie und Biologie des Sonnendachses (*Helictis personata* Gray 1831) in Gefangenschaft. *Zool. Gart. Lpz.* **33**, 225–47
POLDERBOER, E.B. (1942) Habits of the least weasel (*Mustela rixosa*) in northeastern Iowa. *J. Mammal.* **23**, 145–7
POLLACK, E.M. (1951) Observations on New England bobcats. *Ibid.* **32**, 356–8
POOLE, T.B. (1966) Aggressive play in polecats. *Symp. zool. Soc. Lond.* **18**, 23–44
POURNELLE, G.H. (1965) Observations on birth and early development of the spotted hyaena. *J. Mammal.* **46**, 503
POWELL. J.E. (1913) Notes on the habits of the small Indian mongoose (*Mungos auropunctatus*). *J. Bombay nat. Hist. Soc.* **22**, 620
PRATER, S.H. (1935) The wild animals of the Indian Empire. III Carnivora or beasts of prey. *Ibid.* **37**, 112–66
PRIEWERT, F.W. (1961) Record of an extensive movement by a raccoon. *J. Mammal.* **42**, 113
PROCTER, J. (1963) A contribution to the natural history of the spotted-necked otter (*Lutra maculicollis* Lichtenstein) in Tanganyika. *E. Afr. Wildl. J.* **1**, 93–102
PROGULSKE, D.R. (1969) Observations of a penned, wild-captured black-footed ferret. *J. Mammal.* **50**, 619–20
PROVOST, E. (in press) Population dynamics in the bobcat: in *Ecology and conservation of the world's cats*, ed. R.L. Eaton
PRUITT, W.O. (1965) A flight releaser in wolf-caribou relations. *J. Mammal.* **46**, 350–1
PULLIANEN, E. (1965) Studies on the wolf in Finland. *Ann. Zool. Fenn.* **2**, 215–59
QUICK, H.F. (1951) Notes on the ecology of weasels in Gunnison County, Colorado. *J. Mammal.* **32**, 281–90
—— (1952) Some characteristics of wolverine fur. *Ibid.* **33**, 492–3
—— (1955) Food habits of marten (*Martes americana*) in northern British Columbia. *Canad. Field Nat.* **69**, 144–7
RÄBER, H. (1944) Versuche zur Ermittlung des Beuteschemas an einem

Hausmarder (*Martes foina*) und einem Iltis (*Putorius putorius*). *Rev. Suisse Zool.* **51**, 293–332

RAND, A.L. (1935) On the habits of some Madagascar mammals. *J. Mammal.* **16**, 89–104

RAUSCH, R.A. (1967) Some aspects of the population ecology of wolves, Alaska. *Amer. Zool.* **7**, 253–65

RAUSCH, R.L. (1961) Notes on the black bear, *Ursus americanus* Pallas, in Alaska, with particular reference to dentition and growth. *Z. Säugetierk.* **26**, 77–107

RAVEN, H.C. (1936) Notes on the anatomy of the viscera of the giant panda (*Ailuropoda melanoleuca*). *Amer. Mus. Novit.* **877**, 1–23

RAYNAUD, A. (1969) 'Les organes génitaux des Mammifères'; in *Traité de Zoologie*, **16** (6), ed. P.P. Grassé (Masson et Cie: Paris)

REICHSTEIN, H. (1957) Schädelvariabilität europaischer Mauswiesel (*Mustela nivalis*) und Hermeline (*Mustela erminea*) in Beziehung zu Verbreitung und Geschlecht. *Z. Säugetierk.* **22**, 151–82

RENSCH, B. & DÜCKER, G. (1959) Die Spiele von Mungo und Ichneumon. *Behaviour* **14**, 185–213

—— (1963) Haptisches Lern- und Unterscheidungsvermögen bei einem Waschbären. *Z. Tierpsychol.* **20**, 608–15

—— & —— (1969) Manipulierfähigkeit eines Wickelbären bei längeren Handlungsketten. *Ibid.* **26**, 104–12

RHEINGOLD, H.L. (1963) Maternal behavior in the dog: in *Maternal behavior in mammals*, ed. H.L. Rheingold (John Wiley & Sons: New York), 169–202

RICHARDSON, W.B. (1942) Ring-tailed cats (*Bassariscus astutus*): their growth and development. *J. Mammal.* **23**, 17–26

RINKER, G.C. (1944) Os clitoridis from the raccoon. *Ibid.* **25**, 92

LA RIVERS, I. (1948) Some Hawaiian ecological notes. *Wasmann Coll.* **7**, 85–110

ROBERTS, A. (1951) *The mammals of South Africa* (Central News Agency: Cape Town)

ROBINETTE, W.L., GASHWILER, J.S. & MORRIS, O.W. (1959) Food habits of the cougar in Utah and Nebraska. *J. Wildl. Mgmt.* **23**, 261–73

——, —— & —— (1961) Notes on cougar productivity and life history. *J. Mammal.* **42**, 204–17

ROBINSON, R. (1969a) The white tigers of Rewa. *Carnivore Genet. Newsl.* **8**, 192–3

—— (1969b) The breeding of spotted and black leopards. *J. Bombay nat. Hist. Soc.* **66**, 423–9

ROBINSON, W.B. (1952) Some observations on coyote predation in Yellowstone National Park. *J. Mammal.* **33**, 470–6

ROEST, A.I. (1961) Partially albino badger from California. *Ibid.* **42**, 275–6

ROLLINGS, C.T. (1945) Habits, food and parasites of the bobcat in Minnesota. *J. Wildl. Mgmt.* **9**, 131–45

ROMER, A.S. (1968) *Notes and Comments on Vertebrate Paleontology* (University of Chicago Press)

ROSE, J.E. (1968) Discussion following paper by E.F. Evans in *C.I.B.A* Symposium on *Hearing Mechanisms in Vertebrates* (Churchill Ltd: London)

ROSENBERG, H. (1971) Breeding the bat-eared fox, *Otocyon megalotis* at Utica Zoo. *Int. Zoo Yb.* **11**, 101–2

ROTHSCHILD, M. (1942) Change of pelage in the stoat, *Mustela erminea* L. *Nature Lond.* **149**, 78
—— (1944) Pelage change of the stoat, *Mustela erminea* L. *Ibid.* **154**, 180–1
—— (1957) Note on change of pelage in the stoat. *Proc. zool. Soc. Lond.* **128**, 602
ROWBOTTOM, J. (1969) Watching otters in Scotland. *Animals* **12**, 159–61
ROWE-ROWE, D.T. (1969) Some observations on a captive African weasel, *Poecilogale albinucha. Lammergeyer* **10**, 93–6
—— (1971) The development and behaviour of a rusty-spotted genet, *Genetta rubiginosa* Pucheran. *Ibid.* **13**, 29–44
ROWLANDS, I.W. & PARKES, A.S. (1935) The reproductive processes of certain mammals. VIII Reproduction in foxes (*Vulpes* spp.). *Proc. zool. Soc. Lond.* **1935**, 823–41
ROZENZWEIG, M.L. (1966) Community structure in sympatric carnivora. *J. Mammal.* **47**, 602–12
RUST, C.C. (1962) Temperature as a modifying factor in the spring pelage change of short-tailed weasels. *Ibid.* **43**, 323–8
—— (1965) Hormonal control of pelage cycles in the short-tailed weasel (*Mustela erminea bangsi*). *Gen. comp. Endocrinol.* **5**, 222–31
——, SHACKLEFORD, R.M. & MEYER, R.K. (1965) Hormonal control of pelage cycles in the mink. *J. Mammal.* **46**, 549–65
RYDER, R.A. (1955) Fish predation by the otter in Michigan. *J. Wildl. Mgmt.* **19**, 497–8
SAINT GIRONS, M.C. (1962) Notes sur les dates de reproduction en captivité du fennec, *Fennecus zerda* (Zimmermann, 1780). *Z. Säugetierk.* **27**, 181–4
SANDEGREN, F.E., CHU, E. & VANDEVERE, J. (in press) Maternal behavior in the California sea otter.
SANDERSON, G.C. (1950) Methods of measuring productivity in raccoons. *J. Wildl. Mgmt.* **14**, 389–402
SANDERSON, I.T. (1940) The mammals of the north Cameroons forest area. *Trans. zool. Soc. Lond.* **24**, 623–725
SANKHALA, K.S. (1967) Breeding behaviour of the tiger, *Panthera tigris* in Rajasthan. *Int. Zoo Yb.* **7**, 133–47
SARICH, V.M. (1969a) Pinniped origins and the rate of evolution of carnivore albumins. *Syst. Zool.* **18**, 286–95
—— (1969b) Pinniped phylogeny. *Ibid.* **18**, 416–22
SAUNDERS, J.K. (1963a) Food habits of the lynx in Newfoundland. *J. Wildl. Mgmt.* **27**, 384–90
—— (1963b) Movements and activities of the lynx in Newfoundland. *Ibid.* **27**, 390–400
—— (1964) Physical characteristics of the Newfoundland lynx. *J. Mammal.* **45**, 36–47
SCAPINO, R.P. (1965) The third joint of the canine jaw. *J. Morph.* **116**, 23–50
SCHAFFER. J. (1940) *Die Hautdrüsenorgane der Säugetiere* (Urban & Schwarzenberg; Berlin)
SCHALLER, G.B. (1967) *The deer and the tiger* (University of Chicago Press)
—— (1968a) Hunting behaviour of the cheetah in the Serengeti National Park, Tanzania. *E. Afr. Wildl. J.* **6**, 95–100

—— (1968b) Serengeti lion study. *UNESCO Bull. Regional Centre for Science and Technology for Africa* **3**, 43-5
—— (1969a) The hunt of the cheetah. *Animal Kingd.* **72** (2), 2-7
—— (1969b) Food habits of the Himalayan black bear (*Selenarctos thibetanus*) in the Dachigam Sanctuary, Kashmir. *J. Bombay nat. Hist. Soc.* **66**, 156-9
—— (1969c) Life with the king of beasts. *Nat. Geogr.* **135**, 494-519
—— (1970) This gentle and elegant cat. *Nat. Hist.* **79** (6), 30-9
—— & LOWTHER, G.R. (1969) The relevance of carnivore behavior to the study of early hominids. *Southwest. J. Anthropol.* **25**, 307-41
SCHAUENBERG, P. (1969) Le lynx, *Lynx lynx* (L.), en Suisse et dans les pays voisins. *Rev. Suisse Zool.* **76**, 257-87
SCHEFFER, V.B. (1939) The os clitoridis of the Pacific otter. *Murrelet* **20**, 20-1
SCHENKEL, R. (1947) Ausdrucks-studien an Wölfen. *Behaviour* **1**, 81-129
—— (1966) Play, exploration and territoriality in the wild lion. *Symp. zool. Soc. Lond.* **18**, 11-22
—— (1967) Submission: its features and functions in the wolf and dog. *Amer. Zool.* **7**, 319-29
SCHMIDT, F. (1934) Über die Fortpflanzungsbiologie von sibirischen Zobel (*Martes zibellina* L.) und europäischen Baummarder (*Martes martes* L.). *Z. Säugetierk.* **9**, 392-403
—— (1943) Naturgeschichte des Baum- und des Steinmarders. *Monogr. Wildsäuget.* **10**, 1-258
SCHMIDT-NIELSEN, K. (1964) *Desert Animals – Physiological Problems of Heat and Water* (Oxford University Press)
——, BRETZ, W.L. & TAYLOR, C.R. (1970) Panting in dogs: unidirectional air flow over evaporative surfaces. *Science* **169**, 1102-4
SCHNEIDER. D.G., MECH, D.L. & TESTER, J.R. (1971) Movements of female raccoons and their young as determined by radio-tracking. *Anim. Behav. Monogr.* **4**, 1-43
SCHNEIDER, K.M. (1926) (quoted by Matthews 1939) *Pelztierzucht* **2**, 1
SCHOLANDER, P.F., WALTERS, V., HOCK, R. & IRVING, L. (1950a) Body insulation of some arctic and tropical mammals and birds. *Biol. Bull.* **99**, 225-36
——, HOCK, R., WALTERS, V., JOHNSON, F. & IRVING, L. (1950b) Heat regulation in some arctic and tropical mammals and birds. *Ibid.* **99**, 237-58
——, HOCK, R., WALTERS, V. & IRVING, L. (1950c) Adaptation to cold in arctic and tropical mammals and birds in relation to body temperature, insulation and basal metabolic rate. *Ibid.* **99**, 259-71
SCHOONMAKER. W.J. (1938) Notes on mating and breeding habits of foxes in New York State. *J. Mammal.* **19**, 375-6
SCHOONOVER, L.J. & MARSHALL, W.H. (1951) Food habits of raccoon (*Procyon lotor hirtus*) in north-central Minnesota. *Ibid.* **32**, 422-8
SCHUMACHER, G.H. (1961) *Funktionelle Morphologie der Kaumuskulatur* (Fischer: Jena)
SCOTT, J.P. (1962) Critical periods in behavioral development. *Science* **138**, 949-58
—— (1967) The evolution of social behavior in dogs and wolves. *Amer. Zool.* **7**, 373-81
SCOTT, T.G. (1943) Some food coactions of the northern plains red fox. *Ecol. Monogr.* **13**, 427-73

—— & KLIMSTRA, W.D. (1955) Red foxes and a declining prey population. *S. Illinois Univ. Press Monogr. 1* (Carbondale, Ill.)

SCOTT, W.B. (1937) *A history of land mammals of the western hemisphere* (Macmillan: New York)

SEAL, U.S. (1969) Carnivora systematics: a study of hemoglobins. *Comp. Biochem. Physiol.* **31**, 799–811

——, PHILLIPS, N.I. & ERICKSON, A.W. (1970) Carnivora systematics: immunological relationship of bear serum albumins. *Ibid.* **32**, 33–48

SEALANDER, J.A. (1943) Winter food habits of mink in Michigan. *J. Wildl. Mgmt.* **7**, 411–17

SEALY, S.G. (1968) Third upper molar in *Canis latrans. Mammalia* **32**, 712–13

SEAMAN, G.A. & RANDALL, J.E. (1962) The mongoose as a predator in the Virgin Islands. *J. Mammal.* **43**, 544–6

SEARLE, A.G. (1968) *Comparative genetics of coat colour in mammals* (Logos Press: London)

SECHZER, J.A. & BROWN, J.L. (1964) Color discrimination in the cat. *Science* **144**, 427–9

SEGALL, W. (1943) The auditory region of the arctoid carnivores. *Field Mus., Chicago, zool. Ser.* **29**, 33–59

SEITZ, A. (1955) Untersuchungen über angeborene Verhaltensweisen bei Caniden. III Beobachtungen an Marderhunden (*Nyctereutes procyonoides* Gray). *Z. Tirepsychol.* **12**, 463–89

—— (1959) Beobachtungen an handaufgezogenen Goldschakalen (*Canis aureus aligirensis* Wagner 1843). *Ibid.* **16**, 747–71

SELKO, L.F. (1937) Food habits of Iowa skunks in the fall of 1936. *J. Wildl. Mgmt.* **1**, 70–6

SETON, E.T. (1920) Acrobatic skunks. *J. Mammal.* **1**, 140

SHADLE, A.R. (1941) Black bear feeds on 'honey dew' and magnolia leaves. *Ibid.* **22**, 321

BEN SHAUL, D.M. (1962) The composition of the milk of wild animals. *Int. Zoo Yb.* **4**, 333–42

SHELDON, W.G. (1937) Notes on the giant panda. *J. Mammal.* **18**, 13–19

—— (1949) Reproductive behavior of foxes in New York State. *Ibid.* **30**, 236–46

—— (1950) Denning habits and home range of red foxes in New York State. *J. Wildl. Mgmt.* **14**, 33–42

—— (1953) Returns on banded red and gray foxes in New York State. *J. Mammal.* **34**, 125

—— & TOLL, W.G. (1964) Feeding habits of the river otter in a reservoir in central Massachusetts. *Ibid.* **45**, 449–55

SHERMAN, H.B. (1954) Raccoons of the Bahama Islands. *Ibid.* **35**, 126

SHORTRIDGE, G.C. (1934) *The mammals of South West Africa* (Heinemann: London)

SHUFELDT, R.W. (1924) The skull of the wolverene (*Gulo luscus*). *J. Mammal.* **5**, 189–93

SHUMAN, R.F. (1950) Bear depredations on red salmon spawning populations in the Karluk River system, 1947. *J. Wildl. Mgmt.* **14**, 1–9

SICHER, H. (1944) Masticatory apparatus in the giant panda and the bears. *Publ. Field Mus. nat. Hist. Zool.* **29** (4), 61–73

SIEFKE, A. (1960) Baummarder-Paarung. *Z. Säugetierk.* **25**, 178
SIKES, S.K. (1964) The ratel or honey badger. *Afr. wild Life* **18**, 29–37
DA SILVEIRA, E.K.P. (1968) Notes on the care and breeding of the maned wolf *Chrysocyon brachyurus* at Brasilia Zoo. *Int. Zoo Yb.* **8**, 21–3
SILVER, H. & SILVER, W.T. (1969) Growth and behavior of the coyote-like canid of northern New England with observations on canid hybrids. *Wildl. Monogr.* **17**, 1–41
SIMPSON, C.D. (1966) The banded mongoose. *Animal Kingd.* **69**, 52–7
SIMPSON, G.G. (1945) The principles of classification and a classification of mammals. *Bull. Amer. Mus. nat. Hist.* **85**, 1–350
SINHA, A.A., CONAWAY, C.H. & KENYON, K.W. (1966) Reproduction in the female sea otter. *J. Wildl. Mgmt.* **30**, 121–30
SISSON, S. & GROSSMAN, J.D. (1953) *The anatomy of the domestic animals* (W. B. Saunders & Co: Philadelphia)
SMITH, W.P. (1944) Red fox's method of hunting field mice. *J. Mammal.* **25**, 90–1
SMITHERS, R.H.N. (1966a) A southern bat-eared fox. *Animal Kingd.* **69**, 163–7
—— (1966b) *The mammals of Rhodesia, Zambia and Malawi* (Collins: London)
—— (1968a) Cat of the pharaohs. *Animal Kingd.* **61**, 16–23
—— (1968b) Preliminary identification manual for African mammals. 25. Carnivora: Felidae (Smithsonian Institution: Washington)
SMYTHE, N. (1970a) Relationship between fruiting seasons and seed dispersal methods in a neotropical forest. *Amer. Nat.* **104**, 23–35
—— (1970b) The adaptive value of the social organisation of the coati (*Nasua narica*). *J. Mammal.* **51**, 818–20
SNEAD, E. & HENDRICKSON, G.O. (1942) Food habits of the badger in Iowa. *Ibid.* **23**, 380–91
SNOW, C.J. (1967) Some observations on the behavioral and morphological development of coyote pups. *Amer. Zool.* **7**, 353–5
SONNTAG, C.F. (1923) Comparative anatomy of the tongues of the Mammalia. VIII: Carnivora. *Proc. zool. Soc. Lond.* **1923**, 129–53
—— (1925) Comparative anatomy of the tongues of the Mammalia. XII Summary, classification and phylogeny. *Ibid.* **1925**, 701–62
SOOTER, C.A. (1946) Habits of coyotes in destroying nests and eggs of waterfowl. *J. Wildl. Mgmt.* **10**, 33–8
SOSNOVSKII, I.P. (1967) Breeding the red dog or dhole, *Cuon alpinus*, at Moscow Zoo. *Int. Zoo Yb.* **7**, 120–2
SOUTHERN, H.N. (1964) *The handbook of British mammals* (Blackwell: Oxford).
—— & WATSON, J.S. (1941) Summer food of the red fox in Great Britain. A preliminary report. *J. anim. Ecol.* **10**, 1–11
SPECTOR, W.S. (1956) *Handbook of biological data* (W. B. Saunders: Philadelphia & London)
SPERRY, C.C. (1933) Autumn food habits of coyotes, a report of progress. *J. Mammal.* **14**, 216–20
—— (1939) Food habits of peg-leg coyotes. *Ibid.* **20**, 190–4
STARCK, D. (1964) Über das Entotympanicum der Canidae und Ursidae (Mammalia, Carnivora, Fissipedia). *Acta Theriol.* **8**, 181–8

—— (1967) Le crane des mammifères: in *Traité de Zoologie*, **16** (1), ed. P.P. Grassé (Masson et Cie: Paris)

STEBLER, A.M. (1938) Feeding behavior of a skunk. *J. Mammal.* **19**, 374

—— (1944) The status of the wolf in Michigan. *Ibid.* **25**, 37–42

STEGMAN, L.C. (1937) Notes on young skunks in captivity. *Ibid.* **18**, 194–202

STEHLIK, J. (1971) Breeding jaguars, *Panthera onca*, at Ostrava Zoo. *Int. Zoo Yb.* **11**, 116–18

STEINBACHER, G. (1939) Nüsse öffnender Sumpfichneumon. *Zool. Gart. Lpz.* **10**, 228–9

—— (1951) Nüsse öffnender Sumpfichneumon. *Ibid.* **18**, 58

STENLUND, M.H. (1955) A field study of the timber wolf (*Canis lupus*) in the Superior National Forest, Minnesota. *Tech. Bull. Minnesota Dept. Conserv.* **4**

STEPHENS, M.N. (1957) *The natural history of the otter* (UFAW: London)

STERNDALE, R.A. (1884) *Natural History of the Mammalia of India and Ceylon* (Thacker, Spink & Co.; London)

STEVENSON-HAMILTON, J. (1914) The coloration of the African hunting dog (*Lycaon pictus*). *Proc. zool. Soc. Lond.* **1914**, 403–5

STORM, G.L. (1965) Movements and activities of foxes as determined by radio tracking. *J. Wildl. Mgmt.* **29**, 1–13

STORY, H.E. (1945) The external genitalia and perfume gland in *Arctictis binturong*. *J. Mammal.* **26**, 64–6

STUEWER, F.W. (1943) Raccoons: their habits and management in Michigan. *Ecol. Monogr.* **13**, 203–57

SULLIVAN, E.G. (1956) Gray fox reproduction, denning, range and weights in Alabama. *J. Mammal.* **37**, 346–51

SUNQUIST, M.E., MONTGOMERY, G.G. & STORM, G.L. (1969) Movements of a blind raccoon. *Ibid.* **50**, 145–7

SUTCLIFFE, A.J. (1970) Spotted hyaena: crusher, digester and collector of bones. *Nature Lond.* **227**, 1110–13

SWANEPOL. P.D. (1962) Feast of kings. *Afr. wild Life* **16**, 215–24

SWITZENBERG, D.F. (1950) Breeding productivity in Michigan red foxes. *J. Mammal.* **31**, 194–5

TATE, G.H.H. (1931) Random observations on habits of south American mammals. *Ibid.* **12**, 248–56

TAYLOR, A. & DAVEY, M.R. (1968) Behaviour of jaw muscle stretch receptors during active and passive movements in the cat. *Nature Lond.* **220**, 301–2

TAYLOR, M. (1969) Note on the breeding of two genera of viverrids, *Genetta* spp. and *Herpestes sanguineus* in Kenya. *E. Afr. Wildl. J.* **7**, 168–9

—— (1970) Locomotion in some East African viverrids. *J. Mammal.* **51**, 42–51

TAYLOR, S. & WEBB, C.S. (1955) Breeding dwarf mongooses. *Zoo Life* **10**, 70–2

TAYLOR, W.P. (1954) Food habits and notes on life history of the ring-tailed cat in Texas. *J. Mammal.* **35**, 55–63

TAYLOR, W.T. & WEBER, R.J. (1951) *Functional mammalian anatomy (with special reference to the cat)* (Van Nostrand: New York)

TEMBROCK, G. (1957a) Das Verhalten des Rotfuchses. *Handb. Zool.* **8**, 10 (15), 1–20

—— (1957b) Zur Ethologie des Rotfuchses (*Vulpes vulpes* (L.)), unter besonderer Berücksichtigung des Fortpflanzung. *Zool. Gart. Lpz.* **23**, 289–532

—— (1958) Spielverhalten beim Rotfuchs. *Zool. Beitr. Berl.* **3**, 423–96

—— (1960) Spielverhalten und vergleichende Ethologie. Beobachtungen zum Spiel von *Alopex lagopus* (L.). *Z. Säugetierk.* **25**, 1–14

—— (1963) Acoustic behaviour of mammals. In *Acoustic Behaviour of Animals*, ed. R.G. Busnel (Elsevier: Amsterdam), 751–86

TENER, J.S. (1954a) A preliminary study of the musk-oxen of Fosheim Peninsula, Ellesmere Island, N.W.T. *Wildl. Mgmt. Bull. Can.* (1), **9**, 1–34

—— (1954b) Three observations of predators attacking prey. *Canada Field Nat.* **68**, 181–2

TERRES, J.K. (1939) Tree-climbing technique of a gray fox. *J. Mammal.* **20**, 256

TESTER, J.R. (1953) Fall foods of the raccoon in the south Platte Valley of north-eastern Colorado. *Ibid.* **34**, 500–2

TETLEY, H. (1941) On the Scottish fox. *Proc. zool. Soc. Lond.* **111**B, 25–35

THEBERGE, J.B. & FALLS, J.B. (1967) Howling as a means of communication in timber wolves. *Amer. Zool.* **7**, 331–8

THOMAS, O. (1894) On the mammals of Nyasaland: third contribution. *Proc. zool. Soc. Lond.* **1894**, 136–46

THOMPSON, D.Q. (1952) Travel, range and food habits of timber wolves in Wisconsin. *J. Mammal.* **33**, 429–42

THOMPSON, E.T. (1923) The mane on the tail of the gray-fox. *Ibid.* **4**, 180–2

THOMSON, A.P.D. (1951) A history of the ferret. *J. Hist. Med.* **6**, 471

THOMSON, T.S. (1842) Report of meeting of Zoological Society for January 25th, 1842. *Proc. zool. Soc. Lond.* **1842**, 10

THORNTON, I.W.B., YEUNG, K.K. & SANKHALA, K.S. (1967) The genetics of the white tigers of Rewa. *J. Zool.* **152**, 127–35

TIMMIS, W.H. (1971) Observations on breeding the oriental short-clawed otter, *Amblonyx cinerea*, at Chester Zoo. *Int. Zoo Yb.* **11**, 109–11

TINBERGEN, N. (1965) Von den Vorratskammern des Rotfuchses (*Vulpes vulpes* L.). *Z. Tierpsychol.* **22**, 119–49

TODD, N.B. (1970) Karyotypic fissioning and canid phylogeny. *J. theoret. Biol.* **26**, 445–80

—— & PRESSMAN, S.R. (1968) The karyotype of the lesser panda (*Ailurus fulgens*) and general remarks on the phylogeny and affinities of the panda. *Carniv. Genet. Newsl.* **5**, 105–8

TOLDT, C. (1905) Der Winkelfortsatz des Unterkiefers beim Menschen und bei Säugetieren und die Beziehungen der Kaumuskulatur zu demselben. *Sitz. K. Akad. Wiss. Wien Math-Naturw. Kl.* **114**, 315–476

TOMICH, P.Q. (1969) Movement patterns of mongoose in Hawaii. *J. Wildl. Mgmt.* **33**, 576–84

TRAUTMAN, M.B. (1963) Solitary carnivore. *Nat. Hist.* **72**, 12–18

TSCHANZ, B., MEYER-HOLZAPFEL, M. & BACHMANN, S. (1970) Das Informationssystem bei Braunbären. *Z. Tierpsychol.* **27**, 47–72

TURNBULL-KEMP, P. (1967) *The leopard* (Howard Timmins: Cape Town)

TURNER, K. (1968) How to make friends with a bat-eared fox. *Africana* **3** (6), 27–9

ULMER, F.A. (1941) Melanism in the Felidae, with special reference to the genus *Lynx*. *J. Mammal.* **22**, 285–8

—— (1968) Breeding fishing cats, *Felis viverrina*, at Philadelphia Zoo. *Int. Zoo Yb.* **8**, 49–55

URBAN, D. (1970) Raccoon populations, movement patterns and predation on a managed waterfowl marsh. *J. Wildl. Mgmt.* **34**, 372–82
VAN VALEN, L. (1964) Nature of the supernumerary molars of *Otocyon. J. Mammal.* **45**, 284–6
—— (1966) Deltatheridia, a new order of mammals. *Bull. Amer. Mus. nat. Hist.* **132**, 1–126
VANDERPUT, R. (1937) La civette. *Bull. agric. Congo Belge.* **28**, 135–46
VARADAY, D. (1964) *Gara-Yaka, the story of a cheetah* (Collins: London)
VERBENE, G. (1970) Beobachtungen und Versuche über das Flemen katzenartiger Raubtiere. *Z. Tierpsychol.* **27**, 807–27
VERHEYEN, W. (1962) Quelques notes sur la zoogéographie et la craniologie d'*Osbornictis piscivora* Allen 1919. *Rev. Zool. Bot. Afr.* **65**, 121–8
VERSCHUREN, J. (1958) *Ecologie et biologie des grands mammifères. Exploration du Parc National de Garamba*, **9** (Inst. Parcs nat. Congo Belge: Brussels)
VERTS, B.J. (1967) *The biology of the striped skunk* (University of Illinois Press: Urbana)
VINCENT, R.E. (1958) Observations of red fox behaviour. *Ecology* **39**, 755–7
VOGEL, C. (1962) Einige Gefangenschaftsbeobachtungen am weiblichen Fenek *Fennecus zerda* (Zimm. 1780). *Z. Säugetierk.* **27**, 193–204
VOLF, J. (1957) A propos de la reproduction du fennec. *Mammalia* **21**, 454–5
—— (1959) La reproduction des genettes au Zoo de Prague. *Ibid.* **23**, 168–71
—— (1963a) Bemerkungen zur Fortpflanzungsbiologie der Eisbären, *Thalarctos maritimus* (Phipps) in Gefangenschaft. *Z. Säugetierk.* **28**, 163–6
—— (1963b) Einige Bemerkungen zur Aufzucht von Eisbären (*Thalarctos maritimus*) in Gefangenschaft. *Zool. Gart. Lpz.* **28**, 97–108
—— (1965) Trente-deux jeunes de la Genette. *Mammalia* **28**, 658–9
—— (1968) Breeding the European wild cat, *Felis s. silvestris*, at Prague Zoo. *Int. Zoo Yb.* **8**, 38–42
VOSSELER, J. (1928) Beobachtungen am Fleckenroller (*Nandinia binotata* (Gray)). *Z. Säugetierk.* **3**, 80–91
—— (1929a) Vom Binturong (*Arctictis binturong* Raffl.). *Zool. Gart. Lpz.* **1**, 296–302
—— (1929b) Beitrag zur Kenntnis der Fossa (*Cryptoprocta ferox* Benn.) und ihrer Fortpflanzung. *Ibid.* **2**, 1–9
WACKERNAGEL, H. (1968) A note on breeding the serval cat, *Felis serval*, at Basle Zoo. *Int. Zoo Yb.* **8**, 46–7
WAGER, D. (1946) *Umhlanga – a story of the coastal bush of South Africa* (Knox: Durban)
WALKER, A. (1930) The 'hand-stand' and some other habits of the Oregon spotted skunk. *J. Mammal.* **11**, 227–9
WALKER, E.P. (1964) *Mammals of the world, vol. II* (Johns Hopkins Press: Baltimore)
WALLER, G.R., PRICE, G.H. & MITCHELL, E.D. (1969) Feline attractant, cis, trans-nepetalactone, metabolism in the domestic cat. *Science* **164**, 1281–2
WALLMO, O.C. & GALLIZIOLI, S. (1954) Status of the coati in Arizona. *J. Mammal.* **35**, 48–54
WALLS, G.L. (1942) *The vertebrate eye and its adaptive radiation* (Harper: New York)

WALTON, K.C. (1968) The baculum as an age indicator in the polecat, *Putorius putorius*. *J. Zool.* **156**, 533–6

WATERS, J.H. (1964) Red fox and gray fox from New England archaeological sites. *J. Mammal.* **45**, 307–8

—— (1967) Foxes on Martha's Vineyard, Massachusetts. *Ibid.* **48**, 137–8

WAYRE, P. (1967) Breeding Canadian otters, *Lutra c. canadensis*, at Norfolk Wildlife Park. *Int. Zoo Yb.* **7**, 128–30

WEBER, M. (1927) *Die Säugetiere* (Gustav Fischer: Jena)

WEBSTER, D.B. (1962) A function of the enlarged middle-ear cavities of the kangaroo rat, *Dipodomys*. *Physiol. Zool.* **35**, 248–55

WEIGEL, I. (1961) Das Fellmuster der wildlebenden Katzenarten und der Hauskatze in vergleichender und stammesgeschichtlicher Hinsicht. *Säugetierk. Mitt.* **9**, 1–120

WELKER, W.I & SEIDENSTEIN, S. (1959) Somatic sensory representation in the cerebral cortex of the raccoon (*Procyon lotor*). *J. comp. Neurol.* **111**, 469–501

——, JOHNSON, J.I. & PUBLOS, B.H. (1964) Some morphological and physiological characteristics of the somatic sensory system in raccoons. *Amer. Zool.* **4**, 75–94

—— & JOHNSON, J.I. (1965) Correlation between nuclear morphology and somatotopic organisation in ventrobasal complex of the raccoon's thalamus. *J. Anat.* **99**, 761–90

WELLS, M.E. (1968) A comparison of the reproductive tracts of *Crocuta crocuta*, *Hyaena hyaena* and *Proteles cristatus*. *E. Afr. Wildl. J.* **6**, 63–70

WEMMER, C. (1971) Birth, development and behaviour of a fanaloka, *Fossa fossa*, at the National Zoological Park, Washington, D.C. *Int. Zoo Yb.* **11**, 113–15

VAN DER WERKEN, H. (1968) Cheetahs in captivity. *Zool. Gart. Lpz.* **35**, 156–61

WEVER, E.G., VERNON, J.A., RAHM, W. & STROTHER, W. (1958) Cochlear potentials in the cat in response to high-frequency sounds. *Proc. nat. Acad. Sci. Wash.* **44**, 1087–90

WHITEMAN, E.E. (1940) Habits and pelage changes in captive coyotes. *J. Mammal.* **21**, 435–8

WHITFIELD, I.C. (1971) Mechanisms of sound location. *Nature Lond.* **233**, 95–7

WHITNEY, L.F. (1952) *The raccoon* (Practical Science Publishing Co.: Orange, Connecticut)

WICKLER, W. (1964) Vom Gruppenleben einiger Säugetiere Afrikas. *Mitt. Max-Planck Ges.* **6**, 296–309

WIGHT, H.M. (1931) Reproduction in the eastern skunk (*Mephitis mephitis nigra*). *J. Mammal.* **12**, 42–7

WILKE, F. (1957) Food of sea otters and harbor seals at Amchitka Island. *J. Wildl. Mgmt.* **21**, 241–2

WILLIAMS, C.B. (1918) Mongoose. The food habits of the mongoose in Trinidad. *Bull. Dept. Agric. Trin. Tob.* **17**, 167–86

WILLIAMS, C.S. (1938) Notes on the food of the sea-otter. *J. Mammal.* **19**, 105–7

WILSON, K. (1954) The role of the mink and otter as muskrat predators in northwest North Carolina. *J. Wildl. Mgmt.* **18**, 199–207

WILSON, V. & CHILD, G. (1966) Notes on development and behaviour of two captive leopards. *Zool. Gart. Lpz.* **32**, 67–70

WILZ, K.J. (1970) Causal and functional analysis of dorsal pricking and nest

activity in the courtship of the three-spined stickleback, *Gasterosteus aculeatus*. *Anim. Behav.* **18**, 115–24
WIMSATT, W.A. (1963) Delayed implantation in the Ursidae, with particular reference to the black bear (*Ursus americanus* Pallas). In *Delayed Implantation*, ed. A.C. Enders (Chicago University Press), 49–74
WINANS, S.S. & SCALIA, F. (1970) Amygdaloid nucleus: new efferent input from the vomeronasal organ. *Science* **170**, 330–2
WINDLE, C.A. & PARSONS, F.G. (1897, 1898) The myology of the terrestrial Carnivora. Part I *Proc. zool Soc. Lond.*, **1897**, 370–409; part II *Ibid.* **1898**, 152–86
WOOD, A.E. & WOOD, H.E. (1933) The genetic and phylogenetic significance of the presence of a third upper molar in a modern dog. *Amer. midl. Nat.* **14**, 36–48
WOOD, J.E. (1954) Food habits of furbearers of the Upland Post Oak region in Texas. *J. Mammal.* **35**, 406–14
—— (1958) Age structure and productivity of a gray fox population. *Ibid.* **39**, 74–86
WOOD-JONES, F. (1939) The forearm and manus of the giant panda, *Ailuropoda melanoleuca*, M-Edw., with an account of the mechanism of its grasp. *Proc. zool. Soc. Lond.* **B. 109**, 113–29
WOOLPY, J. (1968) The social organisation of wolves. *Nat. Hist.* **72**, 46–55
WRIGHT, B.S. (1960) Predation on big game in East Africa. *J. Wildl. Mgmt.* **24**, 1–15
WRIGHT, G.M. (1934) Cougar surprised at well-stocked larder. *J. Mammal.* **15**, 321
WRIGHT, P.L. (1947) The sexual cycle of the male long-tailed weasel (*Mustela frenata*). *Ibid.* **28**, 343–52
—— (1963) Variations in reproductive cycles in North American mustelids. In *Delayed Implantation* ed. A. C. Enders (University of Chicago Press), 77–94
—— (1966) Observations on the reproductive cycle of the American badger (*Taxidea taxus*). *Symp. zool. Soc. Lond.* **15**, 27–45
—— & RAUSCH, R. (1955) Reproduction in the wolverine, *Gulo gulo*. *J. Mammal.* **36**, 346–55
—— & Coulter, M.W. (1967) Reproduction and growth in Maine fishers. *J. Wildl. Mgmt.* **31**, 70–87
WURSTER, D.H. & BENIRSCHKE, K. (1968) Comparative cytogenetic studies in the order Carnivora. *Chromosoma Berl.* **24**, 336–82
WÜSTEHUBE, C. (1960) Beiträge zur Kenntnis besonders des Spiel und Beuteverhaltens einheimischer Musteliden. *Z. Tierpsychol.* **17**, 578–613
WYMAN, J. (1967) The jackals of the Serengeti. *Animals* **10**, 79–83
YADAV, R.N. (1967) Breeding of the smooth-coated Indian otter, *Lutra perspicillata*, at Jaipur Zoo. *Int. Zoo Yb.* **7**, 130
—— (1968) Notes on breeding the Indian wolf, *Canis lupus pallipes*, at Jaipur Zoo. *Ibid.* **8**, 17–18
YALDEN, D.W. (1970) The functional morphology of the carpal bones in carnivores. *Acta anat.* **77**, 481–500
YEAGER, L.E. (1938) Tree-climbing by a gray fox. *J. Mammal.* **19**, 376
—— (1943) Storing of muskrats and other foods by minks. *Ibid.* **24**, 100–1
—— & WOLOCH, J.P. (1962) Striped skunk with three legs. *Ibid.* **43**, 420–1
YOUNG, S. & GOLDMAN, E. (1946) *The puma, mysterious American cat* (North American Wildlife Institute: Washington)

YOUNG, S.P. (1958) *The bobcat of North America* (Wildlife Management Institute: Washington)

ZANNIER, F. (1965) Verhaltensuntersuchungen an der Zwergmanguste, *Helogale undulata rufula*, im Zoologische Garten Frankfurt am Main. *Z. Tierpsychol.* **22**, 672–95

ZARROW, M.X. & CLARK, J.H. (1968) Ovulation following vaginal stimulation in a spontaneous ovulator and its implications. *J. Endocrinol.* **40**, 343–52

ZELLER, F. (1960) Notes on the giant otter, *Pteronura brasiliensis*, at Cologne Zoo. *Int. Zoo Yb.* **2**, 81

ZEUNER, F.E. (1963) *A history of domesticated animals* (Hutchinson: London)

ZOLLMAN, P. & WINKLEMAN, R.K. (1962) The sensory innervation of the common north American raccoon (*Procyon lotor*). *J. comp. Neurol.* **119**, 149–57

ZUCKERMAN, S. (1953) The breeding seasons of mammals in captivity. *Proc. zool. Soc. Lond.* **122**, 827–950

ZUMPT, I.F. (1968a) The handling, housing and nutrition of captive wild meerkats. *J. S. Afr. vet. med. Ass.* **39**, 105–8

—— (1968b) The feeding habits of the yellow mongoose, *Cynictis penicillata*, the suricate, *Suricata suricatta*, and the Cape ground squirrel, *Xerus inauris*. *Ibid.* **39**, 89–91

Author Index

Species and Subject Index

Both common and scientific names are indexed so as to make it easy to equate one with the other. Thus 'Aardwolf' tells you '*see Proteles cristatus*' and '*Proteles cristatus*' is followed by 'or aardwolf'. The numbers following the names refer to the list in Chapter 10 and make it possible to find out the taxonomic position of the species: thus VI 4, 406 indicates that the animal in question is species number 4 in family VI and is to be found on page 406. Referring to this will show that the aardwolf is classified in the family Hyaenidae, subfamily Protelinae.